Energy Storage Devices for Renewable Energy-Based Systems

Energy Storage Devices for Renewable Energy-Based Systems

Rechargeable Batteries and Supercapacitors

Second Edition

Nihal Kularatna
School of Engineering, University of Waikato,
Hamilton, New Zealand

Kosala Gunawardane
Department of Electrical and Electronic Engineering,
Auckland University of Technology, Auckland, New Zealand

ELSEVIER

ACADEMIC PRESS
An imprint of Elsevier

Academic Press is an imprint of Elsevier
125 London Wall, London EC2Y 5AS, United Kingdom
525 B Street, Suite 1650, San Diego, CA 92101, United States
50 Hampshire Street, 5th Floor, Cambridge, MA 02139, United States
The Boulevard, Langford Lane, Kidlington, Oxford OX5 1GB, United Kingdom

Library of Congress Cataloging-in-Publication Data
A catalog record for this book is available from the Library of Congress

British Library Cataloguing-in-Publication Data
A catalogue record for this book is available from the British Library

ISBN 978-0-12-820778-9

For information on all Academic Press publications
visit our website at https://www.elsevier.com/books-and-journals

Publisher: Brian Romer
Acquisitions Editor: Lisa Reading
Editorial Project Manager: Aleksandra Packowska
Production Project Manager: Prem Kumar Kaliamoorthi
Cover Designer: Victoria Pearson

Typeset by SPi Global, India

Contents

About the authors

Nihal Kularatna is an electronics engineer with more than 45 years of experience in industry and research. He won the NZ Engineering Innovator of the Year 2013 award for his work on supercapacitor applications and power electronics. This is his 10th book, which comes as a new edition of his 7th book on energy storage devices. Nihal is an associate professor in electronic engineering at the School of Engineering, University of Waikato, New Zealand. He leads the Power Electronic Research Group at Waikato, and multiple international patents were granted for his supercapacitor-assisted (SCA) techniques. He earned a Doctor of Science degree in 2015 from the University of Waikato for his research contributions to power electronics, instrumentation, and telecommunications.

Kosala Gunawardane was the first PhD student of Dr. Nihal Kularatna at the University of Waikato to develop the supercapacitor-assisted low dropout (SCALDO) regulator technique. She joined the academic staff of Auckland University of Technology in 2014 and is currently a senior lecturer in electrical and electronic engineering. Kosala has been collaborating with Nihal further developing SCA techniques for more than 7 years. She completed her Bachelor of Engineering degree at the University of Moratuwa, Sri Lanka, in 2005 and moved to New Zealand for her PhD in 2009.

Preface

This book is the new edition of Energy Storage Devices for Electronic Systems published in 2015 and expands a lot on the application knowledge gained by a team of power electronics researchers in New Zealand, directly and indirectly supported by the concerted knowledge and efforts globally.

The wider subject presented in this book is aimed at renewable energy systems for low emissions and minimal carbon footprint. Wider applications of supercapacitors are presented, including very low-frequency interference free power converters useful in renewable energy systems as well as supercapacitor-based surge protection techniques.

The authors strongly believe that supercapacitors will play a major role in future power electronics, and renewable energy systems.

Thank you,

Nihal Kularatna
School of Engineering, University of Waikato, Hamilton, New Zealand
Kosala Gunawardane
Department of Electrical and Electronic Engineering, Auckland University of Technology, Auckland, New Zealand

Acknowledgments

This book is the result of researching and developing converter and protection techniques, useful in renewable energy applications over a period of 15 years in a major teamwork at two universities in New Zealand. In this work two research teams led by the two authors guiding more than 15 PhD students developed different applications of energy storage devices, particularly the supercapacitors, which are very rapidly evolving into a major new component family useful in distributed energy systems.

Given this background, both authors are very grateful to both the University of Waikato and Auckland University of Technology (AUT) senior management, and their respective schools and departments for providing the necessary resources as well as laboratory space and computing facilities. Kiwinet and Callaghan research grants during the period 2009–19 are gratefully acknowledged for supporting several PhD projects.

Our former and current postgraduate students, Dr. Kasun Subasinghage, Dr. Dilini Jayananda, and Thilanga Ariyarathna are gratefully acknowledged for their valuable contributions based on their PhD theses work as book chapters of this book. Dr. Jayathu Fernando contributed to Chapters 7 and 13 based on his work on the patented SCA-Surge Absorber (SCASA) technique. For readers who require detailed information on the SCASA technique, we happily direct them to the 2019 Elsevier publication on Design of Transient Protection Systems.

Dr. Nimesha Wijesooriya is also kindly acknowledged for her massive support in uploading the chapters and for reorganizing the original content of the book Energy Storage Devices for Electronic Systems published in 2015 and assisting us with new figures required. Charlies Hua, Nalin Bandara and Savin Kokuhennadige are also acknowledged for their support in drawing some figures for new chapter versions.

Nihal is most grateful to the advice he received during early 1990s from Prof. John R Pierce (the Bell Labs engineer who named the transistor) on how to write a technical reference book. Nihal is also sincerely grateful to Sir Arthur C. Clarke, who was the patron of Arthur C. Clarke Institute for Modern Technologies, for his continuous encouragement for his research work and the book projects during his midcareer of 16 years in Sri Lanka. Expertise and experience he gained there allowed him to confidently start his research in New Zealand on supercapacitor applications area. Nihal is always thankful to his beloved wife Priyani, daughters Dulsha and Malsha, sons-in-law Rajith and Kasun, for their unstinting encouragement and appreciation of his technical book writing, while the four grandchildren Nethuli, Mineli, Kaavya, and Kavindu keep them entertained during the weekends.

Kosala extends her heartfelt thanks to Nihal for being a great mentor for her academic career of over 10 years specially providing the guidance for the projects from an industry perspective and also providing this golden opportunity to become a coauthor of his 10th book. Without his persistent help, the goal of this book project would not have been realized.

Kosala expresses sincere gratitude to her parents for always believing in her and supporting, loving, and encouraging her to follow her dreams. Thanks also to her sister, Supuni and the family for their constant support and encouragement with their best wishes. Most importantly, Kosala wishes to thank her loving and supportive husband, Asanga, and wonderful little girl, Kiana, for the tolerance and patience, and for putting up with her tech addiction!

Authors are very grateful to the Elsevier team, for their kind support to bring the manuscript content into a high quality presentation for the benefit of the readers. We are specially acknowledging the kind support by Joe Hayton, Lisa Reading, Aleksandra Packowska, Prem Kumar Kailamoorthi, Swapna Praveen and the other members of editorial and production teams of Elsevier.

We both are happy to acknowledge the intended future support for our work on energy storage devices and DC microgrid systems from the Advanced Energy Technology Project grant from the Ministry of Business Innovation and Enterprise in 2020, where five New Zealand Universities will be collaborating to develop the Future Architecture Network (FAN) in the next 7 years.

Nihal Kularatna
School of Engineering, University of Waikato, Hamilton, New Zealand

Kosala Gunawardane
Department of Electrical and Electronic Engineering, Auckland University of Technology, Auckland, New Zealand

Modern electrical power system and the role of distributed generation

1

1.1 Evolution of electricity systems

Until the late 19th century, for electrical power demand was relatively low and generators, distribution, and utilization facilities in places close to cities due to consumer spatial density and the technological development limited to voltaic cells, batteries, and small thermoelectric units (Buriticá-Arboleda and Álvarez-Bel, 2011). This early electricity system was primarily some form of distributed generation based on DC.

Over the years, with the population growth and the expansion of cities, the demand for electricity has significantly increased, and the popular sources such as hydroelectric and thermoelectric generation were discovered to meet the rising demand. However, the environmental concerns of thermoelectric units and the geographical location of water sources moved such generation facilities away from cities. These generation units were clustered into major centers of generation in order to supply electricity to many utilities through transmission and distribution of AC grids. Such a system is known as *centralized generation (CG)* (Fig. 1.1), where central power plants with bulk electric power generation located away from end users is linked with the users via transmission and distribution grid network.

Centrally generated in plant sizes of 1–1000 MW, the most commonly used technologies for centralized electricity are large fossil-fired (oil, gas, and coal) boilers or nuclear boilers to produce steam that drives turbine generators. In some cases, large hydro is also used.

Over the years, the widespread proliferation of CG was driven by the following remarkable factors (Mojumdar et al., 2015):

(1) Economies of scale: With the introduction of steam turbines, the size of the turbines increased while decreasing the marginal cost of electricity production.
(2) Systems with high energy efficiency: Higher efficiency is achievable in handling higher pressures and temperatures of steam used in electricity generation.
(3) Long distance electricity transmission feasibility: AC grid's transmitting scenario of electricity over long distances with a significant loss reduction by the use of transformers.
(4) Environmental constraints: Facility to relocate the generation facilities away from the busy city centers thus replacing pollution concerns away from big cities.
(5) Reliability of interconnected grid: Compensation of plant failures by other plants in the interconnected system.

More details about centralized generation and the principles of development of the operation in generation, transmission, and distribution of such power systems are available in standard power system textbooks such as Grigsby (2012).

Energy Storage Devices for Renewable Energy-Based Systems. https://doi.org/10.1016/B978-0-12-820778-9.00002-4

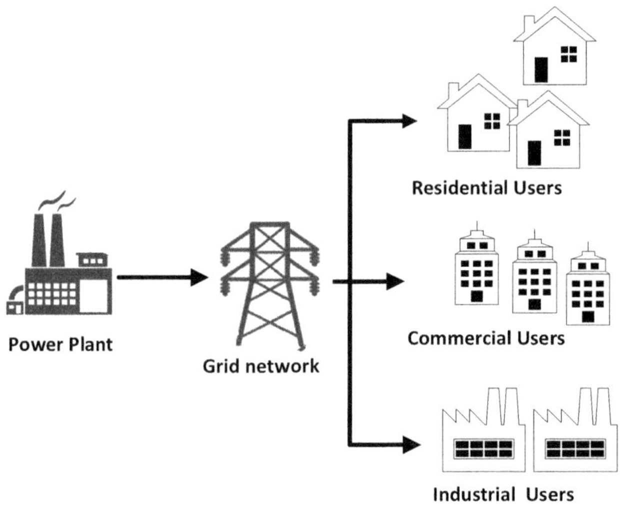

Fig. 1.1 Typical centralized power system.

1.2 Status of current power systems

Electricity has become an essential service in the modern society. Therefore, the overall electricity supply system should be able to assure quality and continuity of supply at any time anywhere. With more socially, economically, and environmentally informed present generation of stakeholders, the modern electrical industry is facing a paradigm shift in the generation, transmission, and distribution of electricity (Barroso and Hugh, 2019).

Over the years, centralized systems have provided the potential for efficient resource allocation and generated substantial economies of scale in the process of building and operating very reliable power delivery networks. However, the energy supply paradigm has been rapidly fading, especially with the depletion of fossil-based primary energy resources and their environmental impact (Bouffard and Kirschen, 2008).

Currently, the electricity supply system is facing the following challenges (Buriticá-Arboleda and Álvarez-Bel, 2011; Bouffard and Kirschen, 2008):

(1) Demand is increasing permanently

The demand for energy on a global scale is increasing exponentially. Economic growth and the population growth are the two major driving forces that continue to cause increase in demand for energy. For example, Fig. 1.2 (Kober et al., 2020) shows the rise in the total global primary energy consumption within the past 40 years (1978–2018), which was mainly covered by fossil fuels. About 85% of the total primary energy consumption was still covered by coal, oil and gas in 2018. As a consequence, the share of fossil fuels in global primary energy consumption was dropped by seven percentage points over these 40 years. It is predicted that if the current global energy consumption pattern continues, the world energy consumption will increase by over 50% before 2030(Suganthia and Samuel, 2012).

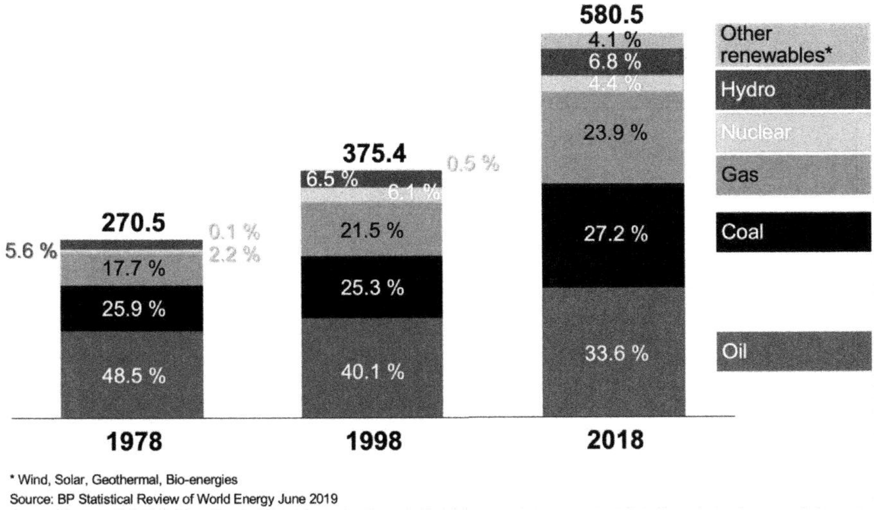

Fig. 1.2 Global primary energy consumption 1979–2018 (Exajoules).
Adopted from Kober, T., Schiffer, H. W., Densing, M., Panos, E., 2020. Global energy perspectives to 2060—WEC's World Energy Scenarios 2019. Energy Strategy Rev. 31, 1–19.

(2) Fossil fuel depletion

Crude oil, coal, and gas are the main sources of fossil-based energy resources whereas coal constitutes approximately 65% of the fossil fuel reserves in the world, with the remaining 35% being oil and gas (Shafiee and Topal, 2009).

The important issue after global consumption of fossil fuels is the price movement. Fig. 1.3 shows the consumption of fossil fuel worldwide from 1965 to 2030. Proven fossil fuel reserves fluctuate according to fossil fuel prices. The energy information administration (EIA) has projected that energy consumption will increase at an average rate of 1.1% per annum (Shafiee and Topal, 2009; Deng et al., 2012). The human civilization is over-exploiting resources faster than they are being renewed. It is expected that the global energy market will continue to depend on fossil fuels for at least the next few decades and the reserves will diminish eventually. The global community is looking for renewable energy instead of nonrenewable energy as a solution to replace fossil fuel resources in the future.

(3) Environmental issues

Extensive fossil fuel consumption has led to some undesirable phenomena such as atmospheric and environmental pollutions, which have not been experienced before. The environmental impact of the centralized energy system is significant due to heavy reliance on fuel, coal, and gas. The electricity sector is primarily responsible for NO_x, CO_2, and SO_2 emissions (Markandya and Wilkinson, 2007). The different forms of power generation have very different contributions to CO_2 emissions as shown in Fig. 1.4. The direct emission of carbon dioxide occurs due to consumption of fossil fuels in machinery involved in power generation operations. Whereas, the indirect emissions occur in other industrial sectors, which supply the power plants inputs consumed in the power generation operations or processes. Global warming, greenhouse effect, climate change, ozone layer depletion, and acid rain are the major consequences of such pollutions.

In order to avoid further impacts of these phenomena, the reliable alternative is to significantly replace fossil fuel as much as possible with environmentally friendly, clean, and renewable energy sources (Deng et al., 2012; Sen, 2004). The growing awareness of the

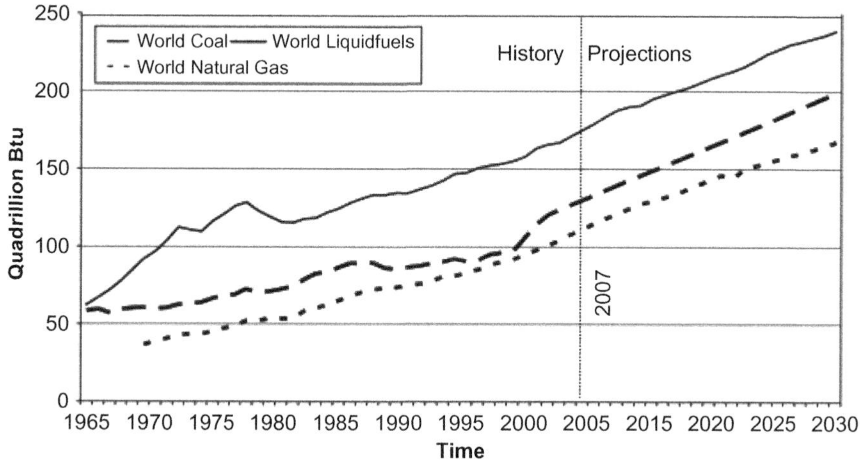

Fig. 1.3 Consumption of fossil fuel worldwide from 1965 to 2030.
Adopted from Shafiee, S., Topal, E., 2009. When will fossil fuel reserves be diminished? Energy Policy 37, 181–189.

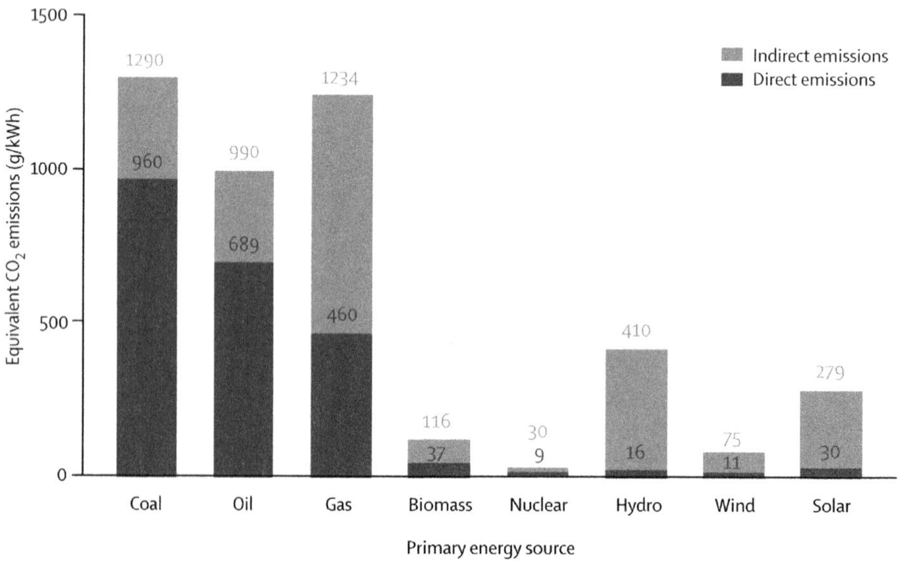

Fig. 1.4 Full energy chain of CO_2 equivalent emissions by primary energy source.
Adopted from Markandya, A., Wilkinson, P., 2007. Electricity generation and health. Lancet 370(9591), 979–990.

environmental concerns about the threats of climate change have led world leaders to agree on a shared, long-term goal of limiting global emissions of greenhouse gases to ensure a 2°C compatible pathway within this century, and pursue efforts to limit global warming to 1.5°C above preindustrial levels (Child et al., 2018).

(4) Penetration of DG in local and regional electricity systems

More recently, within the last decade, varieties of DG technologies have emerged, and are altering the traditional organization of centralized electricity industry. The replacement of additional power demand with decentralized sources such as solar photovoltaics or wind turbines, the energy future may rely on partial shifts from centralized energy generation to distributed energy generation.

Renewable energy sources (RES) are planned to replace a large percentage of fossil fuel electricity generation as shown in Fig. 1.5 in the future (Deng et al., 2012) where 1 EJ is equal to 10 to the power 18 joules. However, they also create opportunities to resolve a broad problem on competitive electricity markets.

Furthermore, in the context of the operational scenarios of power system, there are other main shortfalls in the centralized generation, which explicitly motivate the movement toward distributed generation.

(a) Rural electrification

Providing electricity to the rural communities extending the centralized grid is uneconomical due to high capital expenditures to connect through the long-distance overhead lines (El-Khattam and Salama, 2004). In rural electrification, it is more economical to rely on DG in low population density areas.

(b) High transmission and distribution costs

In the cost of delivering electricity on average, about 30% share (Singh et al., 2014) is due to transmission and distribution costs. This accounts mainly for the line losses and the

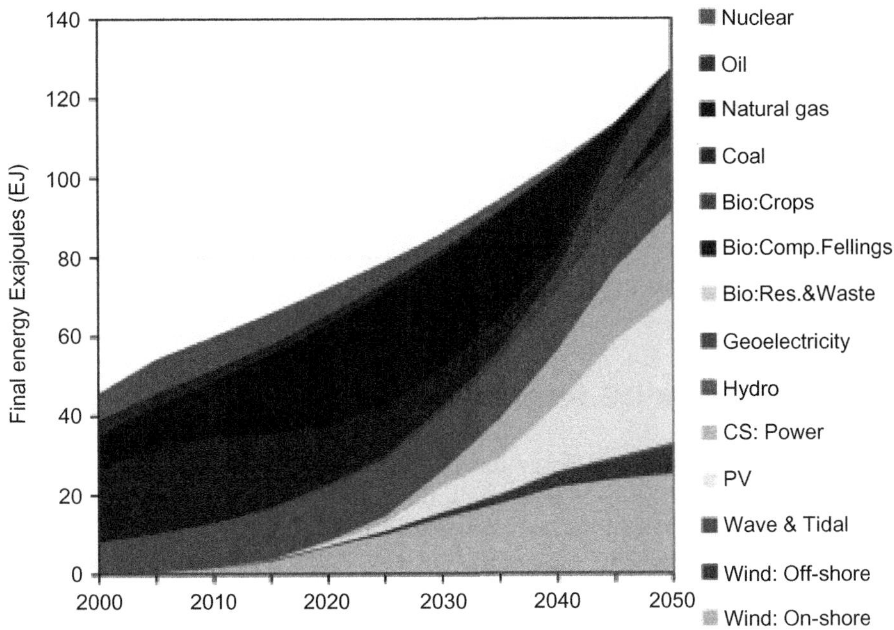

Fig. 1.5 Global Power consumption.
Adopted from Deng, Y. Y., Blok, K., Leun, K. V. D., 2012. Transition to a fully sustainable global energy system. Energy Strategy Rev. 1, 109–121.

conversion losses. In order to reduce these costs DG can be used as a way to bypass the transmission and distribution network costs especially in new developments.

(c) Security and reliability

In the case of operational failures in the main grid, DG can be used as backup system to prevent major interruptions to overall systems. As DG system could always run with auxiliary storage, it has the ability to maintain the continuity of operation through energy storage system. In some areas where the centralized grid is already established, DG has the ability to improve grid resilience by providing reliable electricity during hazard times such as bad weather conditions.

1.3 Distributed generation

The term "distributed generation" (DG) refers to the generation of electricity at or near the utilities as shown in Fig. 1.6. However, there are various definitions available in the literature for DG, which are based on diverse range of factors such as the purpose, the location, the rating of distributed generation, the power delivery area, the technology, the environmental impact, etc. (Ackermann et al., 2001a, b). DGs are originated by both renewable and nonrenewable energy sources. The most attractive DGs are from natural renewable resources such as water, wind, sun, tides, waves, geothermal, and biomass. Such DGs require less area of installation, possess relatively small capacities ranging from few kW to hundreds MW, and also reduce greenhouse gas emissions addressing environment concerns (Montoya-Bueno et al., 2016).

DGs have been classified into four groups based on the capacity as follows (Ackermann et al., 2001a; Ackermann et al., 2001b):

- microdistributed generation (1–5 kW)
- small-distributed generation (5–5 MW)
- medium-distributed generation (5–50 MW)
- large-distributed generation (50–300 MW)

Table 1.1 presents the most commonly used DG technologies and their typical module size (Prakash and Khatod, 2016; Agarwal and Jain, 2019).

1.4 Renewable DG technologies

Sun is the primary source of energy for the entire life on the Earth. Primarily, both forms of energy for the power generation, fossil energy and renewable energy, are derived from sun through various mechanisms as shown in Fig. 1.7. The fossil energy in forms of oil, gas, and coal originated from sun, captured through photosynthesis which take millions of years to form. However, after the consumption it takes further millions of years to replenish for future generations.

There are several common options for DG. Following sections provide a summary with essential details.

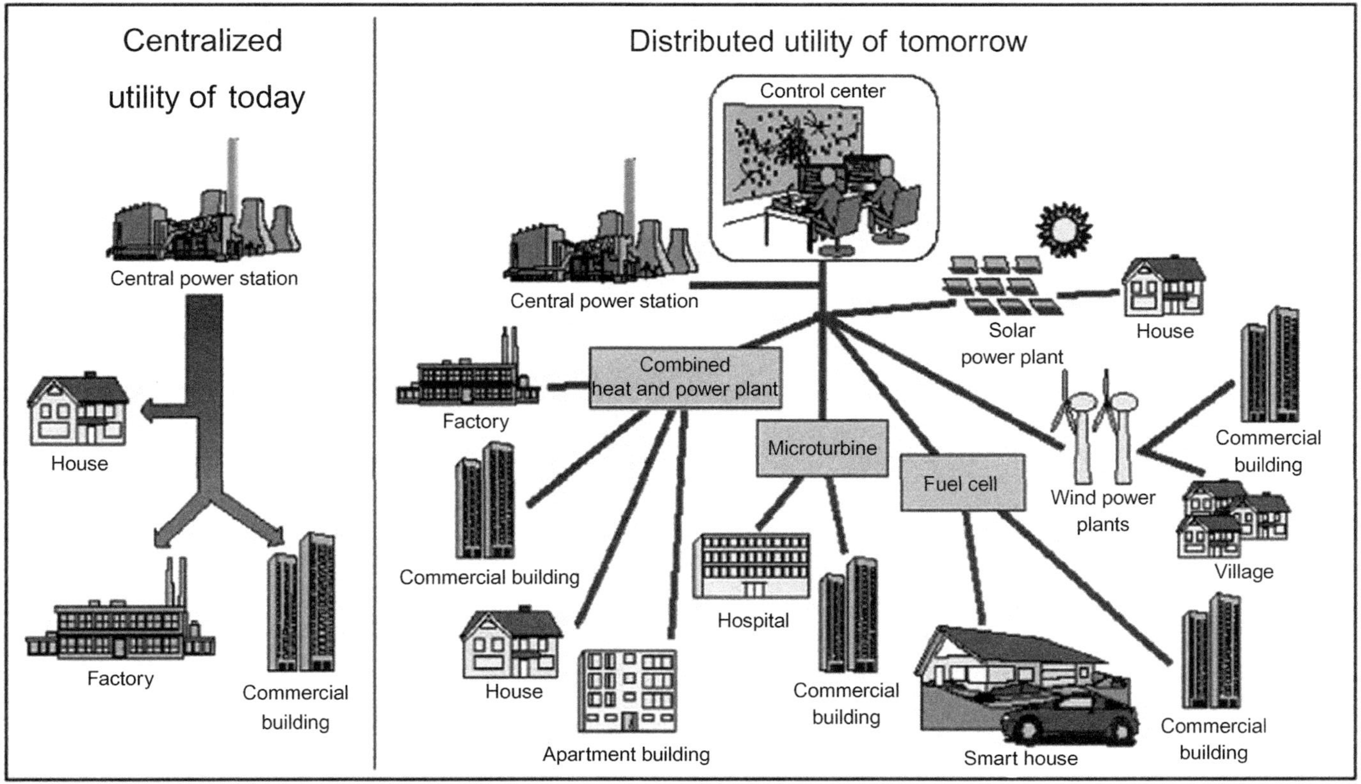

Fig. 1.6 Distributed generation—A typical example.
Adopted from Viral R., Khatod D.K., 2012. Optimal planning of distributed generation systems in distribution system: a review. Renew. Sustain. Energy Rev. 16, 5146–5165.

Table 1.1 Summary of major DG technologies and their characteristics (Prakash and Khatod, 2016).

S. No.	DG technology	Electric power generation range(s)	Merit(s)	Demerit(s)	Energy conversion	Primary source of energy
1	Solar technology solar photovoltaic (SPV) solar thermal power plant solar thermal (Lutz system)	200 W–3000 kW 1–80 MW 10–10 MW	• Easiest and cleanest • Maintenance cost is very less • Fuel free • Overall environmentally friendly	• Require large solar collector • Solar thermal systems are health hazard • Need battery bank for storage • SPV module pose disposal problem • Initial cost is very high	Solar radiation to electric	Sun
2	Small hydro	5 kW–100 MW	• Free and renewable energy source • Less installation cost • Eco-friendly	• Generation depends on water • Affected in flood time • Load demand cannot be meet out	Gravitational potential energy to electrical	Water
	Micro hydro	1 kW–1 MW	• Small size, light weight • Emission free • Less noise	• Cost is more • Low working temperature • Less efficiency • Initial cost is more	Gravitational potential energy to electrical	Water
3	Wind turbine	200 W–5 MW	• Generation cost is very less • No adverse effect on global environment • Fuel free • Saving of land use	• Wind generators are hazards • Noise pollution • Wind speed affect the output • Variable power production highly dependent on wind speed	Wind energy to electrical	Wind

4	Bio-mass energy	100 kW–20 MW	• Renewable energy source • Reduce greenhouse gas (GHG) emission • Lessen dependency on upon conventional fuel • Stops desertification	• Combustion of biomass produces air pollution • Source limitation • Maintenances are expensive • Soil erosion	Chemical to electrical, thermal and in biofuels	Biomass
5	Geothermal energy	5–100 MW	• Economical as stations need small space • No fuel is needed • Renewable source • Its cost does not rise with time	• Discharge of wastewater polluted rivers • Noise pollution • Gases escape into atmosphere • Large-scale withdrawal of underground fluid may cause damage to surface structure	Heat to electrical	Hot Water
6	Tidal energy	0.1–1 MW	• It restricts emission of GHG • Fuel not required • Energy use is forecast able	• Problem faced in placement of machinery • Supply may available for limited time • Machines and supporting equipment are expensive	Kinetic energy to electrical	Ocean water

Continued

Table 1.1 Continued

S. No.	DG technology	Electric power generation range(s)	Merit(s)	Demerit(s)	Energy conversion	Primary source of energy
7	Hydrogen energy system	40–400 MW	• Carbon free • It is a nonconventional energy source • It is an energy carrier	• Initial cost is more • Difficulty in storage, packaging and public distribution • Pressurization is difficult	Chemical to electrical	Water, organic compounds, biomass, and hydrocarbons
8	Ocean wave energy	100–1000 kW	• Carbon free • It is a renewable-based energy source	• Difficult to harness • Very less efficiency • Initial cost is high	Kinetic energy to electrical	Ocean water

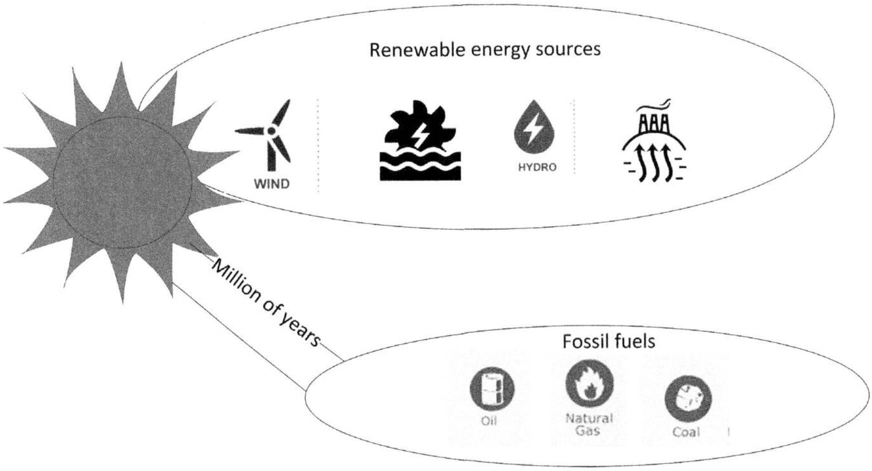

Fig. 1.7 Fossil energy and renewable energy are derived from same primary source—Sun.

1.4.1 PV power generation technology

As a renewable energy source, solar energy is much more attractive, has received increased interest around the globe due to ready availability at any location, and cost effectiveness. However, solar energy has high variability in availability. The intensity and spectral distribution of the solar radiation depend on the geographic position, time, day of the year, climate conditions, composition of the atmosphere, altitude, and many other factors. The availability of PV energy is recorded as solar irradiance (also called as insolation), expressed as the power per unit area (watts per square meter, W/m^2). Fig. 1.8 shows a photovoltaic power potential world map from SOLARATLAS (Global Solar ATLAS, 2019).

There are many useful studies and databases are available to obtain estimation of the annual PV plant energy production for a given location and following are some of such websites:

https://globalsolaratlas.info/
https://solargis.com/
https://www.nrel.gov/gis/solar.html

A solar cell or photovoltaic (PV) cell is a semiconductor device used to convert solar radiation energy to electrical energy by the photovoltaic effect. The elementary building block of a photovoltaic panel is the PV cell that consists of a p-type layer and an n-type layer jointly forming a p-n junction. Fig. 1.9 illustrates the physical structure of a PV cell (Villalva and Gazoli, 2009; Cotfas et al., 2013). When the PN junction is exposed to the solar radiation, the electrons start moving and generate direct current (DC). Typically, the current generated by a single cell is quite small, around 30 mA/cm^2. A PV array may be either a panel or a set of panels connected in series or parallel to form large PV systems. PV panels are generally composed of series cells in order to obtain higher output voltages. The large output currents are achieved by increasing the surface area of the cells or by connecting cells in parallel. A typical

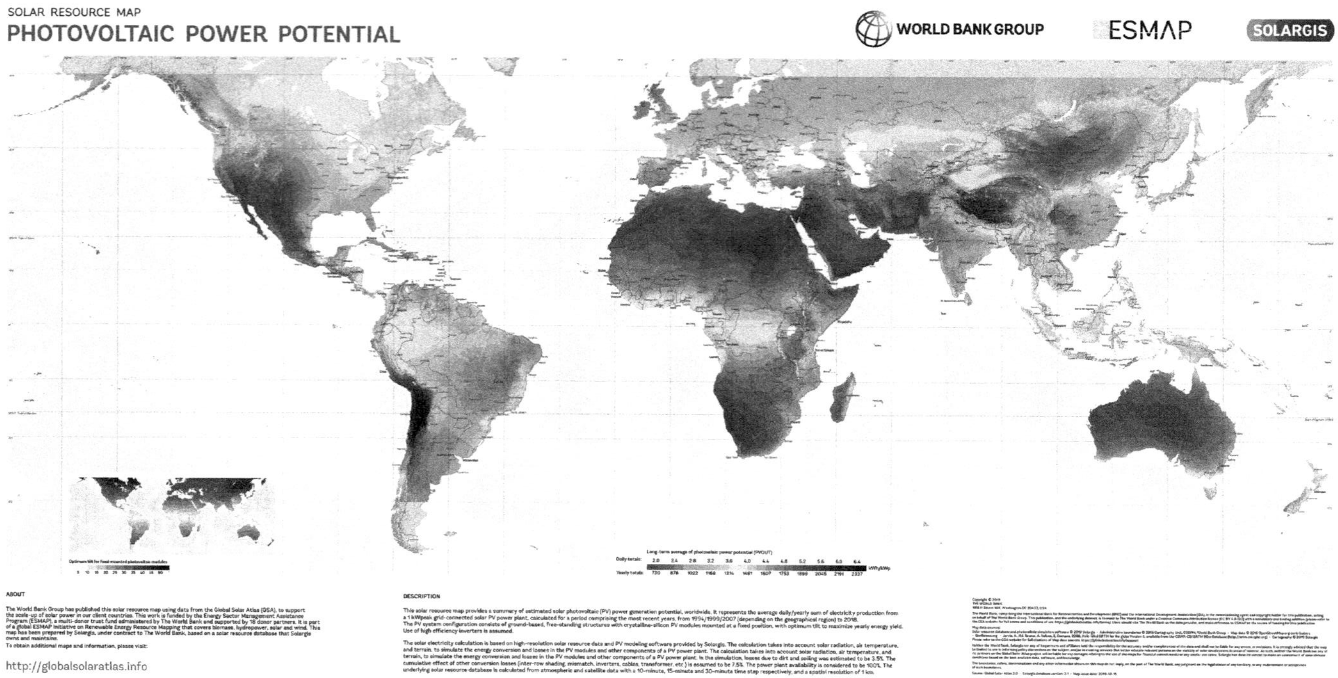

Fig. 1.8 Photovoltaic power potential.
Adopted from SOLARATLAS (Global Solar ATLAS. 2019 https://globalsolaratlas.info/download/world (Accessed)).

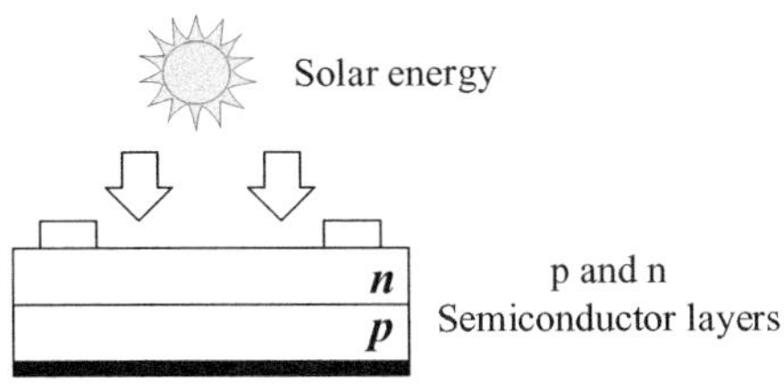

Fig. 1.9 Physical structure of a PV cell.

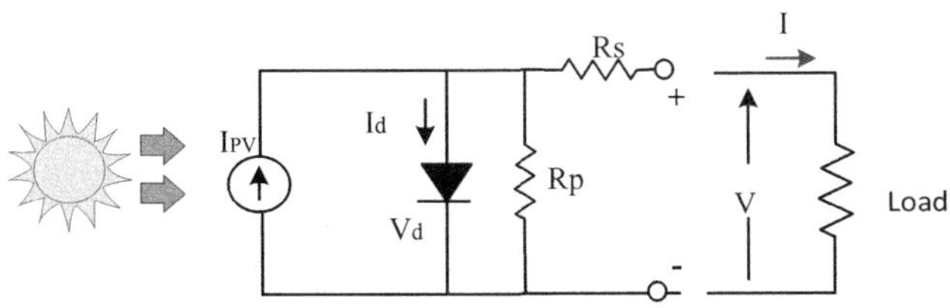

Fig. 1.10 Equivalent circuit model of a PV cell including series and parallel resistances.

PV module is made up of around 36 or 72 cells connected in series, encapsulated in a structure made of aluminum and Tedlar.

The equivalent circuit of the solar cell is shown in Fig. 1.10. This circuit is useful in evaluating the solar cells.

According to Fig. 1.10 the characteristics equation of solar cell is given by

$$I = I_{pv} - I_d \tag{1.1}$$

$$I_d = I_0.e^{\left[\left(\frac{qV}{akT}\right)-1\right]} \tag{1.2}$$

where q is the electron charge ($1.60217646 \times 10^{-19}$ C), k is the Boltzmann constant ($1.3806503 \times 10^{-23}$ J/K), T is the temperature of the p-n junction in Kelvin, and a is the diode ideality constant.

I_{PV} is the current generated by the incident light, which is directly proportional to the irradiation. This can be denoted as a current source equivalent in I-V graph as shown in Fig. 1.11A. I_d is the diode current which is shown in Fig. 1.11B. Then a combination of the two graphs according to Eq. (1.1), originates the standard I-V curve of a solar cell as shown in Fig. 1.11C, represents the operation model of the PV cell.

In a practical PV cell, there is a series of resistance in the current path through the semiconductor material, the metal grid, contacts, and current collecting bus. These resistive

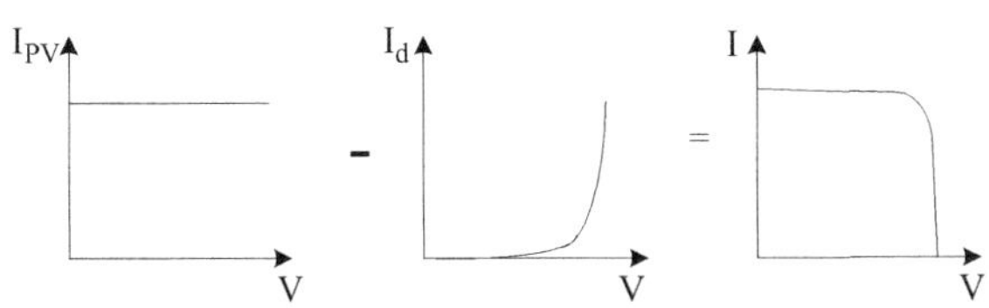

Fig. 1.11 Standard I-V curve of a solar cell: (A) I_{PV} is the current generated by the incident light, which is denoted as a current source equivalent, (B) I_d diode current, and (C) standard I-V curve of the overall solar cell.

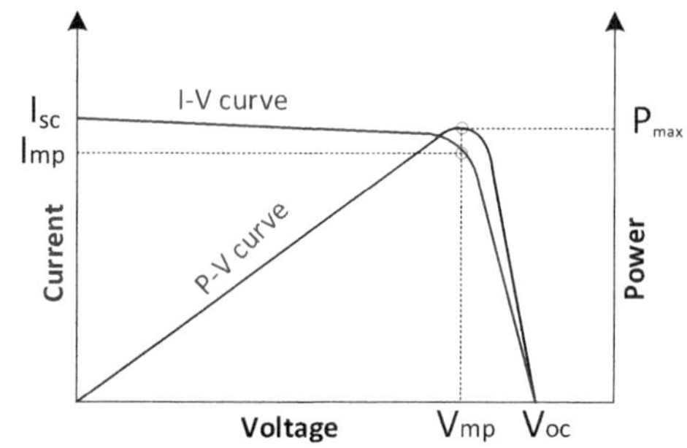

Fig. 1.12 I-V curve together with the P-V (power-voltage) curve for a PV cell

losses are lumped together as a series resister (Rs) as shown in Fig. 1.10. Rp is the loss due to small leakage of current through a resistive path in parallel with the intrinsic device.

Fig. 1.12 shows a typical I-V curve together with the power-voltage (P-V) curve for a PV cell. P_{max} is the maximum power point whereas I_{mp} is the current and V_{mp} is the voltage at the maximum power point. To continuously yield the maximum power from PV panels it needs to operate at the maximum power point (MPP). To accomplish this, PV panels use maximum power point tracking (MPPT) techniques, which is the key to optimum solar system efficiency.

The MPPT methods compel the PV array to operate at or near the maximum power point of the V-P characteristic curve. The algorithms that are used in maximum power point trackers dynamically bring the current or voltage at or near the maximum power point, which operates an electronic system along with the PV module in a manner that allows it to generate the maximum power. Fig. 1.13 shows the typical functional block diagram of MPPT tracker of a PV panel. R_{in} is the input resistance of the converter and R_L is the load resistance.

The purpose or role of various MPPT algorithms is to control the duty ratio (D) of the DC-DC converter used in the design in such a way that R_{in} of the DC-DC converter matches the instantaneous resistance of the solar panel system, which keeps varying with the irradiance.

The most commonly used MPPT algorithms are perturb-and-observe (P&O) method and incremental conductance (IC) method. A good review on various types

Fig. 1.13 Schematic diagram of typical PV system with MPPT technique.

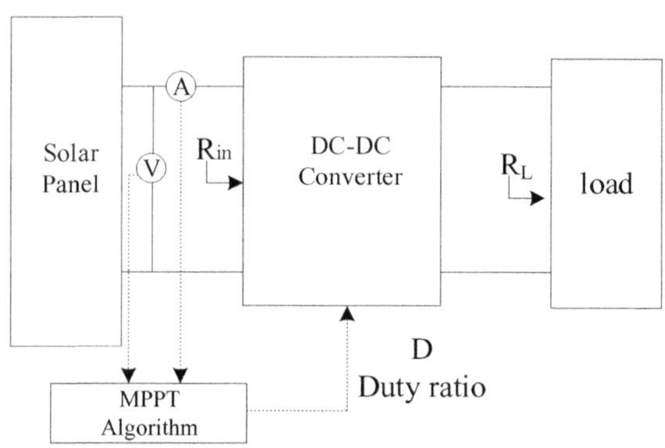

of MPPT algorithms are available in Ref. (Child et al., 2018). An expanded discussion on this subject is available in Chapter 12.

1.4.2 Wind power generation technology

In general terms, wind is atmospheric air in motion. The primary source of wind energy is solar radiation. Solar radiation is differentially absorbed by the Earth's surface and converted through convective processes due to temperature differences which create air in motion. The velocity of the air determines the strength of wind. Air speed is also proportional to the amount of wind energy available in the form of kinetic energy. Fig. 1.14 shows the mean wind distribution with speeds of air across the world.

In a wind power generation system, the kinetic energy of wind is converted into mechanical energy by turbine as shown in Fig. 1.15. The aerodynamic power, P, of the wind turbine is given by

$$P = \frac{1}{2}\rho\pi C_p r^2 V^3 \tag{1.3}$$

where ρ, C_p, r, and V are the air density, the turbine power coefficient representing the power conversion efficiency of a wind turbine, the turbine radius, and the velocity of wind, respectively. The captured power fluctuates as the velocity of input wind changes.

Figure. 1.15 depicts the simplified arrangement of a wind turbine system (Blaabjerg et al., 2004). For high power wind turbines, to optimize with the mechanical speed

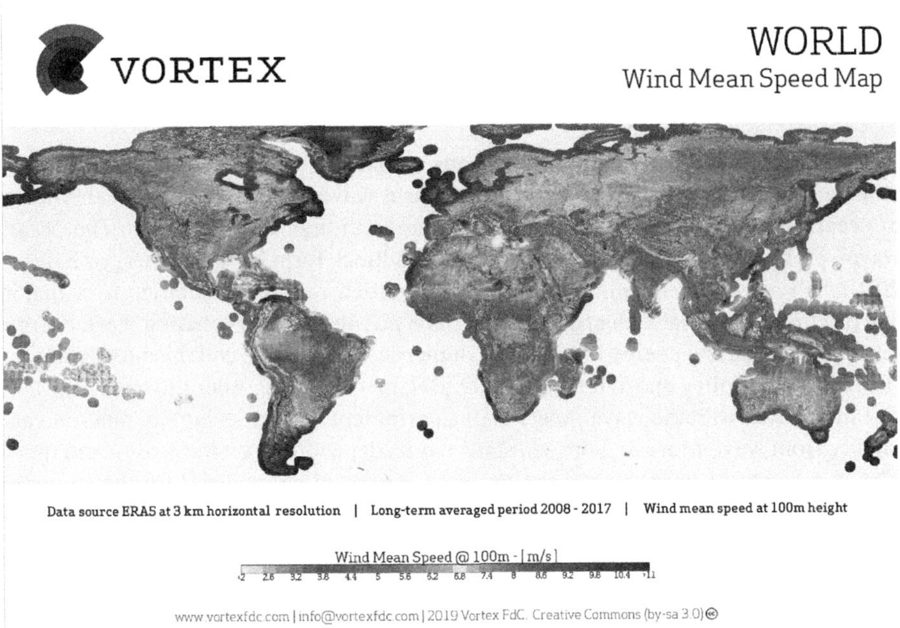

Fig. 1.14 Mean wind distribution with speeds of air across the world. Adopted from https://vortexfdc.com/knowledge/wind-map-world/.

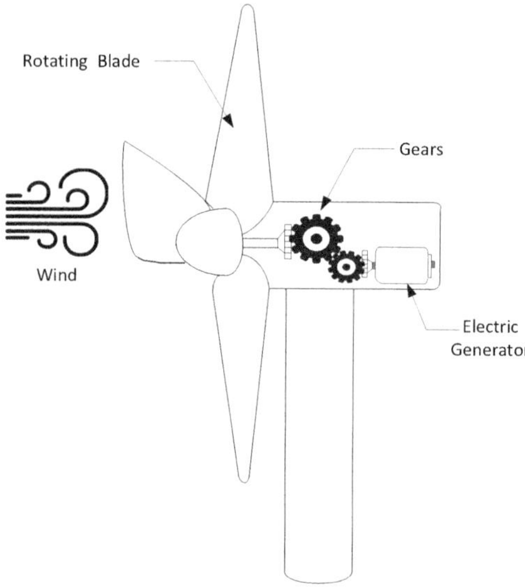

Fig. 1.15 Simplified diagram of a wind generator.

and the torque a gear box is used. With the gear box interfaced between the turbine and the electrical generator, different electrical generators could be chosen from DC generator, synchronous (AC) generator or induction (AC) generator. For optimum operation MPPT scheme based DC-DC converters and battery banks could be used to power the electrical systems in the house-hold. In AC output systems transformer based up or down converters could be used with down stream connections. A more detailed discussions on this topic is available in (Beik and Al-Adsani, 2020).

1.4.3 Ocean energy power generation

Renewable energy sources from the oceans include offshore wind, underwater currents, and wave energy. The energy from ocean waves is the most noticeable form of ocean energy, due to its often-violent nature with high energy effects. The ocean waves produced by wind action are also an indirect form of solar energy (Falcao, 2010). Wave energy is being increasingly regarded in many countries as a major and promising renewable energy resource. The possibility of converting wave energy into usable electrical energy has inspired numerous researchers and inventors globally (Falcao, 2010; Folley and Whittaker, 2009; Bakar et al., 2011; Aderinto and Li, 2018). The main issue with the wave power is its intermittent nature, its largely random variability from wave to wave, with sea state and its dependence on the season and time. The assessment of the wave energy resource is a basic prerequisite for the strategic planning of its utilization and for the design of wave energy devices. The studies aiming at the characterization of the wave energy resource, having in view its utilization, started naturally in those countries where the wave energy technology was developed first. The European Wave Energy Atlas (WERATLAS) was a basic tool developed for wave energy planning in Europe (Folley and Whittaker, 2009). The annual mean wave power density and annual mean best direction of open sea wave distribution around the globe are shown in Fig. 1.16. Seasonal variations are in general considerably larger in the Northern than in the Southern hemisphere, which makes the southern coasts of South America, Africa, and Australia particularly attractive for wave energy exploitation (Falcao, 2010; Gunn and Stock-Williams, 2012; Falnes, 2007).

A sea wave is composed of many elementary waves of different frequencies and directions. Per unit area of sea surface, a stored energy amounting to an average of

$$E = \rho g H_o^2 / 16 \tag{1.4}$$

Eq. (1.4) is associated with wave, where $\rho = 1030\,\text{kg/m}^3$ is the mass density of seawater, $g = 9.81\,\text{m/s}^2$ is the gravitational acceleration, and H_0 is the significant wave height for the actual sea state. This stored energy is equally partitioned between kinetic energy, due to the motion of the water, and potential energy, due to mechanical work performed when the flat-water surface is being deformed to a wavy surface. This work corresponds to water lifted against the gravity force from wave troughs to wave crests (Falnes, 2007).

Theoretically, the energy-extracting device interacts with the waves to reduce the amount of wave energy that is otherwise present in the sea. An oscillating energy-extracting system as shown in Fig. 1.17 is a good wave absorber and also a good wave generator (Falnes, 2007).

Many concepts and designs of wave energy converters have been developed over the years (Falcao, 2010; Aderinto and Li, 2018). Many of such designs remained at the preliminary stages such as laboratory testing or protote designs. Only a few designs have been succsesfully deployed so far. Despite all these studies focusing on wave energy conversion technology and the amount of identified resource potential, the commercial utilization of ocean wave energy is still low compared to other most commonly used renewable sources discussed earlier.

Issues affecting the commercial utlilization of wave energy are as follows.

- Difficulty in integrating wave energy power into the electricity grid due to the high intermeinttent nature of generation.
- Low grid connecting facilities in offshore environments.
- Extreme offshore weather conditions that make the construction, operation, and maintenance of converter systems challenging.

1.5 Energy storages technologies for distributed generation

1.5.1 Summary of energy storage technologies

The output of renewable energy sources is intermittent by nature, fluctuates over time due to reasons such as weather conditions, which ultimately result in a mismatch of energy supply and demand. Currently, in many cases such mismatch is managed by exporting excess renewable production to the grid or importing electricity from the

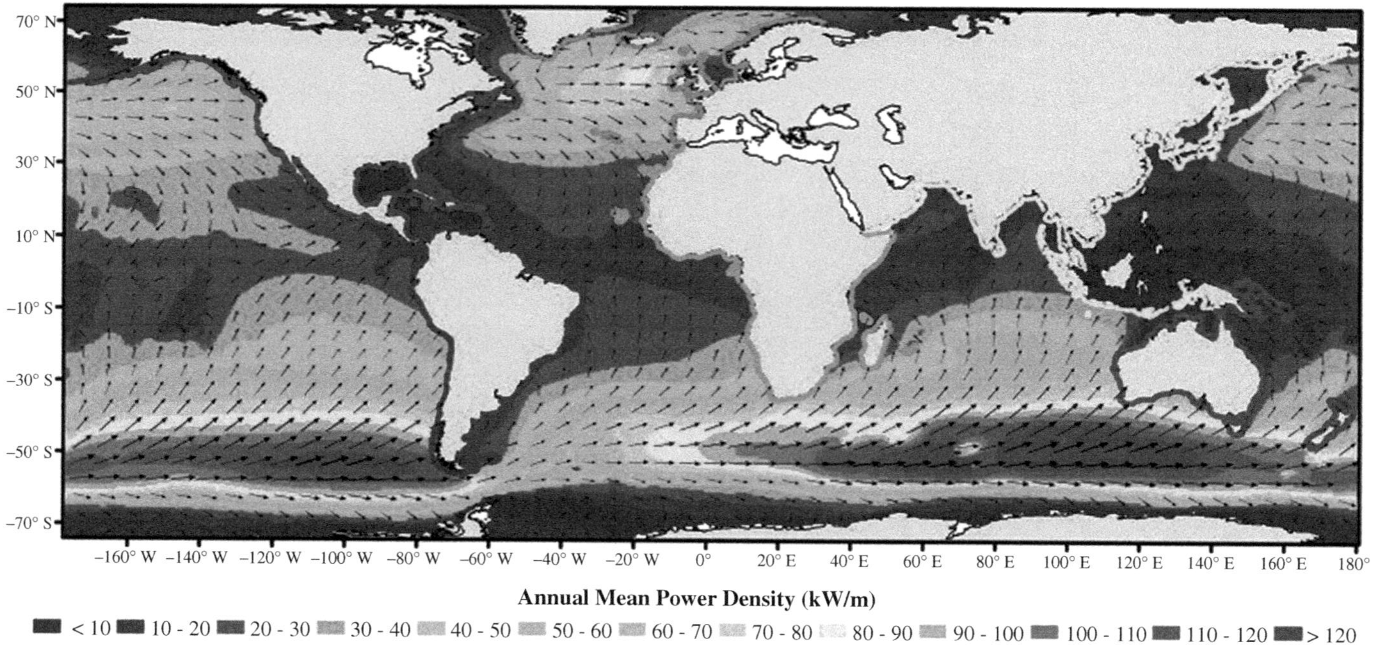

Fig. 1.16 Annual mean wave power density and annual mean wave direction.
Adopted from Gunn, K., Stock-Williams, C., 2012. Quantifying the global wave power resource, Renew. Energy 44, 296–304.

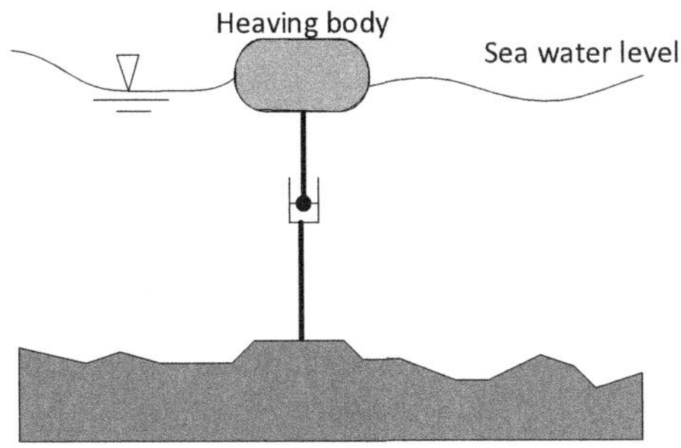

Fig. 1.17 An oscillating wave energy-extracting system.

grid when demand is not met by renewable energy supply. On the other hand, as the scale of renewables in the electricity grid increases, the challenge of matching energy production and demand to ensure a steady supply of electricity becomes increasingly predominant and need to be addressed carefully (Cárdenas et al., 2019; Murray et al., 2018).

In most cases the renewable energy generated on site is fed back to grid at the day time (when solar energy generation is active) based on feed-in tariffs, which is in favor of the utility AC company. If we can store the excess energy generated in an energy storage system (ESS) it could be in favor of the end user, when it is used in the peak load times, when the unit price could be high.

Energy storage refers to a process of converting electrical energy from a power network into a form that can be stored for converting back to electrical energy when needed (Chen et al., 2009). Such a process enables electricity to be produced at times of either low demand, low generation cost, or from intermittent energy sources and to be used at times of high demand, high generation cost, or when no other generation means is available. Energy storages for power generation systems come in many different varieties depending on the application scope. Shifting demand over longer periods of time is vital, as the ability of these systems to generate electricity depends on the location of the user, the surrounding geography, and weather conditions (Murray et al., 2018).

Energy storage systems applicable to renewable generation is classified into following categories (Chen et al., 2009).

- pumped hydrostorage
- rechargeable battery packs (including the cases of electric cars)
- compressed air energy storage system
- flywheel
- supercapacitor banks
- superconducting magnetic energy storage system
- thermal energy storage system
- fuel cell

Pumped hydro and compressed air, which have high storage capacity, are used for energy arbitrage storing of large amounts of cheap energy during off-peak time. Other technologies such as flywheels and battery technologies are applicable to frequency regulation and load, which require an energy storage device to release small-to-medium amounts of stored electricity in seconds on demand. Research on new battery technologies has led to significant improvements in these in recent years. A detailed discussion on selected types of energy storage devices are discussed in Chapter 2.

1.5.2 Hydrogen storage

Currently, lot of inspiring researches are going on utilizing hydrogen storage as a promising long-term storage option (Murray et al., 2018; Chen et al., 2009; Parra et al., 2016). As shown in Fig 1.18, this refers to the use of an electrical current, generated from renewable electricity, to split water via electrolysis process into hydrogen and oxygen. The hydrogen can be stored in compressed tanks or metal hydride storage, used to run fuel cells, can be directly injected into the natural gas grid up to certain concentration limitations, or converted to methane via methanation and used as a substitute for natural gas. This technology can be easily integrated into currently developing DG networks.

Current research trends in hydrogen-related technologies, such as fuel cells, hydrogen storage, and electrolysis, are rapidly growing. A schematic representation of the hydrogen and associated technologies and possible implementation scenarios of a DG model are shown in Fig. 1.19 which consist of application scope in Central gas supply, Localised internal heating, and electricity grid (Murray et al., 2018). Small-wind turbines, small hydro, and solar PV are the renewable generation technologies; heat pumps, electrolysers, fuel cells, gas turbines, and gas boilers are conversion technologies; and battery storage, thermal storage, and hydrogen storage are the storage technologies included in this model. From the hydrogen storage, a limited portion of hydrogen is injected directly into the natural gas grid. The output production of these technologies is fed into the networks and is used to meet the heating and electric demand of the network.

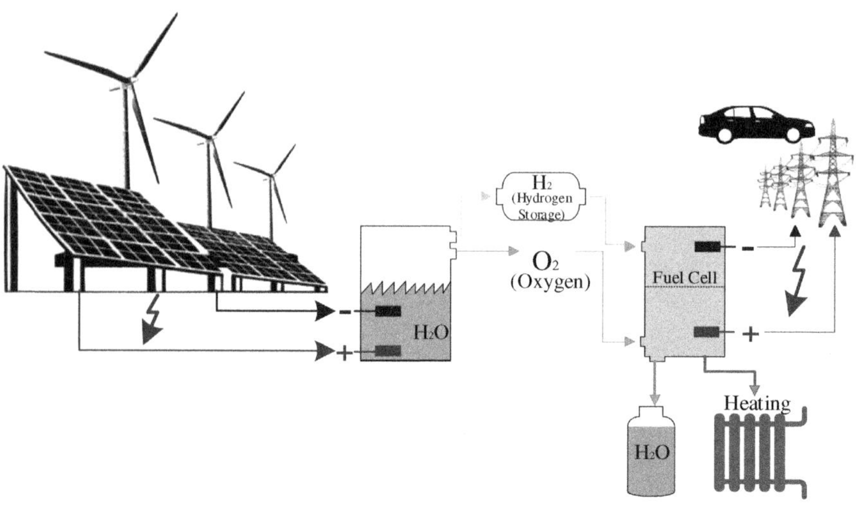

Fig. 1.18 Future prospects of hydrogen technologies together with renewable energy applications.

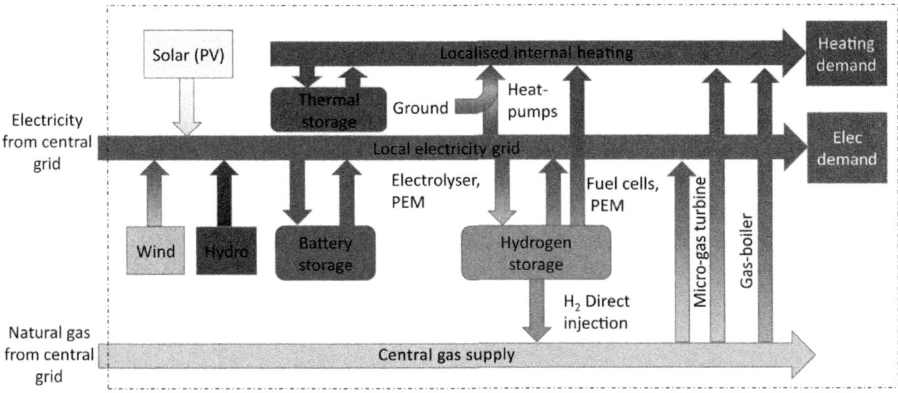

Fig. 1.19 Schematic representation of the hydrogen and associated technologies and possible implementation scenario of DG model.
Adopted from Murray, P., Orehouniga, K., Grosspietschc, D., Carmelieta, J., 2018.
A comparison of storage systems in neighbourhood decentralized energy system applications from 2015 to 2050, Appl. Energy 231, 1285–1306.

1.5.3 Implementation aspects of energy storages

As discussed earlier, most renewable sources have dependency on environmental factors and their power generations show the intermittency in the output. Integrating local energy storage systems to resolve such demand deficit is currently booming in many renewable energy applications. In general, energy storage systems require devices to operate in different configurations, depending on the implementation. There are two commonly adopted configurations:

(i) energy storage installed in parallel—Fig. 1.20A
(ii) energy storage system between the energy generator and the load—Fig. 1.20B

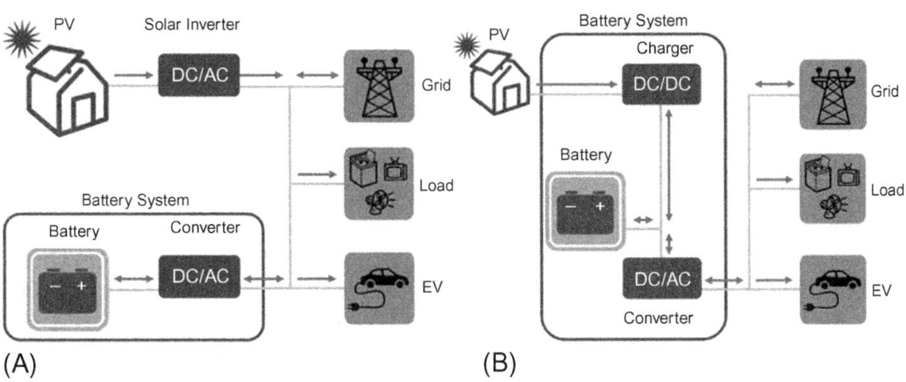

Fig. 1.20 Commonly adopted energy storage configurations for renewable generation systems: (A) energy storage system not directly integrated with renewable energy source and (B) energy storage system integrated with renewable energy source (Speidel and Braunl, 2016)

Table 1.2 Characteristics of energy storage configurations in Fig. 1.20.

Fig. 1.20A Configuration	**Fig. 1.20B Configuration**
Energy storage is run in parallel	Energy storage is installed between the energy producer and the load/grid
Easy installation, cheaper units	Higher installation cost, expensive
No need to match the specifications of the solar system	Need different types and configurations of solar installations
Require DC/AC bidirectional converters	Require DC/DC and DC/AC converter

These two configurations have different characteristics as summarized in Table 1.2 (Speidel and Braunl, 2016).

To store electrical energy many implementation scenarios are adopted in current energy storage technologies to suit various applications.

(a) Integrating energy storage into some devices

Some manufacturers are now introducing energy storage integrated into their solar inverter with the aim of reducing the amount of energy needed from the grid (Speidel and Braunl, 2016). Even for grid-connected households, local energy storage has the ability to maximize the use of renewable energy generated due to the difference in buying and selling prices of electricity. There is also a lot of research aimed solving the issues associated with such systems, such as generation and storage selection optimization (Mohammed et al., 2014).

(b) Developing community energy storage

Another form of bulk energy storage is the community energy storage (CES) systems for end-user applications. The method evaluates the optimum performance, the internal rate of return, and the levelized value of suitable energy storage technologies. An example of community approach where the optimum CES system was calculated as a function of the size of the community. For example, lead-acid and lithium-ion battery technologies can be used when performing PV energy time shift using real demand data from a single home to a 100-home community which could lead to a reduction in cost by 37% and 66% regarding a single home (Murray et al., 2018).

(c) Hybrid power technologies

Emerging technologies of electric power generation is essentially advocating for combinations of different resources for efficient and quality power delivery especially on the utilization of different kinds of renewable sources thereby causing some paradigm shift of attentions from single energy system application for electricity generation to combinations of energy system called hybrid power technologies. In most cases, the end product of a hybrid electrical power system is electricity using a combined facility known as cogeneration. Hybrid electric power system is a mechanism that explores multiple sources of energy connected together as shown in Fig. 1.21 to achieve synchronized power output. Nowadays, the uses of hybrid energy system are tremendously working in favor of renewable energy system exploitation. Notably, such hybrid system has enormous potential to provide quality, secured, and reliable electricity to the consumers especially on off-grid operations for remote locations(Mohammed et al., 2014).

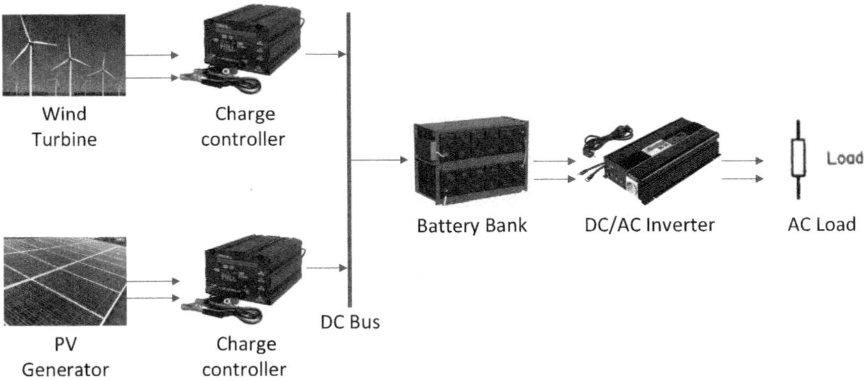

Fig. 1.21 Simple hybrid power system with a wind and PV generator.

1.6 Impact of distributed generation

Currently, the power distribution systems are primarily based on the assumption of the centralized distribution architecture with unidirectional electric power flows from the power system grid to the load. In such a distribution network, bidirectional flow from localized renewable source generators could have a significant impact on the overall distribution system in terms of stability, reliability, voltage profile, power quality, safety, and protection issues. This section is aimed at a discussion on issues in DG systems and its impact on the legacy AC grid.

1.6.1 DG benefits

When talking about the benefits of DG, the most remarkable factor is the utilization of renewable energy sources, facilitating greener production to achieve the sustainable energy goals, thus contributing to lower greenhouse gas emissions. However, DG provides much broader spectrum of benefits to the power system and the community. According to Viral and Khatod (2012), the overall benefits of DG can be categorized as technical, economic, and environmental benefits.

1.6.1.1 Technical benefits

Technical benefits of integrating DG into the current power system include advantages of improvement of some power system attributes as well as reduction in eliminable attributes in the power system as presented in Table 1.3 (Chiradeja and Ramakumar, 2004).

(a) Power losses

The power loss in a power system is significant under heavy load conditions. The inclusion of DG causes a distribution of loading and reduction in power flows within the distribution ends of the AC grid. As a result, the power losses of the AC system will be reduced. Furthermore, DG can significantly reduce the electric line losses due to its close proximity

Table 1.3 Technical benefits of DG.

Improving power system attributes	• Voltage profile • Reliability • Security • Power quality • Energy efficiency
Reducing eliminable attributes	• Line losses
	• Transmission and distribution congestion

to the load centers (Chiradeja and Ramakumar, 2004). Generally, DG units are commissioned in places where they could provide a higher reduction in line losses. On the other hand, DG has impact on both the active and the reactive power. For a particular DG capacity in a power system, an optimum location for the DG can be estimated such that the power losses are at the minimum.

(b) Voltage profile

Introduction of DG has been identified as a possible scenario to eliminate major blackouts in the power system. DG source can provide a portion of the real and reactive power delivered to the load, minimizing the current along a section of the distribution line. As a result it will create a boost in the voltage magnitude at the load side (Shafiee and Topal, 2009; Chiradeja and Ramakumar, 2004). This will improve the voltage profile of the system and maintain the voltage at load terminals to within an acceptable range (Chiradeja and Ramakumar, 2004).

(c) Power quality

The effects of DG can address some of the power quality problems. In a situation where maintaining the voltage profile within acceptable standards is difficult, introduction of DG can lead to a corrective rise or reduction in the voltage in the local network and also assist in any required power factor corrections (Viral and Khatod, 2012).

(d) Reliability

Maintaining reliability of a power system is vital because the cost of power outages cause more severe overall impact (Viral and Khatod, 2012). Traditionally, reliability analysis and evaluation techniques are used mostly in the generation or transmission level outages due to high level of cost impact. However, in general, a large portion of contribution for unavailability of supply comes from distribution system failures. One of the main purposes of integrating DG to distribution system is to increase the reliability as a backup system or as a main supply. DG can also be operated during peak load periods in order to avoid additional charges. This is particularly the case when you have local energy storage options coupled with the DG.

1.6.1.2 *Economic benefits*

- Economic advantages of DG cover lot of aspects including saving on fuel cost, transmission, and distribution cost and wholesale electricity price. The major economic benefits are as follows (Viral and Khatod, 2012; Gil and Jöos, 2006):
  - o Deferred investments for transmission and distribution upgrades of facilities.
  - o Reduced operation and maintenance costs of some DG technologies.
  - o Enhanced productivity.

- o Reduced health-care costs due to improved environment.
- o Reduced fuel costs due to increase in overall efficiency.
- o Reduced reserve requirements and the associated costs.
- o Lower operating costs due to peak shaving.
- o Increased security for critical loads.

1.6.1.3 Environmental benefits

Obviously, the widespread use of DG technologies substantially reduces emissions. Some of the commonly used renewable DG technologies, such as wind turbines, solar photovoltaic panels, and hydro-electric turbines, do not consume fossil fuels at all. On the other hand, DG technologies also create a diverse energy sources which helps to overcome the price shocks, interruptions, and fuel shortages (Viral and Khatod, 2012).

1.6.2 Issues with the DG penetration

With the trends of renewable generation, when the widespread penetration of the DG becomes significant, the original power system dynamics can be hugely affected. According to Viral and Khatod (2012), these issues can been denoted as below.

- **(a)** Reverse power flow: The vast majority of the distribution networks are configured in radial form. This generates unidirectional flow patterns, so the protections system is designed accordingly. The installation of distributed generation changes the flow into bidirectional, as a result of connecting DG in the system causing malfunctions of protection circuits as they are configured at present (Blaabjerg et al., 2017).
- **(b)** Reactive power: Many DG technologies use asynchronous generators that do not supply reactive power to the grid.
- **(c)** System frequency: The increase in the amount of distributed generation affects the system frequency. Deviations from the system nominal frequency are caused by imbalances between supply and demand.
- **(d)** Voltage levels: The installed distributed generation changes the voltage profile of the distribution network because of the change in the magnitudes of power flow. Usually the voltages profile will tend to rise, which is not a problem in congested networks with low voltage problems, as would be in the contrary.
- **(e)** Protection schemes: The installation of distributed generation changes the flow into bidirectional, so necessarily implying new safety equipment and resizing of the network [grounding, short-circuit, breaking capacity, Supervisory Control and Data Acquisition (SCADA) systems, etc. (Blaabjerg et al., 2017)].
- **(f)** Islanding protection is an important security, this is a condition in which a portion of the utility system that contains both load and distributed resources remains energized while isolated from the remainder of the utility system. A DG may be feeding a short circuit, thereby generating the possibility of fire or energizing a certain segment of the network, which may put the lives of workers at risk who come into contact with it when not timely advised of the possibility of it being energized. One way of solution is to place protective equipment such as relays (electronic or mechanical) and transfer switches.

(g) Harmonics injection into the system by asynchronous DG sources that use inverters for interconnection.
(h) Stability issues.
(i) Increased fault currents depending on the location of DG units.
(j) High financial cost/kW generation if the renewable energy is used because these technologies are not much matured.

1.6.3 Distributed generation optimization

DG integration changes the behavior of network to a bidirectional flow. As per the explanation given in Sections 1.6.1 and 1.6.2 DG penetration has its pros and cons in the current power system. In this scenario, proper planning for DG integration into an existing system requires optimal location, size, type of DG, and also network connection (Singh and Sharma, 2017). Inappropriate allocation of DG units will have a reverse effect on all advantages it could bring on offer. Nevertheless, it is very important to allocate DG unit in the optimal location and of appropriate size.

The most obvious positive technical impact of DG is the ability to reduce distribution system losses which is only achievable in an optimum selection. The inappropriate DG allocation may lead to increased system losses and system operation costs. The most of the electrical power losses in electric power systems are dissipated in distribution systems due to heavy currents flowing in primary and secondary feeders. Developing efficient tools that can optimally allocate different DG types in distribution systems could help here thereby reducing losses (Singh and Sharma, 2017; Mahmoud et al., 2016). The selection of the best places for DG installation and the preferable size of the DG units in large distribution systems is a complex optimization problem (Viral and Khatod, 2012). The impact of DG in system operating characteristics, such as electric losses, voltage profile, stability and reliability need to be appropriately evaluated.

The use of an optimization method capable of indicating the best solution for a given distribution network can be very useful for the system planning. The optimization methods have been identified for the planning of distribution systems with DG to minimize losses. The main scope of these methods is the computational robustness of an algorithm that provides near-optimal solutions with intensive computational effort.

The optimization methods can be classified into following three categories (Mahmoud et al., 2016; Sambaiah, 2018):

- analytical-based (AB) method
- numerical-based (NB) method
- heuristic-based (HB) method

Analytical-based methods are performing well for small and simple systems; they are not suitable for a system with large and complex networks. In the literature, great interest is shown toward AB methods, as they are easy to implement and fast. AB methods often follow various strategies to simplify the optimization problem, either by assuming uniformly distributed loads or by allocating only a single DG unit in the entire system (Mahmoud et al., 2016). Such commonly used analytical-based methods are 2/3 rule or Golden rule, Kalman filter, and sensitivity factor analysis (Speidel and

Braunl, 2016). For example, a popular analytical-based method used ~~for~~ only for uniformly distributed loads is 2/3 capacity method in which the incoming generation placing at 2/3 length of the line (Mahmoud et al., 2016; Sambaiah, 2018; Wang and Nehrir, 2004).

The most common examples of NB methods are gradient search (Rau and Wan, 1994; Keane and O'Malley, 2005), linear programming (Keane and O'Malley, 2005), optimal power flow (Harrison and Wallace, 2005), and exhaustive search. These algorithms are considered effective ways for obtaining the optimal DG size at a certain location. Therefore, these methods fail to represent the accurate behavior of a DG optimization problem that involves two continuous variables, both optimal DG size and optimal DG location.

The HB methods are based on utilizing advanced artificial intelligence (AI) techniques, such as genetic algorithms (GAs), particle swarm optimization, harmony search, etc. (Mahmoud et al., 2016).

1.6.4 Impact of larger renewable energy systems on the existing grid

Existing electricity grid is still dominated by large AC synchronous generators, which are located away from urban areas, and these generating houses are coupled to the electricity grid via high-voltage transmission lines and the associated switch gear at high-voltage substations. Historically the network was developed based on these large generators, which work on the principle of locking on to the network's average frequency which is kept stable by all the generators sharing the loads while keeping them synchronized to the average frequency. Synchronous generators at large thermal power plants or the hydrolocations are synchroniszd at the time they are coupled on to the network, by a strict procedures adapted at the generating centers. However, if a sudden failure of large generator connected to the network occurs, a major transmission link failure occurs, or an urban or an industrial center's load move into a sudden rise, the network frequency could enter into an unstable situation. This can create a disconnection of a large generator due to loss of its synchronous speed, governed by the network's average frequency. This kind of an unexpected failure could create what is known as a cascaded failure or a cascading failure, forcing the entire grid or a major part of the grid into a blackout. In such situations, the generating and transmission authorities have the difficult task of starting the system's large generators in a sequential order and synchronizing on to the network, in a step-by-step manner to get the system back in operation.

In general, most large-scale outages can often be traced back to the failure of a single transmission element (Rohden et al., 2016). The initial failure causes secondary failures, which can eventually lead to a whole cascade of failures. Cascading failures have been analyzed in various studies with different models and from different viewpoints (Rohden et al., 2016; Schäfer et al., 2018).

With the modern scenario of renewable generation distributed over geographical locations, many inverter-based connections are established linking into the AC grid, pumping significant amounts of power which could affect the frequency stability of the grid, which is fundamentally important in the case of stability of the large generators. The stability issues are mainly originating from reduction in synchronous inertia, reduction in reactive power reserve, low short-circuit strength of the power network, and fault ride-through(FRT) strategy/capability of the power electronic converters interfaced renewable energy resources (Meegahapola, 2020). Moment these localized renewable sources are dependent on the solar irradiation or the wind, the stability of the power becomes uncertain. In such situations, smart grid-based communication and control infrastructures could play a major role in stabilizing the utility AC grid. A related discussion is available in Ref. (Gungor et al., 2011). How the current progress toward the smart grids will help in these situations as discussed in Ref. (Yu et al., 2011).

1.7 Smart grid

Electricity industry across the world is trying to address numerous challenges, including generation diversification, optimal deployment of assets, demand response, energy conservation, and reduction in the carbon footprint. It is evident that such critical issues cannot be addressed within the confines of the existing electricity grid.

The existing electricity grid is unidirectional in nature. It converts only one-third of fuel energy into electricity, without recovering the waste heat. Almost 8% of its output is lost along its transmission lines, while 20% of its generation capacity exists to meet peak demand only (i.e., it is in use only 5% of the time). In addition to that, due to the hierarchical topology of its assets, the existing electricity grid suffers from domino effect failures (Farhangi, 2010).

The next-generation electricity grid, known as the "smart grid" or "intelligent grid," is to address the shortcomings of the existing grid. In essence, it is to provide the utility companies with full visibility and control over their assets and services. The smart grid is required to be self-healing and resilient to system anomalies, while it needs to empower its stakeholders to define and realize new ways of engaging with each other and performing energy transactions across the system (Table 1.4).

Table 1.4 Comparison of the existing grid with the future smart grid.

Existing grid	Intelligent grid/ smart grid
Electromechanical	Digital
One-way communication	Two-way communication
Centralized generation	Distributed generation
Hierarchical	Network
Few sensors	Sensors throughout
Blind	Self-monitoring
Manual restoration	Self-healing
Failures and blackouts	Adaptive and islanding
Manual check/test	Remote check/test
Limited control	Pervasive control
Few customer choices	Many customer choices

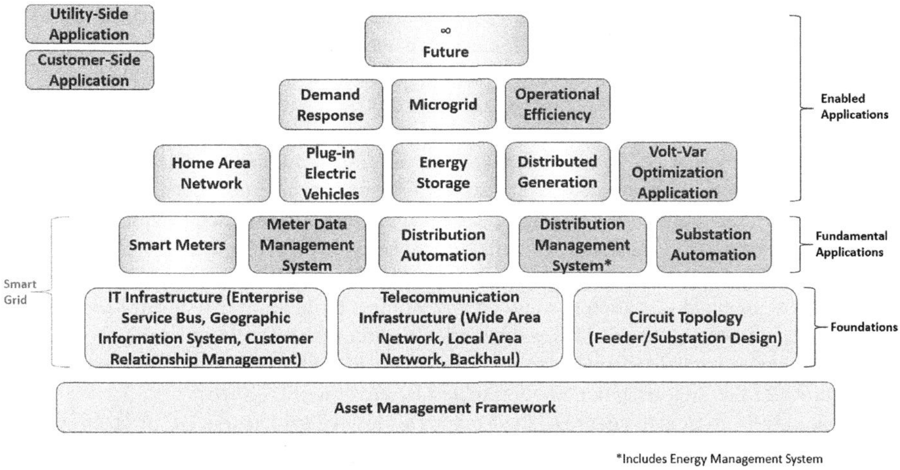

Fig. 1.22 Pyramid of the future smart grid.
Adapted from BC Hydro.

The electricity grid is now the focus of assorted technological innovations. Utilities worldwide are taking steps toward incorporating new technologies in many operations and infrastructure. At the core of this transformation is the need to make more efficient use of current assets. Fig. 1.22 shows a typical utility pyramid in which asset management is at the base of smart grid development. It is on this base that utilities build a foundation for the smart grid through a careful overhaul, information technology, communication, and systems infrastructure (Gungor et al., 2011; Yu et al., 2011; Hossain et al., 2016) provide a good overview of smart grid and its directions.

With the renewable resources at customer premises growing rapidly, while electricity providers move into large solar and wind farms to reduce emissions and the carbon footprint, the integration of wider concepts behind the smart grid as shown in Fig. 1.22 is becoming important. However, the existing inertia of the legacy AC grid, with a profit-based operation, is limiting the rapid progress into fully functional smart grids in the world.

1.8 DG penetration and evolving DC microgrids

1.8.1 Reasons behind DC evolution

As we discussed earlier in this chapter, the global electric infrastructure is dominated by legacy alternating current (AC) grid. However, recently, with the growing interest in DG, renewable energy resources, energy storage, and the availability of high-power semiconductor devices, DC distribution has been receiving much attention among industry, researchers, and community as a competitive alternative to conventional AC distribution. The encouraging reasons to move to DC microgrids are as follows (Patterson, 2012; Brissette et al., 2015; Iyer et al., 2015; Prabhala et al., 2018).

(a) Except few applications, almost all the electric loads around us particularly consumer electronics have become DC dominated loads which natively produce and consume DC in their internal components. Many household electrical appliances such as smart phone, laptop, refrigerator, air conditioner, LED lights, etc., work internally on DC. Modern " inverter driven " white goods are internally operate with single phase AC input directly rectified to create an internal DC bus, which in turn feeds all the building blocks of the appliance. Figure 1.23 depicts the internal block diagram of such an inverter driven white good.
(b) DG sources such as photovoltaics (PV) are in favor of DC. Distributed generation systems, such as photovoltaic cells and fuel cells, produce energy in the form of DC power so it is more efficient and cost effective to connect with DC than converting to AC.
(c) Increasing demand for efficiency and energy storage, through the use of brushless permanent magnet DC motors (BLDC) and batteries, and the uptake in electric vehicles have tipped the scales further toward an end-use environment dominated by DC. For example, in Figure 1.23 the motors shown are dominated by brush-less DC types.
(d) In the current context to adapt DC loads to our legacy AC grid, native DC products require AC-DC power supplies and chargers to rectify and down-convert higher voltage AC power to the lower voltage DC power required by the device. Despite the best engineering efforts, these power conversion stage still turn 10%–30% of AC power into waste heat in the conversion process (Prabhala et al., 2018). The direct use of DC power in a DC grid would eliminate many of these power conversion losses by maximizing self-consumption of the energy produced on site and by reducing imports of electricity from the grid. An illustration of reduced power conversion stages in the future implementation of DC homes compared with present-day AC homes are shown in Figs. 1.23 and 1.24.
(e) With DC power, there is no reactive power loading in the lines, there is no need for phase synchronization, and no harmonic current mitigation required in the DC installation. This reduces the complexity of engineering networks with multiple sources of supply and helps to improve system efficiency.
(f) DC-DC power converters consume far less standby power than conventional AC-DC converters.

1.8.2 DC grid adoption paths

A systematic transition to DC distribution in homes presents a much steeper technology adoption curve than market transitions for stand-alone products. Installing or converting to DC requires not only DC-ready loads, but also infrastructure such as power converters particularly DC-DC converters and DC-AC converters, and DC wiring and DC standards, resulting in higher system costs and the need for a relatively savvy consumer to play an active role in initiating and carrying out the transition. In specific cases of new construction or retrofits, DC may be used in the whole home, but in existing homes, more limited applications could meet particular consumer goals. These goals may range from total energy independence to resiliency during grid outages to the convenience of eliminating external power supplies, in addition to saving money on electricity bills.

Standards and codes play critical roles in moving us to indigenous DC electricity generation, distribution, and use. Product and system standards also play an essential role is supporting the effective deployment of products for alternative energy equipment and systems. Proactive development of the requirements for appropriate

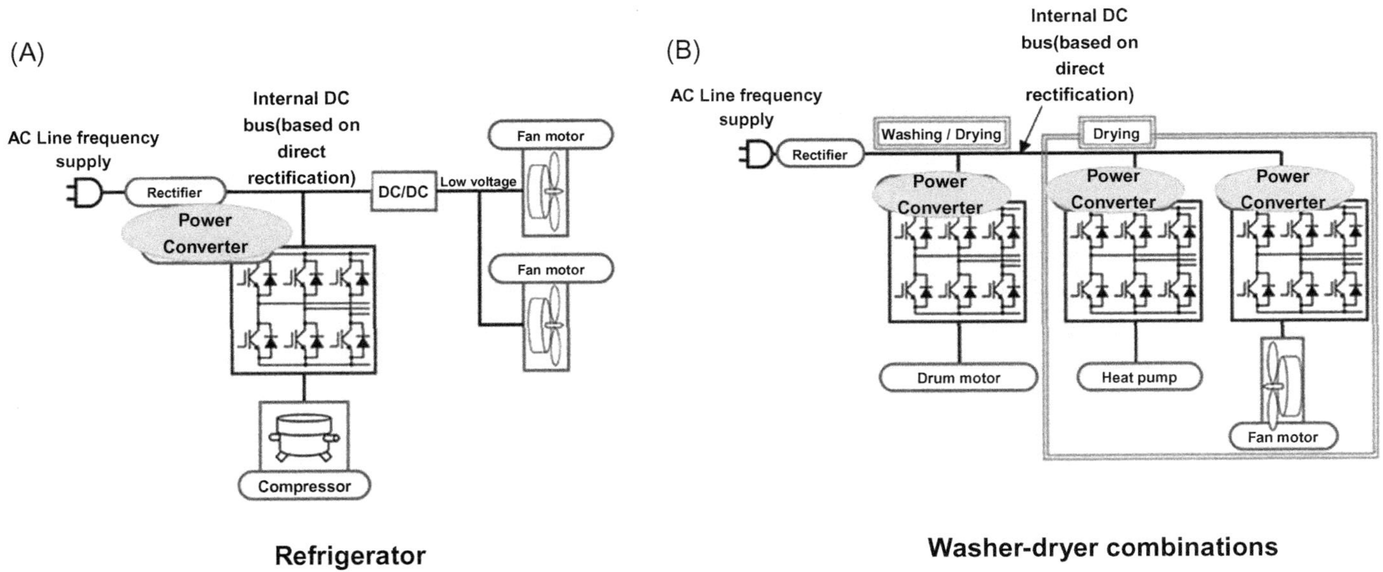

Fig. 1.23 A representative block diagrams for "inverter-driven" white goods (A) refrigerator (B) washer-dryer combination.

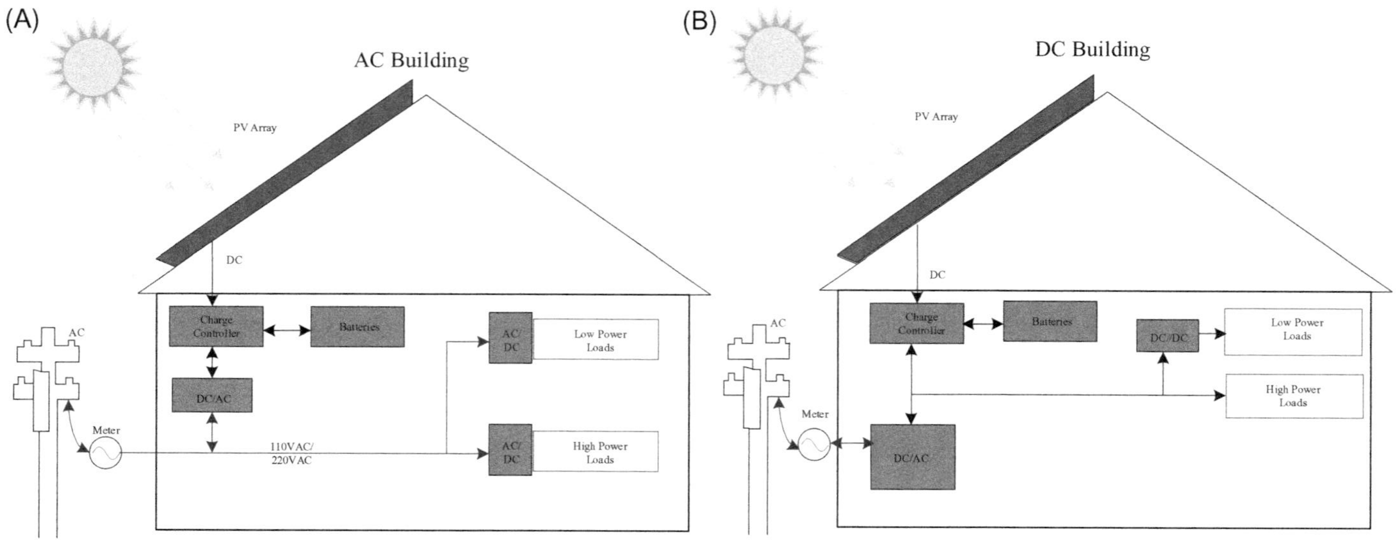

Fig. 1.24 An illustration of conversion stages of (A) AC house and (B) DC house in solar energy harvesting.

application, design and test requirements, code compatibility, and the definition of standardized product interoperability, system attributes, and usage outcomes are all a part of their clarifying roles. In the case of DC power distribution systems organizations such as EA (Emerge Alliance), IEEE (Institute of Electrical and Electronics Engineers), IEC (International Electrotechnical Commission), and ETSI (European Telecommunications Standards Institute) have directly teamed up in a number of formal and informal ways to develop these much-needed standards. Future looks promising with current trends in research innovations in this DC microgrid area.

References

Ackermann, T., Andersson, G., Soder, L., 2001a. Distributed generation: a definition. Electric Power Syst. Res. 57, 195–204.

Ackermann, T., Andersson, G., Söder, L., 2001b. Distributed generation: a definition. Electr. Power Syst. Res. 57 (3), 195–204.

Aderinto, T., Li, H., 2018. Ocean wave energy converters: status and challenges. Energies 1250 (11).

Agarwal, U., Jain, N., 2019. Distributed energy resources and supportive methodologies for their optimal planning under modern distribution network: a review. Technol. Econom. Smart Grids Sustain. Energy 4 (3).

Bakar, B.A., Musirin, I., Othman, M.M., Rahim, M.N.A., 2011. Mathematical model of sea wave energy in electricity generation. In: Presented at the International Power Engineering and Optimization Conference (PEOCO2011), Shah Alam, Selangor, Malaysia.

Barroso, L., Hugh, R., 2019. The future power system centralized, distributed, or just integrated? IEEE Power Energy Mag., 10–12.

Beik, O., Al-Adsani, A.S., 2020. Wind turbine multiphase operational trajectory in an all-DC wind generation system. IET Renew. Power Gener. 14 (15), 2916–2923.

Blaabjerg, F., Chen, Z., Kjaer, S.B., 2004. Power electronics as efficient interface in dispersed power generation systems. IEEE Trans. Power Electron. 19 (5), 1184–1194.

Blaabjerg, F., Yang, Y., Yang, D., Wang, X., 2017. Distributed power-generation systems and protection. Proc. IEEE 105 (7).

Bouffard, F., Kirschen, D.S., 2008. "Centralised and distributed electricity systems," Elsevier Energy Policy, vol. 36, pp. 4504–4508.

Brissette, A., Carr, J., Cui, T., Xu, J., Qi, L., 2015. Analysis of emerging technology for DC-enabled smart homes. In: Presented at the 2015 IEEE First International Conference on DC Microgrids (ICDCM), Atlanta, GA.

Buriticá-Arboleda, C.I., Álvarez-Bel, C., 2011. Decentralized energy: key to improve the electric supply security. In: Presented at the 2011 IEEE PES Confrenece on Innovative Smart Grid Technologies Latin America Medellin.

Cárdenas, B., Swinfen-Styles, L., Rouse, J.P., Hoskin, A., Xu, W., Garvey, S.D., 2019. Energy storage for a high penetration of renewables. In: Presented at the 2019 Offshore Energy and Storage Summit (OSES), BREST, France.

Chen, H., Cong, T.N., Yang, W., Tan, C., Li, Y., Ding, Y., 2009. Progress in electrical energy storage system: a critical review. Prog. Nat. Sci. 19, 291–312.

Child, M., Koskinen, O., Linnanen, L., Breyer, C., 2018. Sustainability guardrails for energy scenarios of the global energy transition. Renew. Sust. Energ. Rev. 91 (321-334).

Chiradeja, P., Ramakumar, R., 2004. An approach to quantify the technical benefits of distributed generation. IEEE Trans. Energy Convers. 19 (4), 764–773.

Cotfas, D.T., Cotfas, P.A., Kaplanis, S., 2013. Methods to determine the dc parameters of solar cells: a critical review. Renew. Sust. Energ. Rev., 588–596.

Deng, Y.Y., Blok, K., Leun, K.V.D., 2012. Transition to a fully sustainable global energy system. Energy Strategy Rev. 1, 109–121.

El-Khattam, W., Salama, M.M.A., 2004. Distributed generation technologies, definitions and benefits. Electr. Power Syst. Res. 71, 119–128.

Falcao, A.F., 2010. Wave energy utilization: a review of the technologies. Renew. Sust. Energ. Rev. 14, 899–918.

Falnes, J., 2007. A review of wave-energy extraction. Marine Struct. 20, 185–201.

Farhangi, H., 2010. The path of the smart grid. IEEE Power Energy Mag. 8 (1), 18–28.

Folley, M., Whittaker, T.J.T., 2009. Analysis of the nearshore wave energy resource. Renew. Energy 34, 1709–1715.

Gil, H.A., Jöos, G., 2006. On the quantification of the network capacity deferral value of distributed generation. IEEE Trans. Power Syst. 21 (4), 1592–1599.

Global Solar ATLAS. 2019 https://globalsolaratlas.info/download/world (Accessed).

Grigsby, L.L., 2012. Electric Power Generation, Transmission and Distribution. CRC Press.

Gungor, V.C., et al., 2011. Smart grid technologies: communication technologies and standards. IEEE Trans. Indust. Inform. 7 (4), 529–539.

Gunn, K., Stock-Williams, C., 2012. Quantifying the global wave power resource. Renew. Energy 44, 296–304.

Harrison, G.P., Wallace, A.R., 2005. Optimal power flow evaluation of distribution network capacity for the connection of distributed generation. IEE Proc. C: Gener. Transm. Distrib. 152 (1), 115–122.

Hossain, M.S., Madlool, N.A., Rahim, N.A., Selvaraj, J., Pandey, A.K., FaheemKhan, A., 2016. Role of smart grid in renewable energy: an overview. Renew. Sust. Energ. Rev. 60, 1168–1184.

Iyer, S., Dunford, W.G., Ordonez, M., 2015. DC distribution systems for homes. In: Presented at the IEEE Power & Energy Society General Meeting, Denver, CO.

Keane, A., O'Malley, M., 2005. Optimal allocation of embedded generation on distribution networks. IEEE Trans. Power Syst. 20 (3), 1640–1646.

Kober, T., Schiffer, H.W., Densing, M., Panos, E., 2020. Global energy perspectives to 2060—WEC's World Energy Scenarios 2019. Energy Strategy Rev. 31, 1–19.

Mahmoud, K., Yorino, N., Ahmed, A., 2016. Optimal distributed generation allocation in distribution systems for loss minimization. IEEE Trans. Power Syst. 31 (2), 960–969.

Markandya, A., Wilkinson, P., 2007. Electricity generation and health. Lancet 370 (9591), 979–990.

Meegahapola, L., Sguarezi, A., Bryant, J.S., Gu, M., Conde D, E.R., Cunha, R., 2020. Power system stability with power-electronic converter interfaced renewable power generation: present issues and future trends. Energies 13, 3441.

Mohammed, Y.S., Mustafa, M.W., Bashir, N., 2014. Hybrid renewable energy systems for off-grid electric power: review of substantial issues. Renew. Sust. Energ. Rev. 35, 527–539.

Mojumdar, M.R.R., Himel, M.S.H., Kayes, G., 2015. A distinctive analysis between distributed and centralized power generation. Int. J. Rec. Res. Electr. Electron. Eng. 2 (4), 1–6.

Montoya-Bueno, S., Munoz-Hernández, J., Contreras, J., 2016. Uncertainty management of renewable distributed generation. J. Clean. Prod. 138, 103–118.

Murray, P., Orehouniga, K., Grosspietschc, D., Carmelieta, J., 2018. A comparison of storage systems in neighbourhood decentralized energy system applications from 2015 to 2050. Appl. Energy 231, 1285–1306.

Parra, D., Gillott, M., Walker, G.S., 2016. Design, testing and evaluation of a community hydrogen storage system for end user applications. Int. J. Hydrog. Energy 41, 5215–5229.

Patterson, B.T., 2012. DC, come home DC microgrids and the birth of the "Enernet". IEEE Power Energy Mag., 60–69.

Prabhala, V.A., Baddipadiga, B.P., Fajri, P., Ferdowsi, M., 2018. An overview of direct current distribution system architectures & benefits. MDPI Energies 11 (2463).

Prakash, P., Khatod, D.K., 2016. Optimal sizing and siting techniques for distributed generationin distribution systems: a review. Renew. Sust. Energ. Rev. 57, 111–130.

Rau, N.S., Wan, Y.-H., 1994. Optimum location of resources in distributed planning. IEEE Trans. Power Syst. 9 (4), 2014–2020.

Rohden, M., Jung, D., Tamrakar, S., Ketteman, S., 2016. Cascading failures in ac electricity grids. Phys. Rev. E 3, 94.

Sambaiah, K.S., 2018. A review on optimal allocation and sizing techniques for DG in distribution systems. Int. J. Renew. Energy Res. 8 (3).

Schäfer, B., Witthaut, D., Timme, M., Latora, V., 2018. Dynamically induced cascading failures in power grids. Nat. Commun. 9.

Sen, Z., 2004. Solar energy in progress and future research trends. Prog. Energy Combust. Sci. 30, 367–416.

Shafiee, S., Topal, E., 2009. When will fossil fuel reserves be diminished? Energy Policy 37, 181–189.

Singh, B., Sharma, J., 2017. A review on distributed generation planning. Renew. Sust. Energ. Rev. 76, 529–544.

Singh, R., Alapatt, G.F., Bedi, G., 2014. Why and how photovoltaics will provide cheapest electricity in 21st century. Electron. Energet. 27 (2), 275–298.

Speidel, S., Braunl, T., 2016. Leaving the grid—the effect of combining home energy storage with renewable energy generation. Renew. Sust. Energ. Rev. 60, 1213–1224.

Suganthia, L., Samuel, A.A., 2012. Energy models for demand forecasting—a review. Renew. Sust. Energ. Rev. 16, 1223–1240.

Villalva, M.G., Gazoli, J.R., 2009. Comprehensive approach to modeling and simulation of photovoltaic arrays. IEEE Trans. Power Electron. 24 (5), 1198–1206.

Viral, R., Khatod, D.K., 2012. Optimal planning of distributed generation systems in distribution system: a review. Renew. Sustain. Energy Rev. 16, 5146–5165.

Wang, C., Nehrir, M.H., 2004. Analytical approaches for optimal placement of distributed generation sources in power systems. IEEE Trans. Power Syst. 19 (4), 2068–2076.

Yu, X., Cecati, C., Dillon, T., Simões, M.G., 2011. The new frontier of smart grids. IEEE Ind. Electron. Mag. 5 (3), 49–63.

Fundamentals of energy storage devices

2

2.1 Introduction

From the invention of the transistor in the late 1940s, electronic products and systems have become an essential part of the modern world with interconnected global communication systems, intelligent white goods, electronic subsystem-dominated automotive, and a plethora of portable devices. With the world population growing beyond 7.8 bilion in 2021, and with portable electronic devices getting popular even in the poorest parts of the world, worldwide energy consumption keeps growing at a steady rate. In 2008, the world's energy consumption was estimated to be around 144,000 TWh and by 2019 this was growing towards 160,000 TWh. Energy supply usually comes from several dominant sources, namely, fossil fuel and nuclear (as nondominant major sources), and from many renewable sources such as hydroelectric, solar, wind, geothermal, biomass, and biofuels.

In the world of energy use, energy storage requirement comes in different forms. In our home and work environments, sometimes we need to store energy in systems such as the uninterruptible power supplies (UPSs) which is used for information systems and other critical facilities such as hospitals, airports, and factories in order to safeguard against power outages. In these situations, different types of energy storage systems (ESSs) such as batteries, supercapacitor (SC) banks, flywheels, and compressed air are used together with suitable electromechanical energy conversion systems. In these systems, we usually deal with a few 100 W to few megawatts order requirements with the outages expected in the range of fractional seconds to few hours.

In our modern hybrid electric vehicles (HEVs) and electric vehicles (EVs), battery banks and fuel cells are used to deal with kilowatt order requirements lasting for few minutes to few hours of driving, where the ESS comes into play. With the inefficiencies of internal combustion engine together with mechanical drive systems in the cars, overall energy efficiency is in the range of 15% to over 20%, encouraging the development of more energy efficient HEVs and EVs. In these applications, where tens of kilowatts of power are required, the capacity of the battery banks in terms of energy storage should be in the range of kilowatt-hours.

Compared to the above scenario, if we consider our portable electronic devices such as cell phones, PDAs, and laptops where battery packs are used, we commonly come across the capacity in terms of milliampere-hours (mAh) or ampere-hours (Ah), with battery packs with nominal voltage values varying from about 2.5 to about 16 V. For power tools with rechargeable battery packs the voltages could be in the range of 10 to 60 V in general. Given these nominal voltage values, assumed to be constant over the operational range, will give us the energy storage capacity range from mWh to Wh. Compared to the two above application areas, if we take much smaller

Energy Storage Devices for Renewable Energy-Based Systems. https://doi.org/10.1016/B978-0-12-820778-9.00005-X

power and energy requirements such as wristwatches, hearing aids, and implants, the battery pack capacity may be specified in microwatt-hours to few milliwatt-hours only. Terminal voltages of these battery packs, mostly coming in the form of one or few single cells, may be in the range of 0.5 V to few Volts per cell.

Given a few common day-to-day examples of battery-based systems, we could practically appreciate that a few commonly identified electrical parameters such as terminal voltage, and the energy storage capacity in terms of Watt-hours or milliwatt-hours allow us to compare the capabilities of the battery packs. If a load consumes power at a steady rate of 1 W for 1 h, the device has consumed total energy of 1 Wh during that period. In standard SI units, this translates to a value of 3600 J, since 1 W dissipated during 1 s is equal to 1 J.

So far, we have considered only batteries as an energy storage device (ESD) family, but in general, many different methods and devices can be used for energy storage suitable for electrical and electronic systems. This chapter provides an overview of energy storage and delivery devices, methods, and the essential fundamentals applicable to energy transfer into electronic systems and devices. With some quantification approach, the contents will assist in comparing the details of energy storage capabilities, and energy delivery capabilities and limitations of ESDs in electronic environments in a practically useful way. The contents of this and the overall work are aimed as a practical guideline to engineers, designers, and researchers working in these areas. The reader is expected to have some essential prior knowledge on circuit theory applicable to simple circuits.

2.2 Simple fundamentals

Let us start with the simple analogy of a water tank at a certain height above the ground level. If we have a hose connected from the bottom of the tank, with a tap connected at the end of the hose, water flow starts when the tap is opened. We can measure the rate of water flow in terms of liters/second. If you consider the tank as the storage device, at this rate we can estimate the time to release the total quantity of water. If we make the tank bigger, we get more quantity of water stored in the tank. If the bottom of the tank is at the same level as before, we will have the same rate of flow at the end of the pipe. We will take a longer time to release that water at that same rate. Now we can raise the bottom level of the tank which will increase the rate of flow. This means the tank will empty faster. We can also increase the diameter of the water hose connected to achieve a higher flow rate of water. This happens since the resistance to water flow will reduce due to the higher diameter of the hose. If you increase the length of the hose, the flow rate will be slower.

The above simple analogy helps us understand ESDs. The level of the water tank (above the reference level of the ground) is analogous to the (positive) terminal voltage of the device with respect to the negative terminal (which is the reference voltage level). The volume of the tank is analogous to the amount of total charge (or energy) available for release into the external circuit. Water flow in the hose is analogous to the current (in Amperes or coulombs/second) in the external circuit. By changing the length or diameter of the hose, we create the analogy of resistance of a conductor (which decreases with the diameter and increases with the length). Sometimes changing the material of the hose, we can notice the different rate of flow for the same diameter/length, which indicates us the analogy of the resistivity of electrical material.

2.2.1 Work, power, and energy

We all have an intuitive sense of the term *work*. The precise definition of work reflects this intuition, and we can define work done, W, by an applied force $\vec{F}$ (a vector quantity, which has an amplitude and a direction) moving an object by a distance Δx,

$$W = \vec{F}\,\Delta x \tag{2.1}$$

In SI units, force is measured in Newton and the distance moved is in meters, giving units of Newton-meter. Newton-meter is given the name Joule. It is important to note that to have distance moved and force applied should be in the same direction. Otherwise, respective directional components need to be used. Work is a scalar quantity and has no direction, but the force and distance both are vector quantities.

Closely related to work, *energy* is the ability to do work. Work and energy have the same unit, Joule. Energy, which is one of the most important concepts in all of physics, comes in different forms. For example, a moving object of mass m moving at the velocity of v m/s carries kinetic energy, K, given by $(1/2)mv^2$. The work-energy theorem indicates that the change in an object's kinetic energy is equal to the net work done on the object:

$$\Delta k = W_{\text{net}} \tag{2.2}$$

In practice, we come across many different units that are used for energy, particularly applicable to different subject areas. Examples are *electron-volt* used in nuclear, atomic, and molecular physics; and the *calorie* used in thermodynamics and chemistry. Older English units include *foot-pound* and the *British thermal unit (BTU)* and the latter is frequently used in heating and cooling systems. Electricity companies use *kilowatt-hours* (kWh). For all these commonly used practical units, one can use conversion factors as in Table 2.1 to convert to Joules.

The next important item in energy conversion and ESDs is the *power*, which is the rate of doing work or rate of change of energy, which is measured in *watts*. Power is defined as the rate of doing work. *Average power*, $\overline{P}$, can be expressed as

$$\overline{P} = \frac{\Delta W}{\Delta t} \tag{2.3}$$

In practice, as the rate of doing work varies with time it is useful to define the *instantaneous power*, P, as

Table 2.1 Conversion coefficients for different units for energy and power.

	To convert from	To	Multiplication factor
Energy	Barrel of oil	Gigajoules (GJ)	≈ 6
	Cubic foot of methane (at standard temperature and pressure)	Megajoules (MJ)	≈ 1
	kiloWatt-hour	Joule	3.6×10^6
	Calorie (mean)[a]	Joule	4.19002
	British Thermal unit (BTU)[b]	Joule	1055.87
	Electron-volt	Joule	1.60206×10^{-19}
Power	Foot lbf/h	Watt	3.7662×10^{-4}
	Horse power (electrical)	Watt	746
	Horse power (mechanical)	Watt	735

[a] There are slightly different conversion factors for calorie used in different cases (see da Rosa, 2013 for details).
[b] Mean value is indicated (see da Rosa, 2013 for details).
Adapted from da Rosa, A.V., 2013. Fundamentals of Renewable Energy Processes, third ed. Academic Press, Oxford, 884 p.

$$P = \lim_{\Delta t\to 0}\frac{\Delta W}{\Delta t} = \frac{dW}{dt} \tag{2.4}$$

Since our subject of the book is ESDs for electrical or electronic engineering environments, we deal with power and energy associated with electrical, electrochemical, or electromechanical systems, and we can measure electrical power consumed or generated based on the terminal voltage and the current passing through an electrical component as

$$P = VI \tag{2.5}$$

Unit of power is Joules/second and this unit is given the name *watt* (W) in honor of James Watt, a Scottish engineer who developed the steam engine. Watt himself defined another unit, *horsepower (hp)*, which is about 746 J/s or 746 W. Table 2.1 provides some useful conversion factors usable in work, energy, and power (da Rosa, 2013).

2.2.2 Impact of the open-circuit voltage and internal resistance of an energy source

Now let us look at a simple energy source, with a constant terminal voltage of V volts, with an energy storage capacity of E watt-hours (Joules ≡ watt-seconds). In any such ESD, there is a finite internal resistance, and if we quantify that by a lumped resistance, r_{int}, we can draw up an equivalent circuit as shown in Fig. 2.1A. Now we

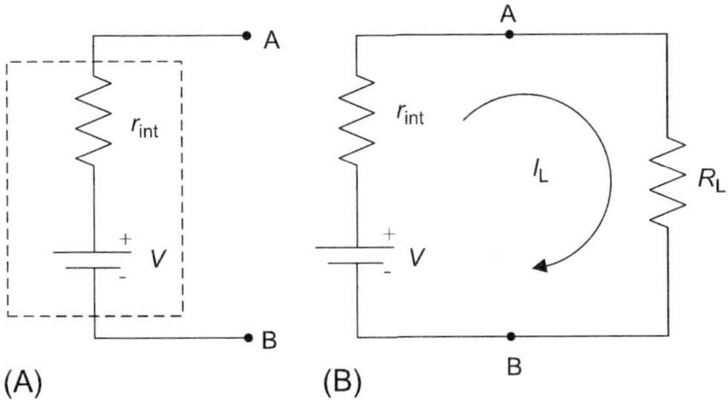

Fig. 2.1 Energy storage device (ESD) and a resistive load connected to its external terminals (A) storage device represented by a constant voltage source and a fixed lumped resistance of value r_{int} (B) closed circuit formed by the external load.

can have a load in the form of a resistor depicted by R_L connected to this device which forms the closed circuit as shown in Fig. 2.1B.

This situation can be simply analyzed by applying the ohms law to the closed circuit and this will give us a current of I_L given by

$$I_L = \frac{V}{R_L + r_{int}} \tag{2.6}$$

The actual voltage across the load is given by

$$V_L = \frac{VR_L}{R_L + r_{int}} \tag{2.7}$$

This can be rewritten as

$$V_L = \frac{V}{1 + \frac{r_{int}}{R_L}} \tag{2.8}$$

Eq. (2.8) indicates that useful voltage appearing at the load end keeps dropping as the internal resistance of the energy source keeps increasing. Therefore, a good energy source should not only have a higher open-circuit voltage, but also a very low internal resistance. Now we can calculate the power dissipated at the load by

$$P_L = V_L I_L = \frac{V}{1 + \frac{r_{int}}{R_L}} \frac{V}{R_L + r_{int}} = \frac{V^2}{R_L}\left[\frac{1}{\left(1 + \frac{r_{int}}{R_L}\right)^2}\right] \tag{2.9}$$

This indicates that the total power deliverable to a load by an ideal power source with zero internal resistance is V^2/R_L, and in practice, the maximum power deliverable to a load is restricted by the internal resistance of the energy source. For instance, in many rechargeable batteries, internal resistance keeps increasing with the discharge. Then even if the electrochemistry in the cell allows us to have a constant open-circuit voltage, as the cell keeps discharging the available power keeps dropping due to the increase of internal resistance. This is one of the reasons for SCs' ability to supply large amounts of power for short durations, compared to batteries. Many modern SC families have very low internal resistance, for example in the order of fractional milliohms to tens of milliohms, and this allows these devices to drive large amounts of power into a load when fully charged up to the rated voltage.

2.2.2.1 Maximum power transfer

Given the relationship in Eq. (2.9), we see that due to the internal resistance of the ESD, part of the energy is converted and wasted as heat generated in the internal resistance, and only a particular percentage is usable at the external load. In many cases, we need a particular level of power to be delivered to the external load and this is limited by the combination of internal resistance and external load resistance. Using differentiation of the function given by Eq. (2.9), with respect to R_L we can show that maximum possible load power occurs when $R_L = r_{int}$. Under this condition, where $R_L = r_{int}$ maximum possible power is given by $V^2/4R_L$.

2.2.3 Energy wasted inside a source and its heating effect

From Eq. (2.6), we can derive the amount of power wasted as heat inside an energy source as

$$P_{loss,\,int} = I_L^2 r_{int} = \left(\frac{V}{R_L + r_{int}}\right)^2 r_{int} \tag{2.10}$$

Using Eq. (2.4), this can be rewritten as

$$P_{loss,\,int} = P_L\left(\frac{r_{int}}{R_L}\right) \tag{2.11}$$

This indicates that while we extract energy from an energy source, a quantifiable share is wasted within the source. At the maximum power transfer condition, internal dissipation becomes equal to the power consumed by the load. This energy wastage is in the form of heat, and it could usually have an adverse effect on the lifetime of the device. In general, most device specifications are temperature sensitive, and higher operating temperatures tend to have adverse effects on the delivery of energy and the lifetime of the device.

2.2.4 Time delays in delivering or transferring energy

In many cases, the amount of useful energy stored in a storage device needs to be transferred into a load within a very short duration. One automotive example is the quick energy required in accelerating a sports car to a particular speed within a very short period of time. More discussion on this is given in Chapter 8 related to Lamborghini Sian model. Another possible domestic example is to heat the stored water in a pipeline between the tap and the central heating system so that when we open the tap, we get warm water instantaneously. In such cases, a designer should be able to estimate and quantify the speeds and delays related to a combination of the ESD and the associated load. In this process, time constants associated with the ESD and the load will come into the picture.

To appreciate this aspect, let us consider a simple example of a capacitor of value C with an equivalent series resistance (ESR) of r_C connected to a load of R_L via a switch as in Fig. 2.2. If the capacitor is initially charged to a DC voltage of V_{C0}, the total energy stored in the capacitor is given by

$$E_{C0} = \frac{1}{2} C V^2_{Co} \tag{2.12}$$

At time $t=0$ when the switch is activated, the instantaneous current through the load resistor is given by

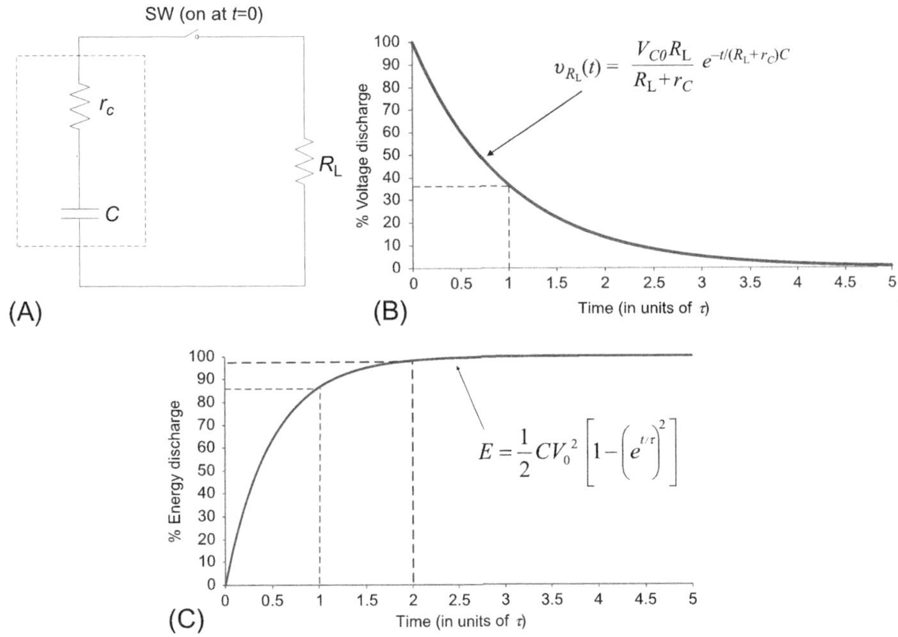

Fig. 2.2 Capacitor as an energy storage device: (A) simplified circuit showing capacitor ESR in the overall circuit; (B) capacitor discharge curve; and (C) percentage energy released to a load as a function of time measured in time constant as a unit.

$$I_{R_{\mathrm{L}},t=0} = \frac{V_{C0}}{R_{\mathrm{L}} + r_C} \tag{2.13}$$

This indicates that the ESR of the capacitor could play a significant role in the maximum possible current, particularly when the load resistance is pretty small. Considering the combination of the capacitance, external load resistance, and the ESR in series, circuit time constant, τ, is given by

$$\tau = (R_{\mathrm{L}} + r_C)C \tag{2.14}$$

Using this time constant and the initial voltage across the capacitor, we can derive the time-varying exponential voltage waveform across the capacitor as

$$v_C(t) = V_{C0}\mathrm{e}^{-t/\tau} = V_{C0}\mathrm{e}^{-t/(R_{\mathrm{L}} + r_C)C} \tag{2.15}$$

Based on Eq. (2.15), in practical situations within about 5 time constants, capacitor voltage discharges down to almost 98% of its original voltage. Given these relationships, the capacitor releases its stored energy based on the following relationship:

$$\begin{aligned} E_{C0} - E_{C,t} &= \frac{1}{2}C\left[V_{C0}^2 - \left(V_{C0}\mathrm{e}^{-t/(R_{\mathrm{L}} + r_C)C}\right)^2\right] \\ &= \frac{1}{2}CV_{C0}^2\left(1 - \mathrm{e}^{-[2t/(R_{\mathrm{L}} + r_C)C]}\right) \end{aligned} \tag{2.16}$$

$E_{C,t}$ is the remaining energy in the capacitor after a time period of t.

The above indicates that the energy discharged within a period of one time constant, E_{D}, τ is given by

$$E_{\mathrm{D},\tau} = \frac{1}{2}CV_{C0}^2\left(1 - \mathrm{e}^{-2}\right) \approx 0.86E_{C0} \tag{2.17}$$

Similarly, we can calculate the discharge of useful energy into an external load and the associated loss in the ESR of the capacitor. Given the above situation, if the ESR is significantly large compared to the external load, transfer efficiency will be quite low. Fig. 2.2B depicts the voltage across the capacitive element of the capacitor versus time and Fig. 2.2C indicates the total energy discharged into the external load and the ESR. From Fig. 2.2C, we see that within less than 2 time constants, nearly all energy stored could be discharged.

The two examples we have considered, namely, an electrochemical cell (with a constant open-circuit voltage over time and temperature) and a capacitor, indicate us that the energy delivery capabilities of these two devices are different. In the case of the capacitor, the amount of energy stored is dependent only on the voltage capability of the capacitor and the value of capacitance. Energy delivery is not constant over time and it tends to deliver over 85% of its energy within the first time constant. Also in an electrochemical device like a battery, to predict the overall stored energy, you need much more information about the device, compared to the electrical parameters applicable to the device operation. That part of the subject is electro-chemistry, which is beyond the scope of this book.

2.2.5 Complex models of ESDs

In the previous section, we considered extremely simplified models for an electrochemical cell or a capacitor with two lumped-parameter elements. However, in the real world of ESDs, the device models are much more complicated than these two sample cases. Fig. 2.3 indicates a more realistic model for a lead-acid battery (Grillo et al., 2011) and Fig. 2.4 indicates a model for a SC (Musolino et al., 2013).

Compared to the simplified model indicated in Fig. 2.1, the equivalent circuit shown in Fig. 2.3 has an electromotive force which varies with the temperature and the state of charge of the battery (Moubayed et al., 2008). Single lumped resistance used to represent the internal resistance of the battery model in Fig. 2.1 really comprises three different components based on terminal resistance, and two other components that depend on the state of charge and the discharge current. More details are available in Moubayed et al. (2008). Two capacitors are shown across the two resistances R_1 and R_2 add time delays to make the voltage and current estimations in external circuits more complex, for both charging and discharging. Later chapters in the book will discuss more details on modeling batteries and SCs.

When we look at the SC equivalent circuit in Fig. 2.4, to estimate time delays and limitations, such a situation calls for much more analytical- or simulation-type approaches, since the mathematics involved with such a problem is much more complex. In developing these complex models, which includes a voltage dependent capacitance (given by $C_{i1}*V_{Ci}$) researchers have combined experimental and simulation work, in order to match the physics in the electric double-layer effect. Present-day SCs or ultra-capacitors, technically mean the same family of electrical double-layer capacitors (Zubieta and Bonert, 2000). We will discuss more details on these capacitors in Chapters 7 & 8.

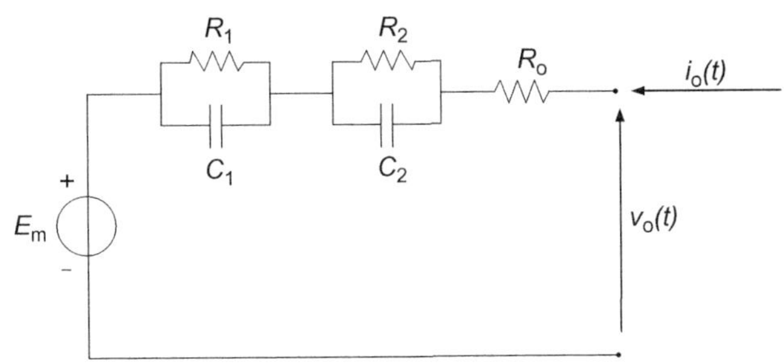

Fig. 2.3 Equivalent circuit for a lead-acid battery.
Adapted from Grillo, S., Martini, L., Musolino, V., Piegari, L., Tironi, E., Tornelli, C., 2011. Management of different energy storage devices using a losses minimization algorithm. In: Proceedings of IEEE-ICCEP Conference, pp. 420–425.

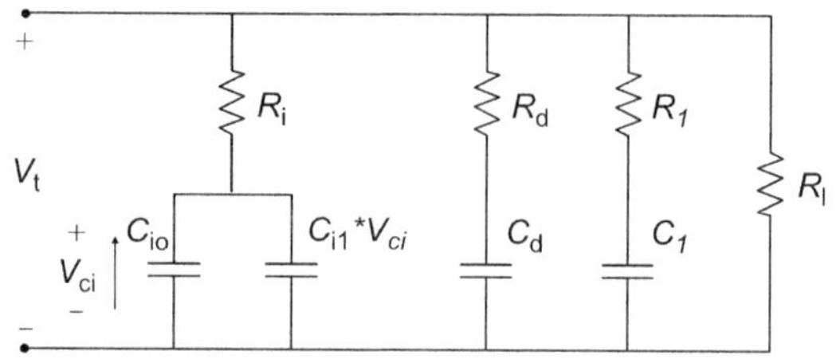

Fig. 2.4 Supercapacitor (SC) equivalent circuit which includes a voltage dependent capacitor in the first branch.
Adapted from Zubieta, L., Bonert, R., 2000. Characterization of double-layer capacitors for power electronic applications. IEEE Trans. Ind. Appl. 36 (1), 199–205.

2.3 Energy storage in electrical systems

In electrical systems and electronic products, energy storage requirements come in two basic forms. One is short-term storage where components in powered electrical circuits store energy in electrostatic or electromagnetic form. Capacitors and inductors are typical examples. The other case is where we use ESDs for longer-term energy supply as a case of a redundant energy source.

When a system is powered by an unreliable energy supply, such as frequently failing AC power line, or, the system is a portable one, we require some form of energy storage to keep the system operational with the expected reliability levels. As an example, if the AC mains supply to a critical industrial load such as a manufacturing plant suddenly fails, there should be a suitable standby power system such as a generator set buffered by a UPS. In this scenario, a UPS is expected to have sufficient energy storage in the form of chemical batteries, or by other means to supply energy until the generator sets installed are started. When the generator is coming into operation, energy storage is in the form of the fuel storage available for the generator set. In critical situations such as semiconductor fabrication facilities, where the AC mains supply is primarily fail-safe, still there are short-term situations where the quality of the AC supply is not up to the expected standard, and a short autonomy UPS system comes into operation to back up the AC mains. In such a situation, the UPS system may have only limited energy storage in the form of an SC bank, ready to back up the needs of a few AC cycle duration. In such cases, other options could be a flywheel or a compressed air supply.

2.3.1 *Basic electrical components as in-circuit energy storage*

When electrons flow through a resistor, energy is dissipated in the resistance and gets converted as heat and no ideal resistor could store energy. However, in an inductor within a closed circuit carrying a current of I, stored energy E_L is given by

$$E_L = \frac{1}{2}LI^2 \tag{2.18}$$

This energy storage occurs due to the magnetic flux generated by the current within the core of the inductor. However, since the energy is stored in the inductor due to the flux lines created by the current, this energy is not physically transportable with the device unless the rest of the circuit responsible for generating this current is kept undisturbed.

In contrast, if we connect a voltage source to a capacitor, its voltage exponentially rises to the value of the voltage source. If the DC source used to charge the device has an open-circuit voltage of V, stored energy in the capacitor of value C is given by

$$E_C = \frac{1}{2}CV^2 \tag{2.19}$$

Compared to the case of an inductor, this energy storage occurs in the form of energy associated with an electrostatic field based on electric-charge storage. For this reason, if the capacitor has no significant leakage (which represents an infinite leakage resistance), the device with the energy stored can be transported after disconnecting the voltage source. However, in practical capacitors, the leakage resistance is finite, and this creates a leakage current through this equivalent resistance, and energy is wasted in the form of heat dissipated in this large resistance. Practical devices such as electrolytic capacitors with capacitance values ranging from nanofarads to few thousand microfarads do retain their charge from microsecond order durations to few minutes, but they cannot be treated as useful ESDs similar to electrochemical devices such as batteries. In modern SC families, capacitance values vary from fractional Farads to few thousands of Farads, and their leakage currents are in the order of few microamperes to milliamperes only. This allows them to be used as short-term energy storage and delivery devices in power electronic systems.

2.3.2 Energy storage options for longer-term and infrequent utilization

In the previous paragraphs, we quantified how common energy storage components such as inductors and capacitors store energy based on current through the device or the voltage across the device. These scenarios are mostly used in circuit level design and on a very short-term basis. However, in situations such as critical electrical facilities at airports, semiconductor manufacturing, and computer server farms, there are few longer-term options for energy storage. They are

1. Electrochemical devices such as batteries
2. SC banks
3. Flywheels
4. Fuel cells and the associated energy source
5. Compressed air supply
6. Superconductive magnetic energy storage (SMES)

Most of these systems are usually used for back up purposes when a power failure or a power quality issue occurs on the incoming AC mains supply. On electrochemical batteries, which allow you to store energy for few minutes to few hours of backup, we will have a detailed discussion in this chapter and Chapter 3.

SC banks are usually combined with battery packs as hybrid devices to get the best performance of batteries and the SCs. Given the simple basics in the previous sections, a capacitor to achieve high stored energy, capacitance should be large or the voltage across the terminals should be large. In general, common forms of capacitors provide the designers with capacitance values in the range of picofarads to several thousand microfarads. Voltage capabilities usually come in the range of 10 V to over 10 kV and the manufacturing technology capabilities to improve these capacitors grow slowly.

Fig. 2.5 Different capacitors: (A) commonly used electrolytic caps; (B) supercapacitors from three different manufacturers; (C) a very large electrolytic capacitor; and (D) an SC module with balancing circuits.
Panel (C) Courtesy of AECS Ltd., Hastings, NZ and (D) Courtesy of ABB Research, Napier, New Zealand.

In contrast to commonly used capacitors, SCs come with very low voltage capability, usually in the range of 2 to less than 4 V. However, their capacitance values range from about 0.1 to over 50,000 F. Photograph in Fig. 2.5A depicts a set of electrolytic capacitors in the range of 10–2200 μF. Photograph in Fig. 2.5B depicts a set of SCs with values varying from 0.2 to 650 F. Photograph in Fig. 2.5C depicts a very large electrolytic capacitor used in Toyota Prius electrical system. Photograph in Fig. 2.5D indicates an SC module with multiple SCs (of 4000 F value) in series combined with a voltage balancing module. Table 2.2 provides a comparison of different capacitor options for the energy storage capabilities varying from 1 to over 10 kJ (Kularatna et al., 2011).

SCs usually have very low internal resistance (in the form of ESR) compared to rechargeable batteries, in general. Also, their ESR is relatively constant over its discharge period or the calendar life. This situation is illustrated from the example in Fig. 2.6 where it indicates the case of discharging these two devices at an approximate current of 1 A and plotting the measured internal resistance against the depth of discharge. Even a small SC such as a 0.6-F value (HS206F from Cap-XX) indicates a constant internal resistance of less than 100 mΩ, while the AA energizer cell indicates a case where its internal resistance keeps increasing with the depth of discharge. Fig. 2.7 depicts a case of discharging different types of SCs at a constant current of 1 A and comparing it with the behavior of an AA-type cell. While the SCs simply

Table 2.2 Comparison of typical electrolytic capacitors and supercapacitors for their ESR values and other useful specifications.

Energy storage limit	Capacitor type	Manufacturer	Parameters			
			Capacitance	**Terminal voltage (V)**	**Short circuit current (A)**	**ESR (mΩ)**
<1 J	Electrolytic	RSS	2200 μF	16	104	153
1–5 J	Supercap	Maxwell	1 F	2.7	3.85	700
		Cap-xx	2.4 F	2.3	115	20
	Electrolytic	Cornell Dubilier	2200 μF	50	704	71
5–50 J	Supercap	Maxwell	10 F	2.5	14	180
		Cap-xx	1.2 F	4.5	112.5	40
		Nesscap	10 F	2.3	33	70
	Electrolytic	Cornell Dubilier	82,000 μF	16	1441	11.1
		VICOR	270 μF	200	325	614
>50 J	Supercap	Maxwell	350 F	2.7	840	3.2
		Nesscap	120 F	2.3	144	16
>10 kJ	Supercap	Samwha Electric	3000 F	3.0 V	13,000 A	0.23

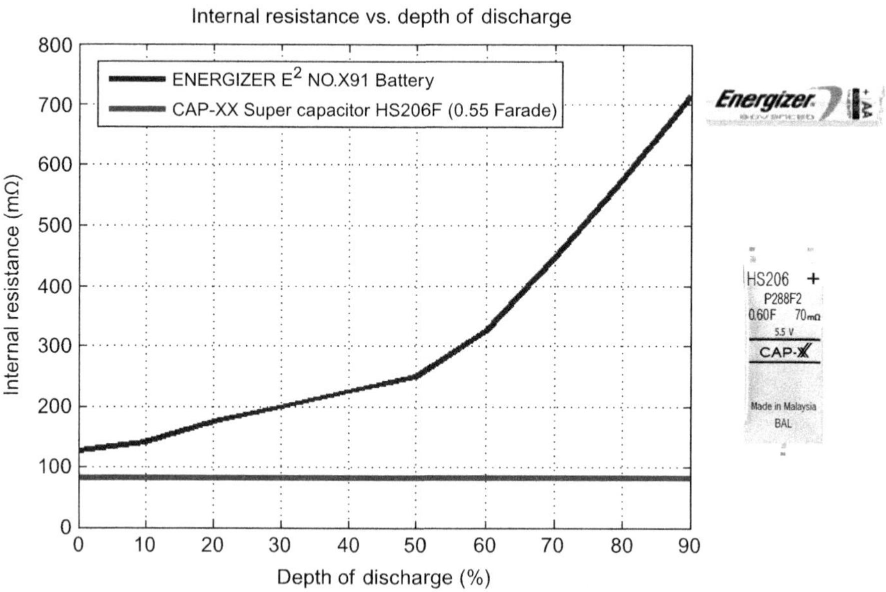

Fig. 2.6 Comparison of internal resistance of a supercapacitor and an energizer type AA cell.

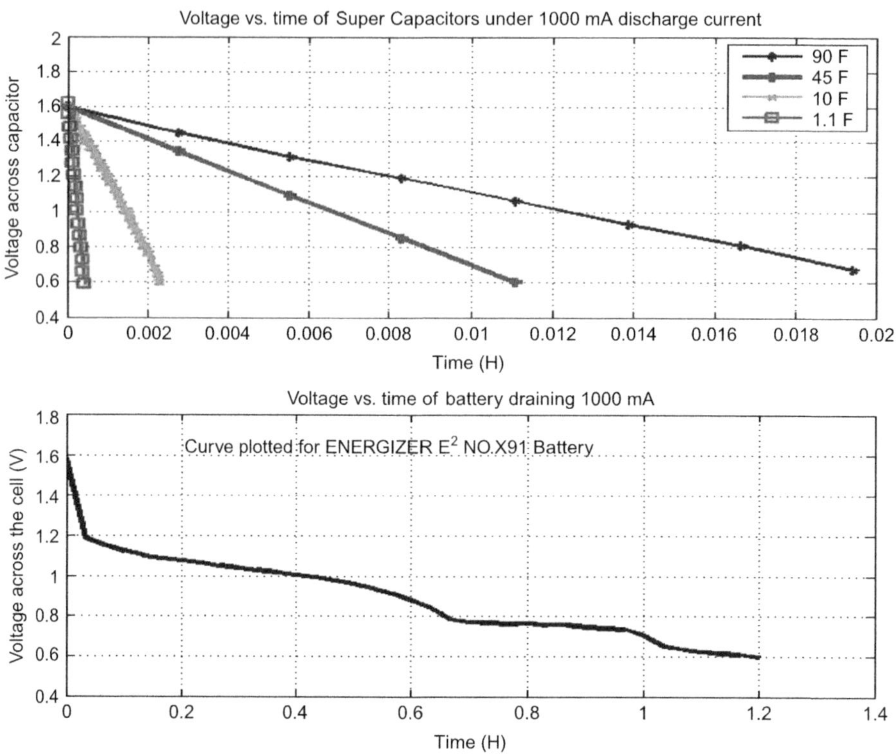

Fig. 2.7 Terminal voltage comparison of supercapacitors and batteries under constant current discharge: (A) supercapacitors and (B) a disposable AA-type cell.

obey the behavior of a capacitor voltage (based on the simple relationship of $\Delta V = \Delta Q/C$) given by

$$\Delta V = \frac{I_L}{C}\Delta t \tag{2.20}$$

However, due to the complex nature of the internal resistance variation of the AA cell as we saw from Fig. 2.6, the terminal voltage variation is not easily predictable by a simple model as we indicated in Fig. 2.1.

In general, SCs do not have very high energy storage capacity compared to electrochemical batteries. However, for short-term high-power requirements, very low and relatively fixed internal ESR of SCs allows them to be used as short-term high-power delivery devices. This complementary nature of the two families allows designers to develop hybrid-architectures for cost-effective backup energy sources.

2.3.3 Flywheel as an ESD in electrical systems

Flywheels are kinetic ESDs where a rotating mass is used to store energy based on the inertia of the rotating mass and the rotational speed of the rotor. The situation is equivalent to the case of a mass of m moving at a speed of v where the kinetic energy is given by $\frac{1}{2}mv^2$. In the case of a flywheel, usually rotating at a very high speed, the kinetic energy stored in the rotor, E_k is given by

$$E_k = \frac{1}{2}J\omega^2 \tag{2.21}$$

where J is the moment of inertia and ω is the angular velocity.

In practice, the rotor is built up in a way that mass is kept low, but the value of J is maximized. Early generations of flywheels used large steel rotors, but the newer systems use carbon-fiber and other composite material in order to keep the weight low. Magnetic bearings of different complexities are used to minimize friction and high-temperature superconductors (HTSC) are used in very high-speed latest designs. A relatively new application of flywheels is in the storage of electrical energy, which is achieved by the addition of an electrical machine and a power converter. The electrical machine may be integrated with the flywheel and operates at variable speed, and the power converter is usually provided by a power electronic variable speed drive. The main feature of flywheel energy storage (FES) systems is that they can be charged and discharged at relatively high speed for many AC cycles in an application such as a UPS. Typical state-of-the-art composite rotors have the capability to store 100 Wh/kg (Ruddell, 2003). There are two basic categories of flywheels, low speed, and high speed.

Low-speed ones are made of steel rotors and speeds reach up to about 600 rev/min. They usually carry conventional bearings. High-speed flywheels coming into commercial use with operating speeds up to 50,000 rev/min use advanced composite material, and magnetic bearings including HTSC techniques (Ruddell, 2003). The main electrical applications of FES are UPS, power quality systems, traction applications, etc. A detailed discussion on these is beyond the scope of this book.

2.3.4 Fuel cells

A fuel cell is a galvanic cell in which the chemical energy of a fuel is converted directly into electrical energy by means of electrochemical processes. The fuel and oxidizing agent are continuously and separately supplied to the two electrodes of the cell, where they undergo a chemical reaction. Fig. 2.8 depicts the basic arrangement of a fuel cell. An electrolyte is placed between two electrodes in order to conduct the ions from one electrode to the other. Fuel is supplied to the anode, where electrons are released from the fuel under the catalyst. The electrons, under the potential difference between these two electrodes, flow through the external circuit to the cathode, where combining positive ions and oxygen, reaction products, or exhaust are produced. This process is different from a battery since we have to supply the fuel and oxidant into the fuel cell to produce electrons which flow into the external circuit. More details on theory are available in Chapter 14 of Eshani et al. (2010).

Fuel cells do not come under the category of electrochemical batteries, since they require an external supply of fuel. Fuel cells can be divided into five subsets: (i) stationery plug-in power (hundreds of megawatts capability), (ii) back-up power (tens to hundreds of kilowatts), (iii) traction power (portable power supplies at 10–100 kW), and (iv) small portable power (1–100 W) and mini- or micropower (10 μW to 1 W). Fuel cells have the potential to provide clean, efficient sustainable power in a range of market segments (Zhao, 2009).

There are many types of fuel cells, with different power capabilities. Six different types are: (i) proton exchange membrane or polymer exchange membrane fuel cells

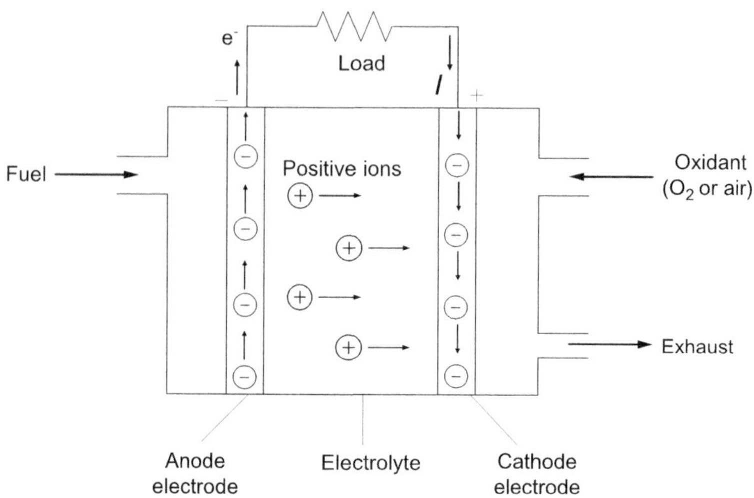

Fig. 2.8 Basic operation of a fuel cell.

Table 2.3 Operating data for different fuel cell systems.

Cell system	Operating temp range (°C)	Electrolyte
PEMFC	60–100	Solid
AFC	100	Liquid
PAFC	60–200	Liquid
MCFC	500–800	Liquid
SOFC	1000–1200	Solid
DMFC	100	Solid

Source: Eshani, M., Gao, Y., Emadi, A., 2010. Modern Electric, Hybrid Electric and Fuel Cell Vehicles. CRC Press, Boca Raton, 534 p.

(PEMFC), (ii) Alkaline fuel cells (AFC), (iii) phosphoric acid fuel cells (PAFC), (iv) molten carbonate fuel cells (MCFC), (v) solid oxide fuel cells (SOFC), and (vi) direct methanol fuel cells (DFMC). Table 2.3 summarizes some operating temperature ranges and the state of electrolyte. PEMFCs were developed in the 1960s to help the US manned space program and they are now developed for automotive applications. With a power density of 0.35–0.6 W/cm^2, these are favorable for EV and hybrid electric vehicles (HEV). Low-temperature operation and fast start capability are the most desirable characteristics of these for EV and HEVs.

Methanol- and hydrogen-based fuel cells seem to be attractive to automotive applications. Fig. 2.9 indicates the characteristics of the hydrogen-oxygen fuel cell. Fig. 2.9A indicates that these fuel cells have terminal voltage ranges varying from 0.5 to 1.0 V. Fig. 2.9B indicates their efficiency and power capability versus current density. More details are available in Baldauf and Preidel (1999), Berlowitz and Darnell (2000), and Wang (2002). For fuel cells using oxygen as the oxidant, air is usually used as the oxidant. Fig. 2.10 depicts the cell voltage, system efficiency, and net power density versus net current density applicable to a hydrogen-air fuel cell.

Usually, a hydrogen-air fuel cell works together with a power electronic DC-DC converter and associated auxiliaries. Fig. 2.11 indicates this case where the fuel cell exchanges air, water, and hydrogen. More details are available in Eshani et al. (2010).

Compared to the use of hydrogen-based fuel cells in larger systems such as EVs and HEVs, another very interesting category of fuel cells is the microfuel cells which are mostly based on DMFC. The applications of these are in microelectronic environments and wireless sensor nodes. Fig. 2.12 depicts the arrangement of a DFMC and its device life versus duty cycle of operation. The graphs in Fig. 2.12B indicate the ability of these devices to last long in micropower applications such as sensor nodes. For details, Zhao (2009) is suggested.

It is important to recognize that fuel cells are strictly not ESDs, but they are energy converters based on a fuel feed. Therefore, a detailed discussion on this subject is beyond the scope of this book.

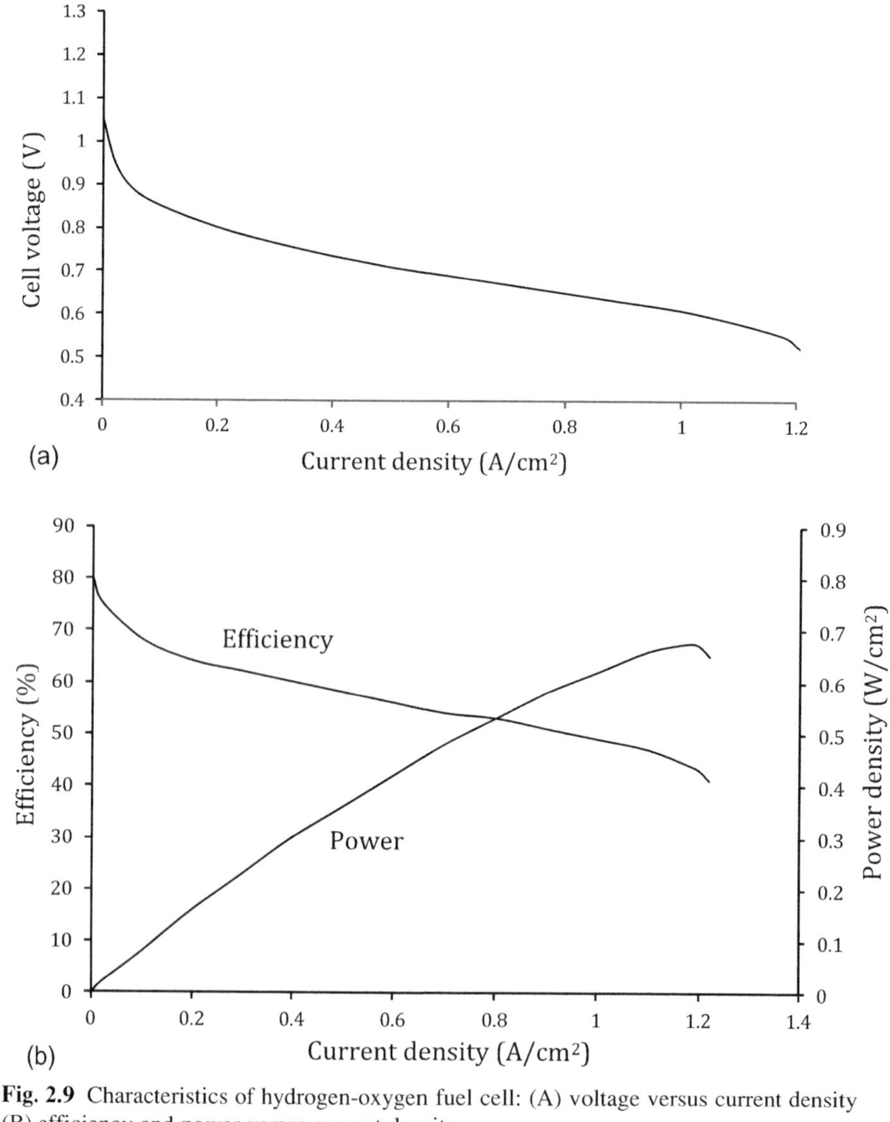

Fig. 2.9 Characteristics of hydrogen-oxygen fuel cell: (A) voltage versus current density (B) efficiency and power versus current density.

2.4 Compressed air energy storage

Compressed air energy storage (CAES) is a technology that stores energy as compressed air for later use. Energy extraction may finally occur at a gas turbine. CAESs have been considered for numerous applications, most notably for electric grid support for load leveling applications. In such systems, energy is stored during periods of low demand and then converted back to electricity when the demand is high. Natural

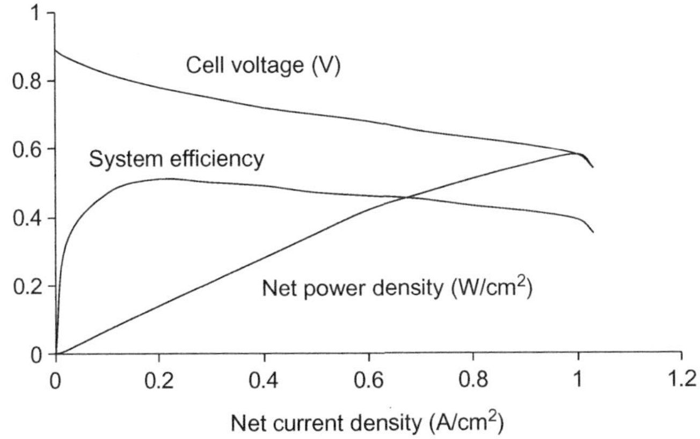

Fig. 2.10 Characteristics of hydrogen-air fuel cell at different current densities.

caverns are often used as air reservoirs in order to store large amounts of energy, as these systems are generally used for power levels varying from 35 to over 300 MW (Vazquez et al., 2010). During the last decade, attempts were made to combine CAES and SCs with the possibility to achieve maximum efficiency point tracking (Lemofouet and Rufer, 2006). This subject is not within the scope of this book.

2.5 Superconductive magnetic energy storage

Superconductive magnetic energy storage (SMES) systems make use of the energy stored in a DC magnetic field created by a direct current flowing through a superconducting coil. The coil is cryogenically cooled to achieve extreme conductivity

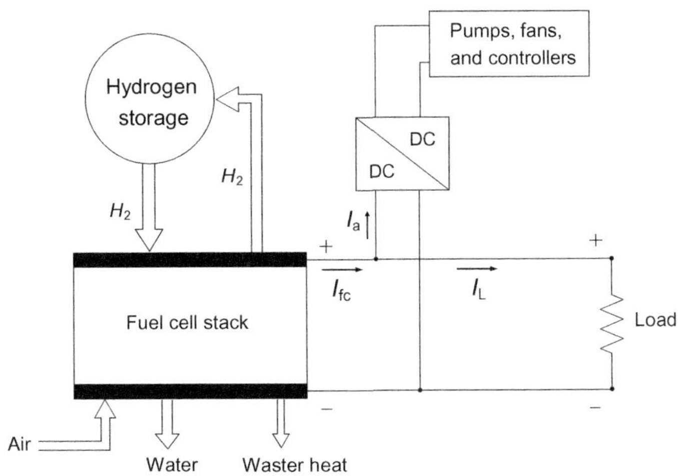

Fig. 2.11 A hydrogen-air fuel cell system.
Adapted from Eshani, M., Gao, Y., Emadi, A., 2010. Modern Electric, Hybrid Electric and Fuel Cell Vehicles. CRC Press, Boca Raton, 534 p.

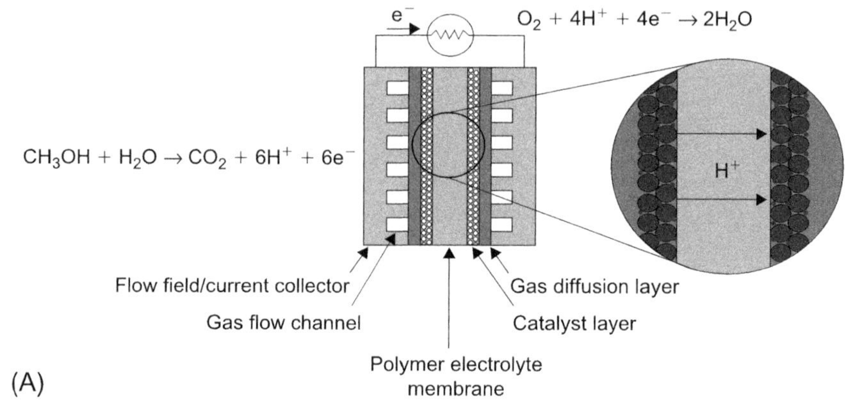

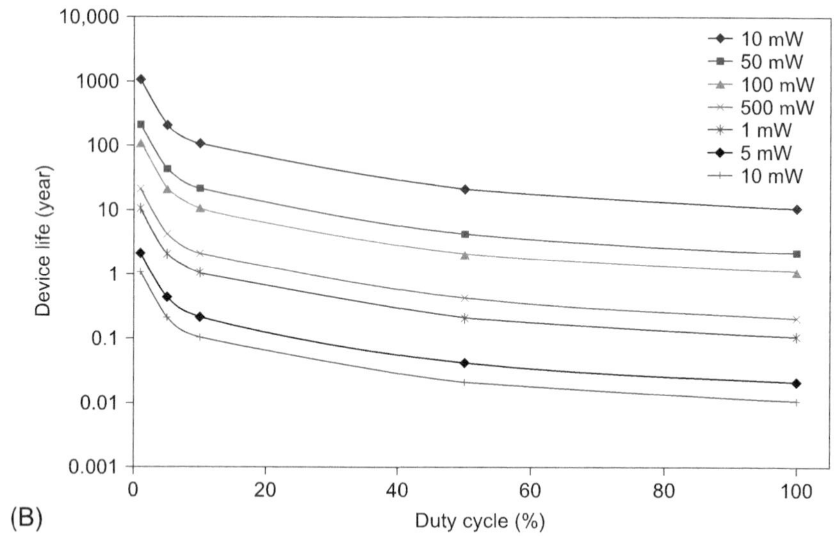

Fig. 2.12 Direct methanol fuel cell (DFMC): (A) simplified arrangement and (B) device life versus duty cycle for 12 M methanol fuel cell carrying 1 cubic centimeter of methanol and discharged at 0.5 V.
Source: Zhao, T.S., 2009. Micro Fuel Cells—Principles and Applications. Academic Press, London, 300 p.

(where it is defined as a superconductor) and no resistive losses. This combination of a DC fed inductor with near-zero resistance can be used as a very fast-acting ESD. Industrial applications include adjustable speed drives, power quality improvement products, and back-up power supplies. Due to the capability of very fast response, these systems are used in many special circumstances, with power levels ranging from few 100 kW to few MW. More details can be found in Buckles and Hassenzabl (2000) and Ali et al. (2010).

2.6 Rapid energy transfer requirements and fundamental circuit issues

In general, an ESD's electrical capabilities can be summarized by few simple specifications such as:

(i) Energy storage capability
(ii) Internal resistance of the device
(iii) Fundamental delays associated with device properties (time constants and related issues)

Energy storage capability is obviously the prime specification, but when it comes to delivering that energy into an external load, the maximum power that can be transferred depends on the internal resistance of the device. This was discussed as a summary in Section 2.2.2. What we can deliver as maximum power is limited by the minimum possible value of the internal resistance of the device.

This situation is well illustrated in the case of modern SCs with very low ESR in the order of 0.3 milliohms to few 10s of milliohms. In a practical case of an SC with a rated voltage around 2.5 V and an ESR of 1 mΩ, the device could deliver an instantaneous maximum power of $V^2/4r_C$, (when $r_C = R_L$, which is the condition for max power delivery) which translates to a value of 1562 W at 2.5 V. Compared to this case, if we are to use an electrolytic capacitor with a voltage capability of 16 V, and with an ESR value of 100 mΩ, the maximum power capability is only 640 W. Table 2.4 summarizes the energy storage capability and maximum power delivery capability of an SC and an electrolytic capacitor.

What is indicated in Table 2.4 provides us an interesting scenario. A 3000 F SC can store over 10 kJ of energy and it can initially supply over 6 kW of power. However, a 2200-μF, 50 V capacitor just cannot store even 3 J of energy. However, its instantaneous power capability is around 8 kW. In a situation like where the two capacitors feed a heating element, SC will do significantly better in heating, despite its low DC voltage rating. Based on the discussion in Section 2.2.4, a capacitor could release a huge instantaneous current, limited only by the total path resistance is indicated in Fig. 2.2B. Also, the graph in Fig. 2.2C indicates that it could release almost over 85% of its charge within a period of one time constant. This in summary indicates us that an SC could deliver a huge percentage of its stored energy within a period of time determined by the time constant of the overall circuit.

Now let us consider another scenario of a low-voltage AC supply based on a transformer feeding a heating coil as in Fig. 2.13A. In a situation such as a mains transformer stepping down to a very low voltage, and supplying a very low resistance heater coil, we can configure the system to deliver a large continuous power into the load on a continuous basis. However, since a transformer's nonideality leads to an equivalent circuit such as in Fig. 2.13B, load sees an equivalent circuit with an inductive impedance together with an AC source of value nV_p, as depicted in Fig. 2.13C. Now with the transformer (turns ratio of $n = N_2/N_1$) secondary providing a voltage source of approximately nV_p, together with an equivalent series internal impedance based on inductance and a resistance representing its nonideal behavior,

Table 2.4 Comparing the capability of supercapacitors and electrolytic capacitors.

Capacitor type	Manufacturer	Capacitance	Voltage rating (V)	ESR (mΩ)	Total energy storage capability	Maximum possible output power (load resistance = ESR) (kW)
Supercapacitor	Maxwell	3000 F	2.7	0.3	10.9 kJ	6.07
Electrolytic capacitor	Cornell-Dubilier	2200 μF	50	71	2.75 J	8.8

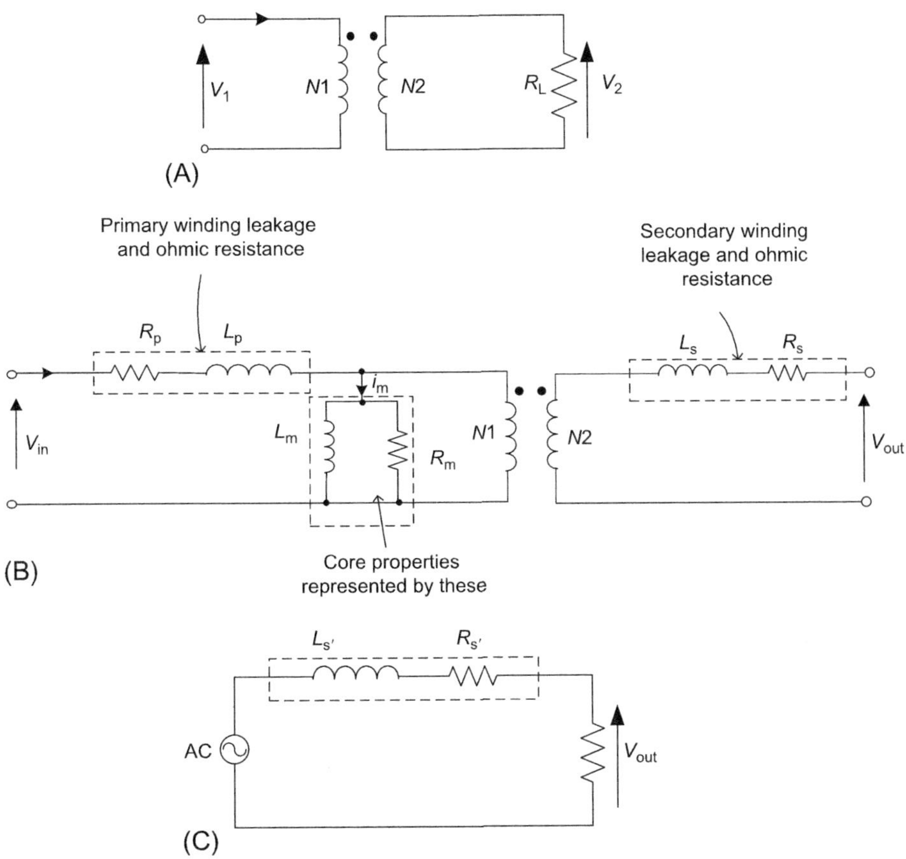

Fig. 2.13 Step down transformer-based supply delivering power into a resistive load: (A) basic circuit; (B) general equivalent circuit of transformer indicating lumped components representing nonideal behavior; and (C) equivalent circuit seen by the resistive load.

the overall equivalent circuit seen by the load will be a combination of a voltage source, an inductance, and a resistance feeding the resistive load.

This creates the opposite of the SC's case where the inductive circuit does not allow a sudden change in current at the initial switch on time. The circuit will behave in such a way that, instantaneous AC voltage developed at the secondary side (approximately nV_p) will be equal to the sum of the drops across the inductance and the resistive components. With an *L-R* combination, circuit transients will behave in an opposite manner compared to the case of a capacitor and will take a finite time to reach the steady state of voltage developed across the resistive load. However, once this stage is reached power can be delivered to the load continuously.

This case can be easily illustrated in a water heating example as in Fig. 2.14. We supplied cold water at a flow rate of 0.1 L/s at the right-hand side of the glass tube and used a series SC bank of 65 F with a maximum voltage capability of over 15 V, and a transformer secondary supplying a similar RMS voltage at the heating coil.

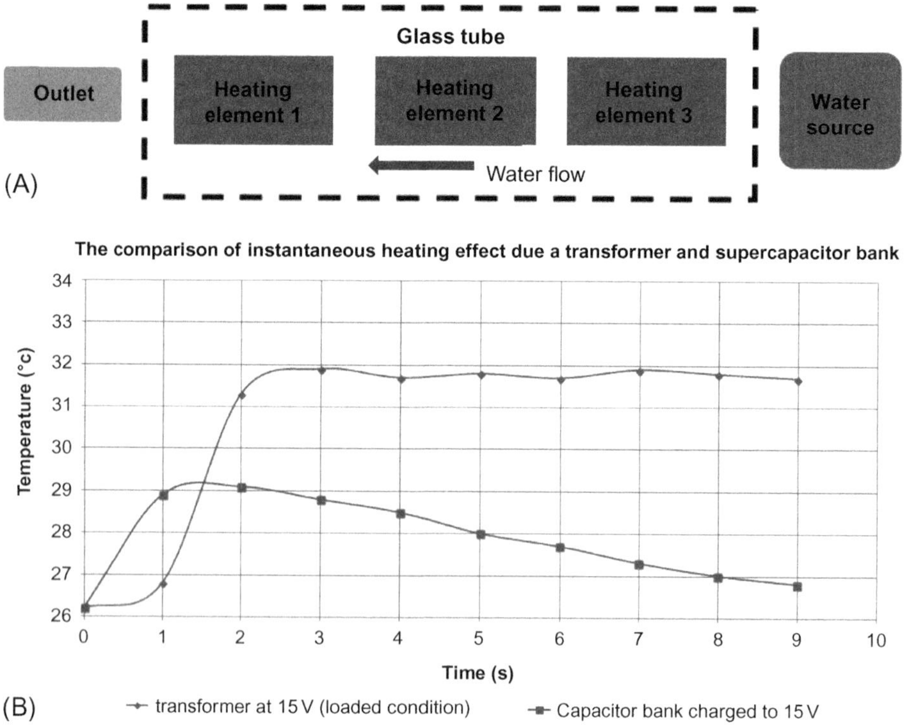

Fig. 2.14 Attempt to heat water from a cold-water supply using prestored energy in a capacitor bank versus a transformer secondary supply with a similar RMS voltage capability: (A) arrangement of heating elements and (B) temperature rise versus time for two cases.

Temperature rise of the water at the left-hand side of the tube is shown in Fig. 2.14B. While the SC bank could dump almost all the energy into water within about 2.5-time constants (in this example case, it was 9 s) and increase the temperature of the flowing water, the transformer was unable to heat the water as fast as the SC bank. However, the transformer supply could maintain the temperature continuously after its transient delay associated with heating. The case simply illustrates the differences of capacitive and inductive AC energy sources, in an application where instant heating is needed. This is a simple and practical illustration of the capability of SCs for instant energy supply requirements.

2.7 Technical specifications of ESDs

ESDs are enabling technologies for many established and new applications of power electronic systems. Well-established areas are power quality improvement products, UPS systems, and portable devices. New application areas include EV, renewable energy systems and power peak shaving, and many more new applications. ESS could be characterized by a set of general specifications and these allow us to quantitatively compare various device family options and also to look at the performance of a particular energy source continuously as it is effectively utilized in a given power electronic environment.

2.7.1 Energy and power density

2.7.1.1 Energy density

Energy density of an ESS is its energy divided by its weight or volume. When weight is used it is called the *gravimetric energy density* and *volumetric energy density* when the volume is used. Terms "energy density" and "specific energy" are sometimes used for volumetric and gravimetric measures, respectively. Units are watt-hour per liter or watt-hour per kilogram.

2.7.1.2 Power density

It is the maximum amount of power an ESS can deliver per unit volume or unit weight. When the volume is used, it is also called the *volumetric power density* and is usually measured in Watts per liter. When weight is used it is called *gravimetric power density* or *specific power* and measured in Watts per kilogram.

2.7.1.3 Cycle life

Cycle life is a measure of an ESS's ability to withstand repetitive deep discharging and recharging using the manufacturer's cyclic charging recommendations and still provide minimum required capacity for the application. Cyclic discharge testing can be done at any of the various rates and depths of discharge to simulate conditions in the application.

2.7.1.4 Cyclic energy density

For purposes of comparison, a better measure of ESS characteristics is a composite characteristic which considers energy density over the service life of the system. A composite characteristic, *cyclic energy density*, is defined as the product of energy density and cycle life at that energy density and has the dimensional units, watt-hour-cycles per kilogram (gravimetric) or watt-hour-cycles per liter (volumetric).

2.7.1.5 Self-discharge rate

The self-discharge rate is a measure of how long an ESS can be stored or left unused and still provide minimum required capacity and be recharged to rated capacity. This commonly is measured by placing an ESS (such as batteries) on a shelf stand at room (or elevated) temperature and monitoring open-circuit voltage over time. In cases such as batteries, self-discharge is measured in terms of percentage capacity loss per month or per year in terms of energy (in watt-hour) lost.

In cases like capacitors, it is usual to indicate a value of a resistor which is placed in parallel to the terminals of the capacitor to denote this issue.

2.7.1.6 Charge acceptance or coulombic efficiency

In ESS such as batteries where the open-circuit voltage is relatively constant, charge accumulated or discharged in terms of $\int_0^t i dt$ is used to discuss the capability of the device to accept and deliver current into a given load. The charge delivered to the load, C_{load} will be usually less than the charge fed into the device, C_{charge}. The ratio of these two quantities, measured in coulombs or Ampere-hours, is called the charge-acceptance or the coulomb-efficiency. This measure usually depends on the charge (or discharge) rate, temperature, age, and the overall condition of the ESS.

The above specifications are generally applicable to many types of ESSs, and in batteries, there are many other specifications which are used in practical circumstances, which are further discussed.

2.8 Ragone plot

ESSs come in different sizes, such as pumped ESSs (used in hydroelectric plants), large and small fuel cells, different types of batteries, SCs, and SMES. In all these systems, two very important specifications used to compare the overall performance are energy density and power density. To compare different energy delivery capability options, a Ragone plot (developed by David V Ragone) is very frequently used.

Fig. 2.15 depicts the concept of the Ragone plot where the vertical axis is volumetric or gravimetric energy density, and the horizontal axis is power density. A Ragone plot reveals at a glance the characteristics of different energy deliveries or storage systems. More capable device means a large energy density as well as a large power

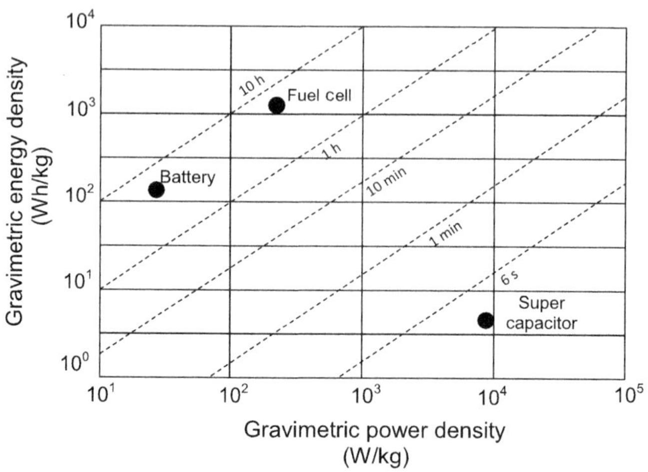

Fig. 2.15 Ragone plot with three specific systems placed at relative locations.

density. One should try to achieve systems placed at the north-east corner of the chart. da Rosa (2013) provides a summary of how to use the data sheet information and calculate results to place the device on the Ragone plot. In Fig. 2.15, the relative positioning of a lead-acid battery (of capacity 80 Ah; internal resistance of 37.5 mΩ, weight 31 kg, volume 0.01 m^3), SC (3000 F, internal resistance of 3.7 mΩ, weight 0.55 kg, volume 0.000475 m^3), and a fuel cell (continuous power capability 110 kW, the maximum current of 440 A, DC voltage of 250 V, weight 120 kg, volume of 82 L). More details on the calculations related to these are in da Rosa (2013).

References

Ali, M.H., Wu, B., Dougal, R.A., 2010. An overview of SMES applications in power and energy systems. IEEE Trans. Sustain. Energy 1 (1), 38–47.

Baldauf, M., Preidel, W., 1999. Status of development of direct methanol fuel cell. J. Power Sources 84, 161–166.

Berlowitz, P.J., Darnell, C.P., 2000. Fuel cell choices for fuel cell powered vehicles. In: Proceedings of Fuel Cell Transportation 2000 (SAE SP-1505), Warrendale, PA, pp. 15–25.

Buckles, W., Hassenzabl, W.V., 2000. Superconducting magnetic energy storage. IEEE Power Eng. Rev. 20 (5), 16–20.

da Rosa, A.V., 2013. Fundamentals of Renewable Energy Processes, third ed. Academic Press, Oxford, p. 884.

Eshani, M., Gao, Y., Emadi, A., 2010. Modern Electric, Hybrid Electric and Fuel Cell Vehicles. CRC Press, Boca Raton, p. 534.

Grillo, S., Martini, L., Musolino, V., Piegari, L., Tironi, E., Tornelli, C., 2011. Management of different energy storage devices using a losses minimization algorithm. In: Proceedings of IEEE-ICCEP Conference, pp. 420–425.

Kularatna, N., Kankanamge, K., Fernando, J., 2011. Supercapacitors enhance LDO efficiency—part 2: implementation. Power Electron. Technol. Mag. 37 (5), 30–33.

Lemofouet, S., Rufer, A., 2006. A hybrid energy storage system based on compressed air and supercapacitors with maximum efficiency point tracking (MEPT). IEEE Trans. Ind. Electron. 53 (4), 1105–1115.

Moubayed, N., Kouta, J., El-Ali, A., Dernayka, H., Outbib, R., 2008. Parameter identification of the lead-acid battery model. In: Proceedings of IEEE-PVSC Conference, pp. 1–6.

Musolino, V., Piegari, L., Tironi, E., 2013. New supercapacitor model with easy identification procedure. IEEE Trans. Ind. Electron. 60 (1), 112–120.

Ruddell, A., 2003. Investigation on Storage Technologies for Intermittent Renewable Energies: Evaluation and Recommended R & D Strategy. CCLRC-Rutherford Appelton Laboratory, p. 30. Report WP-ST6 Flywheel.

Vazquez, S., Lukic, S.M., Galvan, E., Franquelo, L.G., Carrasco, J.M., 2010. Energy storage systems for transport and grid applications. IEEE Trans. Ind. Electron. 57 (12), 3881–3895.

Wang, M., 2002. Fuel cell choices for fuel cell vehicles, well-to-wheel energy and emission impacts. J. Power Sources 112, 307–321.

Zhao, T.S., 2009. Micro Fuel Cells—Principles and Applications. Academic Press, London, p. 300.

Zubieta, L., Bonert, R., 2000. Characterization of double-layer capacitors for power electronic applications. IEEE Trans. Ind. Appl. 36 (1), 199–205.

Bibliography

Christen, T., Carlen, M.W., 2000. Theory of Ragone plots. J. Power Sources 91, 210–216.

Christen, T., Ohler, C., 2002. Optimizing energy storage devices using Ragone plots. J. Power Sources 110, 107–116.

Pell, W.G., Conway, B.E., 1996. Quantitative modeling of factors determining Ragone plots for batteries and electrochemical capacities. J. Power Sources 63, 255–266.

Ter-Garzarian, A.G., 2011. Energy Storage for Power Systems, second ed. IET, London, p. 276.

Rechargeable battery technologies: An electronic circuit designer's viewpoint

3

3.1 Introduction

Since the invention of the lead-acid battery by Gaston Planté in 1859, electrochemistry has progressed steadily over the last one and a half centuries. With the miniaturization of electronic components proliferating steadily over the last half century, the insatiable demand for smaller and lightweight portable electronic equipment has dramatically increased the need for research on battery chemistries as well as semiconductor components for the optimal management of batteries. Battery chemistries come in two different forms, namely, disposable or primary batteries and secondary or rechargeable batteries. The market for batteries can be found in IDTechEx Ltd (2012a, b), marketresearch (2009), and US demand (2009).

Mature rechargeable battery chemistries are: (i) lead acid, (ii) nickel cadmium, (iii) nickel metal hydride, (iv) lithium-ion, (v) lithium-polymer/lithium metal, and (vi) lithium-iron phosphate. With the demand growing from electric vehicles (EV) and portable consumer products, many organizations spend huge amounts of research money on new battery chemistries. Some significant areas include zinc-based chemistries and silicon as a material for improving some properties of batteries (EE Times Asia, 2009). Higher energy density, superior cycle life, environmental friendliness, and safe operation are among the general design targets of secondary battery manufacturers. Primary or disposable batteries are a reasonably mature market and a product chemistry range, but still there are attempts to increase the energy density, reduce the self-discharge rate (in order to increase the shelf life), and also to improve the usable temperature range. To complement these developments, many semiconductor manufacturers continue to introduce new integrated circuit families for battery management.

This chapter describes the characteristics of batteries such as sealed lead acid, NiCd, NiMH, Li-based chemistries, and zinc-air from an engineer's viewpoint together with modern techniques used in battery management integrated circuits (ICs). This is done without detailing the chemical reactions and associated battery chemistries. Concepts and applications related to smart battery systems and related standards, and IEEE 1625/1725 standards for safety of battery-powered systems are also briefly introduced.

Energy Storage Devices for Renewable Energy-Based Systems. https://doi.org/10.1016/B978-0-12-820778-9.00001-2

3.2 Battery terminology and fundamentals

3.2.1 Capacity

Battery or cell capacity means an integral of current over a defined period of time.

$$\text{Capacity} = \int_0^t i\,dt \tag{3.1}$$

This relationship applies to either charge or discharge, that is, capacity added or capacity removed from a battery or cell. The capacity of a battery or cell is measured in milliampere-hours (mAh) or ampere-hours (Ah).

Although the basic definition is simple, many different forms of capacity are used in the battery industry. The distinctions between them reflect differences in the conditions under which the capacity is measured.

3.2.1.1 Standard capacity

Standard capacity measures the total capacity that a relatively new, but stabilized production cell or battery can store and discharge under a defined standard set of application conditions. It assumes that the cell or battery is fully formed, that it is charged at standard temperature at the specification rate, and that it is discharged at the same standard temperature at a specified standard discharge rate to a standard end-of-discharge voltage (EODV). The standard EODV is itself subject to variation depending on the discharge rate as discussed.

3.2.1.2 Actual capacity

When the application conditions differ from the standard conditions, the capacity of the cell or battery changes; the term *actual capacity* includes all nonstandard conditions that alter the amount of capacity the fully charged new cell or battery is capable of delivering when fully discharged to a standard EODV. Examples of such situations might include subjecting the cell or battery to a cold discharge or a high-rate discharge.

3.2.1.3 Available capacity

That portion of actual capacity which can be delivered by the fully charged new cell or battery to some nonstandard EODV is called available capacity. Thus, if the standard EODV is 1.6 V/cell, the available capacity to an EODV of 1.8 V/cell would be less than the actual capacity.

3.2.1.4 Rated capacity

Rated capacity is defined as the minimum expected capacity when a new, but fully formed, cell is measured under standard conditions. This is the basis for C rate (defined later) and depends on the standard conditions used which may vary depending on the manufacturers and the battery types.

3.2.1.5 Retained capacity

If a battery is stored for a period of time following a full charge, some of its charge will dissipate. The capacity which remains that can be discharged is called retained capacity.

3.2.2 Peukert's law and the battery capacity

The capacity of a battery, sometimes referred to as C_{load} or simply C, is a somewhat inaccurate measure of how much charge a battery can deliver to a load. It is an imprecise number because it depends on temperature, age of the cells, state of the charge, and the rate of discharge. It has been observed that two identical, fully charged batteries, under the same circumstances, will deliver different charges to a load depending on the current drawn by the load. In other words, C is not constant and the value of C for a fully charged battery is not an adequate description of the characteristic of the battery unless it is accompanied by additional information, *rated time of discharge*, with the assumption that the discharge occurs under a constant current regime. This discussion is based on Peukert's law, according to which Peukert in 1897 showed that for a lead-acid battery,

$$I^n t = \Lambda = \text{constant} \tag{3.2}$$

where t is the time in hours and I is the current in amperes. Number n, the Peukert Number, is always somewhat larger than 1. For lead-acid batteries this is approximately 1.2. In most practical situations in engineering, we define the capacity of a battery, C, for a constant discharge rate of I amperes as

$$C = It \tag{3.3}$$

Combining the two relationships in Eqs. (3.2) and (3.3),

$$C = \Lambda I^{1-n} \tag{3.4}$$

This provides an explanation of the battery discharge curves we discuss later, where the capacity of a battery is dependent on the discharge current as shown in Fig. 3.1.

It is important to note that, in this form, Peukert's equation (often referred to as a law) presents a unit imbalance. The intent of this equation is to account for the intrinsic losses associated with discharging batteries at elevated currents. In particular, when discharging a battery at increasingly higher currents, the internal cell resistance increases and the recovery rate of the cell decreases (Hausmann and Depick, 2013). Later research demonstrated that the decrease in recovery rate is due to the degradation in the number of active sites in the positive active material as well as the increase in the resistance between the positive material and the electrolyte (Baert and Vervaet, 1999). The exponent, n, known as Peukert's constant, is to account for these losses (Doerffel and Sharkh, 2006). This equation is valid only if the discharge current and the cell temperature are constant. More details have been provided by Hausmann and Depick (2013), Baert and Vervaet (1999), Doerffel and Sharkh (2006), and

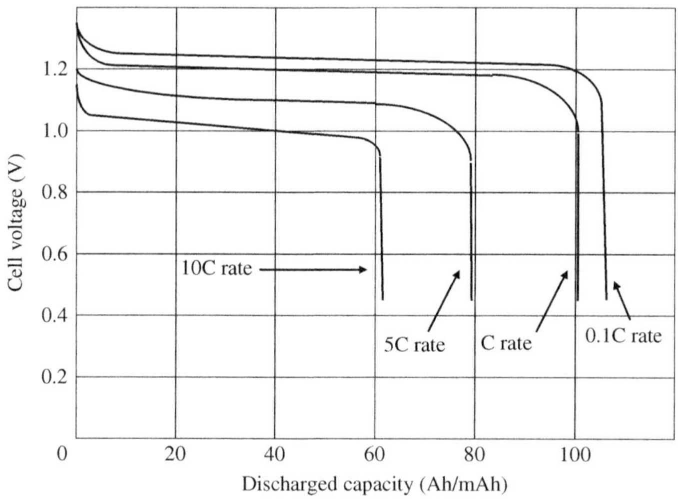

Fig. 3.1 Capacity vs discharge rate of a typical cell.

Guoliang et al. (2010) on how this relationship needs to be modified in practical batteries such as lithium-ion and nickel metal hydride batteries.

3.2.3 C rate

The C rate is defined as the rate in amperes or milliamperes which is numerically equal to the capacity rating of the cell given in Ah or mAh. For example, a cell with a 1.2-Ah capacity has a C rate of 1.2 A. The *C* concept simplifies the discussion of charging for a broad range of cell sizes since the cells' responses to charging are similar if the C rate is the same. Normally a 10-Ah cell will respond to a 1.0-A (0.1C) charge rate in the same manner that a 2-Ah cell will respond to a 0.2-A (also 0.1C) charge rate. The rate at which current is drawn from a battery affects the amount of energy which can be obtained. At low discharge rates, the actual capacity of a battery is greater than that at high discharge rates. This relationship is shown in Fig. 3.1.

3.2.4 Energy density

The energy density of a cell is its energy divided by its weight or volume. When weight is used, it is called the gravimetric energy density, and volumetric energy density when volume is used. The terms "energy density" and "specific energy" are sometimes used for volumetric and gravimetric measures, respectively.

3.2.5 Power density of a battery

It is the amount of power a battery can deliver per unit volume at a specified state of charge (SOC), usually 20%. It is also called the volumetric power density and is usually measured in watts per liter.

3.2.6 Cycle life

Cycle life is a measure of a battery's ability to withstand repetitive deep discharging and recharging using the manufacturer's cyclic charging recommendations and still provide the minimum required capacity for the application. Cyclic discharge testing can be done at any of the various rates and depths of discharge (DODs) to simulate conditions in the application. It must be recognized, however, that cycle life has an inverse logarithmic relationship to DOD.

3.2.7 Cyclic energy density

For purposes of comparison, a better measure of rechargeable battery characteristics is a composite characteristic which considers energy density over the service life of the battery. A composite characteristic, *cyclic energy density*, is defined as the product of energy density and cycle life at that energy density and has the dimensional units, watt-hour-cycles/kilogram (gravimetric) or watt-hour-cycles/liter (volumetric).

3.2.8 Self-discharge rate

Self-discharge rate is a measure of how long a fully-charged can be stored and still provide the minimum required capacity and be fully-charged to the rated capacity. This is commonly measured by placing batteries on the shelf stand at room (or elevated) temperature and monitoring open-circuit voltage over time. Samples are discharged at periodic intervals to determine the remaining capacity (RC) and recharged to determine rechargeability.

3.2.9 Charge acceptance

Charge acceptance is the willingness of a battery or cell to accept charge. Charge acceptance can be defined as the ratio $\frac{C_{\text{load}}}{C_{\text{charge}}}$ where C_{charge} is the total coulombs of charge pumped into the battery. Sometimes this is called the coulombic efficiency. This is affected by cell temperature, charge rate, and the SOC. More on this will be discussed later with graphs.

3.2.10 Depth of discharge

DOD is the capacity removed from a battery divided by its actual capacity, expressed as a percentage.

3.2.11 Battery discharge curves and related terminology

Fig. 3.2 is the typical discharge curve of a secondary battery, applicable to almost all chemistries in general. A fully charged battery usually displays a higher than nominal terminal voltage, which is relatively flat over the discharge curve. However, when a battery is discharged up to about 80% of its rated capacity, the voltage drop becomes

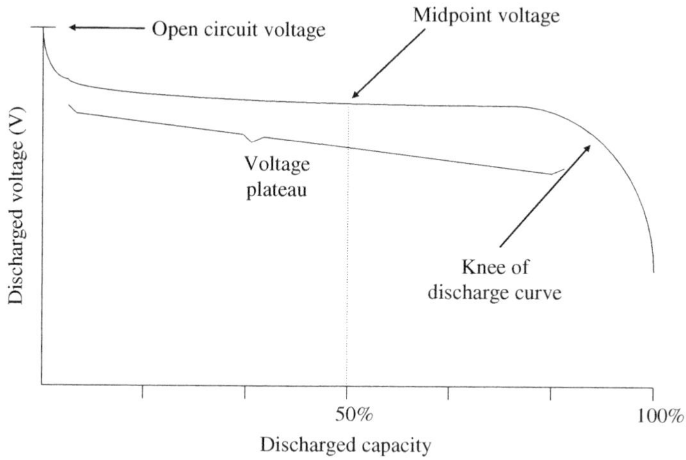

Fig. 3.2 Midpoint voltage and the voltage plateau.

significantly larger, due to its internal resistance increase, etc., and usually displays a sharp drop characterized by a knee.

3.2.11.1 Voltage plateau

Voltage plateau is the protracted period of very slowly declining voltage that extends from the initial voltage drop at the start of a discharge to the knee of the discharge curve. An example is shown for a typical lead-acid cell in Fig. 3.2.

3.2.11.2 Midpoint voltage

Midpoint voltage is the battery voltage when 50% of the actual capacity has been delivered (see Fig. 3.2). This is considered as the nominal cell voltage of the battery for design purposes.

3.2.12 Overcharge

Overcharge is defined as continued charging of a cell after it has become fully charged. When a cell is not yet fully charged, the electrical energy of the charge current is converted to chemical energy in the cell by the charging reactions. But when all of the available active material has been converted into the charged state, the energy available in the charging current goes to produce gases from the cell or to activate other nonuseful chemical reactions. Usually, this results in a temperature rise in the cell.

3.2.13 State of charge (SoC)

The SOC of a battery pack is defined as the percentage of the RC (or RC divided by its rated capacity). In other words, SOC is the ratio between the saved energy in the battery and the total energy that can be saved in the battery. If the rated capacity is measured in coulombs or Ah, RC should also be estimated in the same units. In cases where the remaining or the rated energy content is estimated in watt-hours, SOC can be estimated based on the energy storage capability. The SOC provides the current state of the battery, and enables the battery to be safely charged and discharged at a level suitable for battery life enhancement.

It is common to estimate the SOC percentage using charge transfer in and out of the battery, and in this case typical coulomb-counting methods are used. In this case, we can use the rated capacity of the battery, Q_{rated}, which is not a function of temperature or aging (since it is usually estimated at standard laboratory conditions at a controlled temperature). However, to accommodate the SOC variation with aging, it is necessary to use a different term, $Q_{\text{discharge}}$. This is defined as the maximum total electrical charge, expressed in Ah, that a battery can deliver from the fully charged state (SOC of 100%) to a fully discharged state (SOC of 0%) as indicated in Eq. (3.5),

$$Q_{\text{discharge}} = \int_0^{\text{total}} I_b(t)dt \tag{3.5}$$

Then the SOC can be expressed using the ratio or percentage of the capacity that has been discharged relative to $Q_{\text{discharge}}$

$$\text{SOC}(t) = \text{SOC}(t_0) - \frac{\int_0^{t} I_b(t)dt}{Q_{\text{discharge}}} \tag{3.6}$$

where I_b is the battery current.

In applications where energy transferred to and from the battery is considered, in Wh or kWh, one needs to use the integration of terminal voltage multiplied by the battery current, and this may be useful in applications where the true energy transferred between system elements is important. In applications such as EV, hybrid electric vehicles (HEV), and plug-in HEV (PHEV), since the engine capacities are measured in kW, battery pack SOC estimates are estimated based on the initial energy capacity $E_0(t)$ based on the relationship given in Eq. (3.7).

$$\text{SOC}(t) = \text{SOC}(t_0) - \frac{1}{E_0(0)} \int_0^t P_i(t)\text{d}t \tag{3.7}$$

where $E_0(0)$ and $P_i(t)$ are the initial energy capacity and the instantaneous power fed from the battery into the load, respectively. When $P_i(t)$ is negative, the battery is charged from the external energy source. In this case, the system should have monitoring capability for both battery voltage and current. Kim and Cho (2011) and Ebbesen et al. (2012) provide some details on estimation of this useful parameter, and also some insight into dealing with practical inaccuracies in estimating the SOC and other useful parameters such as state of health (SOH).

3.2.14 State of health

In practical circumstances, designers are challenged to predict the life of a battery pack, and for this purpose SOH is a useful parameter to estimate the remaining useful life of a battery pack.

SOH is the ability of a cell to store energy, source and sink high currents, and retain charge over extended periods, relative to its initial or nominal capacity or capability. The available charge stored within a fully charged cell is expected to fall with cell usage, as active material on cell plates degrades due to multiple reasons within the battery chemistry. Early detection of SOH degradation would allow a "smart" battery pack to take remedial action, thereby restoring cell capacity (Bhangu et al., 2005).

In estimating the fading of the nominal (energy) capacity of a battery over time due to several complex chemical processes taking place within the battery cells, several types of capacity fade models are usable, namely, (i) electrochemical models, (ii) event-based models, and (iii) energy throughput-based models. Electrochemical models are derived from first principles and they are far superior to the other two models. However, to use them in developing battery packs with predictable performance is very difficult in real time due to limitations in computational resources.

In order to estimate the SOH based on a throughput-based capacity model, an assumption is made that, under constant operating conditions, a battery can withstand a certain amount of energy throughput, equivalent to a number of charge/discharge cycles, before it reaches the end of life. *By definition, the end of life of a battery is reached when the battery capacity has dropped by 20% from its initial capacity.* Based on these conditions, similar to the SOC, we can define SOH of a battery as

$$\mathrm{SOH}(t)=\mathrm{SOH}(0)-\frac{1}{2N\cdot E_0(0)}\int_0^t P_{\mathrm{i}}(t)dt \tag{3.8}$$

SOH(0) = 1 and N are the initial SOH and the total number of cycles before end of life, respectively. The factor 2 in the denominator accounts for both charging and discharging. When SOH = 0, end of life for the pack is reached. In general, N is not constant but depends on operating conditions of the battery such as DOD and C rate, etc. More details are available in Ebbesen et al. (2012).

3.3 Battery technologies: An overview

Most commonly, the available primary batteries (the disposable types) offer significantly greater energy density together with very low self-discharge rates. They are usually cheaper than the rechargeable batteries. The most common primary chemistries include: (i) carbon-zinc (Leclanche cells), (ii) alkaline-MnO_2, (iii) lithium-MnO_2, (iv) lithium-sulfur dioxide, (v) lithium-iron disulfide, (vi) lithium-thionyl chloride ($LiSOCl_2$), (vii) silver-oxide, and (viii) zinc-air. Mercury primary batteries have gradually gone out of the marketplace due to toxicity and environmental reasons. More details on these chemistries can be found in Quinnell (1991) and Schimpf (1996).

Many types of rechargeable chemistry are used in electronic systems. Common rechargeable chemistries are mainly based on variations of lead acid, nickel-based and lithium-based systems, while limited zinc-based systems and rechargeable alkaline batteries are also available. The choice of a particular battery technology is limited by size, weight, cycle life, operating temperature range, and cost. Comparison of basic characteristics of major chemistries is depicted in Table 3.1.

NiCd batteries were used to power most rechargeable consumer appliances before 1990. It is a mature, well-understood technology. However, cadmium was coming under increasing regulatory scrutiny (including mandatory recycling in some jurisdictions), and the maturity of NiCd technology also means most of the capacity and life cycle improvements have already been made. NiMH chemistry entered the market in 1989 and offers incremental improvements in energy density both by weight and volume over NiCd. Li-ion is better still, offering over twice the watts per liter and per kilogram of NiCd batteries. As always, this higher performance comes at a higher price.

Using liquid lithium, Li-ion cells exhibit a much higher energy density than Ni-based chemistries, with a relatively higher terminal voltage than Ni-based cells, and are the commonly used energy source for most portable systems after 2000. The advantages of Li-ion chemistry, however, also come at the cost of greater electrical fragility. Li-ion, particularly, is more easily and extensively damaged by less than optimal battery management so much so that fail-safe circuits are required to disconnect the cells from the load under over-current or over-temperature conditions. Therefore, these electronic circuits are usually built into the battery pack.

In general, with an increase of the battery energy densities, and any possibilities of quantum jumps in new chemistries, one important concern is the safety and the transportability. Table 3.2 indicates the energy densities of selected fuels and batteries compared with high explosives with transportation capability. A value of 1000 Wh/kg energy density is essentially the practical maximum with its own oxidant for safe transportation (Hamlen et al., 1995). A conventional battery in general is having its own oxidant, similar to this situation, and therefore the maximums shown in Table 3.2 apply in development options. However, this condition does not apply to metal-air batteries, such as zinc-air, etc., since the oxidant is supplied by oxygen from the air.

With Li-based chemistries with high-energy densities entering the market, and with some fires and explosions occurring in the past, Li-battery manufacturers are not supposed to sell them to uncertified battery pack assemblers (BPA) and strict charge and discharge control practices are in place.

During the last decade, newer lithium-based chemistries entered the market. Important ones are the Li-polymer (sometimes known as Li-ion polymer) and Li-iron phosphate chemistries which have better performance than the Li-ion (Chu, 2009; Kim, 2001). Note that Table 3.1 is only a guide; for more precise details, individual manufacturer data sheets need to be consulted.

Table 3.1 Secondary battery chemistry characteristics.

Parameter	Units/ conditions	Sealed lead acid	NiCd	NiMH	Li-Ion	Li-polymer	Li-iron phosphate	Rechargeable alkaline
Average cell voltage	V	2.0	1.2	1.2	3.6	1.8–3.0	3.2–3.3	1.5
Relative cost	NiCd = 1	0.6	1	1.5–2.0				0.5
Internal resistance		Low	Very low	Moderate	High	High	High	
Self-discharge	%/month	2%–4%	15%–25%	20%–25%	6%–10%	18%–20%		0.3%
Cycle life	Cycles to reach 80% of rated capacity	500–2000	500–1000	500–800	1000–1200		1500–2000	<25
Overcharge tolerance		High	Med	Low	Very low			Med
Internal resistance		Low	Very low	Moderate	High (coke electrode) Highest (graphite electrode)			
Energy by volume (volumetric energy density)	watt-hour/liter	70–110	100–150	200–350	200–330	230–410	200	220
Energy by weight (gravimetric energy density)	watt-hour/kg	30–45	40–60	60–80	120–160	120–210	100	80

Table 3.2 Energy densities for selected fuels and batteries.

Fuel or battery	Volumetric energy density (Wh/kg)
Diesel fuel	10,000
Methanol	5000
High explosive	1000
Primary battery (maximum estimated around 1995)	500
Rechargeable battery (maximum estimated around 1995)	200
$LiSO_2$ battery	175
Alkaline battery	80
NiCd battery	40
Li-ion battery	120

Source: Hamlen, R.P., Christopher, H.A., Gilman, S., 1995. US Army battery needs—present and the future. IEEE AES Mag. 10, 30–33.

Lead-acid batteries are most familiar in automobiles because they are the most economical chemistry for delivering large currents. Lead acid also has a long trickle life and therefore serves well for classic "floating" applications. While flooded lead-acid technology is popular for automobile and similar applications, sealed lead-acid batteries serve the electronic engineering environments. On the downside, lead acid has the least capacity by volume and weight.

3.4 Lead-acid batteries

Lead-acid batteries are the oldest and most widely used rechargeable battery systems today, mostly due to the automotive industry's attachment to lead-acid chemistry. There are basically two types of lead-acid batteries, namely, the (i) flooded lead-acid type and (ii) sealed lead-acid type.

3.4.1 Flooded lead-acid batteries

The flooded lead-acid battery today basically uses the design developed by Faure in 1881. It consists of a container with multiple plates immersed in a pool of dilute sulfuric acid. In most industrial batteries, the cell stack is flooded with high purity, aqueous sulfuric acid at 30%–40% concentration. The traction batteries use more concentrated solutions to increase the energy density. The separators are basically nonconductive, microporous structures with good wettability, small to medium pore size, good strength and flexibility, and low electrical resistance. These can be made of rubber, cellulose, PVC, PE, glass fiber, or microglass. In recent years, leaf separators have often been replaced by envelope separators, which can prevent dendritic growth and improve the retention of active material (Broussely and Pistoia, 2007).

The so-called maintenance-free forms of flooded batteries provide excess electrolyte to accommodate water loss through a normal life cycle. Most industrial applications for flooded batteries are found in motive power, engine starting, and large system power backup. Today, other forms of battery have largely supplanted flooded batteries in small- and medium-capacity applications, but in larger sizes flooded lead-acid batteries continue to dominate.

The applications of lead-acid batteries can be grouped into three main categories: (i) starting, lighting, and ignition (SLI) or automotive; (ii) traction (deep cycle); and (iii) stationary. In Table 3.3, a summary of comparison is given.

By far, the biggest application for flooded batteries is SLI service on automobiles and trucks. Large flooded lead-acid batteries also provide motive power for equipment ranging from forklifts to submarines and provide emergency power backup for many electrical applications, most notably the telecommunications network. More details on different types of lead-acid batteries are available in Broussely and Pistoia (2007).

3.4.2 Sealed lead-acid batteries

Sealed lead-acid batteries first appeared in commercial use in the early 1970s. Although the governing reactions of the sealed cell are the same as other forms of lead-acid batteries, the key difference is the recombination process that occurs in the sealed cell as it reaches full charge. In conventional flooded lead-acid systems, the excess energy from overcharge goes into electrolysis of water in the electrolyte with the resulting gases being vented. This occurs because the excess electrolyte prevents the gases from diffusing to the opposite plate and possibly recombining. Thus,

Table 3.3 Comparison of lead-acid battery applications.

Application	Duty type	Design features
SLI: Cars, motor cycles, trucks, aircrafts, and ships	High-power discharge performance over a wide temperature range. Shallow cycling duty with service life >3 years	Thin plate, mostly 12 V, maintenance free
Traction: Electric trucks, road vehicles, and golf carts	Deep-cycle discharges (1–5 h rates). Service life of 500–2000 cycles	Tubular or flat plate designs, 24–96 V assemblies
Stationary: Telecom, UPS, emergency lighting, and energy storage	Reliability over extended periods. Discharge duty: 1–6 h rates. Service life of 5–25 years on continuous charge. Limited cycling capability	Variety of designs, including tubular, flat plate, Planté, and VRLA cell types

Source: Adapted from Broussely, M., Pistoia, G., 2007. Industrial Applications of Batteries—From Cars to Aerospace and Energy Storage. Elsevier, Amsterdam, pp. 691-736.

electrolyte is lost on overcharge with the resulting need for replenishment. The sealed lead cell, like the sealed nickel-cadmium, uses recombination to reduce or eliminate this electrolyte loss.

Sealed lead-acid batteries for electronics applications are somewhat different from the flooded type commonly found in the automobile. There are two types of sealed lead-acid batteries: the original gelled electrolyte and retained (or absorbed) system. The gelled electrolyte system is obtained by blending silica gel with an electrolyte, causing it to set up in gelatin form. The retained system employs a fine glass fiber separator to absorb and retain liquid electrolyte. Sometimes the retained system is named absorbed glass mat (AGM). AGM is also known in the industry as "starved design." Starved refers to the absorption limits of the glass separator creating a limitation to the AGM design relating to diffusion properties of the separator. In certain cases, the AGM battery must be racked and trayed in a specific position for optimum performance. Both these types, gelled and AGM, are called valve-regulated lead acid (VRLA) systems. Today, sealed lead cells are operating effectively in many markets previously closed to lead-acid batteries. For a detailed account of lead-acid cells, Gates Energy Products Inc (1992), Hirai (1990), Moore (1993), Moneypenny and Wehmeyer (1994), and Nelson (1997) are suggested.

3.4.2.1 Discharge performance of sealed lead-acid cells

The general shape of the discharge curve, voltage as a function of capacity (if the current is uniform), is shown in Fig. 3.2. The discharge voltage of the sealed lead-acid battery typically remains relatively constant until most of its capacity is discharged. It then drops off sharply. The flatness and the length of the voltage plateau relative to the length of the discharge are major features of sealed lead cells and batteries. The point at which the voltage leaves the plateau and begins to decline rapidly is often identified as the knee of the curve.

Sealed lead-acid batteries may be discharged over a wide range of temperatures. They maintain adequate performance in cold environments and may produce actual capacities higher than their standard capacity when used in hot environments. Fig. 3.3 indicates the relationships between capacity and cell temperature. Actual capacity is expressed as a percentage of rated capacity as measured at 23°C. Accounting for the temperature effect is rather difficult, because the capacity also depends on a number of factors: design, method of manufacture, storage, service history, and most importantly current density. However, in a limited temperature range around 30°C, the International Electrochemical Commission has recognized the validity of Eq. (3.9),

$$C_T = C_{30}(1 + 0.008[T - 30]) \tag{3.9}$$

where C_T and C_{30} are respective capacities in Ah at each temperature in Celsius. Also, the effects of low and high discharge rates n (as in Peurkert's equation) vary between 1 and 2.

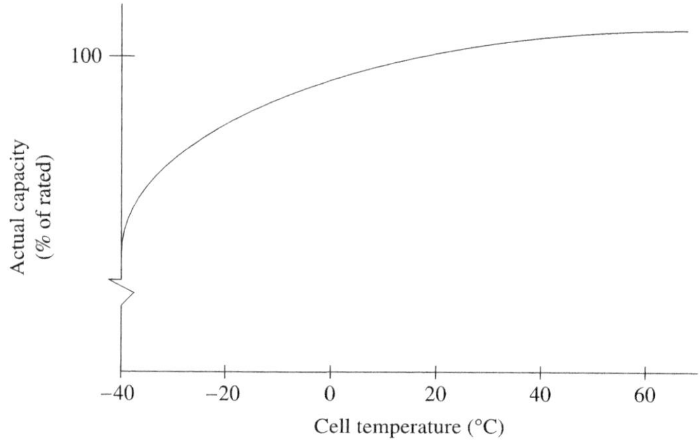

Fig. 3.3 Typical discharge capacity as a function of cell temperature.

3.4.2.2 *Capacity during battery life*

The initial actual capacity of sealed lead-acid batteries is almost always lower than the battery's rated or standard capacity. However, during the battery's early life, the actual capacity increases until it reaches a stabilized value which is usually above the rated capacity. The number of charge-discharge cycles or length of time on float charge required to develop a battery's capacity depends on the specific regime employed. Alternatively, if the battery is on charge at 0.1C, it is usually stabilized after receiving 300% (of rated capacity) overcharge. The process may be accelerated by charging and discharging at low rates.

Under normal operating conditions, the battery's capacity will remain at or near its stabilized value for most of its useful life. Batteries will then begin to suffer some capacity degradation due to their age and the duty to which they have been subjected. This permanent loss usually increases slowly with age until the capacity drops below 80% of its rated capacity, which is often defined as the end of useful battery life. Fig. 3.4 shows a representation of the capacity variation with cycle life that can be expected from sealed lead-acid batteries.

3.4.2.3 *Effect of pulse discharge on capacity*

In some applications, the battery is not called upon to deliver a current continuously. Rather, energy is drawn from the battery in pulses. By allowing the battery to "rest" between these pulses, the total capacity available from the battery is increased. Fig. 3.5 represents typical curves representing the voltage delivered as a function of discharged capacity for pulsed and constant discharge at the same rate.

For the pulsed curve, the upper row of dots represents the open-circuit voltage and the lower sawtooth represents the voltages during the periods when the load is connected. The use of discharged capacity as the abscissa eliminates the rest periods and shows only the periods of useful discharge.

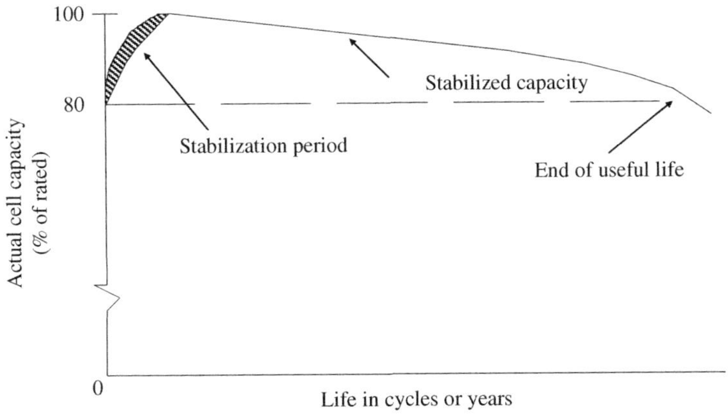

Fig. 3.4 Typical cell capacity during its life.

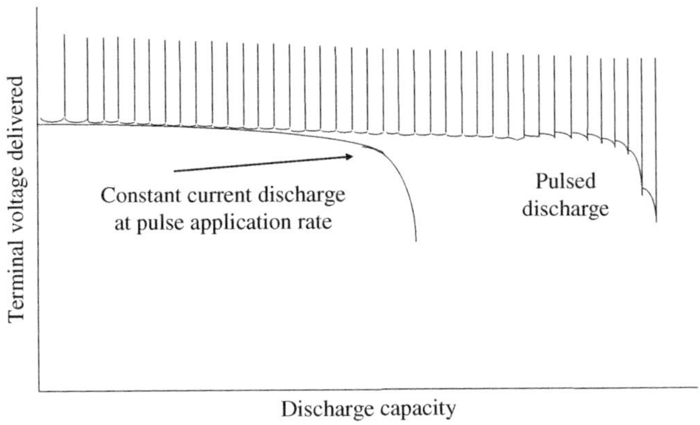

Fig. 3.5 Typical pulsed discharge curve.

3.4.3 Charging

In general, experience with sealed lead-acid chemistry indicates that application problems are more likely to be caused by undercharging than by overcharging. Since the starved-electrolyte cell is relatively resistant to damage from overcharge, designers may want to ensure that the batteries are fully charged, even at the expense of some degree of overcharge. Obviously, excessive overcharge, either in magnitude or duration, should still be avoided.

The charge acceptance of sealed lead-acid batteries in most situations is quite high, typically greater than 90%. A 90% charge acceptance means that for every ampere-hour of charge introduced into the cell, the cell will be able to deliver 0.9 Ah to a load. Charge acceptance is affected by a number of factors including cell temperature, charge rate, cell SOC, age of the cell, and the method of charging.

The SOC of the cell will dictate to some extent the efficiency with which the cell will accept charge. When the cell is fully discharged, the charge acceptance initially is quite low. As the cell becomes only slightly charged, it accepts current more readily and the charge acceptance jumps quickly, approaching 98% in some situations. The charge acceptance stays at a high level until the cell approaches full charge.

As mentioned earlier, when the cell becomes fully charged, some of the electrical energy goes into generating gas, which represents a loss in charge acceptance. When the cell is fully charged, essentially all the charging energy goes to generate gas, except for the very small current that makes up for the internal losses, which otherwise would be manifested as self-discharge. A generalized curve representing these phenomena is shown in Fig. 3.6A. As with most chemical reactions, temperature does have a positive effect upon the charging reactions in the sealed lead-acid cell. Charging at higher temperatures is more efficient than it is at lower temperatures, all other parameters being equal, as shown in Fig. 3.6B.

The starved-electrolyte sealed lead cell charges very efficiently at most charging rates. The cell can accept charge at accelerated rates (up to the C rate) as long as the SOC is not so high that excessive gassing occurs. Also, the cell can be charged at low rates with excellent charge acceptance.

Fig. 3.6C shows the generalized curve of charge acceptance now further defined by charging rates. When examining these curves, one can see that at high states of charge, low charge rates provide better charge acceptance.

3.5 Nickel-cadmium batteries

Nickel-cadmium (NiCd) batteries belong to a group of five rechargeable batteries NiCd, NiMH, Ni-H_2, Ni-Zn, and Ni-Fe, having in common an Ni-based positive electrode and an alkaline solution. Even though some known drawbacks can be associated with the NiCd battery, especially low-energy density and environmental impact (due to the Cd content), still it can advantageously be used in a number of industrial applications. Cd is used as the negative electrode and β-NiOOH as a positive electrode. The electrolyte is a solution of potassium hydroxide (KOH) of 22% concentration with some LiOH for improving the life cycle and temperature performance. Two broad categories of NiCd cells available are vented and sealed types. Vented batteries can have four different electrode constructions, namely, (i) pocket plate, (ii) sintered plate, (iii) fiber nickel cadmium (FNC), and (iv) plastic-bonded plate. Pocket plate type is the oldest with capacities up to 1450 Ah; sintered plate type was developed to increase the energy density by about 50% for high-power requirements; FNC types allow a wide power output capability; and the more recent plastic-bonded type allows reduced weight and volume. Sintered plate types can go up to 100 Ah, while fiber- and plastic-bonded types can reach up to about 500 Ah. More details are found in Broussely and Pistoia (2007).

The operation of the sealed NiCd batteries is similar to VRLA type lead acids. They can come in three different forms: (i) sintered, (ii) fiber, and (iii) foam. The first two types are similar to the cases of vented versions, and in the foam version the Ni electrode is obtained by Ni plating porous fibers followed by pyrolysis. More details are found in Broussely and Pistoia (2007).

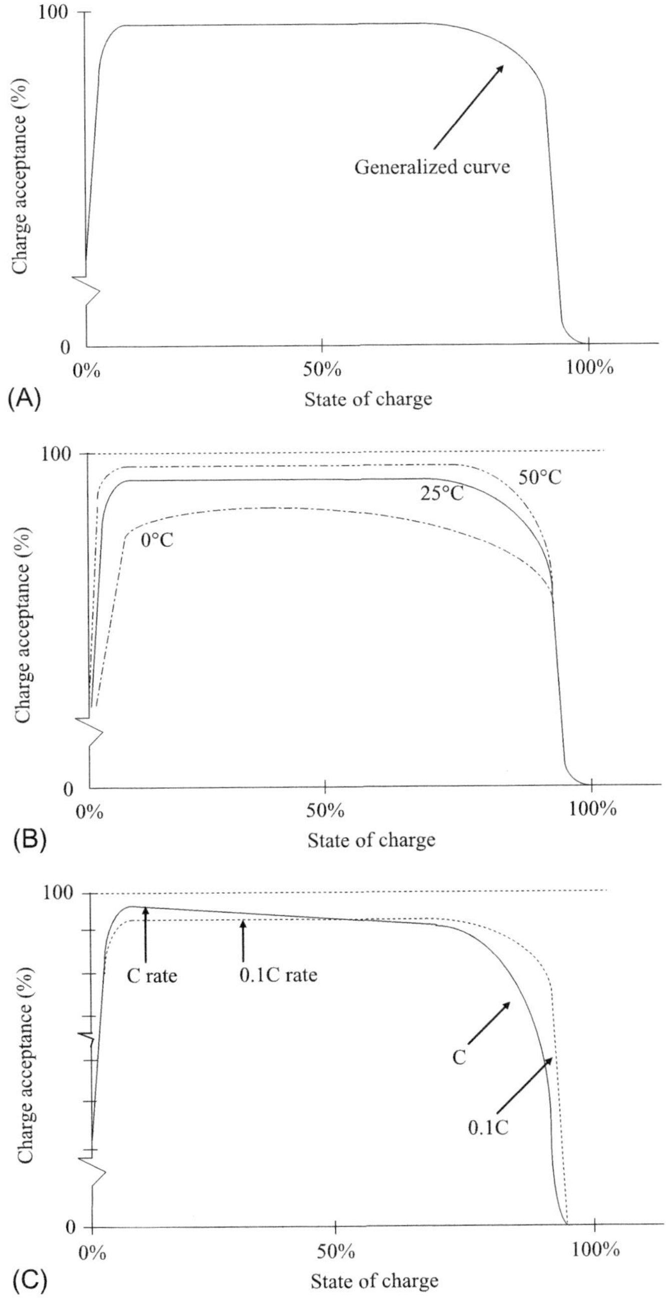

Fig. 3.6 Charge acceptance. (A) Effect of state of charge upon charge acceptance, (B) charge acceptance at various temperatures, and (C) charge acceptance at various charge rates.

Sealed NiCd batteries are well suited to applications where a self-contained power source increases the versatility or reliability of the end product. Among the significant advantages of NiCd families are: higher energy density and discharge rates, fast recharge capability, long operating and storage life, etc.

In NiCd batteries, a reaction at the negative electrode consumes the oxygen gas that generates at the positive electrode during overcharge. The design prevents the negative electrode from generating hydrogen gas, permitting a sealed structure. NiCd batteries mainly adopt cylindrical or prismatic type configurations. Because NiCd batteries contain cadmium, an environmentally hazardous substance, their disposal has become controversial. This has spurred research into other alternative chemistries.

3.5.1 Discharge characteristics

The discharge voltage of a sealed NiCd cell typically remains relatively constant until most of its capacity is discharged. It then drops off rather sharply. The flatness and length of the voltage plateau relative to the length of discharge are major features of sealed NiCd cells and batteries. The discharge curve when scaled by considering the effects of all the application variables provides a complete description of the output of a battery. Differences in design, internal construction, and conditions of actual use of cell affect the performance characteristics. For example, Fig. 3.7A illustrates the typical effect of discharge rate.

3.5.2 Charge characteristics

Nickel-based batteries are easily charged by applying a controlled current. The charging current can be pure direct current or it may contain a significant ripple component such as half-wave or full-wave rectified current.

This section on charging sealed nickel-cadmium batteries refers to charging rates as multiples (or fractions) of the C rate. These C rate charging currents can also be categorized into descriptive terms, such as standard-charge, quick-charge, fast-charge, or trickle-charge as shown in Table 3.4.When a nickel-cadmium battery is charged, not all of the energy input is converting the active material to a usable (chargeable) form. Charge energy also goes to converting active material into an unusable form, generating gas, or is lost in parasitic side reactions.

Fig. 3.8 shows the charge acceptance of NiCd cells. The ideal cell, with no charge acceptance losses, would be 100% efficient. All the charge delivered to the cell could be retrieved on discharge. But nickel-cadmium cells typically accept charge at different levels of efficiency depending upon the SOC of the cell, as shown by the bottom curve of Fig. 3.8.

Fig. 3.8 describes this performance for successive types of charging behavior (zones 1, 2, 3, and 4). Each zone reflects a distinct set of chemical mechanisms responsible for loss of charge input energy.

In zone 1, a significant portion of the charge input converts some of the active material mass into a nonusable form, that is, charged material not readily accessible during

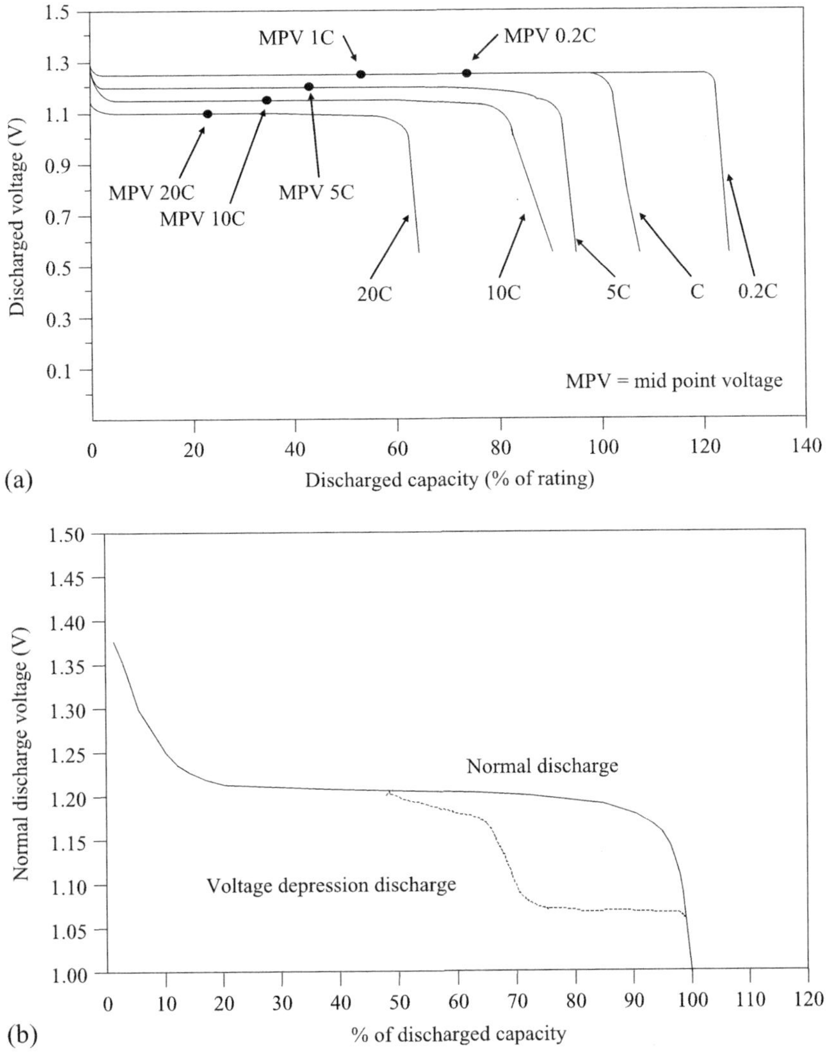

Fig. 3.7 Discharge curves for NiCd cells. (A) Typical curves at 23°C and (B) voltage depression effect.

medium- or high-rate discharges, particularly in the first few cycles. In zone 2, the charging efficiency is only slightly less than 100%; small amounts of internal gassing and parasitic side reactions are all that prevent the charge from being totally efficient. Zone 3 is the transition region.

As the cell approaches full charge, the current input shifts from charging positive active material to generating oxygen gas. In the overcharge region, zone 4, all of the current coming into the cell generates gas. In this zone, the charging efficiency is practically none.

Table 3.4 Definition of rates for charging NiCd cell.

Method of charging	Charge rate Multiples of C rate	Recharge time (h)	Charge control
Standard	0.05	36–48	Not required
	0.1	16–20	
Quick	0.2	7–9	Not required
	0.25	5–7	
	0.33	4–5	
Fast	1	1.2	Required
	2	0.6	
	4	0.3	
Trickle	0.02–0.1	Used for maintaining charge of a fully charged battery	

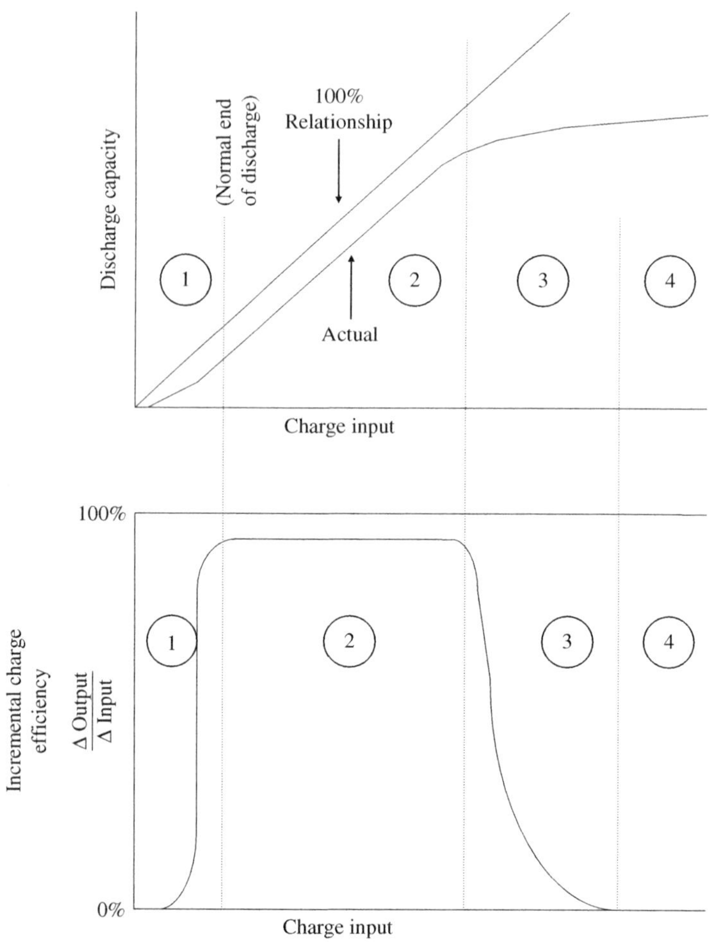

Fig. 3.8 Charge acceptance of a sealed NiCd cell at 0.1°C and 23°C.

The boundaries between zones 1, 2, 3, and 4 are indistinct and quite variable depending upon cell temperature, cell construction, and charge rate. The level of charge acceptance in zones 1, 2, and 3 is also influenced by cell temperature and charge rate. For details, Gates Energy Products Inc (1992) is suggested.

3.5.3 Voltage depression effect

When some NiCd batteries are subjected to numerous partial discharge cycles and overcharging, cell voltage decreases below 1.05 V/cell before 80% of the capacity is consumed. This is called the *voltage depression effect* and the resultant lower voltage may be below the minimum voltage required for proper system operation, giving the impression that the battery has worn out (see Fig. 3.7B). Cells exposed to overcharge, particularly at higher temperatures, are quite common and the voltage may be about 150 mV lower than the normal cell voltage. Voltage depression is an electrically reversible condition and disappears when the cell is completely discharged and charged. This process is sometimes called conditioning. This effect is sometimes erroneously called the "memory effect."

Broussely and Pistoia (2007) provides a significant discussion on NiCd battery chemistry with useful details of construction and applications with chemical reactions.

3.6 Nickel metal hydride batteries

While NiCd battery performance was rapidly advancing after 1980, nickel metal hydride (NiMH) cells introduced in the early 1990s have shown a nearly 170% increase in energy density. By 2000, the volumetric energy density of NiMH cells has increased over 300 Wh/L (Powers, 2000). These extensions of the Ni-based chemistries have become popular with product applications such as notebook computers and cellular phones, etc. The first practical NiMH batteries entered the market in the early 1990s. In these cells, the environmentally unfriendly cadmium negative electrode was replaced by an alloy that could reversibly absorb and desorb hydrogen. This chemistry has been the choice for EV till about 2005 (Powers, 2000; Stempel et al., 1998).

3.6.1 Construction

In many ways, nickel metal hydride (NiMH) batteries are the same as NiCd types, but they use nickel for the positive electrode and a recently developed material, known as a hydrogen-absorbing alloy, for the negative electrode. When a NiMH cell is charged, hydrogen generated by reaction with a cell electrolyte is stored in the metal alloy (M) in the negative electrode. Meanwhile, at the positive electrode, which consists of nickel oxyhydroxide loaded into a nickel foam substrate, a hydrogen ion is ejected and nickel is oxidized (Stempel et al., 1998). With an operating voltage of 1.2 V, they provide high capacity, large energy density characteristics comparable to those of NiCd models.

3.6.2 A comparison between NiCd and NiMH batteries

The NiCd cell is more tolerant of fast recharging and overcharging than NiMH cells. NiCd cells hold their charge longer than do NiMH cells. NiCd cells will withstand between 500 and 2000 charge/discharge cycles compared to about 500–800 cycles for NiMH cells. Further, NiCd cells will withstand a wider temperature range than NiMH cells. On the other hand, NiMH cells do not exhibit the notorious "memory effect" that NiCd cells sometimes do. As with any new technology at the time of entering the market, NiMH's prices were higher than those of NiCds (Briggs, 1994; Small, 1992).

The voltage profile of NiMH cells during discharge is very similar to that of the NiCd cells. NiMH cells' open-circuit voltage is 1.3–1.4 V. At moderate discharge rates, NiMH cells' output voltage is 1.2 V. Both NiCd and NiMH cells have a relatively constant output voltage during their useful service. Fig. 3.9 is a typical graph from a battery company comparing the output voltage of 700-mAh NiCd and 1100-mAh NiMH AA cells while under load. Note that the NiMH cell's greater capacity results in approximately 50% longer service life.

NiMH chemistry advances helped introduce batteries suitable for EV and aerospace applications. These bipolar NiMH batteries were meeting the volumetric and gravimetric energy density needs of these applications during the mid-1990s to mid-2005 (Cole et al., 2000; Reisner and Klein, 1994).

Fig. 3.10 is another typical graph showing that NiCd and NiMH batteries and cells charge in similar fashions as well. However, the little bumps at the end of the two

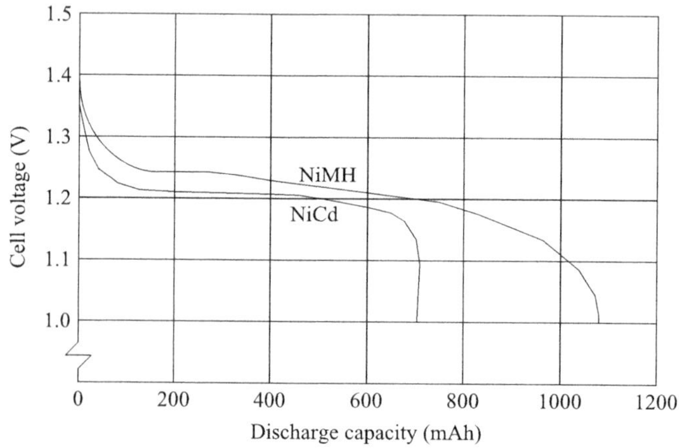

Fig. 3.9 Comparison of discharge characteristics of NiCd and NiMH batteries.

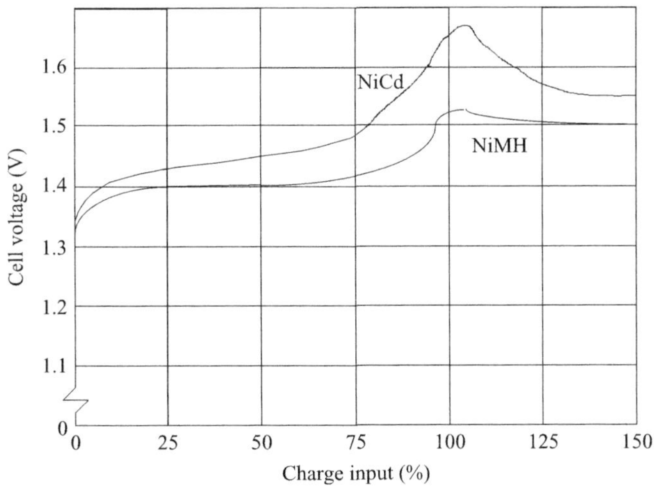

Fig. 3.10 Battery voltage at the achievement of 100% charge.

cells' charge curves bear closer examination. You will always see these negative excursions even though absolute cell voltages vary significantly with temperature.

The negative excursions signal a fully charged cell more or less independently of temperature, a useful quirk that sophisticated battery chargers exploit. Note that the NiCd cell's negative-going voltage excursion after reaching full charge is more pronounced than that of the NiMH cells.

3.7 Lithium-based rechargeable batteries

Toward the latter part of the 1990s, the demand for portable systems was increasing at a dramatic rate. Meeting these goals required improvements in battery technologies beyond traditional NiCd and NiMH systems. Newer lithium-based rechargeable battery systems have overcome the safety and environmental obstacles posted by early efforts and are, in general, the most efficient rechargeable battery packs available. With an energy density by weight about twice that of nickel-based chemistries (Table 3.1), Li-ion batteries can deliver lighter weight packs of acceptable capacity. Li-ion also has about three times the cell voltage of NiCd and NiMH batteries; therefore, fewer cells are needed for a given voltage requirement. Li-ion batteries have become the choice in notebook PCs and many other portable systems because of their high-energy density, declining costs, and readily available management circuits.

Rechargeable lithium cells come in several different chemistries, namely, (i) lithium-ion, (ii) lithium-polymer, (iii) lithium metal, and (iv) lithium-iron phosphate. Most of these chemistries have come into commercial use from 1992 onward (Dan, 1997; Morrison, 2006a, b, 2007). Fig. 3.11A indicates the progress of Li-ion chemistry based on the commonly used 18,650 type cells, which is the most matured out of all the lithium cells. Fig. 3.11B indicates the relative capabilities of the four common rechargeable chemistries.

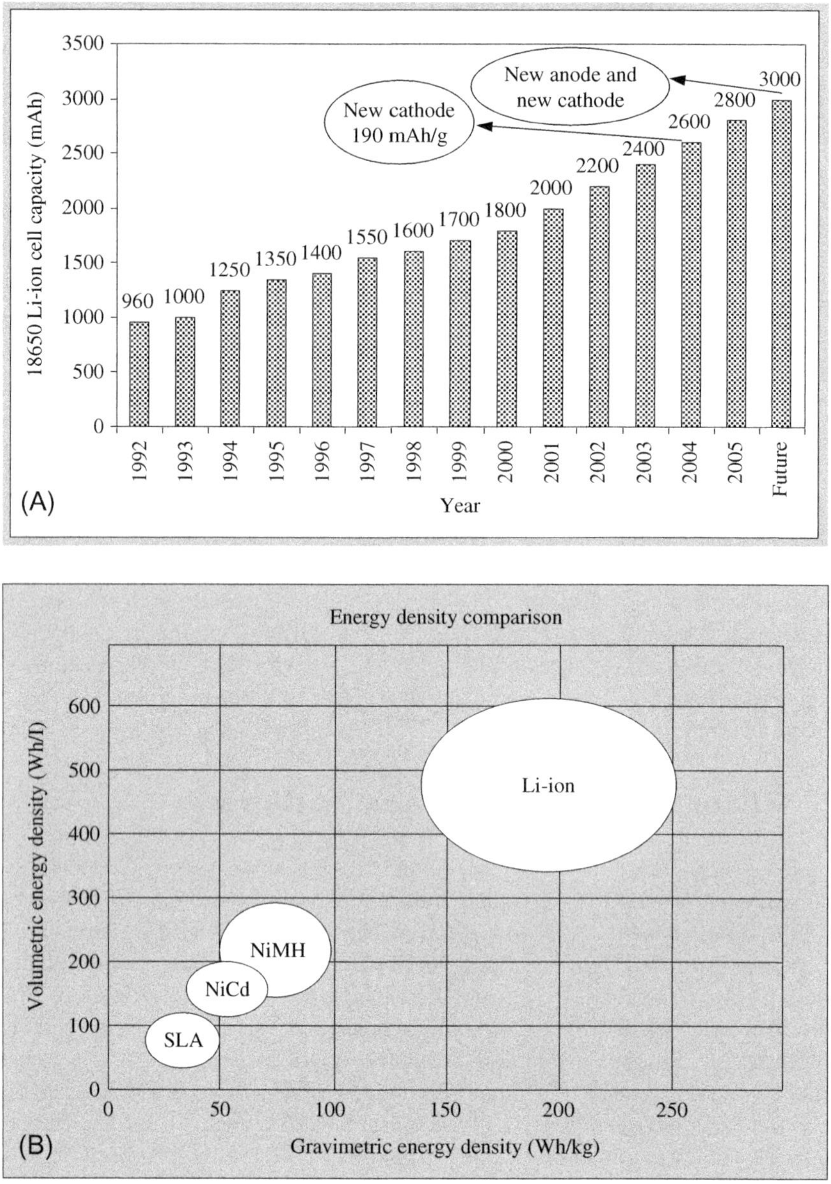

Fig. 3.11 Progress of the Li-ion cells and comparison with other chemistries. (A) mAh capacity increases over the period from 1992 for 18,650 cells and (B) comparison of Li-ion with other chemistries.

Source: Power Electronics Technology (Morrison, D., 2006a. New materials extend Li-ion performance. Power Electron. Technol. 50–52).

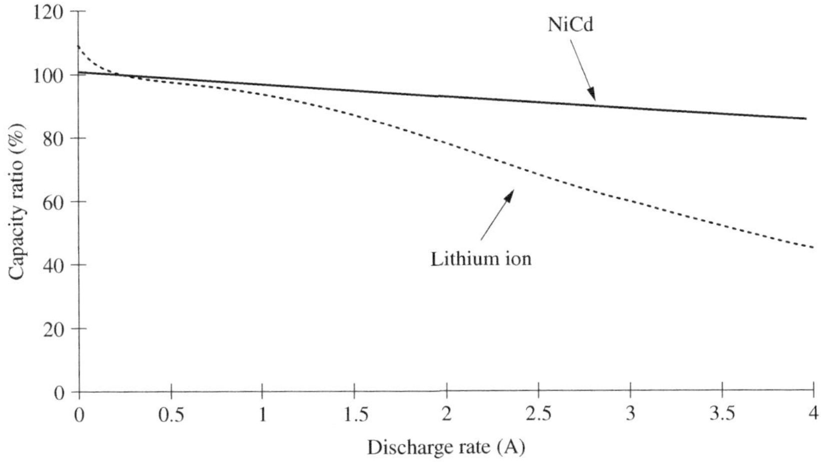

Fig. 3.12 Li-ion and NiCd capacity vs discharge current.

The first noticeable difference between Li-based chemistries and nickel-based chemistries is the higher internal impedance of the lithium-based batteries. Fig. 3.12 shows this by graphing the actual discharge capacity of an Li-ion cell at different discharge currents compared to an NiCd cell. At a 2 A discharge rate (2C), less than 80% of the rated capacity is available for Li-ion compared to nearly 95% of the rated capacity for NiCd (Freeman and Heacock, 1995). For systems with discharge currents greater than 1 A, the capacity realized from the Li-ion battery may be less than expected. Parallel battery stack configurations are often used in Li-ion battery packs to help reduce the severity of this problem. Due to the nature of the lithium chemistries, lithium batteries cannot tolerate overcharge and overdischarge.

Given the comparison of energy densities of fuels, batteries, and explosives (Fig. 3.2), the most important factor in using lithium chemistries in a portable consumer product is safety. Fortunately, battery safety is comprehensively addressed by cell and BPA in concert with the semiconductor manufacturers. For details, Bennett and Brawn (1997) is suggested. Commercially available Li-ion packs have an internal protection circuit that limits the cell voltage during charge to between 4.1 and 4.3 V/cell, depending on the manufacturer. Voltages higher than this rating could permanently damage the cell. A discharge limit of between 2.0 and 3.0 V (depending on the manufacturer) is necessary to avoid reducing the cycle life of the battery and damaging the battery.

3.7.1 Construction

The anode, or negative electrode, in an Li-ion cell is comprised of a material capable of acting as a reversible Li-ion reservoir. This material is usually a form of carbon, such as coke, or graphite, or pyrolytic carbon. The cathode, or positive electrode, is also a material which can act as a reversible lithium-ion reservoir. Due to lithium ions shuttling back and forth between these two reservoirs, these batteries are sometimes known as rocking chair cells (Fuller et al., 1994)

Preferred cathode materials are currently $LiCoO_2$, $LiNiO_2$, or $LiMn_2O_4$ because of their high oxidation potentials of about 4 V versus lithium metal. Commercially available Li-ion cells use a liquid electrolyte made up of mixtures which are predominately organic carbonates containing one or more dissolved lithium salts (Levy, 1995). While $LiCoO_2$ was the preferred cathode material at the early stages of development, in newer systems $LiMn_2O_4$ cathodes are used. Chemical process details are summarized in Levy (1995). In the latest generation of Li-ion cells, a new cathode material, known as Nickel oxide-based New Platform, is used by Panasonic batteries (Morrison, 2006a), extending the commonly used 18,650 series to 3.9 Ah.

In lithium-polymer batteries, the liquid electrolyte is replaced with a polymer in gel or solid form. The polymer electrolyte provides the required electrode stack pressure, so the metal can is no longer required and it becomes possible to package the cells easily. A laminate of aluminum foil and plastic, the pouch occupies less space and weighs less than the metal can. For these reasons, lithium-polymer cells came in compact prismatic cell formats around early 2000 (Morrison, 2000). Also, they are considered safer than liquid Li-ion, as they do not leak when they are punctured. As a result, simplifications are possible for in-pack protection circuits (Morrison, 2000).

$LiFePO_4$ batteries, also known as LFP battery systems, were introduced around the late 1990s, and recently they have gone into mass production, due to demand from EVs, hybrid EVs, electric bikes, and power tools. In these batteries, cathode material, $LiFePO_4$, is environmentally benign, inexpensive, and relatively abundant (Jiayuan et al., 2009).

3.7.2 Charge and discharge characteristics

Today, the predominant Li-ion technologies use coke or graphite for an anode material. Fig. 3.13 illustrates the differences in the two types of cells during discharge. The graphite anode discharge voltage is relatively flat during a majority of the discharge cycle, while the coke anode discharge voltage is more sloped (Juzkow and St. Louis, 1996).

The energy available from the graphite anode cell is higher for a given capacity due to the higher average discharge voltage. This may be useful in systems that need the maximum watt-hour capacity for a given battery size. Also, the charge and discharge cutoff voltages between the two Li-ion systems vary among manufacturers.

Fig. 3.14 shows the typical charge profile for Li-ion batteries. The charge cycle begins with a constant current limit, transitioning to a constant voltage limit, typically specified between 4.1 and $4.3\,V \pm 1\%$, depending on the manufacturer's recommendation. This allows maximum charge capacity without cell damage. Charging to a lower voltage limit does not damage the cell, but the discharge capacity will be reduced. A 100 mV difference could change the discharge capacity by more than 7%.

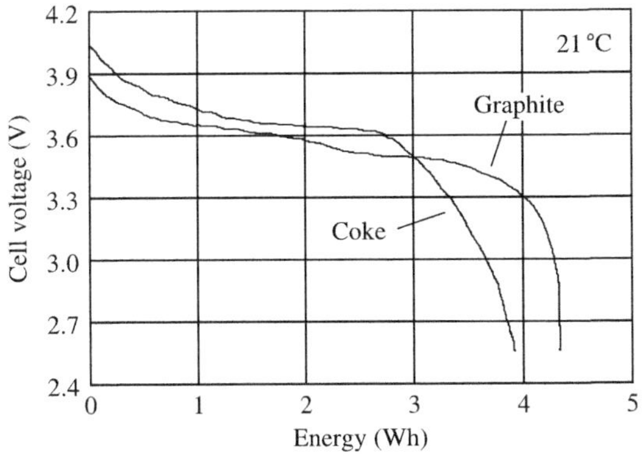

Fig. 3.13 Li-ion discharge profile for different electrodes.
Source: Adapted from (Juzkow, M. W., St. Louis, C., 1996. Designing lithium-ion batteries into today's portable products. In: Portable by Design Conference, pp. 13–22.), Moli Energy Limited, USA.

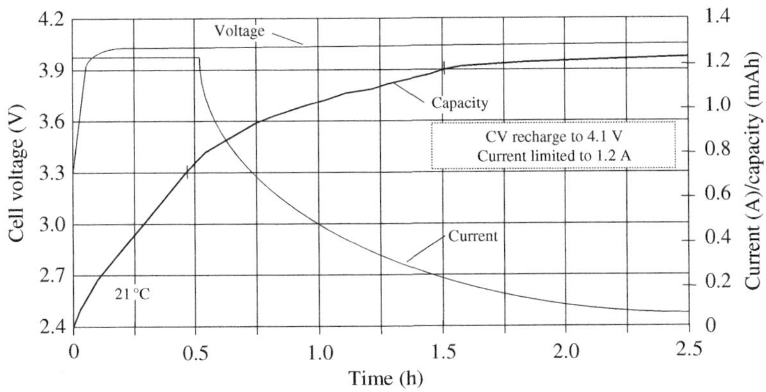

Fig. 3.14 Li-ion charge profile at constant potential charging at 4.1 V and current limited to 1.2 A.
Source: Moli Energy Ltd., USA.

3.7.3 Li-ion microbatteries

Li-ion chemistry has matured vastly during recent years and coin cells of volume less than 1 cm^3 with volumetric energy densities up to about 100 Wh/L are available. Better miniaturization can be achieved by solid-state thin-film batteries (Bates et al., 2000). These devices with solid-state electrodes and a thin-film layer of electrolyte have achieved high cycle stability and specific capacity. However, the overall capacity of these cells is quite small, and to overcome this 3D silicon wafer processing combined with state-of-the art lithium battery electrodes is under development (Hahn et al., 2012). This research ongoing at Fraunhofer IZM, Germany, is expected to yield 2000–10,000 microbatteries on a 300 mm silicon wafer, based on cavities developed on silicon for electrodes, current collectors, and contacts. More details are available in Hahn et al. (2012). These microbatteries are aimed at powering miniaturized sensor nodes, RFID devices, medical devices, etc.

3.8 Reusable alkaline batteries

Alkaline technology has been used in primary batteries for several years. With the development of the reusable alkaline manganese technology, secondary alkaline cells have quickly made their way into many consumer and industrial applications. In many applications, reusable alkaline cells can be recharged from 75 to over 500 times, and initially have three times the capacity of a fully charged NiCd battery. These cells do not compete with NiCds in high-power applications, however.

Intensive research and development activities carried out at Battery Technologies Inc. (BTI), Canada, and at the Technical University in Graz, Austria, in the late 1980s and early 1990s resulted in the successful commercialization of the rechargeable alkaline manganese dioxide zinc (RAM) system. BTI has chosen to sell licenses and production equipment, where necessary, for the manufacturing and worldwide marketing rights of its proprietary RAM technology. For example, Rayovac Corporation, one of the Licensees, launched its line of reusable alkaline products under the name RENEWAL in the USA, Pure Energy Battery Corporation in Canada (PURE ENERGY), and Young Poong Corporation in South Korea (ALCAVA). For details, Nossaman and Parvereshi (1995), Sengupta (1995), and Ivad and Kordesch (1997) are suggested.

The chemistry behind the reusable approach depends on limiting the zinc anode to prevent overdischarge of the MnO_2 cathode. Additives are also incorporated to control hydrogen generation and other adverse effects on charge. The rated cycle life is around 25 cycles to 50% of initial capacity. A longer cycle life is possible depending on the drain rate and DOD. To take advantage of the reusable alkaline cell and increase its life, a special "smart charger" is required.

3.8.1 *Cumulative capacity*

Using reusable alkaline cells can drive down the total battery cost to the consumer. This cost saving can be determined by looking at the cumulative capacity of a reusable cell versus the onetime use of a primary alkaline cell. Fig. 3.15 illustrates the capacity of AA cells being discharged down to 0.9 V at 100 mA. It shows that although the initial use of the reusable alkaline is almost that of primary alkaline, the reusable one can be recharged for continued use. Table 3.5 shows the increase in cumulative capacity by limiting the DOD and achieving more cycles.

Overcharging also affects the cycle life of reusable alkaline. Reusable alkaline is not tolerant of overcharge and high continuous charge currents, and may be damaged if high current is forced into them after they have reached a partially recharged state. Proper charging schemes should be used to prevent an overcharged condition.

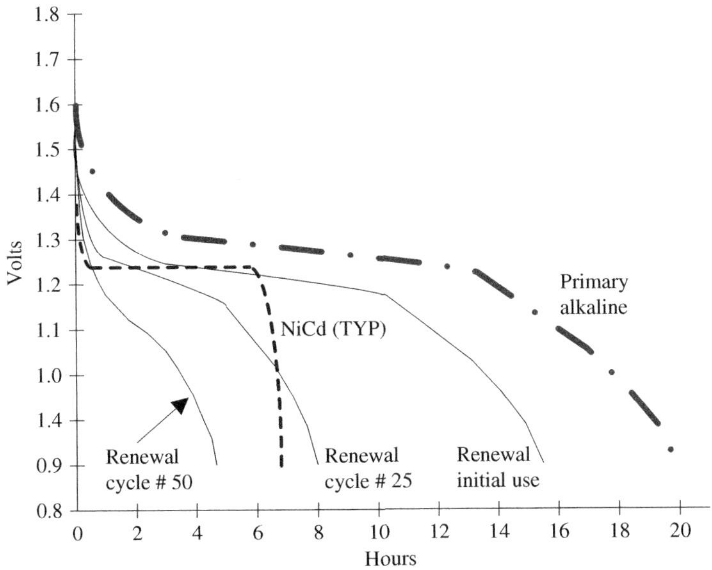

Fig. 3.15 100 mA discharge curve comparison for NiCd, primary alkaline, and reusable alkaline.
Source: Benchmarq Microelectronics/BTI Technologies.

Table 3.5 Capacity of "AA" cells at various depths of discharge (DODs) (values in mAh).

	125 mA to 0.9 V		
Condition	**100% DOD**	**30% DOD**	**10% DOD**
Cycle 1	1500	450	150
Cycle 50	400		
Cumulative 50	33,000	22,000	7000
Cumulative 100		44,000	15,000
Cumulative 500			73,000

3.9 Zn-air batteries

Primary Zn-air batteries have been in existence for over 50 years with applications such as hearing aids and harbor buoys. The light weight and high-energy content in Zn-air technology have promoted research on Zn-air rechargeable chemistry by companies such as AER Energy Resources, USA, in the late 1990s and around 2009 by companies such as RWE Innogy (Germany) and ReVolt Technology AS (RWE Innogy Invests, 2009). The focus is on EV and portable appliances. Another application is solar powered rural telecom systems.

Rechargeable Zn-air technology is an air-breathing technology where the oxygen in ambient air is used to convert zinc into zinc oxide in a reversible process. Cells use an air-breathing carbon cathode to introduce oxygen from air into a potassium hydroxide electrolyte. The cathode is multilayered with a hydrophilic layer and the anode comprises metallic zinc.

The characteristic voltage of zinc-air systems is nominally around 1 V. For example, during discharge, they will operate within a voltage between 1.2 and 0.75 V. The current and power capability of the system is proportional to the surface area of the air-breathing cathode. For more current and power, a larger surface area cell is required. For less current and power, a smaller cell may be used. Compared to other rechargeable chemistries, Zn-air chemistry needs an air manager for the intake and exhaust of air to allow the chemical process.

Fig. 3.16 compares the performance of recent developments on zinc-air chemistry (RWE Innogy Invests, 2009) with other common rechargeable chemistries. This clearly indicates that Zn-air batteries require less weight and volume. Discharge and charge characteristics of Zn-air batteries by AER Energy Resources are shown in Fig. 3.17.

The cells exhibit a flat voltage profile over the discharge cycle. A typical charge voltage is 2 V/cell using a constant voltage/current taper approach. Life cycle varies between 50 and 400, depending on the DOD. Cost per Wh is apparently the lowest compared to Ni- and Li-based chemistries. For details, Cutler (1997) is suggested.

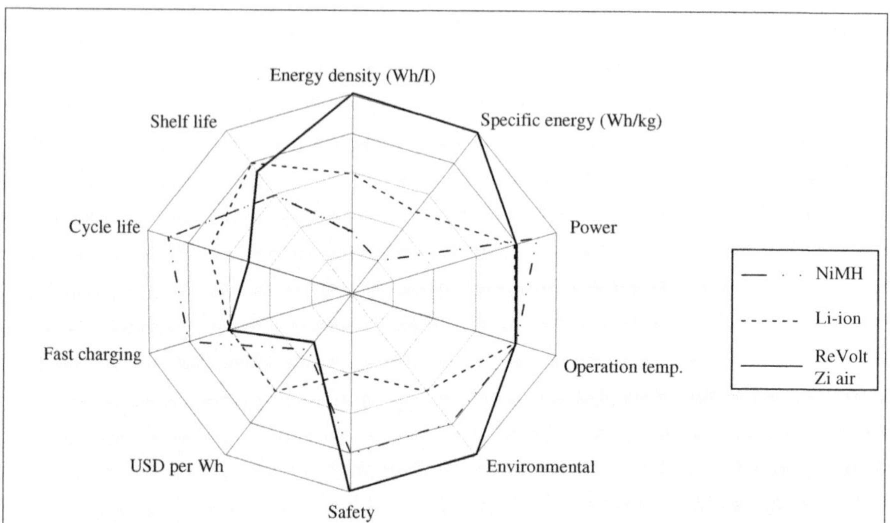

Fig. 3.16 Performance comparison of rechargeable Zn-air chemistry with other chemistries. *Source*: RWE Innogy Invests, 2009. !5.5M in ReVolt; Rechargeable Zinc-air Storage Systems, http://www.greencarcongress.com/2009/09/revolt-20090901.html./Green Car Congress.

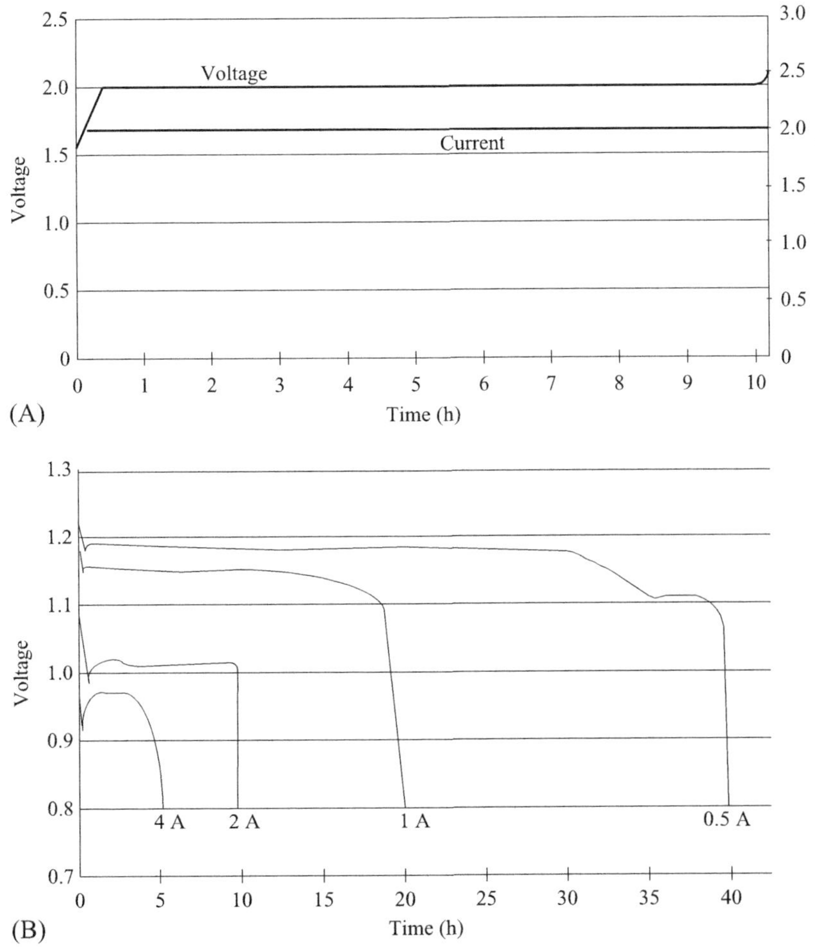

Fig. 3.17 Charge/discharge characteristics of a Zn-air chemistry. (A) Charge and (B) discharge. Courtesy: AER Energy Technologies.

3.10 Rechargeable batteries versus supercapacitors

During the last two decades, a new form of large capacitors, now known as supercapacitors, have come into the world of power electronics. Supercapacitors in general work similar to the normal capacitors, where energy is stored in an electric field. They come in several different commercial variations such as (i) symmetrical electrical double layer capacitors (EDLCs), (ii) hybrid types with one of the electrodes similar to lithium ion batteries, and (iii) capa-batteries (or battery-caps) where pseudo-capacitance property dominates.

After 2015, these device families have advanced significantly, and the capa-batteries today have nearly reached the energy density of lead-acid chemistry. One important property of supercapacitors is that their internal resistance (or equivalent series

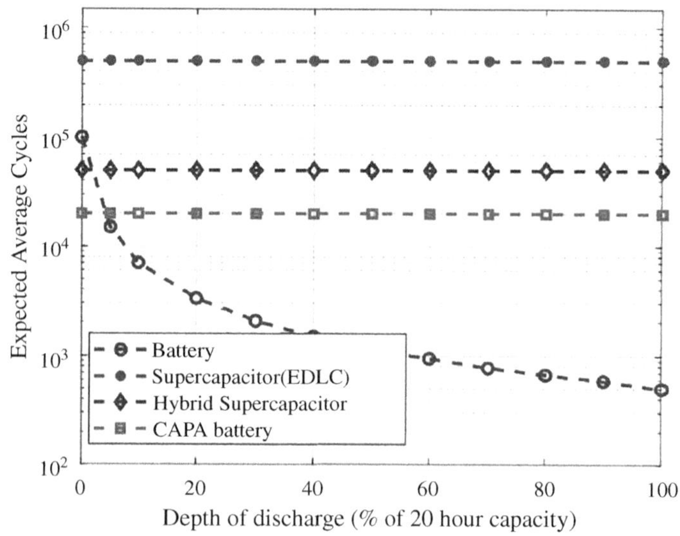

Fig. 3.18 Life cycle vs depth of discharge of batteries and supercapacitors.

resistance, ESR) is not dependent on the DOD. ESR values of supercapacitor families are generally much lower and constant over the DOD, compared to batteries. Another fact is that their cycle life is not dependent on the DOD. This is illustrated in Fig. 3.18 (Jayananda et al., 2020). According to market search reports, supercapacitors utilization in electronic systems will be seen a rapid growth in the near future (Harrop, 2020).

From Chapter 7 onwards, this subject will be discussed in detail.

References

Baert, D., Vervaet, A., 1999. Lead-acid battery model for the derivation of Peukert's law. J. Power Sources 44, 3491–3504.

Bates, J., et al., 2000. Thin-film lithium and lithium-ion batteries. Solid State Ionics 135 (1-4), 33–45.

Bennett, P.D., Brawn, G.W., 1997. Introduction to applying Li-ion batteries. In: Proceedings of Portable by Design Conference, USA, pp. 125–134.

Bhangu, B.S., Bentley, P., Stone, D.A., Bingham, C.M., 2005. Nonlinear observers for predicting state-of-charge and state-of-health of lead-acid batteries for hybrid electric vehicles. IEEE Trans. Veh. Technol. 54 (3), 783–794.

Briggs, A., 1994. NiMH technology overview. In: Portable by Design Conference Proceedings, pp. BT-42–BT-45.

Broussely, M., Pistoia, G., 2007. Industrial Applications of Batteries—From Cars to Aerospace and Energy Storage. Elsevier, Amsterdam, pp. 691–736.

Chu, B., 2009. $LiFePO_4$ batteries help consumer devices come to life. Power Electron. Technol. 35, 10–15.

Cole, J.H., Eskra, M., Klein, M., 2000. Bipolar nickel-metal hydride batteries for aerospace applications. IEEE AES Mag. 15 (1), 39–45.

Cutler, T., 1997. Rechargeable zinc-air design options for portable devices. In: Proceedings of Portable by Design Conference, USA, p. 112.

Dan, P., 1997. Recent advances in rechargeable batteries. Electron. Des. 45 (3), 112–116.

Doerffel, D., Sharkh, S.A., 2006. A critical review of using the Peukert equation for determining the remaining capacity of lead-acid and lithium-ion batteries. J. Power Sources 155 (2), 395–400.

Ebbesen, S., Elbert, P., Guzzella, L., 2012. Battery state-of-health perspective energy management for hybrid electric vehicles. IEEE Trans. Veh. Technol. 61 (7), 2893–2900.

EE Times Asia, 2009. Silicon-Air Battery Touts Unlimited Shelf Life. http://www.eetasia.com/login.do?fromWhere=/ART_8800590556_765245_ NT_17302c65.HTM.

Freeman, D., Heacock, D., 1995. Lithium-ion battery capacity monitoring within portable systems. In: HFPC Conference Proceedings, pp. 1–8.

Fuller, T.F., Doyle, M., Newman, J., 1994. Simulation and optimization of the dual lithium ion insertion cell. J. Electrochem. Soc. 141 (1), 1–10.

Gates Energy Products Inc, 1992. Rechargeable Batteries Applications Handbook. Butterworth-Heinemann, Boston.

Guoliang, W., Rengui, L., Chunbo, Z., Chan, C.C., 2010. Apply a piece-wise Peukert's equation with temperature correction factor to NiMH battery state of charge estimation. J. Asian Electr. Veh. 8 (2), 1419–1423.

Hahn, R., Hoppner, K., Eisenreich, M., Ferch, M., Wilke, M., Lang, K.D., 2012. Development of rechargeable micro batteries based on micro channel structures. In: IEEE International Conference on Green Computing and Communications, pp. 619–623.

Hamlen, R.P., Christopher, H.A., Gilman, S., 1995. US Army battery needs—present and the future. IEEE AES Mag. 10, 30–33.

Harrop, P., 2020. Supercapacitor Markets, Technology Roadmap, Opportunities 2021-2041, IDtechEX report. ISBN: 9781913899011.

Hausmann, A., Depick, C., 2013. Expanding the Peukert equation for battery modeling through inclusion of a temperature dependency. J. Power Sources 235, 148–158.

Hirai, T., 1990. Sealed lead-acid batteries find electronic applications. PCIM, 47–51.

IDTechEx Ltd, 2012a. Batteries & Supercapacitors in Consumer Electronics 2013-2023: Forecasts, Opportunities, Innovation., p. 345 (SKU: CGAQ4888235).

IDTechEx Ltd, 2012b. Traction Batteries for Electric Vehicles Land, Water & Air 2013-2023., p. 333 (SKU: CGAQ4862363).

Ivad, J.D., Kordesch, K., 1997. In-application use of rechargeable alkaline manganese dioxide/zinc (RAM™) batteries. In: Proceedings of Portable by Design Conference, USA, pp. 119–124.

Jayananda, D., Kularatna, N., Steyn-Ross, D.A., 2020. Supercapacitor-assisted LED (SCALED) technique for renewable energy systems: a very low frequency design approach with short-term DC-UPS capability eliminating battery banks. IET Renew. Power Gen. 14 (9), 1559–1570.

Jiayuan, W., Zechang, S., Xuezhe, W., 2009. Performance and characteristic research in $LiFePO_4$ battery for electric vehicle applications. In: IEEE Vehicle Power and Propulsion Conference, 2009 (VPPC '09), pp. 1657–1661.

Juzkow, M.W., St. Louis, C., 1996. Designing lithium-ion batteries into today's portable products. In: Portable by Design Conference, pp. 13–22.

Kim, S., 2001. Lithium-ion polymer batteries promise improved size, safety, energy density. PCIM, 30–39.

Kim, J., Cho, B.H., 2011. State-of-charge estimation and state-of-health prediction of Li- ion degraded battery based on an EKF combined with a per-unit system. IEEE Trans. Veh. Technol. 60 (9), 4249–4260.

Levy, S.C., 1995. Recent advances in lithium ion technology. In: Portable by Design Conference Proceedings, pp. 316–323.

marketresearch, 2009. Advanced Rechargeable Battery Market: Emerging Technologies and Trends Worldwide., p. 245. marketresearch.com. (Pub. ID: SB1933124).

Moneypenny, G.A., Wehmeyer, F., 1994. Thinline battery technology for portable electronics. In: HFPC Conference Proceedings, pp. 263–269.

Moore, M.R., 1993. Valve regulated lead acid vs flooded cell. In: Power Quality Proceedings, pp. 825–827.

Morrison, D., 2000. Thinner Li-ion batteries power next generation portable devices. Electron. Des., 95–106.

Morrison, D., 2006a. New materials extend Li-ion performance. Power Electron. Technol., 50–52.

Morrison, D., 2006b. Li-ion cells build better batteries for power tools. Power Electron. Technol. 32, 52–54.

Morrison, D., 2007. Cathode modeling builds better batteries for power tools. Power Electron. Technol. 52.

Nelson, B., 1997. Pulse discharge and ultrafast recharge capabilities of thin-metal film technology. In: Proceedings of Portable by Design Conference, USA, pp. 13–18.

Nossaman, P., Parvereshi, J., 1995. In systems charging of reusable alkaline batteries. In: Proceedings of HFPC Conference, USA.

Powers, A.R., 2000. Sealed nickel cadmium and nickel metal hydride cell advances. IEEE AES Mag. 15 (12), 15–18.

Quinnell, R.A., 1991. The business of finding the best battery. EDN, 162–166.

Reisner, D.E., Klein, M., 1994. Bipolar nickel-metal hydride battery for hybrid vehicles. IEEE AES Mag. 9 (5), 24–28.

RWE Innogy Invests, 2009. !5.5M in ReVolt; Rechargeable Zinc-air Storage Systems. http://www.greencarcongress.com/2009/09/revolt-20090901.html.

Schimpf, M., 1996. Choosing lithium primary-cell types. Electron. Des. 44, 141–144.

Sengupta, U., 1995. Reusable Alkaline™ Battery Technology: applications and system design issues for portable electronic equipment. In: Portable by Design Conference, pp. 562–570.

Small, C.H., 1992. Nickel-hydride cells avert environmental headaches. EDN, 156–161.

Stempel, R.C., Ovshinsky, S.R., Gifford, P.R., Corrigan, D.A., 1998. Nickel metal hydride: ready to serve. IEEE Spectr. 35 (11), 29–34.

US demand, 2009. US demand for batteries to reach $16.8 billion in 2012. Power Electron. Technol. 35, 9.

Dynamics, models, and management of rechargeable batteries

4

4.1 Introduction

In Chapter 3, we had a simple discussion on battery chemistries and their basic performance specifications from an electronic engineer's viewpoint. To get the best out of any rechargeable battery chemistry, a battery management circuit designer should have an intrinsic understanding of the battery chemistry and the internal behavior from an electrical engineer's perspective. In a modern application, such as a wireless communication product or an electric vehicle, device power consumption fluctuates widely and the energy source has to deal with highly varying current deliveries to electronic loads, and this demands a fast responding battery pack with an optimally designed battery management system (BMS). In addition to these characteristics determined by the load behavior, in a rechargeable battery pack, state of charge (SOC), state of health (SOH), and end of life (EOL) become important overall quality parameters of an installed pack. In determining optimal management criteria for the longest lifetime and the longest runtime of the battery pack, circuit designers have to clearly understand battery dynamics and use optimal battery models to design the correct BMSs.

This chapter provides some fundamentals of battery dynamics, different types of modeling to achieve usable and practical equivalent circuits, and an overview of management systems for different applications. In this chapter, electrochemistry-based modeling approaches are highlighted to better understand the behavior of popular rechargeable chemistries, without going into mathematics. For details of electrochemistry-related modeling, the chapter cites adequate references.

4.2 Simplest concept of a battery

As discussed in Chapter 2, simplest model of a battery (single-cell case) is based on a constant voltage source representing the chemical behavior of the electrode-electrolyte combination together with a constant resistance, and this Thevenin equivalent circuit provides us with an excellent starting point. However, this equivalent circuit as shown in Fig. 2.1 is an oversimplified case, and it does not provide us with any idea of practical battery behavior, particularly applied to situations of charging and discharging with time-varying currents occurring between the charging system and the connected load, respectively.

Energy Storage Devices for Renewable Energy-Based Systems. https://doi.org/10.1016/B978-0-12-820778-9.00009-7

4.3 Battery dynamics

To understand the fundamentals of battery dynamics, we should start with a simple concept applied to dynamic systems, where variables within the system are heavily dependent on time. Such a dynamic system can be represented by a simple block diagram shown in Fig. 4.1, where the system input receives a time-dependent excitation signal and when it is propagated through the system it gets modified and/or converted to another output parameter that is also time dependent. If the system is excited with a time-dependent signal $u(t)$, with the system modifying the input signal, based on a transfer function that may be mathematically defined, and this situation creates a time-varying output signal $y(t)$. The dynamic system can be described by a set of differential equations:

$$\dot{y}(t) = f[y(t), u(t)] \tag{4.1}$$

Input and output signals $u(t)$ and $y(t)$ are generally vectors, and if the system is nonlinear, Eq. (4.1) will also be nonlinear.

For batteries, it is common to define the current and ambient temperature as input signals. The output signals could be (i) battery terminal voltage, (ii) battery temperature, (iii) SOC, and (iv) SOH. While other parameters such as internal resistance and open-circuit voltage (OCV) are also useful output signals, from a dynamic viewpoint these additional output signals yield only limited additional information. As discussed by Jossen (2006), a simple example of the dynamic characteristic in a battery pack is shown in Fig. 4.2, where a NiMH battery undergoes a pulsed discharge after a charging period. The discharge regime is in-line with the GSM standard with a pulse duration of 577 μs and a period of 4.81 ms. Pulse current changes from a 0.2 A during the rest period to a maximum of 2 A during the transmission pulse. The discharge voltage shows a ripple voltage of 40–50 mV and if this is assumed to be due to internal resistance, the battery internal resistance can be estimated as 22–28 mΩ.

In a situation like this, since the discharge current is in pulse mode, compared to a constant DC discharge current operation, we may face additional questions such as

- How can we quantify the transient voltage fluctuations with short-term ripple superimposed on the longer term terminal voltage profile?

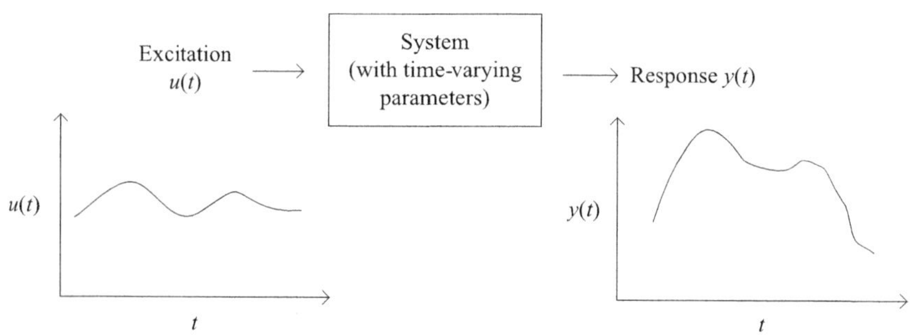

Fig. 4.1 A dynamic system.

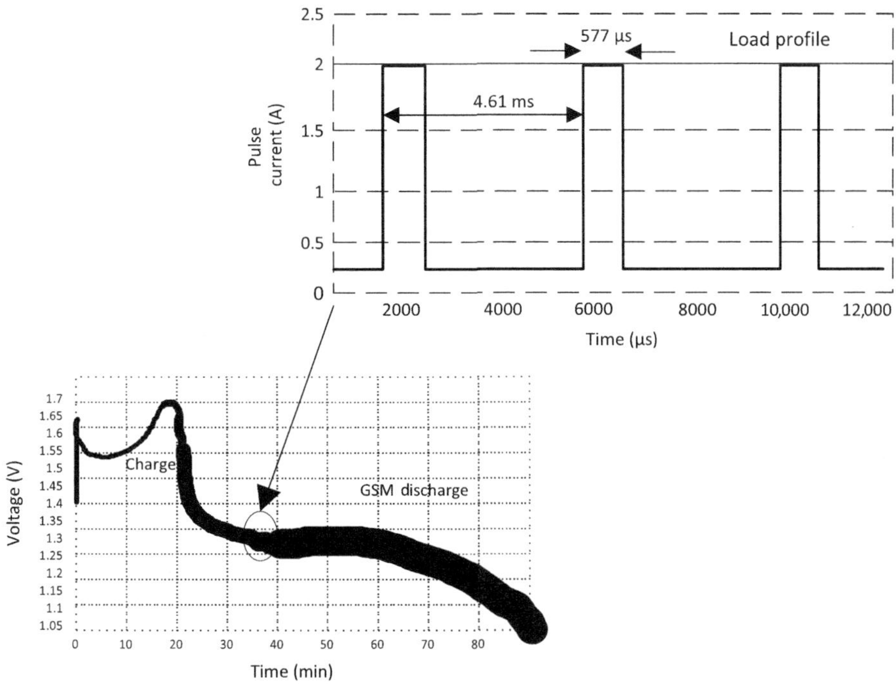

Fig. 4.2 Battery discharge with a pulsed current in a typical GSM cellular phone.
Source: Jossen, A., 2006. Fundamentals of battery dynamics. J. Power Sources 154, 530–538.

- Any added heat generated inside the battery?
- What will be the performance of the battery and how does it affect the lifetime estimated?

A battery's dynamic performance can be influenced by both internal parameters (such as SOC, SOH, DC/AC components of internal resistance, and other design parameters) as well as external parameters (such as ambient temperature, DC current, and the short- and long-term history of the battery). Conversely, the dynamic characteristics contain information on these parameters. As indicated by Jossen (2006), there has been significant amount of research work done on battery dynamics applicable in SOC/SOH determination, and the dynamic behavior of a battery is a combination of several effects, which occurs within microsecond to several years.

As depicted in Fig. 4.3 different dynamic effects in batteries occur within a wide range of time and is a combination of several effects such as

(i) Electrical and magnetic
(ii) Electric double layer
(iii) Mass transport
(iv) Cycling and SOC
(v) Reversible
(vi) Aging

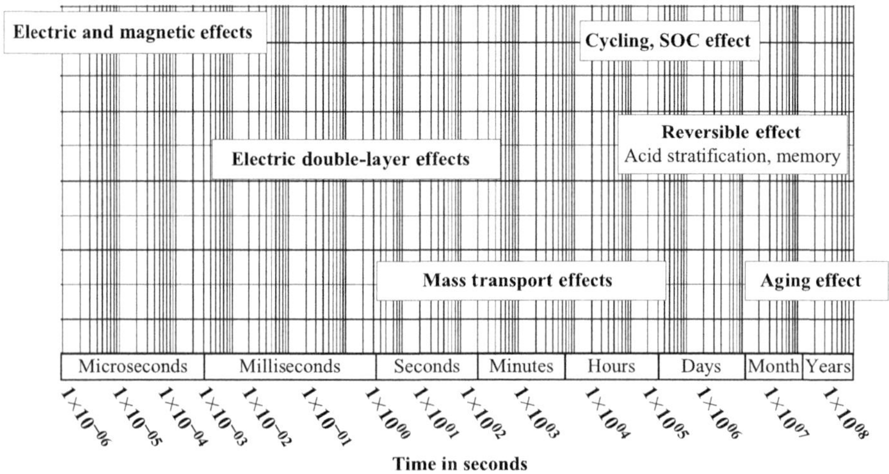

Fig. 4.3 Typical time ranges or dynamic effects within batteries.
Source: Jossen, A., 2006. Fundamentals of battery dynamics. J. Power Sources 154, 530–538.

Fig. 4.3 depicts only different time domains applicable to these effects, which are quite dependent on chemistry, battery design, temperature, SOC, and SOH. Temperature effects are not shown in the figure, and the dynamics of the battery temperature depends on the heat capacity of the pack, dissipated heat, and the heat generated within the battery. Dynamics of these temperature variations can be in the range of seconds to several hours in general.

4.3.1 Long-term effects

Aging, reversibility, cycling, and SOC effects are considered long-term phenomena, and as we see from Fig. 4.3, their dynamics occur within minutes to years of time durations. These are related to the operation regime of the battery.

4.3.1.1 Aging effects

Aging influences the battery performance and hence the output parameters significantly, and the time domain for battery aging is typically in the range of months to years. For primary cells, aging is considered due to storage, but through the discharge process.

4.3.1.2 Reversible effects

Some electrochemical storage systems show reversible effects. These occur during cyclic operations and can be regenerated by special charge-discharge regimes. For example, vented lead-acid batteries show acid stratification that can be removed by an extended charge. In NiCd cells, memory effect is another example, which can be cleared by applying one or more full charge-discharge cycles. Time domain for such reversible effects lies between the values of aging effects and the cycling effects, as shown in Fig. 4.3. Few hours to 1 year time constants are effective in these processes.

4.3.1.3 Cycling and SOC effects

If a battery is charged or discharged, its SOC is changed and this itself is a dynamic characteristic of a battery. Additionally, the open circuit or the terminal voltage itself is a function of SOC. Time domain of SOC change or cycling depends on the operational conditions with time domain from several minutes to days.

During the cycling process, the battery can heat up due to internal heat sources, which can be due to ohmic heating or any other chemical reactions. The generated ohmic heat can be quite influenced by the charge/discharge current profiles. If we assume that there will be only ohmic heating, heating power can be estimated by

$$P_{\text{ohmic}} = R_{\text{int}} I_{\text{eff}}^2 \tag{4.2}$$

where P_{ohmic} is the power dissipated due to ohmic resistances due to an overall internal effective resistance of R_{int} and the I_{eff} is the effective current based on the wave shape. In the example we cited in Fig. 4.2, for the GSM waveform, RMS value of the current is estimated as 0.73 A and the average current is 0.42 A. With a form factor of the waveform of 1.73 approximately, the heat generated due to the RMS current of the GSM pulse transmission, it generates three times bigger power, calculated based on the average current of the waveform. More discussion on the subject of waveform parameters discussed here was provided by Kularatna (2008).

4.3.2 Mass transport effects

Within a chemical system such as a battery, the transport of ions is due to diffusion and migration. Diffusion is caused by the gradient in concentration, while migration is caused by forces of electric field present within the chemical system of the electrodes and the electrolyte. Two cases may have different directions. Migration is often hindered by solvated molecules that cover the ion.

In most cases, diffusion is largely responsible for mass transport. Diffusion can occur at different inside locations of an electrochemical cell such as the following (Jossen, 2006):

1. In the free electrolyte or the separator, due to ions produced on one electrode are consumed by the opposite polarity electrode.
2. With a porous electrode, electrochemical reaction can occur anywhere within the active mass (AM) of the porous electrode.
3. The reaction products can move within the AM to their final location by diffusion.
4. If there is a film on the electrode (e.g., in Li-ion cell there is a film, the solid electrolyte interface (SEI) on the surface of the anode), diffusion can happen through the SEI.

Inside a battery, there are three separate primary parts which are mainly responsible for the chemical reactions. They are the positive electrode, the negative electrode, and the electrolyte. In a battery, there is a separator which is in the middle of the battery in between these two electrodes. Fig. 4.4A depicts a typical case of a Li-ion cell where a porous composite negative electrode made of graphite and a porous composite positive electrode with a polypropylene separator. The two electrodes and the separator are immersed in the electrolyte. More details of such a simple schematic, used in the estimation of capacity fade models for Li-ion batteries, are discussed in a study by Ramadass et al. (2004).

Fig. 4.4B portrays the three locations where diffusion can take place. As the battery has two electrodes, the diffusion within the porous electrode and within the AM is in both electrodes. In such a case, diffusion occurring can be described by Fick's first law (Jossen, 2006), which is dependent on the diffusion coefficient and the rate of change of concentration along the direction of the mass transfer. The diffusion coefficient is material-specific; in gases, it is typically 10^{-1} cm^2/s, in liquids 10^{-5} cm^2/s, and in solids 10^{-10} cm^2/s. Temperature is a key influencing factor in the diffusion coefficient.

The limitations of the diffusion of the ions result in locally changed ion concentrations, and from an electrical engineer's perspective, it results in an overpotential caused by reduced or increased ion concentration at the location of charge transfer. Time constants representing the diffusion depend on electrode thickness and the structure to a great degree, with typical time constants in the range of seconds to minutes (Jossen, 2006). Impedance spectroscopy, or measuring the response of a battery to an excitation of a small amplitude signal of a voltage or a current to estimate the small signal impedance of the cell, is frequently used to analyze the dynamic behavior of the batteries due to mass transport effects (Huet, 1998; Jossen, 2006). More discussion on the subject of impedance spectroscopy is presented in Section 4.4.

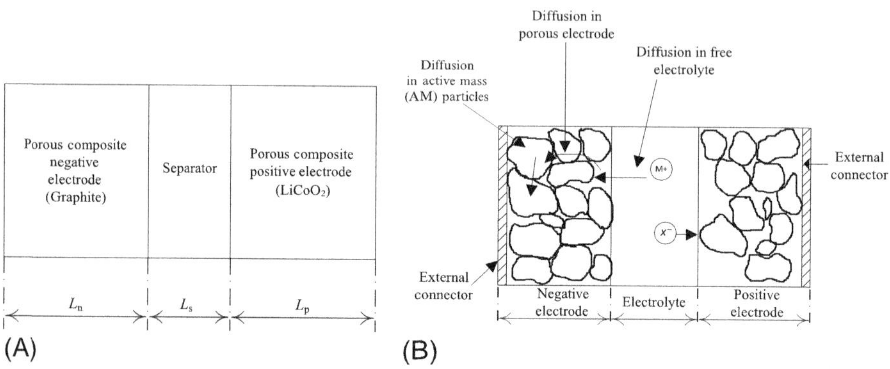

Fig. 4.4 Schematic of typical Li-ion cell sandwich: (A) basic structure and (B) diffusion process in such a cell.
Adapted from Ramadass, P., Haran, B., Gomadam, P.M., White, R., Popov, N., 2004. Development of first principles capacity fade model for Li-ion cells. J. Electrochem. Soc. 151, A196–A203 and Jossen, A., 2006. Fundamentals of battery dynamics. J. Power Sources 154, 530–538.

4.3.3 Double-layer effects

The electrical double-layer effect is a phenomenon present both in electrical double-layer capacitors (more commonly known as supercapacitors) and in electrochemical batteries. Fig. 4.5 depicts the basic idea of a double-layer capacitor in relation to a parallel plate capacitor. Fig. 4.5A shows the case of simple parallel plate capacitor where capacitance is given by kA/d, where k is the dielectric constant, A is the area of the plates, and d is the distance between the two parallel plates. When two electrodes are placed across a liquid electrolyte, due to the action of the electrolyte, two charged layers are present on both ends of the electrolyte, and these are equivalent to two capacitors in series, and this is a liquid double-layer capacitor. In a case like a rechargeable battery, a porous separator is also placed in the middle, and if the active material used in the cell, which is the anode/cathode element, it can be large-area material used similar to supercapacitors (Lai and Rose, 1992). Some materials, known as large surface area materials, typically usable in supercapacitors, have 50–1500 m^2/g

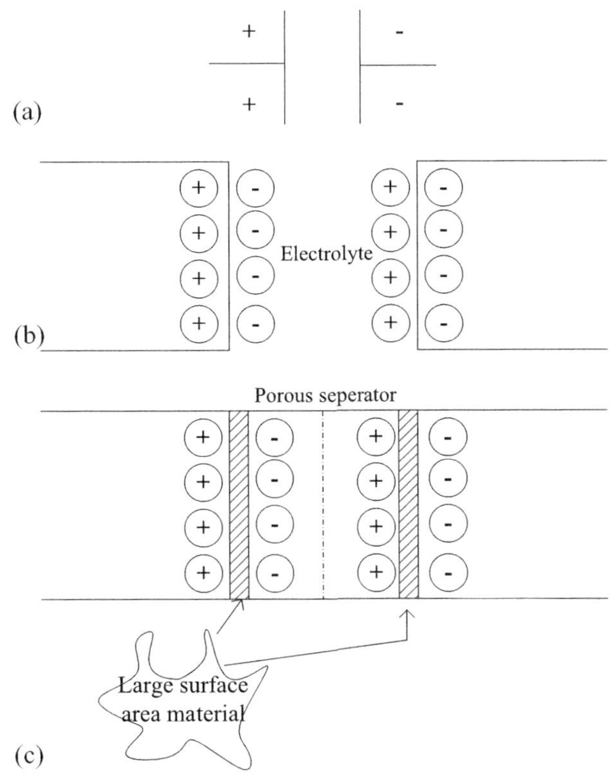

Fig. 4.5 Simplified concept of electrical double-layer capacitance: (A) parallel plate capacitor, (B) liquid double-layer capacitor, and (C) double-layer capacitor formed by a porous separator in an electrolyte.
Adapted from Lai, J.S., Rose, M.F., 1992. High energy density double layer capacitors for energy storage applications. IEEE AES Mag. 7 (4), 14–19.

surface density. When such a material with extremely low thickness is combined with suitable electrodes, a large capacitor is formed due to the large *A*/*d* ratio. Inside batteries, such a high capacitance surface layer is formed next to electrodes and creates a double-layer effect.

In a rechargeable battery, due to large surface area in porous electrodes, combined with the short distance between the electrodes and electrolyte, the electrochemical double layer is formed similar to the example provided in Fig. 4.5C. This creates a double-layer capacitance on the electrode surface, with a significantly large value, in parallel to the electrochemical charge-transfer reaction.

Recent research publications (Li et al., 2014) and the citations there indicate that electrochemistry-based impedance models for battery chemistries such as Li-ion can be developed to achieve a better understanding of battery dynamics. In this process, the charge-transfer reaction happening on the electrode/electrolyte surface can be simplified by several assumptions such as

- **(i)** Electrodes consist of multiple spherical particles of identical size and dynamics.
- **(ii)** The material inside the particle is spatially homogeneous.
- **(iii)** Electrodes have perfect conductivity (i.e., no voltage drop inside a particle or among particles).
- **(iv)** The SEI film growth caused by aging only happens on the anode surface.
- **(v)** The electrolyte, separator, and current collectors are treated as lumped pure resistors.

Such a set of simplifications allow us to consider the cell behavior based on identical multiple spherical particles as shown in Fig. 4.6A which can lead to a simplified analysis based on single-particle pair separated by the separator used in a case like Li-ion chemistry.

This kind of analysis allows the developing of a set of mathematical relationships between the internal dynamics and external measurements, to include the electrical double-layer effect with other effects discussed. As the two electrodes of a battery are not equal, the dynamic characteristics of both electrodes are also different. In a lead-acid battery typical double-layer capacitance values are 7–70F/Ah and 0.4–1F/Ah for the

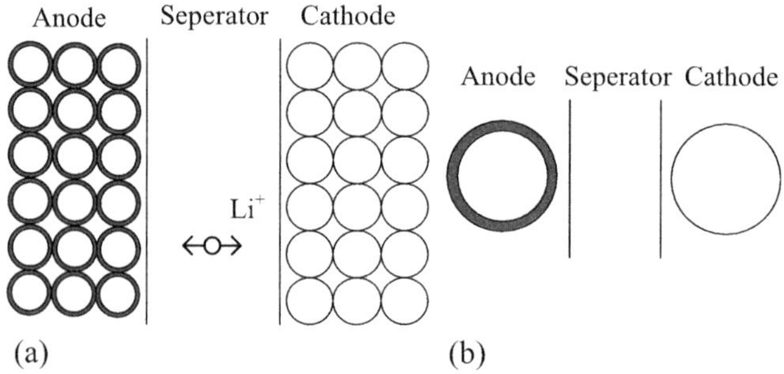

Fig. 4.6 Simplification of Li-ion battery electrodes: (A) based on identical multiple particles and (B) single particle at each electrode.
Source: Li, S.E., Wang, B., Peng, H., Hu, X., 2014. An electrochemistry-based impedance model for lithium-ion batteries. J. Power Sources 258, 9–18.

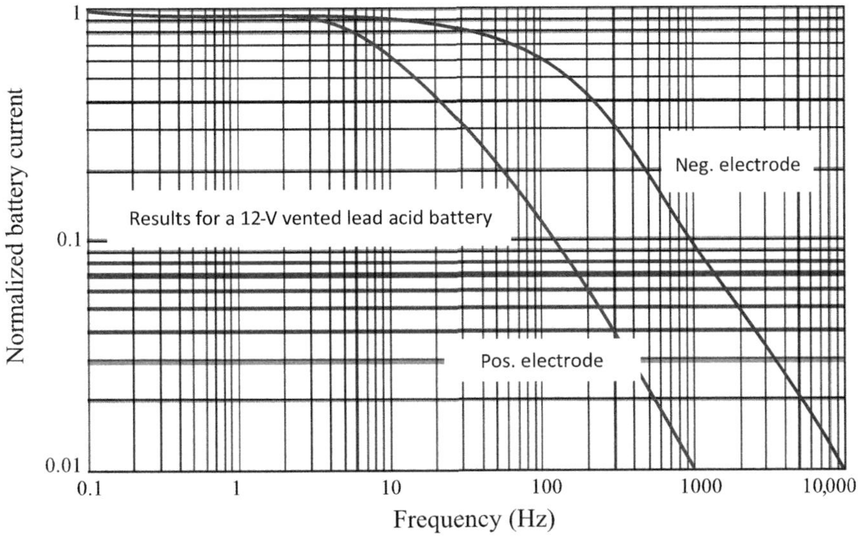

Fig. 4.7 Frequency response of the two electrodes of an automotive lead-acid battery caused by the double-layer effect.
Source: Jossen, A., 2006. Fundamentals of battery dynamics. J. Power Sources 154, 530–538.

positive and negative electrodes, respectively. Lot more analytical details are provided in a study by Li et al. (2014) and an overview of different dynamic effects is discussed in a study by Jossen (2006). Fig. 4.7 depicts the frequency response of the two electrodes of a lead-acid battery caused by the double-layer effect.

As depicted in Fig. 4.7, the cutoff frequency for the positive electrode is approximately 10 Hz and for the negative electrode it is 100 Hz. This means that the alternative currents with a frequency above 100 Hz do not flow through the charge-transfer reaction since they are filtered by the double-layer capacitor.

4.3.4 Effects caused by porous electrodes

Porous electrodes as used in batteries are complex components. The behavior is different from planar electrodes and they have a significant influence on the dynamic characteristics. The diffusion of ions through the porous electrodes is one of the main limiting factors in batteries. Additionally, the porous structure changes during discharging as the volume of the discharged AM in most systems is much higher than the volume of the charged AM. This results in a reduction in the porosity during discharging, which in turn leads to lower diffusion. In many systems, the conductivity of the discharged AM differs from that of the charged AM. This changes the ohmic resistance of the AM during discharging. These effects result in an inhomogeneous discharge of porous electrode, especially at higher currents. The effects of porous distributed electrodes indicate that the dynamic characteristic is more blurred for the porous effect in comparison to other effects.

4.3.5 Electric and magnetic effects

Compared to the slow effects of mass transport and the electrochemical double layer, much faster phenomena are found in a rechargeable battery. The ohmic resistance of the cell is the sum of the electrolyte resistance, resistance of the current collector, the AM, and the transition resistance between the AM and the current collector. In addition to this, each cell has a serial inductance based on the geometry of the cell. For a lead-acid battery values between 10 and 100 nH/cell for 100 Ah, cells are reported (Jossen, 2006). This is in addition to the wiring of the cell. This inductance limits the slew rate of the current, particularly in large batteries such as lead-acid systems. In large lead-acid battery packs, this inductance can be of interest above 1 kHz frequencies. For small batteries, with much smaller inductance, much higher frequencies such as 10–100 kHz show the effect of these inductive components.

With increased frequencies, the penetration depth of the ions in the porous structure decreases and the electrodes start behaving like planar electrodes. At high frequencies, electrode pair becomes a simple plate capacitor equivalent. A typical value of this high-frequency interelectrode capacitance for a lead-acid cell is around 10 nF/cell (Jossen, 2006).

Another important high-frequency effect is the skin effect. Caused by electromagnetic field effects, the penetration depth of AC current in conductive materials is limited. This penetration depth d for cylindrical materials is calculated by $d = 1/\sqrt{k\mu\pi f}$, where k is the conductivity of the material and μ is the permeability of the material. Depending on the depth of materials used for current collectors, this effect can come into play. More details on materials are given in a study by Jossen (2006).

4.3.6 Battery equivalent circuits based on various dynamic effects

For different dynamic effects, a battery pack may be represented by an electrical equivalent circuit, as if only that effect is in charge of the battery dynamics. For example, Fig. 4.8A shows the equivalent circuit for diffusion effects, Fig. 4.8B shows the equivalent circuit applicable to double-layer effect, Fig. 4.8C shows the equivalent circuit applicable for porous electrode effect, and Fig. 4.8D indicates the high-frequency equivalent circuit of a battery.

4.4 Electrochemical impedance spectroscopy for batteries

Electrochemical impedance spectroscopy (EIS) is a widely used technique in several areas such as nondestructive diagnostics, prognostics, and modeling in electrochemical systems. The EIS involves empirically determining the small signal impedance of an electrochemical material or a device such as a battery, supercapacitor, or a fuel cell.

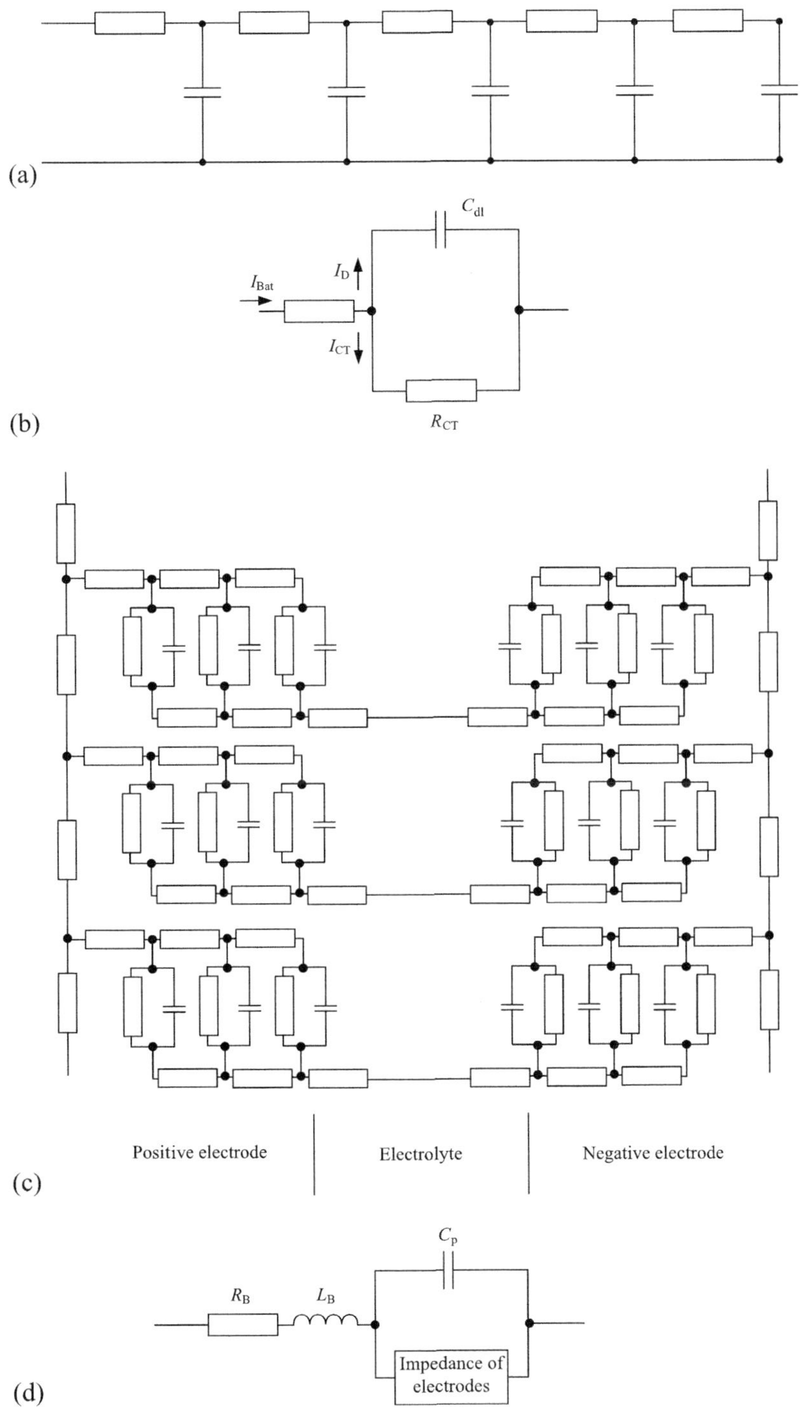

Fig. 4.8 Equivalent circuits applicable to different dynamic effects of the batteries: (A) diffusion effects; (B) double-layer capacitor effect; (C) porous electrode case; and (D) high-frequency operation.

In recent years, the technique has been applied to a host of next-generation direct energy conversion technologies such as fuel cells, photovoltaic systems, as well as energy storage devices such as batteries and supercapacitors (Lindahl et al., 2012).

In EIS, an electrochemical device such as a battery or a supercapacitor is perturbed from steady-state operation by a potentiostatic (constant voltage) or galvanostatic (constant current) excitation signal over the desired frequency range. The resulting current and voltage response is then analyzed based on the two signals' interrelated magnitude, phase and frequency content determining the localized impedance of the device, on small signal basis, as a function of the frequency. The attraction of EIS lies in the direct correlation between the observable electrical performance (at the terminal level) and its nonobservable electrochemical and physicochemical processes. Thus EIS allows bulk monitoring of the internal processes of an electrochemical device with the results indicative of the device's dynamic performance such as the SOC and the overall health such as SOH (Huet, 1998; Lindahl et al., 2012).

4.4.1 Simple basics used in EIS and sample results for different chemistries

The electrochemical impedance (or simply the AC impedance) of a battery characterizes its dynamic behavior around a selected operating DC voltage across its terminals or at a particular DC current (charging or discharging). In general, any type of excitation signal (e.g., sine wave, step, noise, etc.) can be used. Most commonly a sine wave is used and in galvanostatic mode where a DC current of I (polarization current), which could be charging or discharging the battery and this sinusoidal excitation current can be written as

$$\Delta I = I_{\text{peak}} \text{Sin}\,(2\pi ft) \tag{4.3}$$

where I_{peak} is the maximum value of the superimposed excitation current and f is the frequency. This excitation yields a sinusoidal voltage response of ΔV in the form

$$\Delta V = V_{\text{peak}} \text{Sin}\,(2\pi ft + \varphi) \tag{4.4}$$

where V_{peak} is the maximum value of the superimposed excitation voltage around a DC terminal voltage of V and φ is the phase angle related to two superimposed voltage and the current. Similar two relationships can be developed also for potentiostatic mode. Eqs. (4.3) and (4.4) define an electrochemical impedance for the battery which is given by

$$Z(f) = \frac{V_{\text{max}}}{I_{\text{max}}} e^{j\varphi} \tag{4.5}$$

In general, voltage amplitude must not exceed about 10 mV to ensure that the impedance measurements are performed under linear conditions. In that case, the excitation and response signals are actually sine waves, and the value of *Z(f)* does not depend on the amplitude of the excitation signal. With large capacity batteries with very low internal impedances, AC current required may be in the order of several amperes (Huet, 1998). Several standard signal processing techniques, including frequency response analysis, fast Fourier transform, and phase-sensitive detection may be used (Lindahl et al., 2012). For all these techniques, fundamentals of the impedance-extraction process are quite similar and more details can be found in a study by Orazem and Tribollet (2008). The value of $Z(f)$ usually depends on the temperature and the SOC. Also due to mass-transport effects, dynamic behavior of a battery, represented by $Z(f)$, could be quite different during continuous charging and discharging.

In representing the dynamic behavior of an electrochemical system such as a battery or a supercapacitor, based on the value of $Z(f)$ *at* different frequencies, either a Bode plot or a Nyquist plot can be used. It is more common to use the Nyquist plot with the opposite of the imaginary part of the impedance on the ordinate axis, so that the capacitive loops appear in the upper quadrants. Fig. 4.9 shows the typical impedance spectra of a Li-ion battery at different SOC and a constant temperature of 25°C (Buller et al., 2005).

In Fig. 4.9, impedance data have been recorded for different frequencies, and at approximately 286 Hz, impedance becomes a pure resistance of approximately 42 mΩ, with the graphs showing the intersection of the real axis. More discussion on this is provided in a study by Buller et al. (2005). Fig. 4.10 depicts the typical EIS results for a lead-acid automotive battery (Buller et al., 2003).

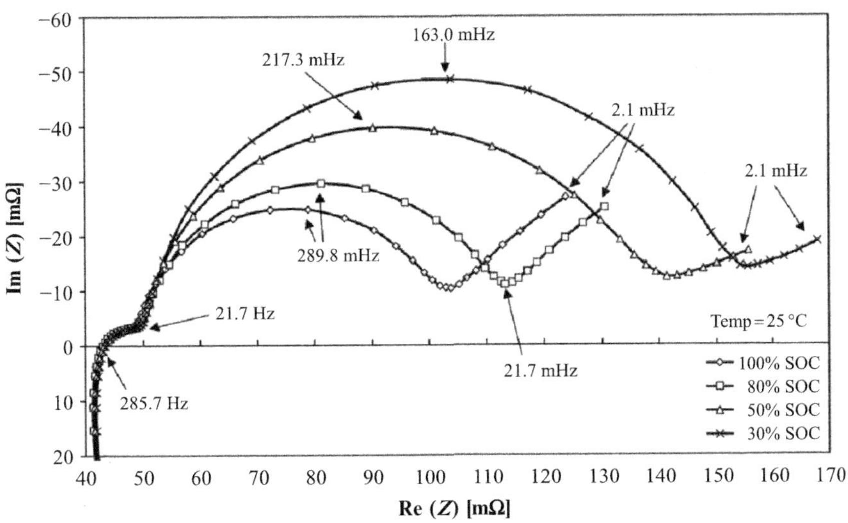

Fig. 4.9 EIS results: typical impedance of a Li-ion battery of the type SAFT LM 176065, 3.6 V/5 Ah at $I_{dc} = 0$ A.
Source: Buller, S., Thele, M., Doncker, R.W., Karden, E., 2005. Impedance based simulation models of supercapacitors and Li-ion batteries for power electronic applications. IEEE Trans. Ind. Appl. 41 (3), 742–747.

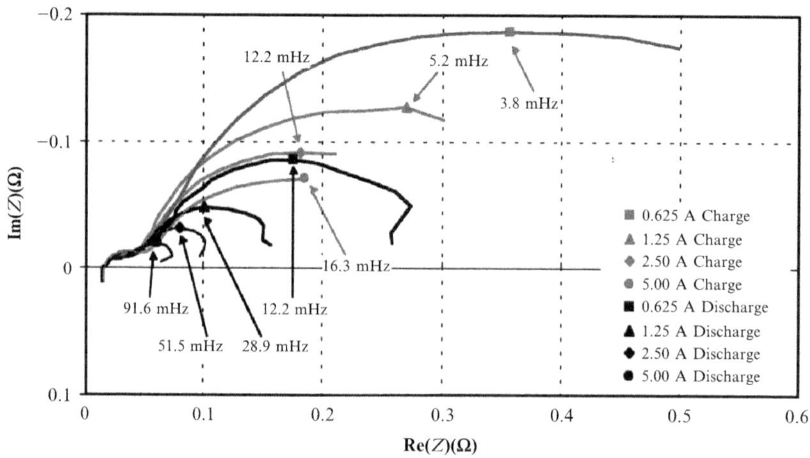

Fig. 4.10 EIS Results for a 36-V-AGM-type lead-acid battery at 70% of SOC at 25°C. *Source*: Buller, S., Thele, M., Karden, E., Doncker, R.W., 2003. Impedance based non-linear dynamic battery modeling for automotive applications. J. Power Sources 113, 422–430.

A discussion on the use of EIS techniques to predict SOC for a NiMH battery is available in a study by Bundy et al. (1998).

4.4.2 Specific requirements for EIS and its limitations

Problems that arise when applying the methods of EIS to batteries may be grouped into three categories (Karden et al., 2000) as follows:

- (i) Nonlinearity of the transport and reaction process occurring at large currents.
- (ii) Nonstationary nature of the process—since batteries change their structure while charging or discharging, where at least one species of reactants deplete.
- (iii) Nonideality of a practical battery (compared to laboratory cell) due to geometry, reaction kinetics and mass transport in porous electrodes.

The above problems make clear that the scope of possible results of an EIS performed is restricted with commercial batteries, compared to experimental cells. However, EIS provides a unique tool for analysis of the dynamical behavior of the batteries. More details are available in a study by Karden et al. (2000).

4.5 Battery equivalent circuit models and modeling techniques

In an electronic engineer's viewpoint, it is most useful to have an equivalent circuit for a battery with any variations of important parameters due to process variables such as temperature and SOC. However, due to various dynamic effects we have discussed in Section 4.3, it is very hard to propose simple equivalent circuits, with linear circuit elements. Following discussion is an attempt to discuss various components within a battery equivalent circuit and how they relate to internal processes occurring within the cells.

4.5.1 Randles' equivalent circuit

First of all, every electrochemical cell is formed by pairing two suitable electrodes and placing them within a suitable electrolyte, and with the necessary separator to get the optimum conversion of chemically stored energy in the cell in the form of electrical energy. For this reason, every battery can be treated as two electrochemical half cells with relatively identical circuit elements, but with different values, which are dependent on several parameters. Randles (1947) proposed a circuit with four components as shown in Fig. 4.11A to show the combination of electrode-electrolyte reaction, double-layer capacitance, and the electrolyte resistance. In an electrochemical cell, there are two electrodes separated by a single electrolyte, and hence we can apply the Randles circuit to each half cell as shown in Fig. 4.11B. In this case, double-layer capacitance is indicated by C_d, diffusion-related performance is indicated by the charge-transfer resistance R_d combined with the Warburg diffusion element (Jossen, 2006; Warburg, 1899) to model the diffusion process occurring at the electrode.

The *Warburg diffusion element* is a common diffusion circuit element (Jossen, 2006) that can be used to model semiinfinite linear diffusion, that is, unrestricted diffusion to a large planar electrode. A Warburg impedance element can be difficult to recognize because it is nearly always associated with a charge-transfer resistance and a double layer but is common in all battery chemistries. The Warburg diffusion element (Z_w) is a constant phase element, with a constant phase of 45 degree (phase independent of frequency) and with a magnitude inversely proportional to the square root of the frequency by

$$Z_W = \frac{\sigma}{\sqrt{\omega}} - j\frac{\sigma}{\sqrt{\omega}} \tag{4.6}$$

where σ is the Warburg coefficient (or Warburg constant) and ω is the angular frequency ($2\pi f$). Amplitude of the Warburg element is given by

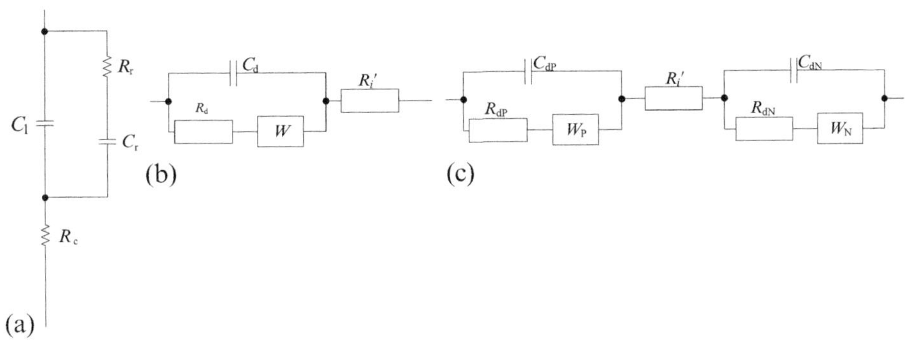

Fig. 4.11 Randles' model extended to two electrode half cells: (A) Randles circuit; (B) basic equivalent circuit of the electrochemical half cell; and (C) full cell with two electrodes.

$$|Z_w| = \sqrt{2}\frac{\sigma}{\sqrt{\omega}} \tag{4.7}$$

The presence of the Warburg element can be recognized if a linear relationship on the log of a Bode plot ($\log|Z|$ vs $\log(\omega)$) exists with a slope of value $-1/2$. In Fig. 4.11B, R_i' indicates the electrolyte resistance of the half cell. Fig. 4.11C shows the combination of the two half cells where the suffixes N and P show negative and positive electrode effects. By EIS methods (Karden et al., 2000), one can estimate values for these parameters of the cell.

4.5.2 More detailed models based on electrochemistry

It is important to understand that the electrochemistry process within a battery is a complex process where several physical phenomena are combined in a complex way. As shown in Fig. 4.6, in a case such as a Li-ion cell with several assumptions as discussed in Section 4.4, current flows and the effects at the anode and cathode could be indicated by Fig. 4.12, where discharging of the cell is indicated. In Fig. 4.12, $J_{FD,i}$ represents the Faradaic current density due to charge-transfer reactions, $Z_{FD,i}$ is the Faradaic impedance of electrodes caused by charge-transfer reactions ($i=n$ for anode and $i=p$ for cathode), determined from the Faradaic current density and the potential across the interface, $\Phi_{1,i}$ and $\Phi_{2,i}$ represent the potential on the two sides of electrode/electrolyte interface, $J_{\text{dl,i}}$ represents the double-layer current density, $C_{\text{dl},i}$ represents the double-layer capacitance, C_{film} represents the capacitance due to the SEI film insulation on the anode, J_{film} is the current density through the SEI film, and R_{film} is the SEI film resistance. More detailed discussion and the mathematical derivations based on Butler-Volmer equation, Nernst equation, and Fick's second law related to the process are available in a study by Li et al. (2014). A detailed discussion on these is beyond the scope of this chapter. For a detailed physical explanation on the processes within an electrochemical cell and related mathematical relationships, article by Grimnes and Martinsen (2008) is suggested.

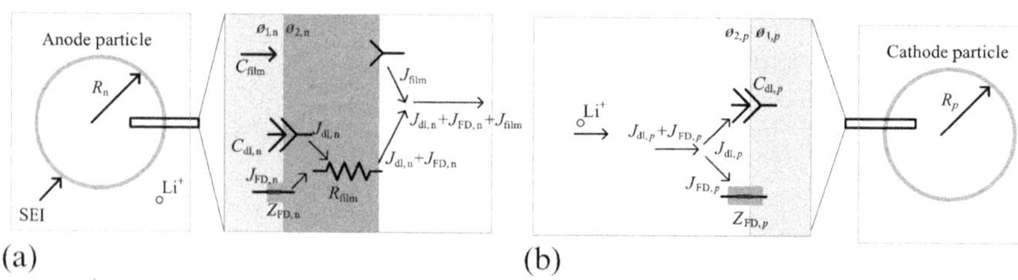

Fig. 4.12 Current flows at the anode and the cathode of a Li-ion cell: (A) anode and (B) cathode. *Source*: Li, S.E., Wang, B., Peng, H., Hu, X., 2014. An electrochemistry-based impedance model for lithium-ion batteries. J. Power Sources 258, 9–18.

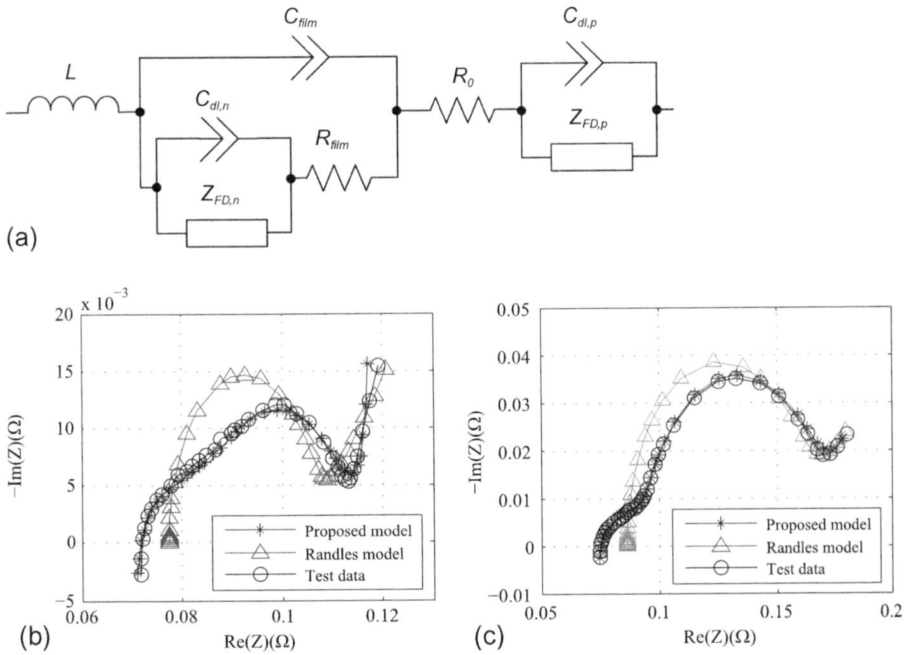

Fig. 4.13 Overall circuit structure of a Li-ion cell and its Nyquist plots: (A) overall impedance model of the cell; (B) Nyquist plots after 200 charge cycles; and (C) Nyquist plots after 800 charge cycles.
Source: Li, S.E., Wang, B., Peng, H., Hu, X., 2014. An electrochemistry-based impedance model for lithium-ion batteries. J. Power Sources 258, 9–18.

Fig. 4.13 shows the circuit structure of the whole battery. The structure has anode impedance and cathode impedance connected in series. The impedance of electrolyte, separator, and current collector is represented by a pure ohmic resistor R_0, placed between anode and cathode. At high frequency, battery current collectors and cables introduce inductance, which is represented by a pure inductor L. As per work by Li et al. (2014), Fig. 4.13B and C indicates the Nyquist plots for the cell after 200 and 800 cycles of charge, respectively, based on models related to the circuit structure shown in Fig. 4.13A. The graphs compare results from the proposed model and test data in relation to data generated from a Randles' model.

4.5.3 Frequency domain behavior of battery models and components

In general, product and systems designers do all their measurements in the time domain, but most individual model elements as discussed in Section 4.5.2 provide us with more information on the frequency domain. Also, EIS techniques are primarily

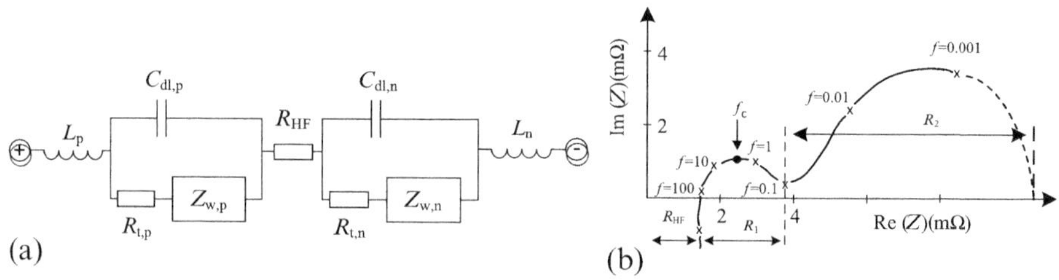

Fig. 4.14 Lead-acid battery: (A) comprehensive equivalent circuit and (B) the Nyquist Pot. *Source*: Huet, F., 1998. A review of impeadance measurements for determination of the state-of-charge or state-of-health of secondary batteries. J. Power Sources 70, 59–69.

the frequency domain methods where we fix the DC operating current in the cell at a constant value and do estimations of the behavior of components in the frequency domain. For example, in Figs. 4.9 and 4.10, we indicate the battery behavior with real and imaginary components together to provide an overview of frequency domain performance. Fig. 4.14A indicates the equivalent circuit based on the combination of the two half cells with Warburg elements, and Fig. 4.14B shows the shape of the Nyquist diagram of a lead-acid battery (Huet, 1998). Frequencies indicated are in Hz and the plot easily identifies the values of ohmic resistance of the battery (R_{HF}) and the effect of a small capacitive loop of size R_1 of the battery. R_{HF} indicates the effect of connections, the separator, electrolyte resistivity, and the effect of surface coverage by crystallized lead sulfate. Small capacitive loop is related to the porosity of the electrodes. A large loop of size R_2 at lower frequencies depends on sulfation reactions controlled by mass transport of Pb^{2+} ions (Huet, 1998).

Fig. 4.14 depicts the battery impedance of a 940-mAh Li-ion battery at different cycles of charge-discharge and full charged conditions. In this, an electrochemistry-based model as indicated in Fig. 4.13 is compared with Randles' model and the test data. In lower frequency regions on the right-hand ends of the figures, the effect of Warburg elements is prominent at lower frequencies.

4.5.4 Practical simplifications of battery models for engineering applications

As we have indicated, most of the dynamic effects inside batteries are quantified based on many assumptions and simplifications. In commercial batteries, many of these criteria are only partially valid and most of the components in the models cannot be accurately estimated. Also with nonlinear components such as the Warburg element and the double-layer capacitor, it is difficult to use these models directly in practical systems to estimate SOC, SOH, and remaining useful life (RUL) of the battery pack. In order to achieve much simplified and less-algorithm-intensive models, many simplifications are used with electrochemistry-based models.

4.5.4.1 Simplified models for lead-acid chemistry

In addition to the automotive area, lead-acid chemistry is still frequently used in electric vehicles such as airport light transport systems. One example is the Urban Light Transport (ULTRA) system used in London Heathrow airport (http://www.ultraglobalprt.com/?page_id=24). This vehicle relies on $4 \times 12\,V$ lead-acid VRLA batteries of capacity nominally rated at 45–50 Ah (Gould et al., 2009). In this kind of a case, the battery pack could be discharging at a high rate such as 150 A and maximum charging current may be around 50 A. In predicting the state of function (SOF) of such batteries online, modified and remapped equivalents of Randles' model are used. SOF estimation (combined SOC and SOH) is done on a real-time basis in light transport systems such as ULTRA. Fig. 4.15 shows the modified version of Randles' circuit shown in Fig. 4.11A. In this five-component-based simplified Randles' circuit description of a 12-V lead-acid battery, C_b is considered as the main charge storage and it is equivalent to a $1 \times 10^5\,F$ capacitor when the battery is fully charged. R_d represents the self-discharge resistance with an approximate value of $5\,k\Omega$. R_i models the battery's terminals and intercell connections (between 5 and $100\,m\Omega$), whereas R_t (in the range of 10–500 $m\Omega$) and C_t (1000–20,000 F) describe mass transport effects.

The voltage across the capacitor C_b is considered to be a suitable indicator for SOC, whereas SOH is inferred by observing a significant change in C_b over time due to aging effects such as AM degradation (Gould et al., 2009). Chemical aspects of lead-acid battery degradation are discussed by Ruetschi (2004). According to this paper, major aging processes leading to a gradual loss of performance, and eventually to the end of service life, are (i) anodic corrosion, (ii) positive plate AM degradation and loss of adherence to grid, (iii) irreversible formation of lead sulfate in the AM, (iv) short circuits, and (v) loss of water.

Randles' model can be remapped into a different equivalent circuit such as in Fig. 4.15B using star-delta transformation (Gould et al., 2009). It can be shown that

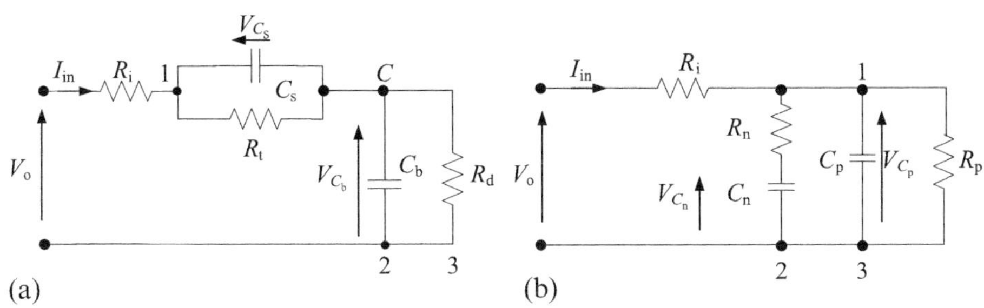

Fig. 4.15 Modified Randles' circuits for lead-acid battery: (A) basic model and (B) remapped circuit.
Source: Gould, C.R., Bingham, C.M., Stone, A., Bentley, P., 2009. New battery model and state-of-health determination through subspace parameter estimation and state observer techniques. IEEE Trans. Veh. Technol. 58 (8), 3905–3916.

$$C_{\mathrm{n}}=\frac{C_{\mathrm{b}}^{2}}{C_{\mathrm{b}}+C_{\mathrm{s}}} \tag{4.8}$$

$$C_{\mathrm{p}}=\frac{C_{\mathrm{b}}C_{\mathrm{s}}}{C_{\mathrm{b}}+C_{\mathrm{s}}} \tag{4.9}$$

$$R_{\mathrm{n}}=\frac{R_{t}(C_{\mathrm{b}}+C_{\mathrm{s}})^{2}}{C_{\mathrm{b}}^{2}} \tag{4.10}$$

$$R_{\mathrm{p}}=R_{\mathrm{d}}+R_{\mathrm{t}} \tag{4.11}$$

By using such a transform to remap the battery parameters into a more useful model and applying Kalman filter (KF), one can get good estimates of SOC and SOH of lead-acid batteries (Gould et al., 2009). More details on these techniques such as KF approach will be presented as an overview later.

4.5.4.2 Hybrid models to consider mass transport effects

Commonly used impedance-based models assume that the battery is in a quasistationary state during measurements. However, the battery packs such as in electric vehicles are subjected to very high continuous, quasicontinuous, or medium-term discharge processes. In situations such as these, to accurately estimate the SOC/SOH, you require improved impedance-based nonlinear simulation models. In these situations, more accurate electrolyte transport models are used, which describe the generation and transport of the sulfuric acid inside the porous electrodes (Thele et al., 2005). In a study by Thele et al. (2005), Matlab/Simulink-based implementation and parameterization in the time domain are given for high Ah throughput automotive and similar battery packs. In this situation, the hybrid modeling approach is used which combines the impedance-based model and a basic electrolyte transport model. Equivalent circuit commencing from the Randles' model is extended to show the effects of (i) generalized capacitance elements, (ii) a generalized DC voltage source, and (iii) a parallel current path for gas as shown in Fig. 4.16.

For the parameter determination, several EIS measurements are performed at different working points considering the dependencies not only on the SOC and the temperature but also on the current rate and the nonlinear resistances (Thele et al., 2005). In a case like this, measurements in Fig. 4.10 can be used. The model extension discussed by Thele et al. (2005) replaces the generalized DC source shown in Fig. 4.16 using a voltage source based on the actual acid concentration in both electrodes. This discussion is beyond the scope of this book that aimed at electronic engineers.

In general, at thermodynamic equilibrium, the terminal voltage of a lead-acid battery adjusts to an equilibrium voltage that depends mainly on the electrolyte concentration and thus on the SOC. Use of EIS for dynamic modeling of a lead-acid battery to estimate this equilibrium voltage which depends on SOC is discussed by Mauracher and Karden (1997). Analysis of uncertainties in an EIS measurement for lead-acid battery is discussed by Stevanatto et al. (2014). Also in a study by Stevanatto et al. (2014),

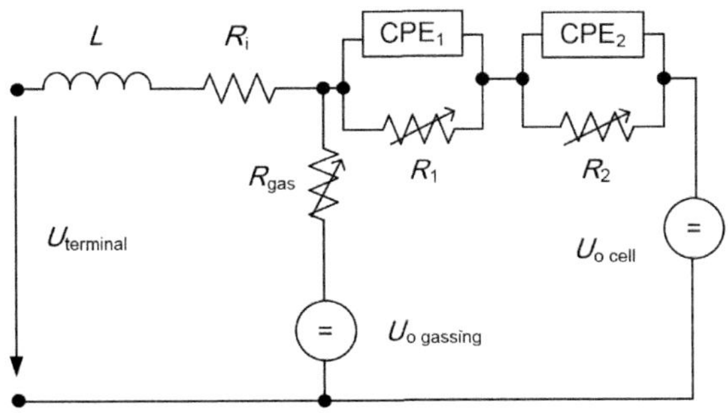

Fig. 4.16 Equivalent circuit of a lead-acid battery to include the effect of gassing and the generalized open-circuit voltage combined with impedance elements.
Source: Thele, M., Buller, S., Sauer, D.U., Doncker, R.W., Karden, E., 2005. Hybrid modeling of lead acid batteries in frequency and time domain. J. Power Sources 144, 461–466.

a discussion on Randles' parameter variations over service life is available. In general, while resistance in a Randles' equivalent circuit keeps increasing, double-layer capacitance keeps decreasing over service life (Stevanatto et al., 2014).

4.5.5 Nickel metal hydride battery models

The NiMH chemistry is an attractive replacement for the NiCd chemistry due to its high energy density, high-current capability, long cycle life, and lack of poisonous materials. NiMH is a popular option for batteries in electric vehicles and this demands relatively accurate models for the chemistry.

A NiMH cell is a dual-intercalation electrochemical system in which proton insertion in the positive solid nickel electrode and the hydrogen de-insertion in the negative metal hydride electrode occur during discharge and vice versa during charge. Fig. 4.17 schematically illustrates these processes (Gu et al., 1999). The open-circuit potentials of both electrodes are functions of local SOC, and the cell behavior is strongly dependent on the utilization of the active materials. In addition, discharge or charge of a NiMH cell is controlled by a number of factors, including the finite rate of solid-state diffusion inside active material particles of either one or both electrodes, charge-transfer kinetics at the electrode/electrolyte interface, and ohmic drop through the electrolyte phase. Any of these factors may also result in low cell potential before the active materials loaded in the cell are completely used up. As a result, the inefficient utilization of active materials becomes a critical issue, particularly for electric and hybrid electric vehicle (HEV) batteries (Gu et al., 1999).

Operation of a sealed NiMH cell is also accompanied by oxygen reactions. Oxygen may be generated at the nickel electrode and cause an excessive buildup of pressure during charge and overcharge. This gassing strongly affects the performance and life of NiMH batteries as well as controls the charge acceptance at high charging rates. It also raises safety concerns in practical applications. A battery must be carefully designed so that the oxygen is effectively transported to and recombined at the metal

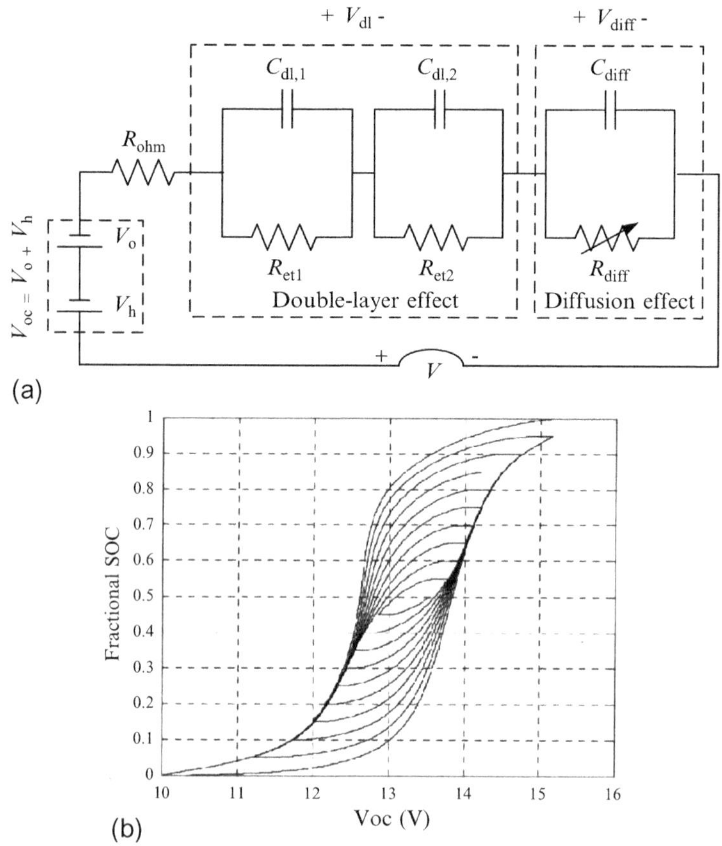

Fig. 4.17 NiMH battery equivalent circuit for hysteresis estimation: (A) circuit used and (B) measured hysteresis behavior of a 12-V NiMH battery module (COBASYS Series 1000) at C/20 rat.
Source: Tang, X., Zhang, X., Koch, B., Frisch, D., 2008. Modeling and estimation of nickel metal hydride battery hysteresis for SOC estimation. In: Proceedings of IEEE International Conference on Prognostics and Health Management, pp. 1–12.

hydride electrode, preventing the cell pressure from rising drastically. It is thus imperative to develop predictive capabilities for not only the active material utilization but also the cell conditions resulting from the oxygen reactions in advanced NiMH batteries (Gu et al., 1999). Modeling of the NiMH devices uses different approaches such as impedance model (Valoen et al., 1997), a combination of numerical and experimental approaches (Mauracher and Karden, 1997), nonisothermal model (Wu et al., 2001), isothermal models (Pan et al., 2002), fuzzy logic (FL) modeling (Singh et al., 2004), and artificial neural networks (ANNs) (Fang et al., 2012).

In general, the number of details to be considered in a NiMH cell will determine the complexity of the model to be achieved. For NiMH cells, the following hierarchy of models could be developed (Wu et al., 2001):

(i) Empirical model—fitted equations from experimental data.
(ii) Lumped model—equations of mass balance, charge balance, electrochemical kinetics, and energy balance.
(iii) Distributed model (with 1D or 2D porous electrode models)—case in item (ii) and equations for the distributed process.
(iv) Coupled electrochemical/thermal model—a combination of (i)-(iii) plus heat transfer equations.

In a study by Wu et al. (2001), a lumped model for NiMH cell is developed with attention to many important characteristics, including the hysteresis potential behavior.

While battery models and equivalent circuits similar to the case of lead-acid chemistry are developed with various complexities in these publications, one important characteristic of NiMH cells for electric vehicles is the hysteresis behavior (Tang et al., 2008). As discussed by Tang et al. (2008), using a simplified model such as shown in Fig. 4.17, NiMH hysteresis can be developed to include both the double-layer effect and the diffusion effect.

General Motors, USA has developed battery state estimator (BSE) algorithms suitable for HEVs and these are based on two different SOC estimates combined: a current-based SOC and a voltage-based SOC. The current-based SOC is based on conventional coulomb integration, which offers limited accuracy, and is sensitive to the initial value of SOC (Tang et al., 2008). In contrast, voltage-based SOC inferred from OCV is independent of the initial SOC and more robust to aging effects. The BSE algorithms are based on an equivalent electrical circuit model that represents the dynamic relationship between the battery terminal voltage and current. For NiMH batteries, as illustrated in Fig. 4.17, the electrical circuit model involves eight parameters, depicting a static resistance R_{ohm}, a static potential V_{OC} that represents the OCV, and two dynamic processes that model the double-layer effect and diffusion effect, respectively. The V_{OC} is the indicator of SOC. Two RC pairs are used to describe the double-layer effect for cathode and anode, respectively. The double layer is a dynamic process with a time constant ranging from hundreds of milliseconds to several seconds. Additional RC pair with a varying diffusion resistance is used to approximate the diffusion effect, whose time constant can be several seconds to hours. Using this model, Tang et al. (2008) present a phenomenological model based on the Preisiah Operator which is a widely used mathematical tool to model the hysteresis in many areas such as transformers, sensors, relays, and smart materials (Tan and Baras, 2005). Fig. 4.17B depicts the behavior of the sample battery used in this exercise. For more details, articles by Tang et al. (2008) and Tan and Baras (2005) are suggested.

In developing more useful equivalent circuits to relate the dynamic behavior, as well as the energetic behavior of a battery pack, Khun et al. (2006) used Cauer and Foster structures. These structures are well adapted to predict the behavior of batteries in real time. The work begins with a modified Randles' equivalent circuit as shown in Fig. 4.18, where R_{Ω} indicates the electrolyte and connection resistances, parallel pair of R_{tc} and R_{dl} double-layer effect-based charge-transfer phenomenon. A Warburg impedance Z_w stands for the diffusion phenomenon. As shown in Fig. 4.18B in a NiMH battery, charge-transfer and diffusion phenomena occur in distinct frequency areas (Khun et al., 2006). Nyquist plot in Fig. 4.18B shows the approximate semicircle

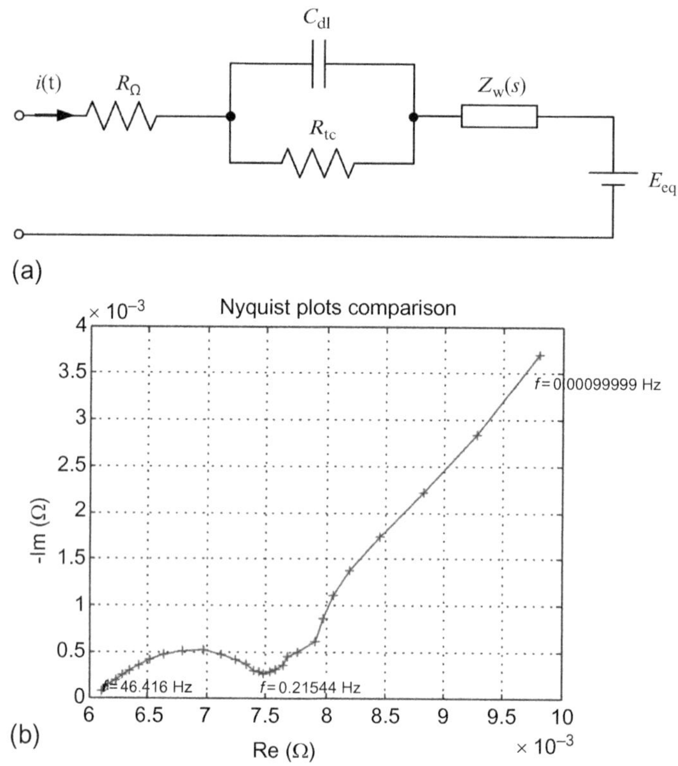

Fig. 4.18 Randles' circuit adapted for a NiMH battery and its Nyquist plot: (A) modified Randles' circuit and (B) Nyquist plot.
Source: Khun, E., Forgez, C., Lagonotte, P., Friedrich, G., 2006. Modeling NiMH battery using Cauer and Foster structures. J. Power Sources 158, 1490–1497.

for the double-layer effect and the approximate 45° slope for the diffusion phenomenon represented by a Warburg element. In a study by Khun et al. (2006), the Cauer and Foster structures together with Mittag-Leffler theorem are used to develop the equivalent circuit components and to compare the dynamic behavior of the battery at different SOC values based on theory vs experimental results.

4.5.6 Li-ion battery modeling, equivalent circuits, and aging issues

Li-ion batteries usually consist of a metal-oxide positive electrode (cathode), a carbon-based negative electrode (anode), and an organic electrolyte based on dissolved lithium salts. Cathode materials used in early versions were based on $LiCoO_2$, $LiMn_2O_4$, or $LiNiO_2$ and with the electric vehicle technology advancing fast, there were many variations in these base materials which are used in batteries, aimed at automotive industry (Belt et al., 2011; Bloom et al., 2006; Gao and Manthiram, 2009;

Stiaszny et al., 2014). Although there is a large variety of possible material combination for electrodes in this chemistry, combining cycle stability (with higher cycle life) and high volumetric or gravimetric energy density reduces the choices to a small number (Stiaszny et al., 2014). Promising candidates for cathodes are layered oxides such as $Li(Ni_{1/3}Co_{1/3}Mn_{1/3})O_2$, which exhibits high specific capacity (Belt et al., 2011), but discharge at high rates is limited. To increase safety and rate capability in the cell, $LiMn_2O_4$ is added to the cathode because it exhibits the fast kinetics that is necessary for high power applications (Belt et al., 2011) while at the same time reducing the battery's cost. In contrast, Mn-rich compounds are preferred compared to Ni- or Co-rich materials, although their specific capacity is low and $LiMn_2O_4$ suffers from manganese dissolution. The best anode material considering both specific capacity and cycle life is graphite (Stiaszny et al., 2014).

Since their introduction in 1991, Li-ion batteries have found many uses in consumer applications, where smaller battery packs are required. With the US government and the US council for automotive research establishing a partnership for new generation electric vehicles (PNGV) in 1993 (Nelson et al., 2002), Li-ion chemistry received a boost for developing higher capacity battery packs and several US national laboratories started collaborating in studying the long-term performance of Li-ion chemistry (Bloom et al., 2002, 2005, 2010; US Department of Energy, 2010; Wright et al., 2002).

As discussed and summarized in Section 4.5.2, Li-ion chemistry current flows in anode and cathode are based on several complex processes, and developing a comprehensive model requires expert knowledge in electrochemistry. It is commonly known that the lifetime of lithium-ion batteries is limited by unwanted side reactions (Stiaszny et al., 2014). These side reactions may affect all parts of a battery including the electrolyte, the active material, binder, conducting agents, current collectors, and the separator, which results in capacity decrease and/or increase of the overall cell impedance (Stiaszny et al., 2014). Understanding the substantial aging mechanisms in lithium-ion batteries is of great importance to make correct lifetime predictions. To study the aging phenomena, several analytical methods as well as EIS are used. More details can be found in a study by Stiaszny et al. (2014).

4.5.6.1 Modeling of aging of high power Li-ion batteries

For plug-in hybrid-electric vehicle (PHEV) applications, performance targets for rechargeable batteries indicated in US department of energy (DOE) documents (US Department of Energy, 2010) include calendar lifetimes of 15 years, with survival temperatures ranging from −46°C to +66°C. In these PHEV battery packs, specifications are aimed at defining the conditions for *recharging* (from a continuous power source for sustained replenishment of energy) as well as *regen* (or regeneration) where short duration high-charge current is fed back to a battery pack, typically during braking. During a regen process during braking, vehicle's kinetic energy is absorbed into the battery pack and this can only persist for a short time due to physical limitations such as heating (US Department of Energy, 2010).

Given this kind of PHEV requirements specified by PNGV, in PHEV battery testing a specific test profile such as in Fig. 4.19A is recommended. The procedures outlined by the US DOE supported US Advanced battery Consortium test manuals such as in US Department of Energy (2010) are intended to show the promise of a technology vs a set of performance and cost targets. No knowledge of the actual battery chemistry is needed. Thus, the evaluation of cells concentrates on their performance and life and how their life is affected by SOC, time, temperature, and type of test. Calendar and cycle life tests were performed to determine the aging characteristics of the blended cathode material under PHEV testing conditions (US Department of Energy, 2010). It should be noted that the calendar tests in the PHEV and HEV (Department of Energy, 2003) manuals are essentially the same. The cycle life test is more complex, depending on what is to be learned. Instead of just a single cycling mode of operation, charge sustaining (CS), as in an HEV, the PHEV operates in charge depleting (CD) as well as CS modes, as shown schematically in Fig. 4.19B. In the CD mode of operation, the battery powers the vehicle directly; the internal combustion engine is not used at all. After a while, the battery energy becomes exhausted. At this low SOC, the PHEV will operate in CS mode, similar to that of an HEV.

Given the above summary of operational details of a PHEV driving cycle and BMS requirements and the required longer calendar life of a vehicle battery pack, designers have to look into battery behavior in terms of capacity degradation and parameter deterioration. As detailed by Belt et al. (2011), Fig. 4.20A depicts the hybrid pulse power characterization (HPPC) of a 10 kW capable pack at 30°C based on the test profile as per Fig. 4.19A. Fig. 4.20B and C indicates the typical test profiles for CS- and CD-type operations as detailed in the US Department of Energy (2010).

Li-ion test cells aimed at PHEV-type applications were tested as per details in a study by Belt et al. (2011), and overall results indicate that in Li-ion chemistry the cell resistance keeps increasing with age and the number of cycles, while the deliverable capacity reduces with age and number of cycles. This is shown in Fig. 4.21. Dotted lines in Fig. 4.21 show the case for CS cycle.

Comparing these results with those obtained from a calendar life test as detailed by Belt et al. (2011) shows that most of the resistance increases (Fig. 4.21A) are due to calendar life effects at all temperatures. At 30°C and 40°C, there were no significant differences between the calendar and CS cycle curves. Only at 50°C and 60°C were significant differences observed; the cells from the CS cycle test displayed about 9%–19% additional resistance increase. Fig. 4.21B compares the capacity fade results in a calendar life test to those found in a CS cycling imposed on the cells.

In all cases, CS cycling increased capacity fade and, in most cases, CS cycling changed kinetic behaviors. Capacity fade in the calendar life test tended to follow Arrhenius-like, linear-kinetics in the temperature range of 30–50°C; those from the CS cycle test exhibited $t^{1/2}$ kinetics. Only at 60°C in the calendar test were parabolic kinetics observed. A detailed discussion on these degradation aspects is found in a study by Belt et al. (2011) and Bloom et al. (2006).

Fig. 4.22 presents a lumped parameter battery equivalent circuit suitable for depicting the impedance rise in high power Li-ion cells as discussed by Bloom et al. (2002). In this equivalent circuit, a large capacitor $I_L/(dV_{oc}/dt)$ is also used to depict the changes in diffusion process in the positive electrode. This lumped

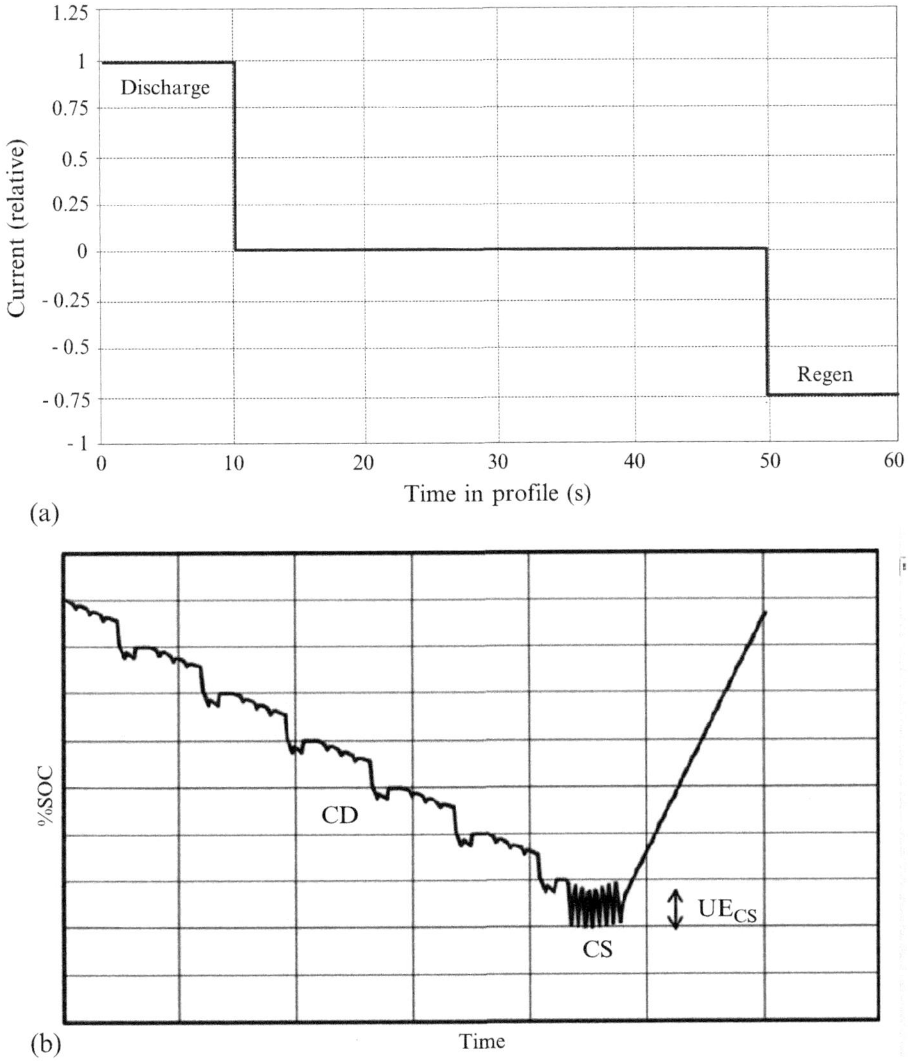

Fig. 4.19 PHEV battery charging-regen recommendations and operating modes: (A) test profile and (B) operating modes indicating charge depletion (CD) and charge sustaining (CS) modes.

parameter model (LPM) described in PNGV documents attempts to linearize the behavior and response of a battery during a simple repeatable current profile such as CS in the US Department of Energy (2010).

The equivalent circuit used in the LPM is shown in Fig. 4.22, where V_{oc} is the ideal OCV; R_o is the internal resistance due to ohmic, interfacial, and kinetic impedances in the cell; R_p is the concentration polarization of the electrolyte and electrodes; C is a capacitance included to simulate the time constant during the polarization of the

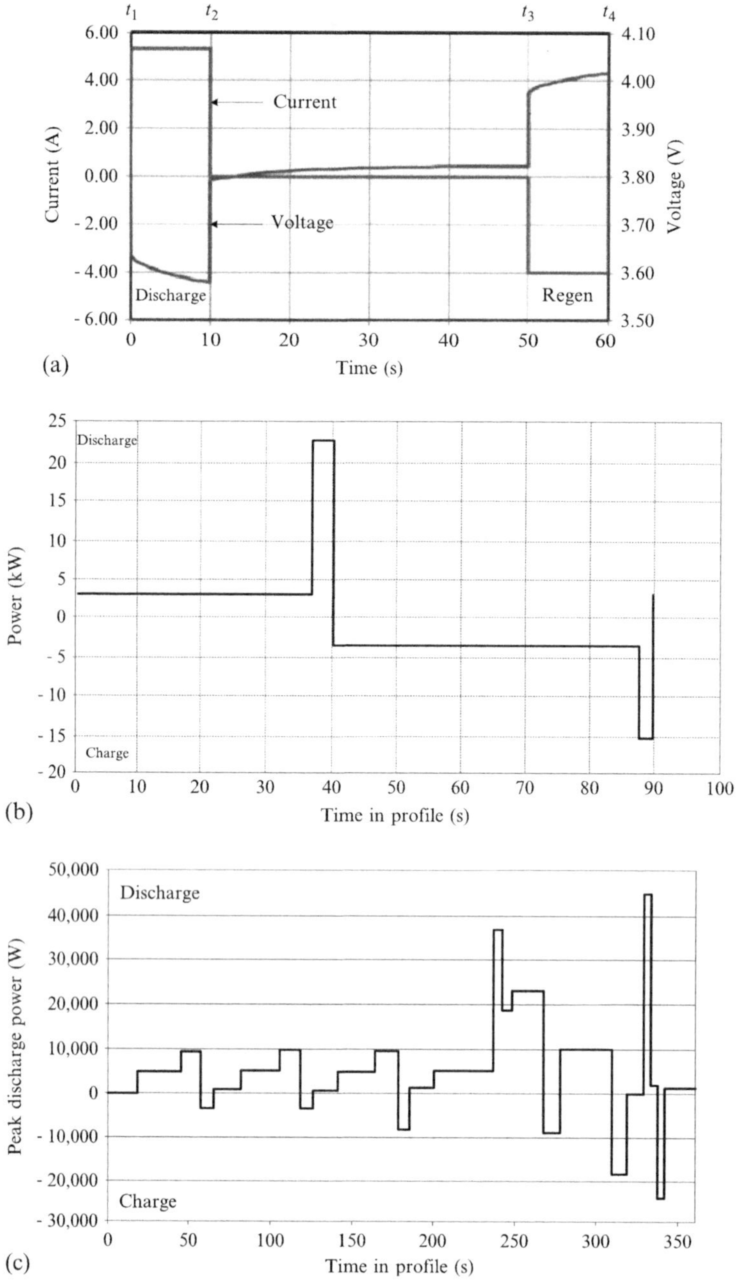

Fig. 4.20 Hybrid pulse power characterization (HPPC) testing of a pack of Li-ion test cells and PNGV specified test profiles (Belt et al., 2011) for PHEV battery packs: (A) HPPC profile and voltage response of a typical Li-ion cell; (B) typical recommended test profile for charge sustaining (CS) mode; and (C) typical recommended test profile for charge depleting (CD) mode.

Source: (A and C) Belt, J., Utgikar, V., Bloom, I., 2011. Calendar and PHEV cycle life aging of high energy, li-ion cells containing blended spinel and layered-oxide cathodes. J. Power Sources 196, 10213–10221.

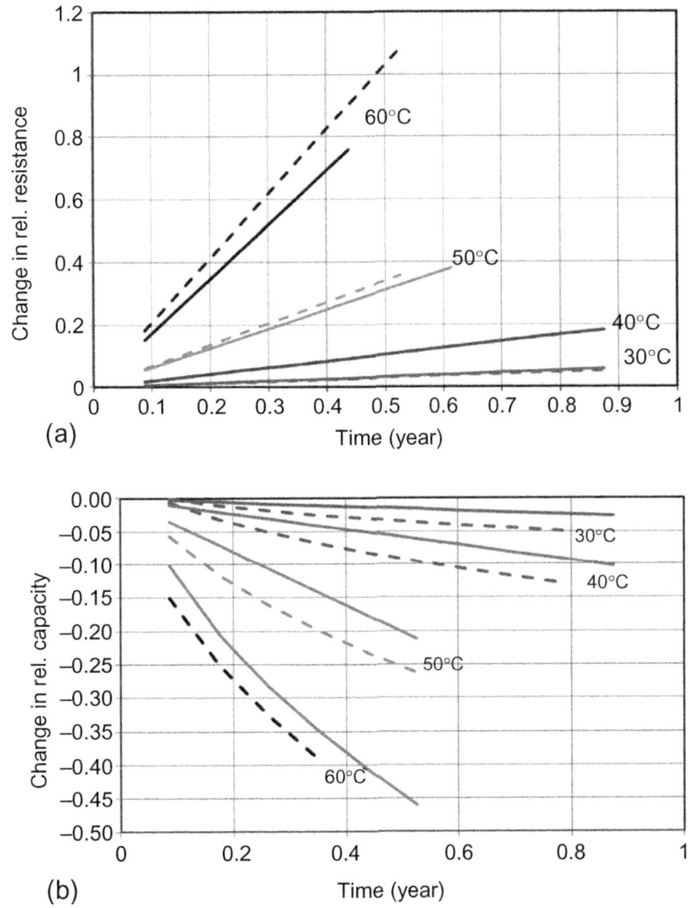

Fig. 4.21 Battery performance vs time: (A) cell resistance increase with time and (B) cell capacity drop with time [*Dotted lines* show CS test and continuous lines show calendar test]. *Source*: Belt, J., Utgikar, V., Bloom, I., 2011. Calendar and PHEV cycle life aging of high energy, li-ion cells containing blended spinel and layered-oxide cathodes. J. Power Sources 196, 10213–10221.

electrolyte and electrodes; I_L is the load current; $I_p(t)$ is the current through the polarization resistance; V_L is the load voltage; R_L is the load resistance; t is the time, and $I_L/(dV_{oc}/dt)$ is related to the changes in the diffusion coefficient of Li+ in the electrodes with time. The relationship between V_L and other parameters could be described by

$$V_L = V_{oc} - [1/I_L][dV_{oc}/dt]\left[\int I_L dt\right] - R_o[I_L] - R_p[I_p(t)] \tag{4.12}$$

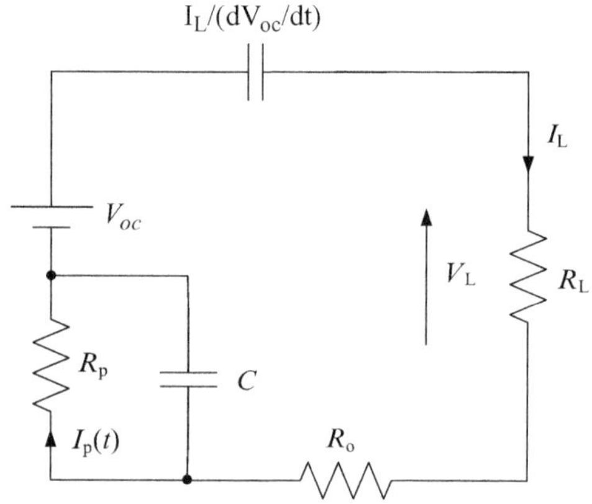

Fig. 4.22 Li-ion battery lumped parameter model (LPM).
Source: Bloom, I., Jones, S.A., Polzin, E.G., Battaglia, V.S., Henriksen, G.L., Motloch, C.G., Wright, R.B., Jungst, R.G., Case, H.L., Doughty, D.H., 2002. Mechanisms of impedance rise in high-power, lithium-ion cells. J. Power Sources 111, 152–159.

More details on combining this relationship with linear regression function (LINEST) in Microsoft Excel to calculate the parameters are given in a study by Bloom et al. (2002). Pattipati et al. (2011) also discuss an estimation procedure for SOC/SOH of a battery pack used for EVs and PHEVs based on the same model as in Fig. 4.22. More discussion on aging of Li-ion is available in a study by Bloom et al. (2001). In Urbin et al. (2010), a model similar to the case by Bloom et al. (2002) is used for energetical model based on a distributed parameter model similar to the model of a transmission line.

4.5.7 Lithium ferro phosphate batteries

With the automotive world getting inclined toward electric drives, the engineering community sees the prime barrier—a long-lasting (cycle life-based), safe, and energy-dense battery packs for energy storage. On another front, green energy world with renewable energy resources getting integrated with the grid, energy storage systems require similar capability battery packs with longer calendar life. In order to meet this demand, Li-based rechargeable chemistries such as lithium ferro phosphate ($LiFePO_4$, also known as lithium iron phosphate) and lithium titanate (LTO) have been developed.

$LiMO_2/Li_4Ti_5O_{12}$-based anode in LTO or Li_xFePO_4-based cathode in lithium ferro phosphate (LFP) batteries offers a long lifetime, high charge/discharge rates, high intrinsic safety for automotive applications and power tools, etc. Table 4.1 provides a comparison of different Li-ion battery chemistries based on cathode and anode material options (Swierczynski et al., 2014).

Fig. 4.23 shows the schematic description of the model of an LFP cell. As pictorially indicated in Fig. 4.23, negative active particle radius is about 140 times bigger than the positive active particle radius (Marcicki et al., 2013). Electrochemistry-based modeling to predict the performance of this chemistry can be found in a study by Wang et al. (2011), Forgez et al. (2010), Kassem and Delacourt (2013), and Kassem et al. (2012).

Similar to the models developed based on electrochemistry, thermodynamic, or otherwise for other chemistries, for LFP cells EIS-based techniques can be developed to predict the SOC/SOH of the battery. Fig. 4.24 depicts the equivalent circuit model based on EIS data and the associated Nyquist plot (Greenleaf et al., 2013) for a commercially available Powerizer LFP-RCR 123 A LFP battery. This Powerizer cell is comprised of a Li_xFePO_4 cathode and a graphite anode with a nominal voltage of 4.2 and a rated capacity of 450 mAh as per manufacturer's datasheet.

Table 4.2 shows the component values based on EIS measurements at different SOC values.

It is interesting to see that the value of almost all resistive components keeps increasing as the SOC goes down. This tells us the reason for a badly depleted cell showing very high internal resistance in general and this is true for almost all battery

Table 4.1 Technical and economic comparison of Li-ion chemistries.

	Cathode					**Anode**
Performance parameter	**LCO (lithium cobalt oxide)**	**NMC (lithium nickel manganese cobalt oxide)**	**NCA (lithium nickel cobalt aluminum oxide)**	**LMO (lithium manganese oxide)**	**LFP (lithium iron phosphate)**	**LTO (lithium titanate)**
Lifetime	Low	Good	Very good	Low	Very good	Very good
Cost per cycle	Average	Average	High	Average	Low	Low
Fast response	Very good	Very good	Very good	Very good	Very good	Very good
Performance	Good	Good	Good	Low	Good	Very good
Safety	Low	Good	Low	Good	Very good	Very good
Self-discharge	Very good	Very good	Very good	Very good	Good	Very good

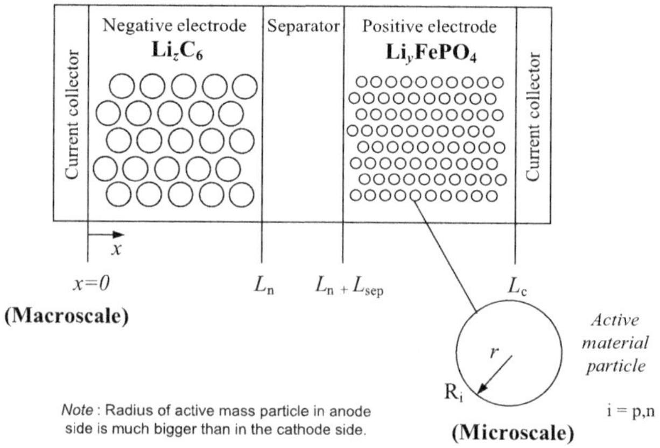

Fig. 4.23 Schematic representation of the LFP cell.

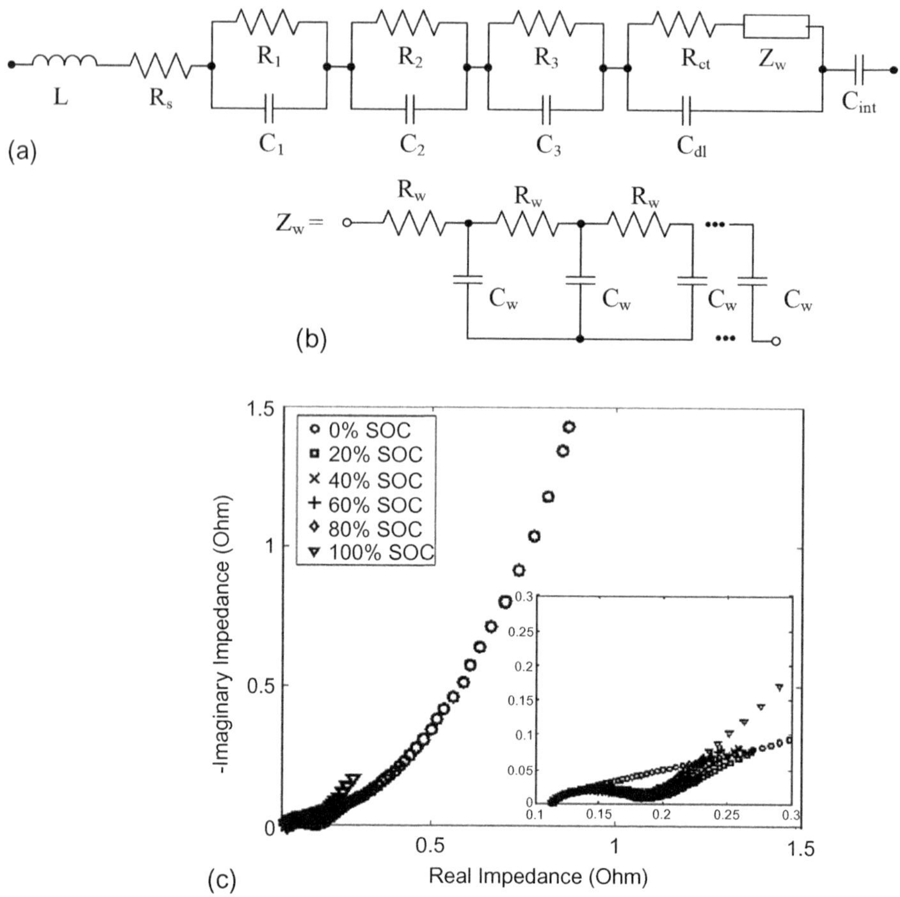

Fig. 4.24 Powerizer LFP battery equivalent circuit and Nyquist plot: (A) total equivalent circuit, (B) expanded Warburg impedance (Z_w), (C) Nyquist plot from 10 mHz to 2 kHz. *Source*: Greenleaf, M., Li, H., Zheng, J.P., 2013. Modeling of LIXFEPO4cathode Li-ion batteries using linear electrical circuit model. IEEE Trans. Sustain. Energy 4 (4), 1065–1070.

chemistries. Frequency-dependent Warburg impedance components that describe the diffusion of Li-ion through the porous electrode can be expressed at a particular measurement frequency as

$$Z_W(\omega) = R_W \frac{\tanh\sqrt{(j\omega T_W)}}{\sqrt{(j\omega T_W)}} \tag{4.13}$$

which is the same as the relationships in Eqs. (4.6) and (4.7).

The premise of the model shown in Fig. 4.24 is intended to relate the electrochemical and phenomenological processes observed in the energy storage devices to their analogous electronic mechanisms, thereby providing a valuable set of data to quantify the electrical behavior of the cell. In order to describe the battery's performance in the time domain, the frequency domain circuit components shown in Fig. 4.24 can be transformed using Fourier transform techniques and the result will be the circuit shown in Fig. 4.25. Component values in Fig. 4.25 and the details of the transformation

Table 4.2 List of parameter values for LFP based on EIS measurements at different SOC values.

		SOC percentage		
Parameter	**Description**	**0**	**40**	**100**
L (μH)		0.78	0.86	1.17
R_s (mΩ)	Ohmic resistance of solution, separator, and contacts	118	116	115
R_1 (mΩ)	Components representing multilayer surface film created during initial cycling and further thickened by successive cycles (SEI)	53	4	6
C_1 (F)		1.39	100	153
R_2 (mΩ)		1540	11	24
C_2 (F)		20.7	2.0	12.6
R_3 (mΩ)		43.5	28.4	24.1
C_3 (F)		0.192	0.126	0.440
R_{ct} (mΩ)	Charge-transfer kinetics at electrolyte/electrode boundary	301	30	36
C_{dl} (F)		0.024	0.019	0.023
C_{int} (F)	Intercalation/deintercalation of Li-ion from active mass	700	675	156.5
R_W (Ω)	Warburg impedance components. T_W is the time constant of the branch circuit	1.07	0.25	0.15
$T_W = R_W * C_W$ (s)		23.4	127.8	125
OCV (V)	Open-circuit voltage	2.53	3.31	3.41

Adapted from Greenleaf, M., Li, H., Zheng, J.P., 2013. Modeling of LIXFEPO4cathode Li-ion batteries using linear electrical circuit model. IEEE Trans. Sustain. Energy 4 (4), 1065–1070.

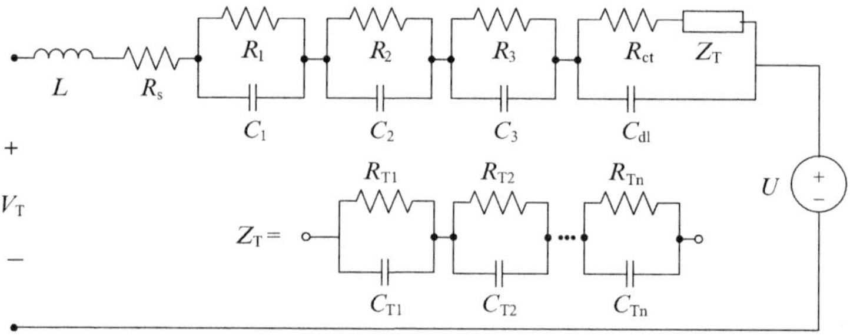

Fig. 4.25 Time domain transformation of frequency domain circuit model shown in Fig. 4.24.

process are detailed in a study by Greenleaf et al. (2013). A similar discussion will be found later in Chapter 5 related to supercapacitors, which has similar processes such as diffusion.

R_{Tn} and C_{Tn} values in Fig. 4.25 can be obtained from a simple conversion into a transmission line model (Greenleaf et al., 2013) as

$$R_{Tn} = \frac{8R_w}{(2n-1)^2\pi^2} \tag{4.14}$$

$$C_{Tn} = C_w/2 \tag{4.15}$$

Albertus et al. (2008) provide a discussion on capacity vs electrode combinations possible for electric vehicle battery packs, based on some electrode material options indicated in Table 4.1, including an electrode pair based on LTO and LFP materials. A detailed discussion on these is beyond the scope of this book since these subjects require significant knowledge components on the electrochemistry of batteries. Most references provided in this chapter (Albertus et al., 2008; Belt et al., 2011; Bloom et al., 2001, 2002, 2005, 2006, 2010; Buller et al., 2003, 2005; Bundy et al., 1998; Department of Energy, 2003; Fang et al., 2012; Forgez et al., 2010; Gao and Manthiram, 2009; Greenleaf et al., 2013; Grimnes and Martinsen, 2008; Gould et al., 2009; Gu et al., 1999; Huet, 1998; Jossen, 2006; Karden et al., 2000; Kassem and Delacourt, 2013; Kassem et al., 2012; Khun et al., 2006; Kularatna, 2008; Lai and Rose, 1992; Li et al., 2014; Lindahl et al., 2012; Marcicki et al., 2013; Mauracher and Karden, 1997; Orazem and Tribollet, 2008; Pan et al., 2002; Pattipati et al., 2011; Ramadass et al., 2004; Randles, 1947; Ruetschi, 2004; Singh et al., 2004; Stevanatto et al., 2014; Stiaszny et al., 2014; Swierczynski et al., 2014; Tan and Baras, 2005; Tang et al., 2008; Thele et al., 2005; US Department of Energy, 2010; Valoen et al., 1997; Wang et al., 2011; Warburg, 1899; Wright et al., 2002; Wu et al., 2001; Urbin et al., 2010; http://www.ultraglobalprt.com/?page_id=24) are work done by many electrochemists and physicists and therefore electronic engineers probably require a basic course on electrochemistry to appreciate these discussions. In Sections 4.6–4.8, a summary of battery equivalent circuits and BMSs in an electronic engineers viewpoint is provided.

4.6 Battery management in practical applications

About three decades ago, battery management was having a reliable, fast, and safe charging methodology to be selected for a battery bank, together with the monitoring facilities for detecting the discharged condition of the battery pack. With modern battery technologies emerging, the demands from the cost-sensitive portable product market, as well as the medium power range products such as UPS and telecom power units and the automotive sector (in EV, HEV, and PHEV sectors), attributes of a modern BMS may include

- Practical battery modeling suitable for electronic systems design.
- Battery charging methods and charge control.
- Determination of SOC/SOH/RUL and the end of discharge (EOD).
- Gas gaging.
- Monitoring battery health issues.
- Communication with the host system/or power management subsystems.
- Battery safety.

Sections 4.6 and 4.7 provide a discussion on some concepts and techniques related to simplified and practical modeling of batteries for best performance extraction, managing given chemistry for best runtime and the longest life, safety of battery packs, and prognostics for health management of battery packs.

4.6.1 Practical modeling of a cell to reflect its electrochemistry—an electronic engineer's viewpoint

As engineers we tend to work with device datasheets, leaving the device design and its behavior to be decided by the device expert, and a team of electrochemists/physicists to give us a mass-producible cell. This electrochemical cell is a multidisciplinary system, which involves complex chemical, electrical, and thermodynamic processes. As we discussed in Section 4.5, while modeling a battery requires a lot of details, as electronic engineers in an oversimplified equivalent circuit, we tend to oversimplify a battery to a voltage source (with OCV) and a series resistance. However, practical voltage profiles of batteries during discharge and charge do not exactly reflect the accuracy of this simplified assumption. Fig. 4.26 indicates the overly simplified chemical process within a battery as it discharges its stored chemical energy to a load.

As we have seen in Fig. 2.1, battery capacity decreases as the discharge rate increases. Also as per Fig. 2.5, we see that a cell can have high pulse discharge if we allow resting between pulses. These kinds of behavior are simply explained in Fig. 4.26. Fig. 4.26A shows the case of a fully charged cell with maximum concentration of active species. When a load is connected, the load current causes this active species to be consumed at the electrode surface (Fig. 4.26B) and replenished by the diffusion process from the bulk of the electrolyte. A higher load current causes a higher concentration gradient, and thus a lower concentration of active species near

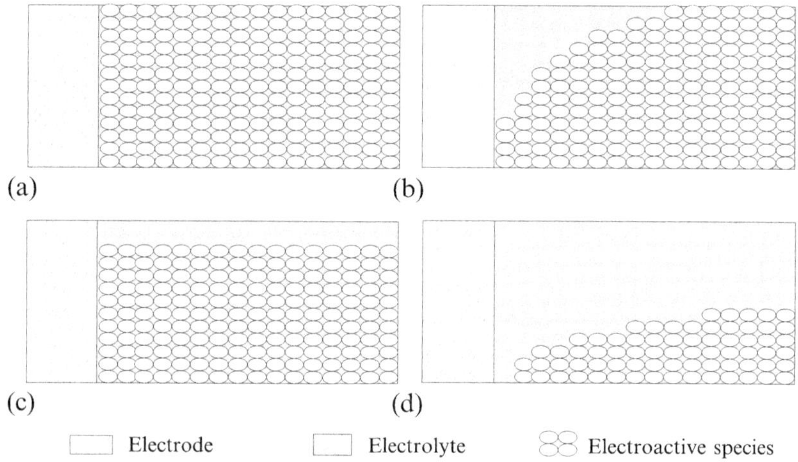

Fig. 4.26 Behavior of a symmetric electrochemical cell: (A) fully charged state; (B) before recovery; (C) after recovery; and (D) discharged state.

the electrode surface. When this concentration falls below a particular threshold, a voltage cutoff occurs. However, the unused charge is not physically lost and it is simply unavailable at the electrode surface due to lag between reaction and diffusion rates. Decreasing the discharge rate effectively reduces this lag as well as the concentration gradient in the vicinity of the electrode. When the recovery occurs (Fig. 4.26C), a high current discharge is possible until the active species are fully consumed as in Fig. 4.26D. For more details and related modeling of batteries, an article by Rao et al. (2003) is suggested.

This simplified discussion on the process of charge-transfer and diffusion process etc. leads to the curiosity toward more complex chemical behavior of a cell, which was summarized in Section 4.5 for different chemistries. Studying such complex chemical processes inside a cell leads us to develop a more accurate circuit model suitable for modeling a battery based on simplified engineering information extracted from a datasheet. NiMH or Li-ion cells can be modeled using modern finite element methods (Gao et al., 2002; Renhart et al., 2008; Schweighofer et al., 2003) for temperature behavior, while automated test systems can be used to extract information for simplified models (Schweighofer et al., 2003) to reflect the cell behavior. For much more simplified requirements, extracting data from cell manufacturer datasheets is also possible (Gao et al., 2002).

As discussed in the literature cited in Sections 4.4 and 4.5, based on different approaches there are several possible battery models as listed below:

(i) Physical models
(ii) Empirical models
(iii) Abstract models
(iv) Mixed models

In general, microscopic-level models are important for a battery designer, while macroscopic-level models are useful for the BMS designer or the battery user. A user-level model should capture only the gross characteristics of the battery behavior.

Physical models are also known as electrochemical models (Rao et al., 2003) and they are very accurate, but require detailed consideration of electrochemical processes, thermodynamic process, physical considerations for both charging and discharging characteristics. These are computationally expensive and practical application of them in the engineering of BMS is limited.

Empirical models are easy to configure and are represented by mathematical expressions with less number of parameters. However, they fail to provide an accurate estimation of battery SOC/SOH, etc. under varying load conditions.

Abstract models are simplified equivalent circuit representations such as PSPICE models or based on stochastic processes (Chiasserini and Rao, 2001).

Mixed models are based on high-level abstraction, avoiding excessive details of physical laws (i.e., electrochemical process) governing the battery characteristics which leads to the derivation of simplified analytical expression. Sometimes these are called analytical models (Rong and Pedram, 2006a).

An example of an empirical lead-acid battery model for charge/discharge and lifetime estimation is given by Agrawal et al. (2010). This empirical model can be extended to other chemistries, and in this work, Rakhmatov and Virdhula model (Rakhmatov et al., 2003) is used as the base. This analytical model, which is useful for a portable product BMS area, allows for a trade-off between the accuracy and the amount of computation performed. A fast prediction model for SOC/SOH estimation with about 5% accuracy for Li-ions is presented by Rong and Pedram (2003).

4.6.2 Application-specific approaches to battery modeling

Dominating battery chemistries used in current systems are lead-acid, NiMH, and Li-based chemistries. With the rapid growth of electric vehicles, larger capacity batteries are developed for HEV, PHEV, and EV (or sometimes called battery electric vehicles) applications and fractional Ah to few Ah packs are developed for portable devices. Table 4.3 depicts a summary of typical battery characteristics expected in HEV, EV, and portable electronics with the expected accuracies in SOC/SOH estimations.

In BMS systems, development designers prefer to use electrical analog battery models such as the ones discussed by Li et al. (2012). For off-line estimation of battery SOC/SOH, these analog battery models with simplified equivalent circuits are commonly used (Chen and Rincon-Mora, 2006; Gao et al., 2002; Husseni et al., 2011; Li et al., 2012; Zhang et al., 2010). As discussed by Li et al. (2012), an analog battery model such as in Fig. 4.27 can be developed, where the impedance Z shown in Fig. 4.27A is a combination of RC sections and the series resistance we discussed in Figs. 4.24 and 4.25. Fig. 4.27B shows the dynamic response of a battery such as Ultralife UBBL 10 Li-ion pack with 6.8 Ah capacity used in EV-type applications.

Table 4.3 Typical characteristics envisaged in three different battery environments.

Characteristic/ parameter	Hybrid electric vehicle (HEV)	Electric vehicle (EV)	Portable electronics (PE)
Maximum charge or discharge rate (C)	±20	±5	±3
Rate profile	Very dynamic	Moderate	Piecewise constant
SOC estimation required	Very precise	Precise	Crude
Predict available power	Yes	Yes	No
SOH estimation	Required	Required	Not essential
Cell balancing	Continuous	Continuous or on charge mode only	On charge only
Lifetime expected (years)	10–15	10–15	<5

Source: Plett, G.L., 2004a. Extended Kalman filtering for battery management systems of LiPB-based HEV battery packs: part 1. Background. J. Power Sources 134, 252–261.

Series resistance in Figs. 4.24 and 4.25 is responsible for the instantaneous drop, and RC sections are responsible for the longer duration transient response. More details are in a study by Li et al. (2012).

4.6.2.1 Battery equivalent circuits and practical ways to estimate parameter values

Battery equivalent circuits are useful for modeling the behavior of a battery pack, to predict its short-term behavior as well as the long-term performance with regard to discharge rates, temperature changes, and life cycle prediction, as well as pulse discharge performance. There are many different models available ranging from a simple voltage source with internal resistance to equivalent circuits modeling the effects of charge state, temperature, pressure, aging, and many more aspects of the battery pack.

To predict the performance of a battery in the short term, simplified equivalents such as shown in Fig. 4.28A is possible. This model is valid for time scales ranging from milliseconds up to a few seconds at a constant SOC/discharge and temperature.

In Fig. 4.28A

- U_B and I_B are the terminal voltage and current.
- U_o is the OCV.
- R_i is the constant part of the internal resistance (connector, electrodes, and electrolyte).
- R_D and C_D are the effects on the surface of the electrode (double-layer capacitance effect).
- R_k and C_k reflect the effect of the diffusion process in the electrolyte.

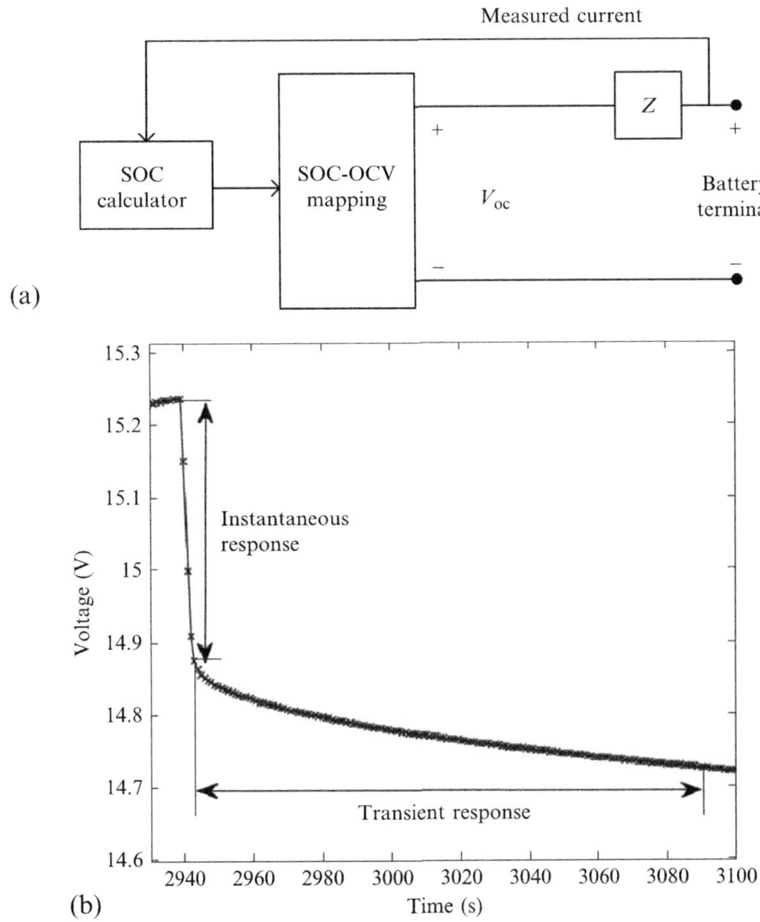

Fig. 4.27 Analog battery model useful in electric vehicle battery pack developments: (A) model and (B) dynamic response of Ultralife UBBL 10 Li-ion battery module.
Source: Li, J., Mazzola, M., Gafford, J., Younan, N., 2012. A new parameter estimation algorithm for an electrical analogue battery model. In: Proceedings of IEEE APEC 2012 Conference, pp. 427–433.

To deal with longer prediction times or different temperature situations, the values of components have to be adapted according to the changes inside the battery (Bernardi and Carpenter, 1995; Gu et al., 1995; Landfords et al., 1995). Schweighofer et al. (2003) discuss the details of a test setup that can be used to extract the parameters for the equivalent circuit model shown in Fig. 4.20A. Fig. 4.20B shows some details of such a test setup. Table 4.4 indicates the component values related to a 9-Ah NiMH battery with 270 A current capacity. Given the time constants for the surface effects of the electrodes, R_D and C_D, and the time constants related to the

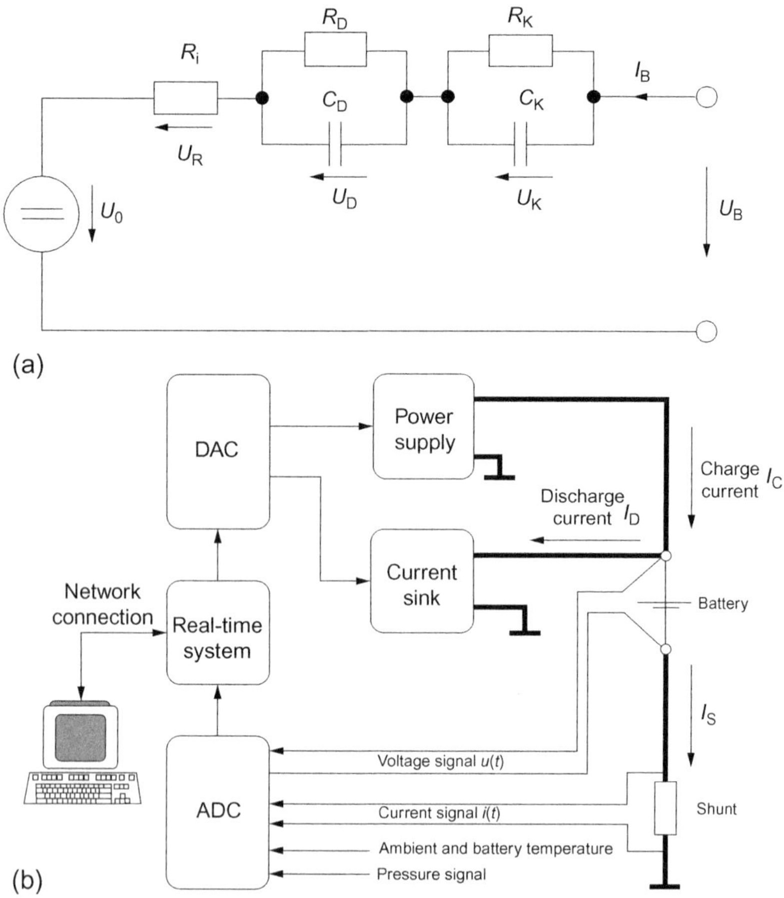

Fig. 4.28 Battery equivalent circuit models and parameter extraction: (A) for short-term behavior and thermal performance and (B) test setup for extracting parameters of model. *Source*: Schweighofer, B., Raab, K.M., Brasseur, G., 2003. Modeling of high power automotive batteries by the use of an automated test system. IEEE Trans. Instrum. Meas. 52 (4), 1087–1091. Reproduced with the permission of IEEE, USA.

Table 4.4 Model parameters for a typical 9-Ah NiMH cell.

Components of cell internal resistances (mΩ)		**Capacitances**	
R_i	**1)**		
R_D	0.35	C_D	171 F
R_k	1.6	C_k	16,000 F

Adapted from Schweighofer, B., Raab, K.M., Brasseur, G., 2003. Modeling of high power automotive batteries by the use of an automated test system. IEEE Trans. Instrum. Meas. 52 (4), 1087–1091.

diffusion process, R_k and C_k, one should be able to predict the battery behavior to match the experimental data available in datasheets. For details, article by Schweighofer et al. (2003) is suggested.

In case a much simplified model is used, public information available on battery datasheets can be used to arrive at a model (Gao et al., 2002) shown in Fig. 4.29. In this case, the battery voltage curves based on different discharge rates, measured in multiples or fractional C rates, can be used using simplified graphical and calculative approaches. This three-component model is based on

- OCV (equilibrium potential), E.
- An internal resistance R_{int} having two components R_1 and R_2.
- An effective capacitance that characterizes the transient response of charge double layers in porous electrodes.

More details related to this procedure are discussed by Gao et al. (2002). For designing of battery packs for portable systems, an accurate prediction of battery models for runtime and V-I performance Cadence simulation can be practically used. More details and various equivalent circuits suitable for this purpose are discussed by Chen and Rincon-Mora (2006), in a design engineer's viewpoint. A set of battery models suitable for Cadence-type simulation work as proposed by Chen and Rincon-Mora (2006) is shown in Fig. 4.30. This uses two sets of parameters where Fig. 4.30A indicates a capacitor- and a current-controlled current source inherited from run-time-based models. RC network-based model in Fig. 4.30B based on Thevenin models simulates the transient response (Chen and Rincon-Mora, 2006).

Another important condition in a battery is the internal battery temperature and the ambient temperature of the battery pack. By using finite element models (FEM), it is

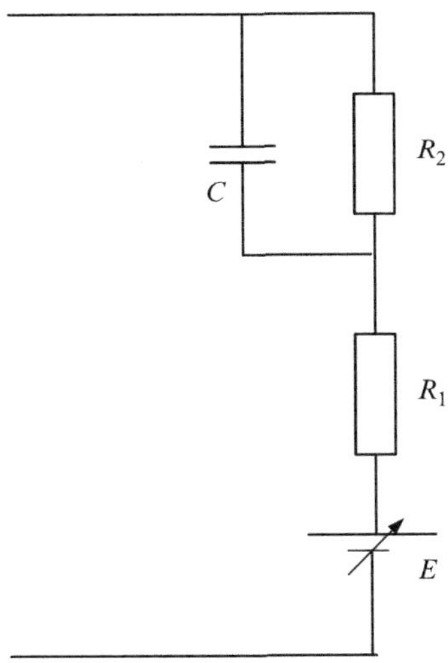

Fig. 4.29 A simplified equivalent circuit model for a Li-ion cell usable for datasheet parameter extraction.

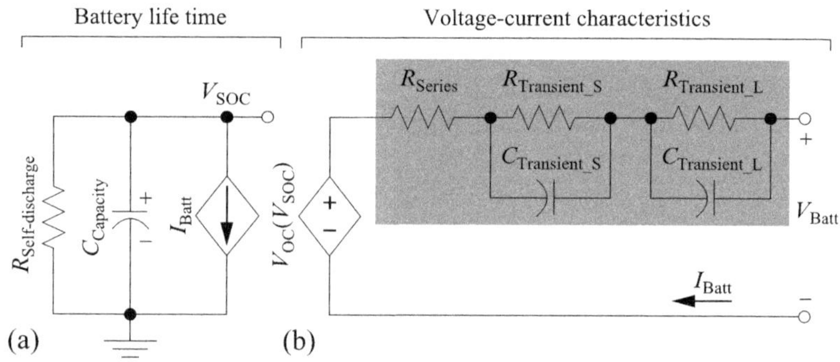

Fig. 4.30 A battery model suitable for Cadence simulation: (A) model suitable for capacity, SOC, and runtime estimation and (B) model suitable for transient response estimation. *Source*: Chen, M., Rincon-Mora, G.A., 2006. Accurate electrical battery model capable of predicting run time and I-V performance. IEEE Trans. Energy Conv. 21 (2), 504–511. Reproduced with the permission of IEEE, USA.

possible to analyze and predict the thermal behavior of a battery pack or a cell using modern FEM software combined with thermal models of the battery (Renhart et al., 2008).

For standby power system requirements such as UPS and telecommunication power supplies, where VRLA-type lead-acid cells are used, reserve life estimation is very useful. In such situations, reserve life estimation is based on SOH measurements on the battery pack. To estimate the SOH in such situations, several parameters and measurements can be utilized:

(a) Float voltage and float current measurement.
(b) Temperature rise in cells in a battery bank.
(c) Number of charge-discharge cycles.
(d) Impedance, conductance, and resistance measurements.
(e) Coup de fouet techniques.

While (a)–(d) are simple enough to understand, coup de fouet (translated as "crack of the whip") requires a brief discussion. In general, start-of-discharge of a VRLA battery is dominated by two transient voltage responses: response associated with batteries internal series resistance and the inductance; and, a more complex phenomenon that is attributed to the electrochemical reactions within the battery known as coup de fouet. As shown in the example of Fig. 4.31 for a Hawker 2H1275-type cell (Pascoe and Anbuky, 2002, 2005), coup de fouet region for a lead-acid battery occurs whenever a fully charged cell is discharged, where the terminal voltage drops to a minimum, or trough, and then recovers back to the plateau value in a relatively short time. A detailed discussion on this subject is found in a study by Pascoe and Anbuky (2002), and Pascoe and Anbuky (2005) provide an overview on how the above parameters or battery behaviors (a)–(e) could be used to estimate the reserve life or the RUL of a lead-acid battery pack.

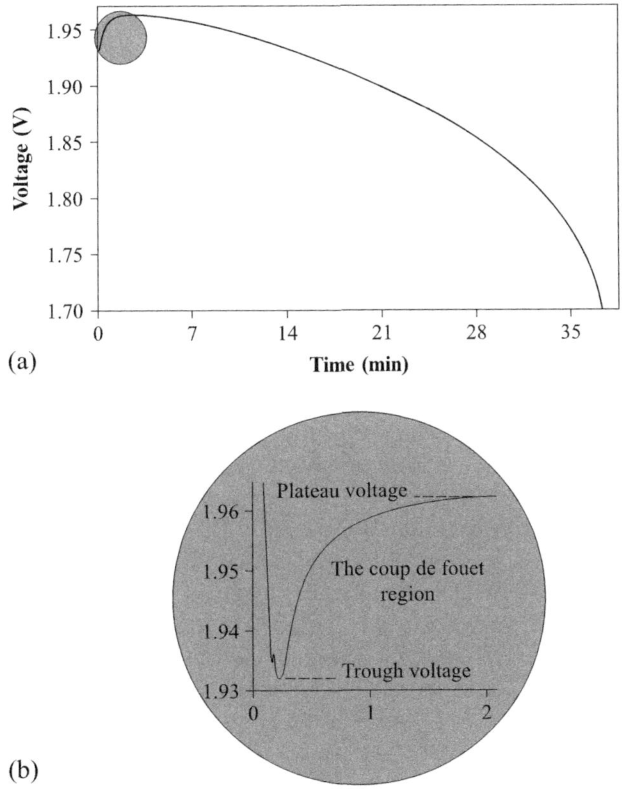

Fig. 4.31 Coup de fouet phenomenon in lead-acid VRLA battery—an example from Hawker 2H1275 battery after a full charge: (A) voltage vs time with highlighted coup de fouet region and (B) expanded coup de fouet region.
Source: Pascoe, P.E., Anbuky, A.H., 2002. The behavior of the coup de fouet of valve-regulated lead-acid batteries. J. Power Sources 111, 304–319.

4.6.2.2 Online parameter estimation and techniques

With battery-powered systems such as electric vehicles and portable products introduced into the market keep growing, in many situations designers of BMS wish to estimate the SOC/SOH dynamically while the battery pack is powering the electronics or electrical systems. In these cases, parameters such as SOC and SOH should be estimated online and few mathematically intensive techniques are currently considered to achieve this target. Few well-documented techniques are

(i) Kalman filter (KF) based
(ii) Neural network (NN) based
(iii) Markov chain based
(iv) Fuzzy logic (FL) based

In all techniques, the modeling of a battery pack based on either electrochemical models or equivalent circuit models is used with intensive mathematical approaches. With high power, high discharge capability battery packs used in different types of electric vehicles, where the driver should be kept informed of battery pack SOC/SOH and other important parameters, the above mathematically complex approaches are considered as possible solutions, for online estimation of the parameters.

KF-based approach

KFs are an intelligent—and sometimes optimal—means for estimating the present value of the time-varying "state" of a dynamic system. An additional benefit of the KF is that it automatically provides dynamic estimation error bounds on these estimates as well, and this fact was exploited to give online performance from a battery pack, without fear of causing damage by overcharge or overdischarge. By modeling a battery system to include the wanted unknown quantities in its state description, in a three-part series of papers, Plett (2004a, b, c) details the use of KF to estimate their values. A brief tutorial to assist applying KFs in BMS is found in a study by Plett (2004a), with a specific case of extended Kalman filtering (EKF) to take care of the nonlinear nature of the battery parameters. The second part of the papers (Plett, 2004b) describes some mathematical cell models that may be used with this method. It also gives an overview of other modeling methods in the literature and shows how an EKF may be used to adaptively identify unknown parameters in a cell model, in real time, given the cell input-output data. The third part of the paper (Plett, 2004c) covers the parameter estimation problem; namely, how to dynamically estimate SOC, power fade, capacity fade, and so forth with an EKF used in conjunction with the cell model. The cell model may be fixed or may itself have adaptable parameters so that the model tracks cell aging effects.

Fig. 4.32 shows a simple block diagram for the algorithm function, and the HEV BMS performs many tasks, including communicating with the vehicle controller, measure cell physical quantities of interest (e.g., cell voltage, current, and temperature), and manage cell balancing. Here, we are only interested in the algorithmic considerations as motivated by the requirements imposed by the environment and the vehicle. In relation to Fig. 4.32, following processes should happen in the algorithm, based on a regular sample collection of battery terminal voltage $V_{k,}$ temperature T_k, and the battery current I_k.

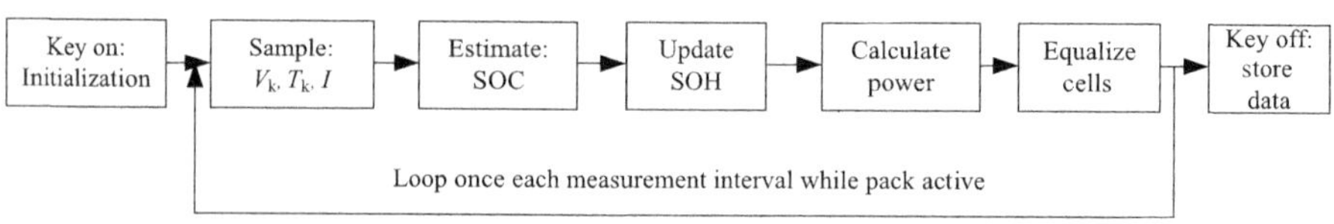

Fig. 4.32 Algorithms in an HEV battery management system.
Source: Plett, G.L., 2004a. Extended Kalman filtering for battery management systems of LiPB-based HEV battery packs: part 1. Background. J. Power Sources 134, 252–261.

- *Initialization.* When the vehicle is turned on, the algorithms must be initialized. The predominant dynamics of a cell while at rest is simply "self-discharge." If the level of self-discharge is too high, the SOH should be flagged as a warning or fault condition.
- *SOC update.* Once in every measurement interval, the voltages, temperatures, and module current are measured. The cell/pack SOC estimate must be updated based on these measurements.
- *SOH update.* Battery capacity and other parameters change over the lifetime of the pack. These must be continuously estimated in order to maintain safety and to obtain maximum performance from the pack.
- *Maximum available power.* Based on the SOC estimate and its uncertainty, and a dynamic cell model, the BMS must be able to estimate the maximum dis/charge power available at any time that will not cause voltage, SOC, or other design limits to be violated.
- *Equalization.* Series strings of cells with unequal capacities (as all are) will become unbalanced. That is, even if the SOC of all cells starts with the same value, they will drift slowly apart as the system operates. The BMS must determine which cells must have their levels of charge altered to keep the pack balanced.

Many of the algorithmic requirements just described prescribing estimating parameters of a battery pack that may not be directly measured. It is well known that Kalman filtering provides an elegant and powerful solution to situations such as this. Kalman filtering is an established technology for dynamic system state estimation that is in common use in many fields including target tracking, global positioning, dynamic systems control, navigation, and communication but is not widely known in the field of BMS until the work done during the last decade. The KF comprises a set of recursive equations that are repeatedly evaluated as the system operates. The reader is referred to Kalman's original paper and several textbooks listed under bibliography for further derivation details together with the content of the three-part paper (Plett, 2004a, b, c) cited here.

NN-based techniques

The inspiration for developing NN came from examination of central nervous systems. In an ANN, simple processing nodes, called "neurons," "neurodes," "processing elements," or "units," are connected together to form a network which mimics a biological NN. NNs are frequently employed to classify patterns by learning from examples. Pattern statistics are determined from a set of training samples, and then new patterns are classified based on these statistics. In a variation called probabilistic NNs (PNN), heuristic approaches are often used to discover the underlying class statistics. They require long computation times for training. Incremental adaptation approaches are susceptible to the convergence of an error function to the local minima. The network structure of PNN is similar to that of backpropagation, but the activation function is replaced by one of the class of functions, such as exponential. Training PNN requires only a single pass. It is based on the Bayesian decision boundaries as the number of training samples grows. After the PNN is trained, all training samples are stored and then used in classifying new patterns. The computation time of PNN is much faster than that of backpropagation types. PNN can be programmed easily into processors or chips to achieve even higher speeds through the use of parallel hardware.

A discussion on how this is applied in battery SOC/SOH estimations is available in a study by Lin et al. (2013), Shen et al. (2005), and Shahriari and Farrokhi (2013).

4.6.2.3 Markov chain and FL-based approaches

Recently, Markov chain and FL-based approaches have been attempted by the researcher community involved with the BMS. Micea et al. (2011), Moura et al. (2013), Gholizadeh and Salmasi (2014), Lin et al. (2012), and Hsieh et al. (2001) provide some insight into these approaches.

4.7 Prognostics in battery health management

Prognostics attempt to estimate the remaining component lifetime when an abnormal condition has been detected. The key to useful prognostics is not only an accurate remaining lifetime estimate but also an assessment of confidence of the uncertainty estimate. The phrase "battery health monitoring" has a wide variety of connotations, ranging from estimating the approximate voltage and other parameters to fully automated online monitoring of various measurements and estimated battery parameters. Prognostics of battery health could be very critical in an electric vehicle, aerospace, and space exploration applications. For example, a catastrophic battery failure has occurred in NASA's Mars Global Surveyor, which stopped operating in November 2006. It was revealed later that when the spacecraft was commanded to go into safe mode, the radiator of the batteries was positioned toward the sun, which created a temperature issue on the battery system (Goebel et al., 2008).

Prognostics and health management (PHM) in batteries for EVs and HEVs is a growing topic (Gao et al., 2002; Goebel et al., 2008). PHM dynamic models use sophisticated reasoning schemes applied with the goal of estimating the SOC, SOH, and the state of life (SOL). However, it remains a difficult task to predict the EOL for a battery pack (Goebel et al., 2008). In estimating battery SOC, SOH, SOL, etc., monitoring the variation in battery impedance becomes very useful, as this parameter has a significant variation as the battery packages. Liejen, P et al (2018) provides details of developing a prognostic system for the Toyota Prius NiMH battery system.

4.7.1 Battery impedance and its time variation estimation as a prognostic parameter

The impedance of a battery is one of the key parameters vary as the battery ages. Due to various electrochemical reasons within a cell, as the cell ages, impedance keeps increasing in general. Fig. 4.33 indicates this situation for NiMH and Li-ion cells. Fig. 4.33A is a simple DC resistance variation for a well-used vs a badly used (overcharged repeatedly) NiMH cell. Fig. 4.33B indicates the results of measuring the impedance of an 18,650-type Li-ion cell in a PHM exercise (Goebel et al., 2008), and Fig. 4.33C indicates the variation in the Li-ion cell impedance vs frequency after different numbers of discharge cycles (Karfthoefer, 2005). In Fig. 4.33C, it clearly

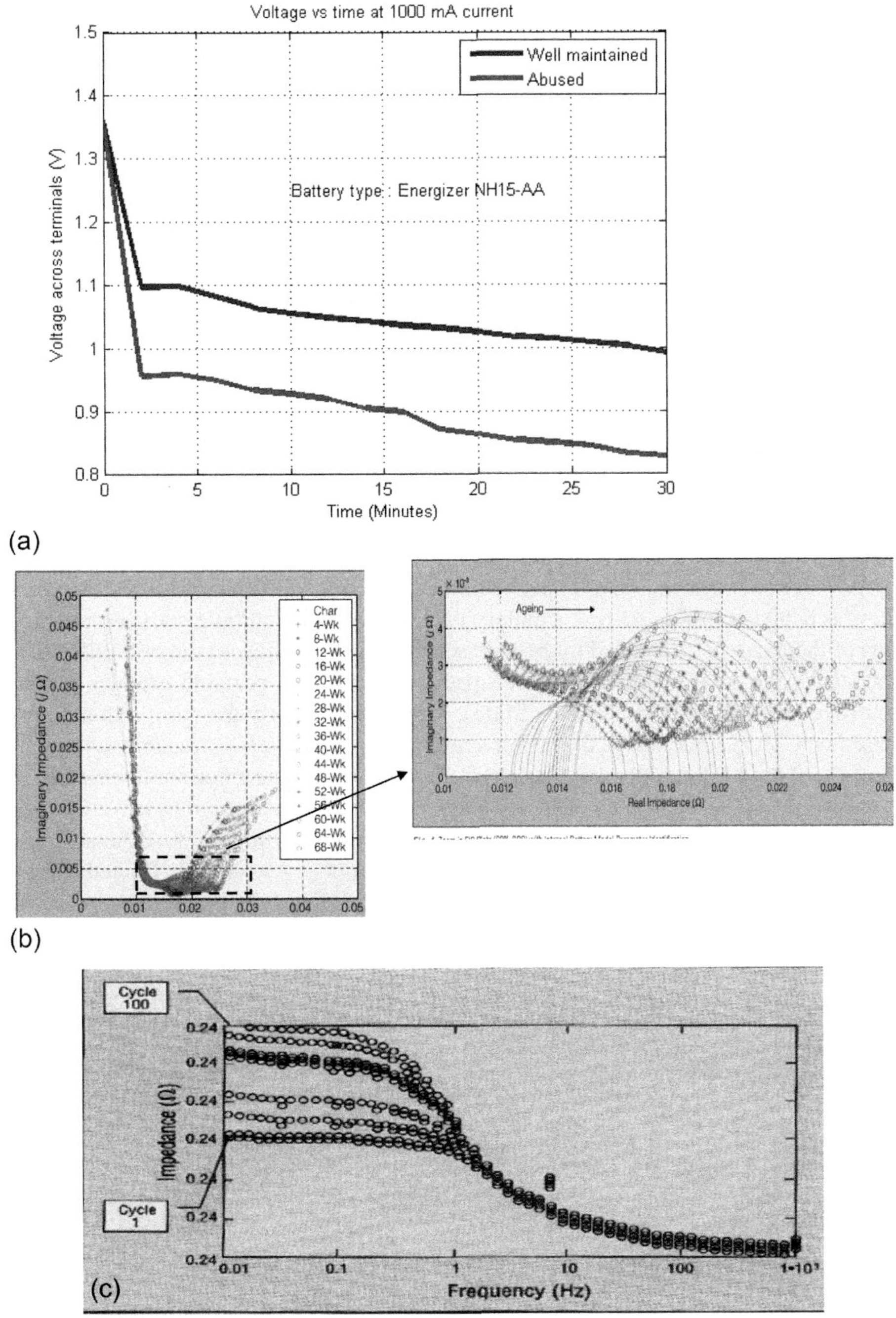

Fig. 4.33 Battery impedance variation with age and abuse: (A) DC impedance comparison of a well-used NiMH cell and an abused one; (B) electrochemical impedance spectrometry (EIS) data for a 18,650-type Li-ion cell with aging (Goebel et al., 2008); and (C) typical variation of Li-ion cells with life cycle at different frequencies (Karfthoefer, 2005).

shows that at lower frequencies below 10 Hz, as the discharge cycles are increased the impedance gradually increases. There are new battery management ICs that monitor this behavior in gas gaging (discussed later) (Karfthoefer, 2005).

In large standby telecommunication battery banks or UPS installations, large numbers of series and parallel connected VRLA cells [sealed lead-acid (SLA) cells] are used and any single or multiple misbehaving cells could in the longer run create expensive replacement issues. In these cases of large and expensive battery banks, battery impedance testing is used as a prognostic technique. By automatic or manual measurement of individual cell voltages over the lifetime of the bank, test engineers can easily predict the potential failure of individual cells. In these situations, typical VRLA cells behave as per Fig. 4.34 in general (Langnan, 2000). A detailed discussion of the impedance variations and terminal voltages is available in a study by Langnan (2000).

4.7.2 Energy-aware battery modeling concepts for best runtime

With the availability of techniques for computationally feasible mathematical models of cells, "energy-aware systems design" is another new concept currently developing, to get the best out of a battery pack, with multiple cells or batteries connected to a given electronic system. Given the chemical structure and the processes inside a battery, as depicted in Fig. 4.26, if a battery pack can contain multiple cells, where each one of them can be connected to the load selectively depending on the best utilization model for the total energy within the pack, runtime of a portable product can be improved significantly (Rao et al., 2003). This is particularly the case with systems based on CMOS logic where the power supply source voltage V_{DD} and threshold voltage, V_{th} of the FETs are the key variables for best power consumption. In case, these parameters can be optimized with a battery efficiency model, an interleaved power supply system (with multiple batteries selectively supplying the load conditions) can provide the best runtime, given the case that a battery pack has only limited energy after an optimal charging process. This concept is shown in a simplified case of Fig. 4.35, with two batteries that can be selectively connected to a suitable DC-DC converter, which supplies the VLSI circuit.

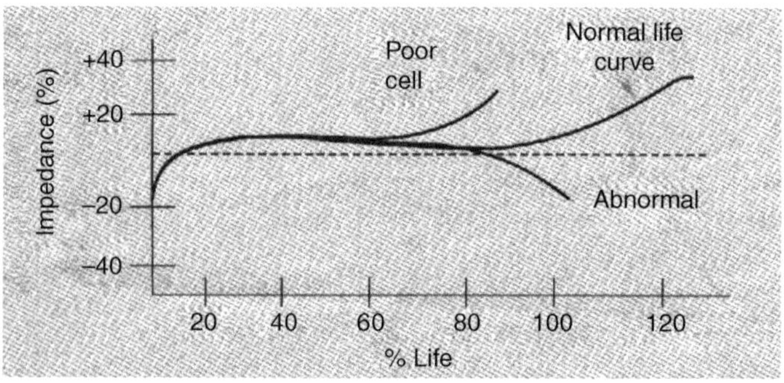

Fig. 4.34 VRLA battery impedance behavior over lifetime.
Adapted from Langnan, P.E., 2000. VRLA battery impedance analysis: cell evaluation via changes in the slope of the impedance curve. Power Qual. Assur., 24–29.

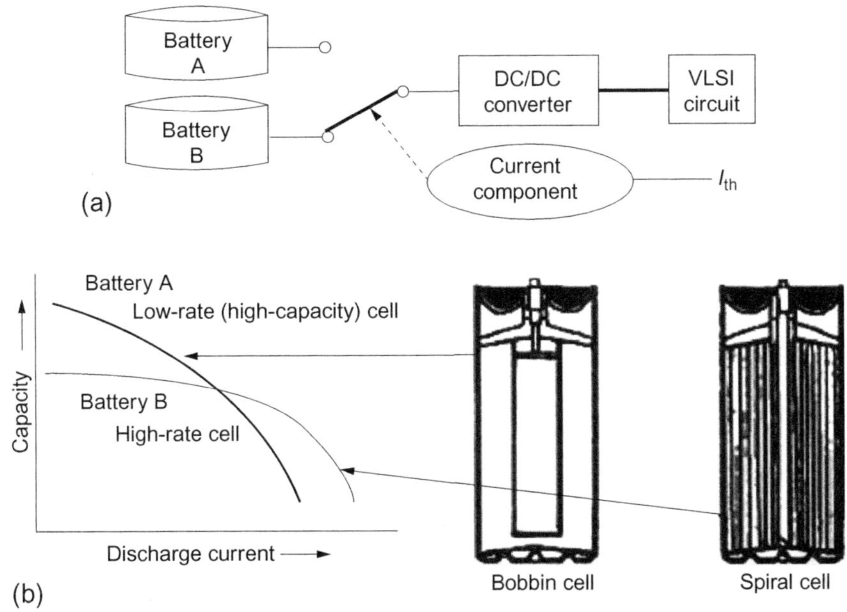

Fig. 4.35 An interleaved power supply system with two batteries with battery switching scheme for best runtime: (A) basic concept and (B) an example of implementation with a high-rate spiral cell and a low-rate bobbin cell.
Source: Rao, R., Vrudhala, S., Rakhmatov, D.N., 2003. Battery modeling for energy aware system design. IEEE Comput. 36 (12), 77–87, p. 86. Reproduced with the permission of IEEE, USA.

In this example, where an interleaved dual battery concept is used as in Fig. 4.35B, the working principle is simple. When the discharge current is lower than a threshold value, I_{th}, the system uses battery A as the energy source and vice versa with battery B. Battery A is low-rate high-capacity cell, while the other one is a high-rate low-capacity cell. In this process of switching based on suitable algorithms, a suitable battery model is taken into account for optimum discharge behavior (Wu et al., 2000). This kind of dynamic power management of battery-powered electronics can be used for battery life optimization as well (Benini et al., 2003). Energy-efficient communication is another application of these concepts with electrochemical battery models (Chiasserini and Rao, 2001).

4.8 Fast charging of batteries

With electric vehicles and portable products proliferating fast charging of rechargeable batteries have become a valuable topic for R & D. However, in fast charging, if the charger cannot enter into a cutoff mode, the battery pack is subjected to a significant overcharge and the life of the battery pack could be reduced, or the pack may even fail.

Table 4.5 Representative charging recommendations for different batteries.

	NiCd or NiMH	Sealed lead acid	Li-ion
Charging current (C)	1	1.5	1
Voltage per cell (V)	1.80	2.5	4.20±0.05
Charging time (h)	~3	~3	2.5–5.0
Method for optimum fast-charge termination point	See Table 4.6	Current cutoff	Typically a timer[a]
Back-up charge termination method	See Table 4.6	Timer	–
"Top off" rate	0.1 C	0.002 (trickle)	
Temperature range (°C)	10–40 (NiCd) 15–30 (NiMH)	0–30	0–40

[a] Depending on manufacturers recommendation.

While all major battery families can accept either a standard (16–24 h) or a fast (2–4 h) charge, most of the discussion here is limited to fast charging methods. Slower charging schemes tend to be found in simpler, price-sensitive applications, which do not need (or cannot afford) much beyond a simple charger and a low-battery indicator. The objective of fast charging a battery is to cram as much energy as it takes to bring the battery back to fully charged state in the shortest possible time without damaging the battery or permanently affecting its long-term performance.

Nickel-based chemistries prefer constant-current-type charging, while lead-acid and lithium cells prefer constant voltage (current limited) charging. For the constant-current cells (NiCd and NiMH), a 1 C charge rate will typically return more than 90% of the battery's usable discharge capacity within the first hour of charging. The constant-potential (CP) cells (lead-acid and lithium-ion) are a bit slower to reach the 90% mark but can generally be completely recharged within 5 h. See Table 4.5 for representative and typical charging recommendations for different chemistries. The most appropriate method should be selected in consultation with the manufacturer.

Fast charging has compelling benefits but places certain demands on the battery system. A properly performed fast charge, coordinated to the specifications of a battery rated for such charging, will deliver a long cycle life. The high charging rates involved, however, cause rapid electrochemical reactions within the cells of the battery. After the battery goes into overcharge, these reactions cause a sharp increase in internal cell pressure and temperature. Uncontrolled high-rate overcharge quickly causes irreversible battery damage. Thus, as the battery approaches full charge, the charging current must be reduced to a lower "top-off" level, or curtailed entirely.

4.8.1 Charge termination methods

If a rapid charge is applied to a battery pack, it is necessary to select a reliable method to terminate charging at the fully charged position. Two practical approaches for charge termination are *temperature termination method* and *voltage termination method*.

4.8.1.1 Voltage termination methods

Four commonly available voltage termination methods are maximum voltage (V_{max}), negative delta voltage ($-\Delta V$), zero slope, and inflection point (d^2V/dt^2). The maximum voltage method senses the increase in battery voltage as the battery approaches full charge. However, this is accurate only on a highly individualized basis. It is necessary to know the exact value at the voltage peak; otherwise the batteries may be over or undercharged.

Temperature compensation is also required because of the negative temperature coefficient of battery voltage. The maximum voltage will increase if the batteries are cold, causing an undercharge because the charging voltage will reach the maximum voltage trip point early. If the batteries are hot, the maximum voltage may never be reached and the batteries will be destroyed due to overheating. Therefore, the V_{max} method is generally not recommended for fast-charge rates.

The negative delta voltage is the most popular of the fast-charge termination schemes. It relies on the characteristic drop in cell voltage that occurs when a battery enters overcharge, as shown in Fig. 4.36. With most NiCd cells, the voltage drop is a very consistent indicator, and the $-\Delta V$ method is fine for charge rates up to 1 C.

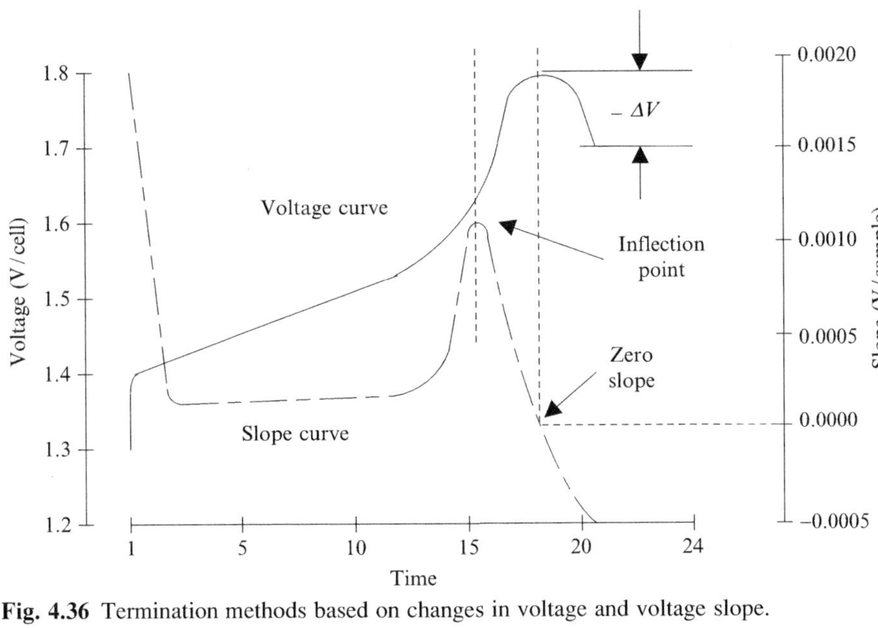

Fig. 4.36 Termination methods based on changes in voltage and voltage slope.

An inherent problem with this method is that the batteries must be driven into the overcharge region to cause the voltage decrease. Pressures and temperatures rise very rapidly at fast-charge rates beyond 1 C. In cyclic applications, the battery must be able to endure that continual abuse.

Another concern is that cells such as NiMH types do not always have the characteristic decrease in voltage, compared to NiCds as shown in Fig. 3.10. This creates a problem of forward compatibility when moving from NiCd to NiMH cells. Most manufacturers of NiMH cells do not advocate the $-\Delta V$ method of charge termination.

The zero slope method monitors the point where the slope of the battery voltage reaches zero. This method is reliable for rapid charge rates up to 4 C and is less susceptible to noise on the voltage sense lines. However, a few types of batteries such as the NiCd button cells may have a voltage slope that never quite reaches zero. Therefore, the zero slope method is better suited as a backup method.

In the inflection point (d^2V/dt^2) method, the system monitors the change in voltage over time and is the most sensitive indicator for preventing overcharge. The inflection point method relies on the changes in the voltage slope, shown in Fig. 4.36, which occurs during charge and is an excellent primary termination method for up to 4 C charge rates.

The change in the voltage slope is an extremely reliable and repeatable indication of charge. It does not rely on the decrease in voltage that may not always occur. Instead, this method looks for the flattening of the voltage profile as the battery reaches full charge. By monitoring the relative change in the steepness of the voltage slope, this method avoids having to use absolute numbers.

4.8.1.2 Temperature termination methods

Temperature is the main cause of failure in a rechargeable cell, so it makes sense to monitor the cell temperature to determine when to shut off charge to a battery. Three methods of charge termination, based on temperature, are common: maximum temperature cutoff (MTC), temperature difference (DT), and temperature slope (dT/dt). The MTC system is the easiest and cheapest to implement but is the least reliable. Using a bimetallic thermal switch or a positive temperature coefficient thermistor, a simple, low-cost circuit can shut down a charging current at an appropriate temperature.

The DT method measures the difference in ambient and cell temperatures to compensate for a cool environment. It requires monitoring two temperature sensors, one for the battery temperature and one for the ambient. This method may be unsuitable if the difference between the cell and ambient temperature is very large.

The DT method can become unreliable with a quickly changing ambient temperature unless an equal thermal mass is attached to the ambient sensor. This means that the DT method is suitable for a primary charge termination at lower charge rates up to $C/5$ if the ambient temperature is not going to change often. The DT method also provides an excellent backup charge termination scheme.

The dT/dt method, a more sophisticated temperature termination scheme, measures the change in temperature over time. This method uses the slope of the battery temperature and therefore is less dependent on changes in ambient temperatures or in large differences between ambient and battery temperatures. The dT/dt method works very well when accurately adjusted to a particular pack and with careful attention paid to the type and placement of the temperature sensor. This dT/dt method is suitable for charge rates up to 1 C and provides an excellent backup method.

4.8.2 NiCd and NiMH fast-charge methods

Nickel-based batteries such as NiCd and NiMH types are mature chemistries. Although it is not correct to consider the NiCd and NiMH electrochemistries or charging regimens as being interchangeable, they are similar enough that they can be discussed together.

There is no one best way to fast charge a NiCd or NiMH battery. Variables introduced by the allowable cost and size of the end application, the choice of charge termination method(s), and the specific battery vendor's recommendations will all influence the final choice of charging technique.

Fig. 4.37A shows the voltage, pressure, and temperature characteristics of a NiCd cell being charged at 1 C rate. Fig. 4.37B shows similar data for NiMH cell.

These curves illustrate the need for a reliable termination of the high-current portion of the charge cycle and assist in understanding the various fast-charge termination methods outlined in Table 4.5. For both electrochemistries, the ideal fast-charge termination point is at 100%–110% of returned charge. The charging current is then reduced to the top-off value for 1–2 h, to bring the cell into a state of a slight overcharge. This compensates for the inefficiencies of the charging process (e.g., heat generation). If the specific application will have the battery on standby for more than several weeks, or at high temperatures, the top-off charge is followed by a continuous, low-level "trickle" charge to counter the self-discharge characteristics of NiCd and NiMH cells.

Under certain conditions, particularly following intervals of storage, a NiMH battery may give an erroneous voltage peak as charging commences. For this reason, the charger should deliberately disable any voltage-based charge termination technique for the first 5 min of the charging interval. Table 4.6 summarizes the fast-charge termination methods for NiCd and NiMH cells.

4.8.3 Charging sealed lead-acid batteries

Unlike nickel-based batteries, SLA batteries are charged using a "CP" regimen. CP charging employs a voltage source with a deliberately imposed current limit (a current-limited voltage regulator). A significantly discharged battery undergoing CP charging will initially attempt to draw a high current from the charger. The current-limiting function of the CP regimen serves to keep the peak charging current within the battery's ratings.

Following the current-limited phase of the charging profile, an SLA battery exposed to a constant voltage will exhibit a tapering current profile as shown in

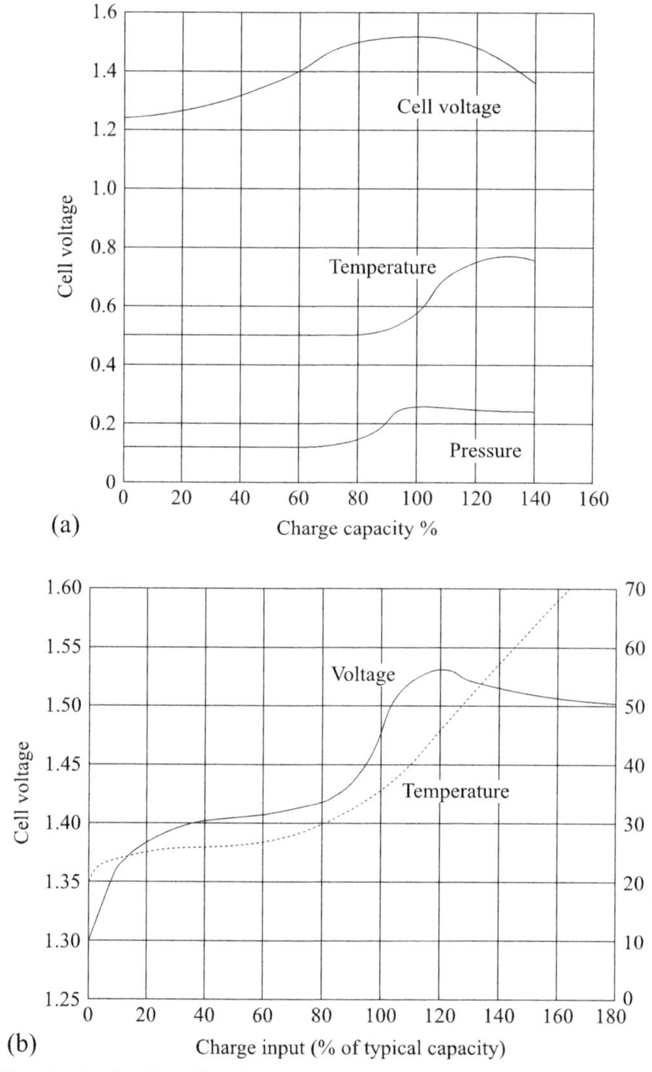

Fig. 4.37 Charging indications for nickel-based batteries at 1 C charge rate; (A) cell voltage, temperature, and pressure for a typical NiCd cell at 1 C rate charging and (B) NiMH voltage and temperature characteristics.

Fig. 4.38. When the returned charge reaches 110%–115% of rated capacity, allowing a dischargeable capacity of 100% of nominal, the charge cycle is complete.

The specifics of fast-charging SLA batteries are more vendor-dependent than those of NiCd or NiMH units. The information provided in Table 4.5 uses data from GS Battery (USA) Inc. The primary termination method, "current cutoff," looks at the absolute value of the average charging current flowing into the battery.

Table 4.6 Fast-charge termination methods for NiCd and NiMH batteries.

Charging technique	Description
Negative $\Delta V(-\Delta V)$	Looks for the downward slope in cell voltage which a cell exhibits (≫30–50 mV for NiCd, 5–15 mV for NiMH) upon entering overcharge. Very common in NiCd applications due to its simplicity and reliability
Zero ΔV	Waits for the time when the voltage of the cell under charge stops rising, and is "at the top of the curve" prior to the downslope seen in overcharge. Sometimes preferred over-ΔV for NiMH, due to NiMH's relatively small downward voltage slope
Voltage slope (dV/dt)	Looks for an increasing slope in cell voltage (positive dV/dt) which occurs somewhat before the cell reaches 100% returned charge (prior to the zero ΔV point)
Inflection point cutoff (d^2V/dt^2, IPCO)	As a cell approaches full charge, the rate of its voltage rise begins to level off. This method looks for a zero or, more commonly, slightly negative value of the second derivative of cell voltage with respect to time
Absolute temperature cutoff (TCO)	Uses the cell's case temperature (which will undergo a rapid rise as the cell enters high-rate overcharge) to determine when to terminate high-rate charging. A good backup method, but too susceptible to variations under ambient temperature conditions to make a reliable primary cutoff technique
Incremental temperature cutoff (ΔTCO)	Uses a specified increase in the cell's case temperature, relative to the ambient temperature, to determine when to terminate high-rate charging. A popular, relatively inexpensive, and reliable cutoff method
Delta temperature/delta time ($\Delta T/\Delta t$)	Uses the rate of increase of a cell's case temperature to determine the point at which to terminate the high-rate charge. This technique is inexpensive and reliable once the cell and its housing have been properly characterized

When that current drops below 0.01 C, the battery is fully charged. If it will be in standby for a month or more, a trickle current of 0.002 C should be maintained. The backup termination method, according to the vendor's recommendations, should be a 180-min time-out on the charging cycle (Schwartz, 1995).

To satisfy more stringent charge control recommendations, where the battery temperature, voltage, and current need be sampled, many dedicated charge controller ICs are available in the market. Bq2031 lead-acid fast-charge IC from Texas/Benchmarq and UC3906 (SLA charger) from Texas/Unitrode Integrated Circuits are examples. The UC3906 battery charge controller contains all of the necessary circuitry to optimally control the charge and hold cycle for SLA batteries. These integrated circuits

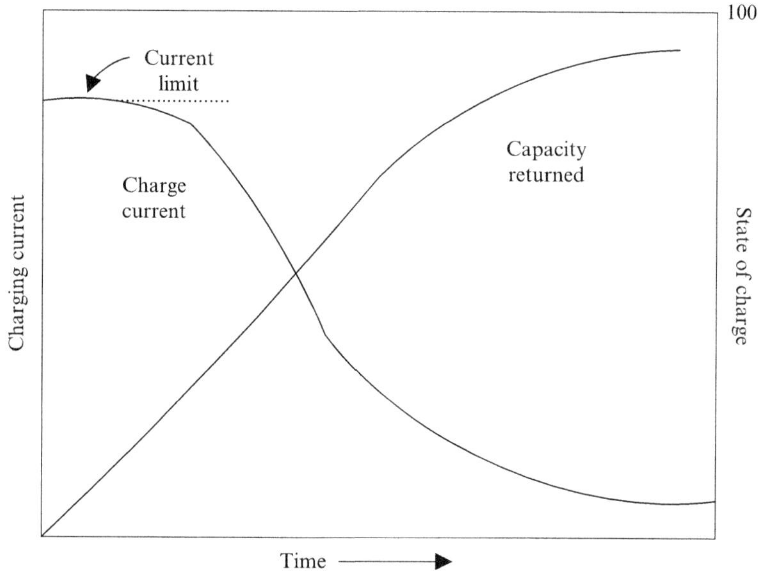

Fig. 4.38 Typical current and capacity returned *Vs* charge time for CP charging in sealed lead-acid batteries.

monitor and control both the output voltage and current of the charger through three separate charge states; a high-current bulk-charge state, a controlled overcharge, and a precision float-charge, or standby state. Sacarisen and Parvereshi (1995) and Application Note U-104 (Unitrode Inc, 1997) provide details about SLA charge control using UC3906.

4.8.4 Li-ion chargers

Lithium-Ion batteries require a constant potential charging regimen, very similar to that used for lead-acid batteries. Because the voltage of Li-ion cell under constant-current charge could continue to rise to the point of cell destruction, absolute voltage limits are required (Bentley and Heacock, 1996). Typical recommendations for Li-ion fast charge are indicated in Table 4.7. As with lead-acid batteries, a Li-ion cell under charge will reduce its current draw as it approaches full charge. Fig. 4.39 indicates this condition. However, if the charger has poor regulation on the output voltage and creates a lower output voltage, it will decrease the capacity as depicted in Fig. 4.39B. The area between the graphs of the current indicates the lost charge.

If the cell vendor's recommendation for charging voltage (generally $4.20\,V \pm 50\,mV$ at 23°C) is followed, the cells will be able to completely recharge from any "normal" level of discharge within 5 h. At the end of that time, the charging voltage should be removed. Trickle current is not recommended. If the voltage on a

Table 4.7 Comparison of recharge requirements of different chemistries.

Parameter	SLA	NiCd	NiMH	Li-ion
Standard charge				
Current (C rate)	0.25	0.1	0.1	0.1
Voltage per cell (V)	2.27	1.5	1.5	$4.1 \pm 50\,mV$
Time (h)	24	16	16	16
Temperature range (°C)	0–45	5–40	5–40	5–40
Termination	None	None	Timer	None
Fast charge				
Current (C rate)	1.5	1	1	1
Voltage per cell (V)	2.45	1.5	1.5	$4.1 \pm 50\,mV$
Time (h)	1.5	3	3	2.5
Temperature range (°C)	0–30	15–40	15–40	10–40
Primary termination method	I_{min}[a], ΔTCO	dT/dt, $-\Delta V$	Zero dv/dt, $-\Delta V$, d^2V/dt^2, ΔTCO	I_{min}+timer, dT/dt, dT/dt
Secondary termination method	Timer, ΔTCO	TCO, timer	TCO, timer	TCO, timer

[a] Minimum current termination threshold.
Adapted from Israelsohn, J., 2001. Battery management included, EDN, 65-74.

Li-ion cell falls below 1.0 V (or to an abnormal level), recharging of that cell should not be attempted. If the voltage is between 1.0 V and the manufacturer's nominal minimum voltage (typically 2.5–2.7 V), it may be possible to salvage the cell by charging it with a 0.1 C current limit until the voltage across the cell reaches the nominal minimum, followed by a fast charge.

Due to special characteristics of Li-ion batteries, most Li-ion manufacturers incorporate custom protection circuits into their battery packs to monitor the voltage across each cell within the battery and to provide protection against overcharge, battery reversal, and other major faults. These circuits are not to be confused with charging circuits. For example, MC 33347 protection circuit is such an IC from Motorola (Alberkrack, 1996).

4.8.5 Portable chargers and comparison of recharge requirements for different chemistries

With the availability of charge controllers, protectors, and other battery management ICs today, a designer can find many commercial ICs and supporting reference designs (Israelsohn, 2001). Choosing the best chemistry requires a deliberative comparison between battery attributes and the required power source specifications in general. As summarized above, the four common chemistries require different recharge

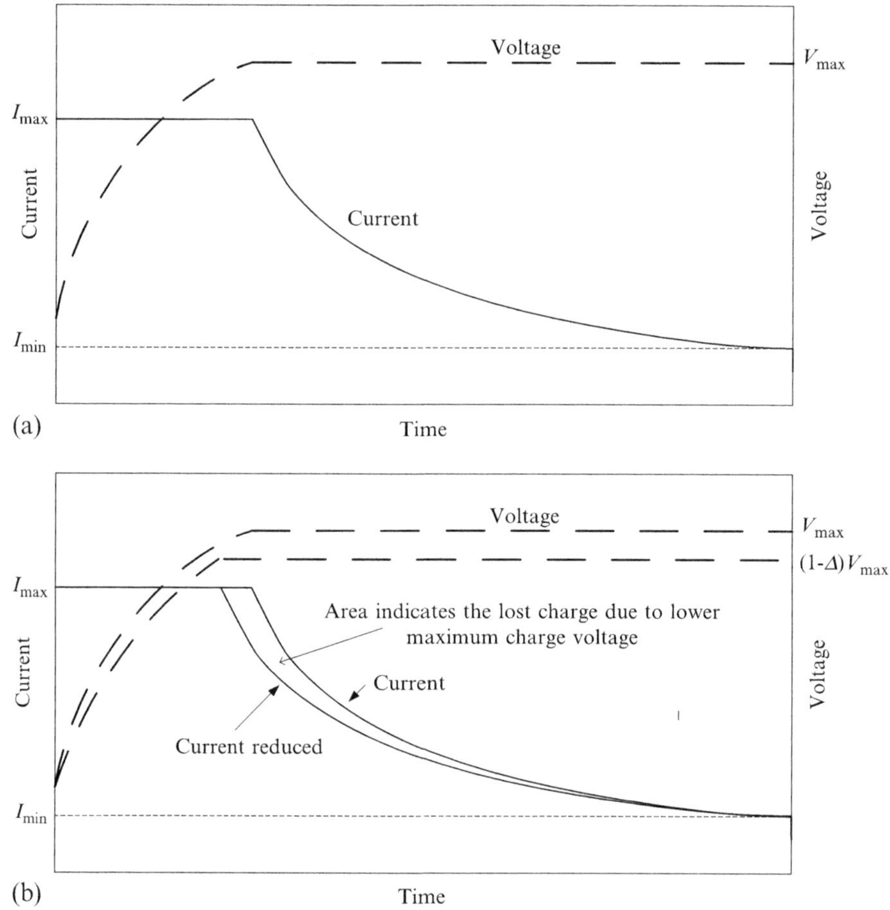

Fig. 4.39 Voltage-regulated charging profile for Li batteries: (A) current-limited voltage-regulated charging profile and (B) charge dependence on voltage regulation.
Adapted from Bentley, W.F., Heacock, D.K., 1996. Battery management considerations for multichemistry systems. In: Proceedings of the 11th Annual Battery Conference, Long Beach, CA.

algorithms and give different indications when they have completed the discharge cycle. Table 4.7 provides a summary, as adapted from the source (Israelsohn, 2001).

Most secondary batteries tolerate trickle charging for long periods. Simplest charging strategy, therefore, uses a simple linear regulator IC or a pulse width modulator IC in conjunction with a series transistor and a currents sense resistor. Circuits such as these are available from many vendors and with a range of auxiliary features for charging single cells or nickel chemistry stacks (Israelsohn, 2001). They sometimes use an adaptive method that adjusts the charger's behavior according to the battery's SOC.

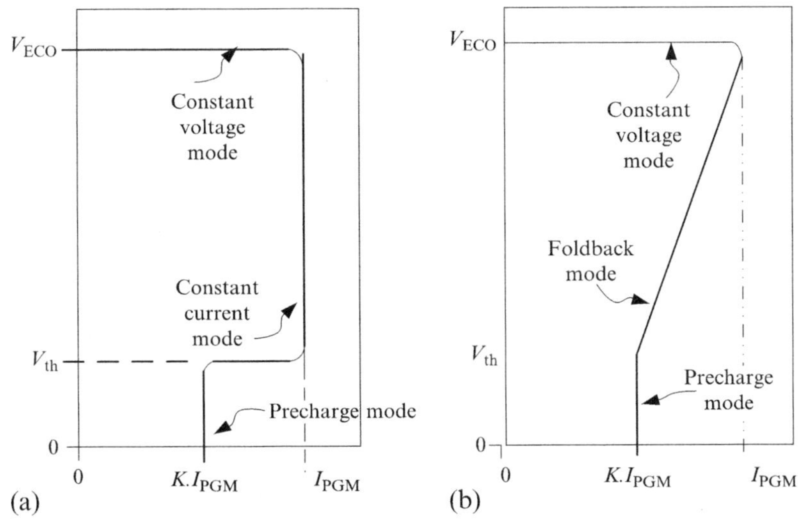

Fig. 4.40 Charger voltage-current algorithms: (A) constant-current/constant-voltage case and (B) current foldback case.
Source: Israelsohn, J., 2001. Battery management included, EDN, 65-74.

A charger of this type may start by testing the battery for deep discharge, which is determined by comparing the battery's terminal voltage with a threshold value. If the battery is deeply discharged, the charger enters a precharge mode in which it limits its current to some fraction, K, of resistor-programmed maximum charge current, I_{PGM}. Fig. 4.40 gives representative options of charging algorithms. Fig. 4.40A is a constant-current/constant-voltage charging algorithm, while Fig. 4.40B is a current foldback mode.

4.8.6 EOD determination

Determination of the point at which a battery has delivered all of its usefully dischargeable energy is important to the longevity of the cells which form that battery. Discharging a single cell too far will often cause irreversible physical damage inside the cell.

If multiple cells are placed in series, unavoidable imbalances in their capacities can cause the phenomenon known as "cell reversal," in which the higher capacity cells force a backward current through the lowest capacity cell. Knowing the EOD point provides a "zero capacity" reference for coulometric gas gaging.

The actual determination of the EOD point is typically done by monitoring the cell voltage. For the most accurate determination of EOD when the load is varying, correction factors for load current and the battery's SOC should be applied, especially to SLA and Li-ion batteries. The essentially flat discharge profiles of NiCd and NiMH make these corrections a matter of user discretion for most load profiles. Table 4.8 presents voltages commonly used to indicate the EOD point for the four battery types.

Table 4.8 Typical end of discharge voltages.

Cell type	EODV (V)	Comments
Lead acid	1.35–1.9 (1.8 V typical)	Depend on loading, state of charge, cell construction, and manufacturer
NiCd	0.9	Essentially constant
NiMH	0.9	Essentially constant within recommended range of discharge rates
Li-ion	2.50-2.70	Depend on manufacturer, loading, and state of charge

4.8.7 Gas gaging

The "gas gaging or fuel gaging" concept discussed here does not refer to the gases which may be evolved by the battery reactions, but rather to the concept of the battery as a fuel tank powering the product. Gas gaging, therefore, involves the real-time determination of a battery's SOC, relative to the battery's nominal capacity when fully charged.

It is possible to make an inexpensive and moderately useful SOC measurement from a simple voltage reading if the battery being used has a sloping voltage profile. Hence, Li-ion batteries and, to a lesser extent, SLA batteries, should be amenable to such an approach. In practice, the results are less than optimal: cell voltages are dependent on loading, internal impedance, cell temperature, and other variables. This reduces the attractiveness of the simple voltage method of gas gaging; the fact that it is not suitable for NiCd or NiMH batteries, due to their essentially flat voltage profile, makes it commercially untenable. A clever and effective alternative is the "coulometric" method (or earlier discussions referred to as coulomb counting).

Coulometric gas gaging, as its name implies, measures the actual charge ($\int I dt$) going into and out of the battery. By integrating the difference of current in and current out, it is possible to determine the charge status of the battery at any given time. There are of course real-world details which must be observed in the actual implementation of such a gas gage; some of the most important of these are:

- **(a)** It is necessary to have an accurate starting point for the integrator, corresponding to a known SOC in the battery. This is often resolved by zeroing the integrator when the battery reaches its EODV.
- **(b)** Compensation for the temperature. The actual capacity of lead-acid batteries increases with temperature, and that of nickel-based batteries decreases as battery temperature rises.
- **(c)** Appropriate conversion factors should be applied for the particular charge regimen and discharge profile used. Under conditions of highly variable battery loading, dynamic compensation may be advisable.

4.8.8 Battery health in a practical viewpoint

SOC is a widely used concept to represent the remaining capacity of a battery in a charge/discharge cycle. However, due to cycle-aging phenomenon, using SOC alone may result in large errors because the full charge capacity (FCC) may be significantly less than the design capacity (DC) that denotes the FCC of a new battery.

Considering the aging effect, SOH of a battery is defined as the FCC of cycle-aged battery to its DC. Analytical modeling of batteries can be used to get theoretically accurate values for SOC and SOH (Rong and Pedram, 2006b). With battery health defined as a battery's actual capacity relative to its rated capacity, the health of the battery can be determined and maintained in three steps:

(i) Discharge the battery to the EOD point, preferably into a known load.
(ii) Execute a complete charge cycle, while gas gaging the battery.
(iii) Compare the battery's measured capacity to its rated capacity.

This sequence will simultaneously "condition" the battery (which means to overcome the so-called "memory effect" of the capacity of NiCd batteries.) and will indicate the capacity of the battery after conditioning. The information so obtained can be used to ascertain whether the battery is in good shape, or is approaching the end of its useful life. Online SOC estimation with a monitoring system is discussed by Kutluay et al. (2005).

4.9 Battery communication and related standards

In the early 1990s, power management and battery management were rarely used in portable products. More advanced processors with very high power consumption and relatively high load currents of tens of amperes were introduced around the mid-1990s, which placed a burden on the available rechargeable battery chemistries due to rapid increase of discharge current. To meet system runtimes expected by the users, innovative power management concepts had to be employed. For this reason, around mid-1990s new industry standards were proposed to standardize the battery and power management subsystems within portable products, where concepts such as Advanced Configuration and Power Interface by Intel were utilized. Following were the standards proposed:

(i) System Management Bus (SMBus) Specification.
(ii) Smart Battery Data Specification.
(iii) Smart Battery Charger Specification.
(iv) Smart Battery Selector Specification.

The above specifications form the smart battery systems (SBS) specification. This specification presents a solution for many of the issues related to batteries used in portable equipment such as laptop computer systems, cellular telephones, or video cameras. Fundamental to the SBS system is the concept that the battery contains all of the necessary components to determine the battery's SOC, predict time to full and time to empty, specify the charging voltage and current, and determine when the battery is fully charged to fully discharged.

Concept of an SBS is shown in Fig. 4.41A. Fig. 4.41B indicates details. It consists of an AC-DC converter (unregulated), power switch, system power supply, smart battery charger, smart battery selector, all of which communicate with the system host

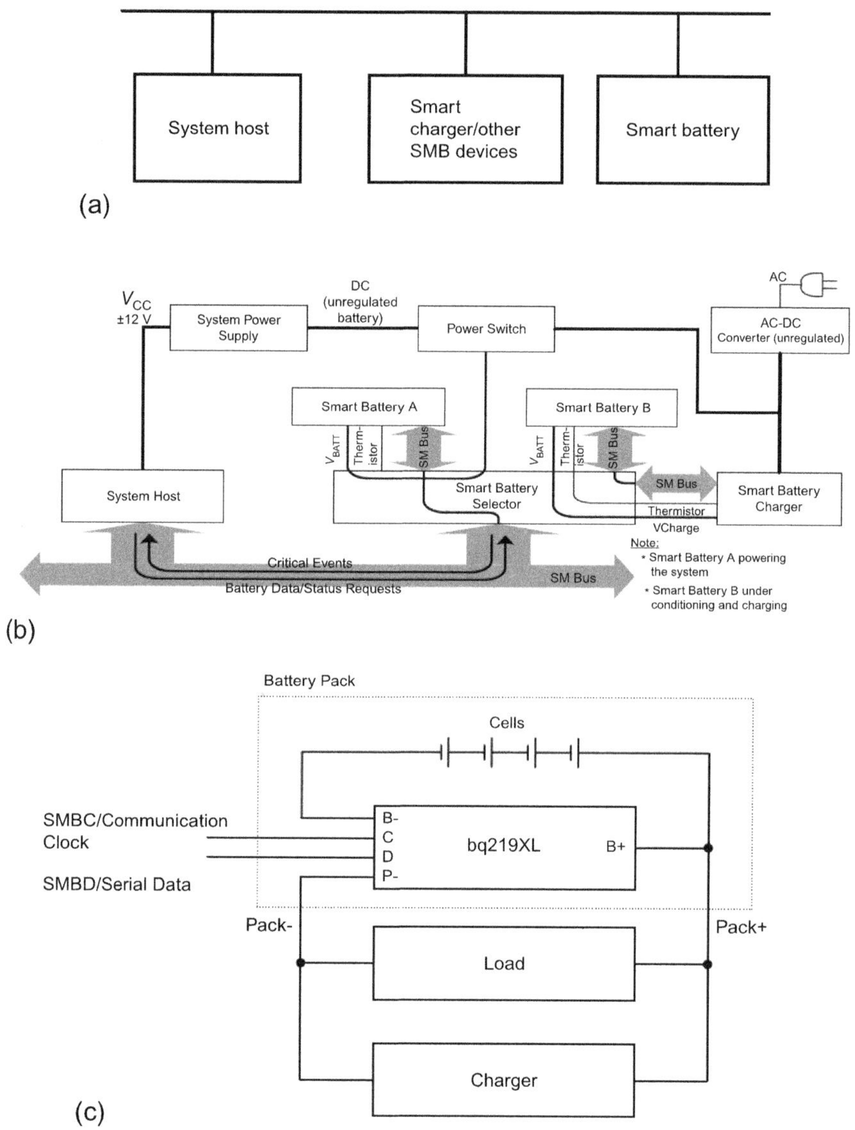

Fig. 4.41 Smart battery system and a SMBus module: (A) concept; (B) a typical multiple smart battery system; and (C) typical control IC (bq 219XL) and its connections.

and the system elements themselves via SMBus. In this case, smart battery A powers the system while battery B is getting conditioned and/or charged.

The System Management Bus (SMBus) is a two-wire interface through which simple power-related chips can communicate with the rest of the system. It uses I^2C as its backbone (Phillips Semiconductors, 2003). A system using SMB passes messages to and from devices instead of tripping individual control lines. Removing the individual control lines reduces pin count. Accepting messages ensures future expandability. With SMBus, a device can provide manufacturer information, tell the system what its mode/part number is, save its state for a suspended event, report different types of errors, accept control parameters, and return its status. The SMBus may share the same host device and physical bus, provided that an appropriate electrical bridge is provided between the respective devices. For details, articles by Benchmarq Microelectronics Inc et al. (1996a, b) and Heacock (1998) are suggested.

4.10 Battery safety

With the introduction of Li-based batteries, with high gravimetric energy density, the safety of the battery-powered portable products has become a concern during the last decade. In the early 2000s, some laptop fires scared the public and promoted widespread recalls of Li-ion cells. During the year 2013, new aircrafts such as the Boeing Dreamliner was grounded due to battery pack issues. Since these events, individual companies have taken many steps to ensure that such disasters do not recur, and the industry has joined forces to unify the standards. By 2004, IEEE has published two standards relevant to the design and manufacture of portable systems: The IEEE 1625 Standard for Rechargeable Batteries for Portable Computing, and a related IEEE 1725 for Rechargeable Batteries for Cellular Phones.

The standard covers design approaches that ensure reliable operation and that minimize the occurrence of faults leading to hazards in portable computing devices and other rechargeable battery-powered systems. The standard guides the system and subsystem design through five major areas:

- systems integration,
- cell,
- pack,
- host device, and
- total system reliability.

Also covered are the critical operational parameters and how they change with time and environment, the effect of extremes in temperature, and the management of component failure. Fig. 4.42 indicates the case of a portable computing device broken into subsystems and within the environment.

In general portable equipment battery packs usually contain an analog front end (AFE) which is monitoring the conditions of the cells, and it usually provides the first level of protection to the pack. A battery management unit BMU (similar to the example in Fig. 4.42C) usually communicates with the cell stack to monitor and communicate with the host system via SMBus. Referring to Fig. 4.42B, in the 1625 standard, it proposes to move the second level of protection into the AFE, which communicates directly with the BMU (Elder, 2004). An industry-wide methodology such as design failure modes and effects analysis (DFMEA), which can be used to highlight and prioritize the possible roots of faults and hazards, is to be used with the 1625 conditions in

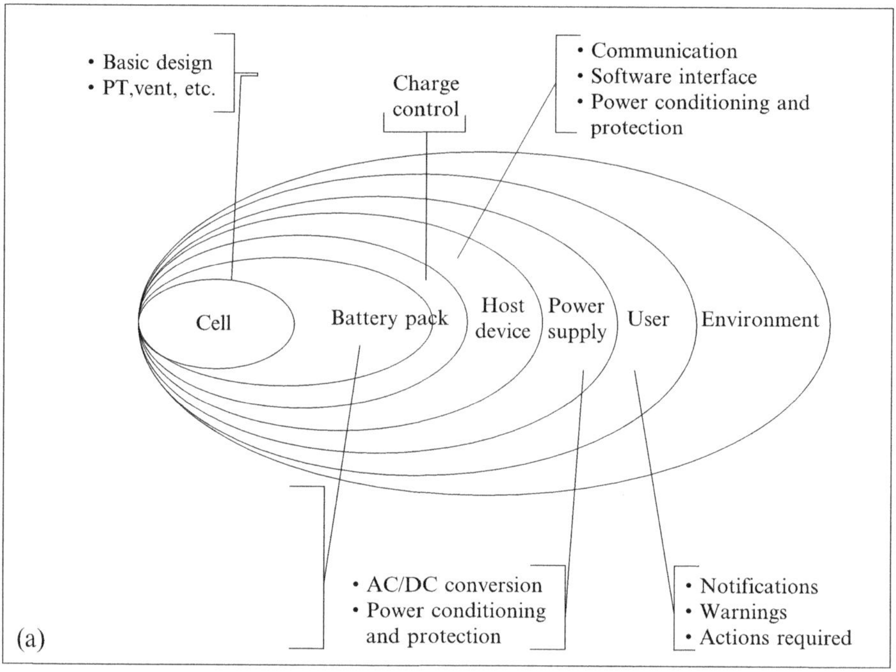

• Basic design
• PT,vent, etc.
Charge control
• Communication
• Software interface
• Power conditioning and protection
Cell
Battery pack
Host device
Power supply
User
Environment
• AC/DC conversion
• Power conditioning and protection
• Notifications
• Warnings
• Actions required
(a)

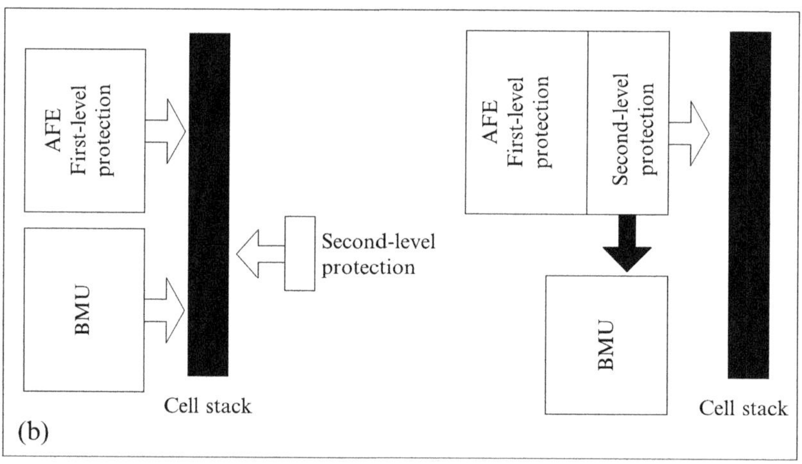

AFE First-level protection
BMU
Second-level protection
Cell stack
AFE First-level protection
Second-level protection
BMU
Cell stack
(b)

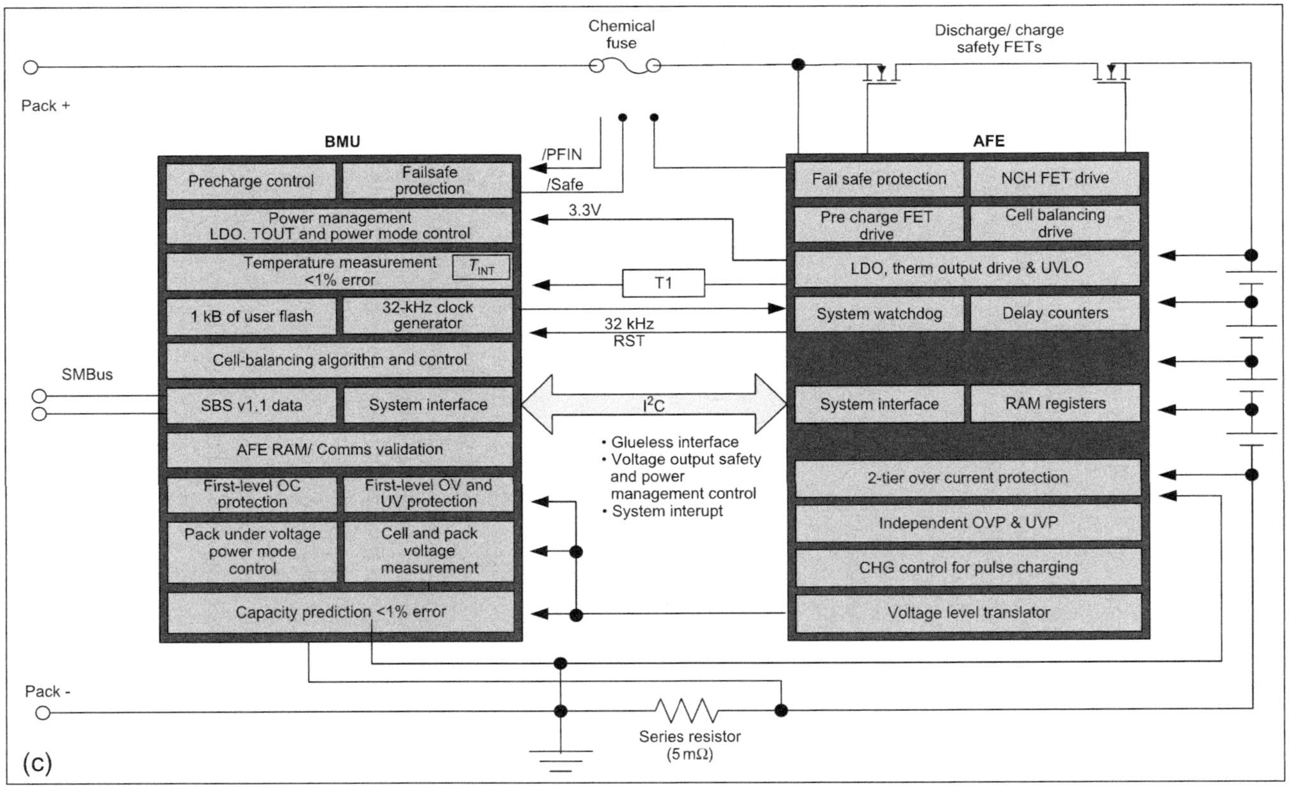

Fig. 4.42 A portable computing device and its environment within the case of IEEE 1625: (A) portable computing device broken out into subsystems; (B) proposed level of safety compared with older battery packs; and (C) implementation of IEEE-1625 concepts.

order to get the benefits proposed in the standard. Fig. 4.42C indicates the block diagram of a battery pack compliant with the SBS 1.1 with 4 series Li-ion cells (Bernardi and Carpenter, 1995). This change in architecture helps clarify the DFMEA analysis. Compared to the older architecture, in the new architecture (Fig. 4.42B), BMU and AFE monitor each other as well as the cell stack. Theoretically, this allows the second-level protection to be integrated with the first-level protection device.

As in Fig. 4.42C, the new architecture consists of a BMU, an AFE-integrated circuit, current sense resistor, safety FET, and a chemical fuse. BMU could be a 3.3-V microcontroller with measurement capabilities and a programmable flash. The AFE provides a high-voltage interface to the cell stack for voltage measurements, along with cell-balancing control, over current protection, and a low dropout regulator to power the BMU. Based on possible problem levels, for discrete components, BMU and AFE interaction DFMEA tables can be prepared similar to the case in Table 4.9 and evaluate each item and score according to the following criteria:

- Severity of the fault (SEV).
- Probability of occurrence (OCC).
- Difficulty of detection (DET).

Typically, evaluation and scoring are based on experience and statistical reliability data. However, these data may not be always available and guidance in 1626 standard may be helpful. For more details, article by Elder (2004) is suggested.

4.11 Future

With consumer electronics, EVs, and portable product markets are growing, newer battery technologies may enter the market. One recent indication in the battery market research reports (www.sdle.co.il) is that high-voltage Li-ion chemistries with better anode and cathode materials will come into the market with higher gravimetric energy densities. In addition, new technologies to combine batteries and supercapacitors may be available. At the time this was finalized newer supercapacitor families have entered the commercial domain, with energy densities closer to lead acid chemistry and the predictions are that they will reach the energy density of Li-ion batteries in the near future.

Major application growth areas for batteries are the transportation systems. Battery packs can challenge the designer because they are no longer simple configuration of cells, but could contain so much safety features, intelligence, energy-aware models and selective batteries feeding the host product, and serial data communication and even recycling suggestions.

In all these situations, a simple and useful concept that should be remembered by the designers is that "batteries are like human beings," and they need care and useful intelligence.

Table 4.9 An example of a DFMEA table for discrete components.

Problem	Problem level	SEV	Issue	OCC	Protection features	DET	APN
BMU to AFE I^2C latch up	Low safety	6	BMU voltage measurements, RAM verification, and AFE control nonfunctional	2	BMU will detect and count I^2C failures and optionally blow the fuse. Fail counter needs to be able to be periodically cleared based on normal operating conditions. First-level current protection still available	1	12
AFE to BMU VCC < VCC (min)	See Table 4.3						
AFE RST output latched high	Severe customer experience	4	BMU is held in reset	1	BMU no longer functions, so the WDI input will stop and the AFE turn off the FETs and optionally blow the fuse	1	4
AFE RST output latched low	Severe customer experience	5	BMU will never be reset on POR or AFE watchdog fault	2	BMU will not be reset on watchdog fault, but watchdog will still turn off FETs and optionally blow the fuse	5	50
AFE TOUT output latched high	Low customer experience	1	Thermistor will always be powered	1	AFE will consume extra current, but there is no safety issue	7	7
AFE TOUT output latched low	Severe safety	9	Thermistor will never be powered	1	BMU will measure out-of-range temperature, which will cause the fuse to blow. All current and voltage protection methods are still fully functional	1	9

Continued

Table 4.9 Continued

Problem	Problem level	SEV	Issue	OCC	Protection features	DET	APN
BMU to AFE CLK < CLK (min)	Severe customer experience	5	AFE will not function correctly	1	The AFE watchdog will timeout, causing a BMU reset, the FETs to turn off, and optionally blow the fuse	2	10
BMU sense of CELL latched high	Moderate safety	7	BMU voltage measurements will be 0 V	2	First-level voltage protection is not valid, but second-level overvoltage is a completely separate circuit; so, it is fully operational. All current- and temperature-based protections are fully functional. Additional measurement validation checks can allow the BMU to blow the fuse	2	28
BMU sense of CELL latched low	Severe customer experience	5	BMU voltage measurements will be full scale (high)	2	The BMU will read all cells at overvoltage; so, pack is not able to be charged but can be discharged. Current and temperature protections still functioning. Additional measurement validation checks can allow the BMU to blow the fuse	2	20
AFE XALERT latched high	No issues	1	XALERT will never activate	2	BMU polls STATUS for updates and checks XALERT, so no issues	2	4
AFE XALERT latched low	Low customer experience	2	XALERT will always be activated	2	BMU will always try to check status and clear XALERT, so will consume more power. All protection functions are functioning	2	8

Source:Elder, G., 2004. IEEE 1625 helps promote safety and reliability. Power Electron. Technol., 34–43. Reproduced with permission from Power Electronics Technology Magazine.

References

Agrawal, V., Uthaichana, K., DeCarlo, R.A., Tsoukalas, L.H., 2010. Development and validation of a battery model useful for discharging and charging power control and lifetime estimation. IEEE Trans. Energy Conv. 25 (3), 821–835.

Alberkrack, J., 1996. A programmable in-pack rechargeable lithium cell protection circuit. In: HFPC Conference Proceedings, pp. 230–237.

Albertus, P., Couts, J., Srinivasan, V., Newman, J., 2008. A combined model for determining capacity usage and battery size for hybrid and plug-in hybrid vehicles. J. Power Sources 183, 771–782.

Belt, J., Utgikar, V., Bloom, I., 2011. Calendar and PHEV cycle life aging of high energy, li-ion cells containing blended spinel and layered-oxide cathodes. J. Power Sources 196, 10213–10221.

Benchmarq Microelectronics Inc, Duracell Inc, Energizer Power Systems, Intel Corporation, et al., 1996a. Smart Battery System Specifications—Smart Battery Charger Specifications. Revision 1.0,.

Benchmarq Microelectronics Inc, Duracell Inc, Energizer Power Systems, Intel Corporation, et al., 1996b. Smart Battery System Specifications—Smart Battery Selector Specifications. Revision 1.0,.

Benini, L., Bruni, D., Macii, A., Macii, E., Poncino, M., 2003. Discharge current steering for battery life optimization. IEEE Trans. Comput. 92 (8), 985–995.

Bentley, W.F., Heacock, D.K., 1996. Battery management considerations for multichemistry systems. In: Proceedings of the 11th Annual Battery Conference, Long Beach, CA.

Bernardi, D.M., Carpenter, M.K., 1995. A mathematical model of the oxygen recombination lead acid cell. J. Electrochem. Soc. 142 (8), 2631–2642.

Bloom, I., Cole, B.W., Sohn, J.J., Jones, S.A., Polzin, E.G., Battaglia, V.S., Henriksen, G.L., Motloch, C., Richardson, R., Unkelhaeuser, T., Ingersoll, D., Case, H.L., 2001. An accelerated calendar and cycle life study of Li-ion cells. J. Power Sources 101, 238–247.

Bloom, I., Jones, S.A., Polzin, E.G., Battaglia, V.S., Henriksen, G.L., Motloch, C.G., Wright, R. B., Jungst, R.G., Case, H.L., Doughty, D.H., 2002. Mechanisms of impedance rise in high-power, lithium-ion cells. J. Power Sources 111, 152–159.

Bloom, I., Jansen, A.N., Abraham, D.P., Knuth, J., Jones, S.A., Battaglia, V.S., Henriksen, G.L., 2005. Differential voltage analyses of high-power, lithium-ion cells: 1. Technique and application. J. Power Sources 139, 295–303.

Bloom, I., Potter, B.G., Johnson, C.S., Gering, K.L., Christophersen, J.P., 2006. Effect of cathode composition on impedance rise in high-power lithium-ion cells: long-term aging results. J. Power Sources 155, 415–419.

Bloom, I., Walker, L.K., Basco, J.K., Abraham, D.P., Christophersen, J.P., Ho, C.D., 2010. Differential voltage analyses of high-power lithium-ion cells. 4. Cells containing NMC. J. Power Sources 195, 877–882.

Buller, S., Thele, M., Karden, E., Doncker, R.W., 2003. Impedance based non-linear dynamic battery modeling for automotive applications. J. Power Sources 113, 422–430.

Buller, S., Thele, M., Doncker, R.W., Karden, E., 2005. Impedance based simulation models of supercapacitors and Li-ion batteries for power electronic applications. IEEE Trans. Ind. Appl. 41 (3), 742–747.

Bundy, K., Karlsson, M., Lindbergh, G., Lundqvist, A., 1998. An electrochemical impedance spectroscopy method for prediction of the state of the charge of a nickel-metal hydride battery at open circuit and during discharge. J. Power Sources 72, 118–125.

Chen, M., Rincon-Mora, G.A., 2006. Accurate electrical battery model capable of predicting run time and I-V performance. IEEE Trans. Energy Conv. 21 (2), 504–511.

Chiasserini, C.B., Rao, R.R., 2001. Energy efficient battery management. IEEE J. Selected Areas Commun. 19 (7), 1235–1245.

Department of Energy, 2003. Freedom CAR Battery Test Manual for Power-Assist Hybrid Electric Vehicles. DOE/ID-11069, 14 pages.

Elder, G., 2004. IEEE 1625 helps promote safety and reliability. Power Electron. Technol., 34–43.

Fang, k., Mu, D., Chen, S., Wu, B., Wu, F., 2012. A prediction model based on artificial neural network for surface temperature simulation of nickel metal hydride battery during charging. J. Power Sources 208, 378–382.

Forgez, C., Do, D.V., Friedrich, G., Morcrette, M., Delacourt, C., 2010. Thermal modeling of a cylindrical LiFePO4/graphite lithium-ion battery. J. Power Sources 195 (9), 2961–2968.

Gao, J., Manthiram, A., 2009. Eliminating the irreversible capacity loss of high capacity layered Li $[Li_{0.2}Mn_{0.54}Ni_{0.13}Co_{0.13}]O_2$ cathode by blending with other lithium insertion hosts. J. Power Sources 191, 644–647.

Gao, L., Liu, S., Dougal, R.A., 2002. Dynamic lithium-ion battery model for system simulation. IEEE Trans. Compon. Packag. Technol. 25 (3), 495–505.

Gholizadeh, M., Salmasi, F.R., 2014. Estimation of state of charge, unknown nonlinearities, and state of health of a Li-ion battery based on a comprehensive unobservable model. IEEE Trans. Ind. Electron. 61 (3), 1335–1344.

Goebel, K., Saha, B., Saxena, A., Celaya, J.R., Christopherson, J.P., 2008. Prognostics in battery health management. IEEE I&M Mag. 11 (4), 33–40.

Gould, C.R., Bingham, C.M., Stone, A., Bentley, P., 2009. New battery model and state-of-health determination through subspace parameter estimation and state observer techniques. IEEE Trans. Veh. Technol. 58 (8), 3905–3916.

Greenleaf, M., Li, H., Zheng, J.P., 2013. Modeling of LIXFEPO4cathode Li-ion batteries using linear electrical circuit model. IEEE Trans. Sustain. Energy 4 (4), 1065–1070.

Grimnes, S., Martinsen, O.G., 2008. Bioimpedance and Bioelectricity Basics. Elsevier, Academic Press, Amsterdam. 471 pages.

Gu, W.B., Wang, C.Y., Liaw, B.Y., 1995. Numerical modelling of coupled electrochemical and transport processes in lead-acid batteries. J. Electrochem. Soc. 144 (6), 2053–2061.

Gu, W.B., Wang, C.Y., Li, S.M., Geng, M.M., Liaw, B.Y., 1999. Modeling discharge and charge characteristics of nickel-metal hydride batteries. Electrochim. Acta 44, 4525–4541.

Heacock, D., 1998. Enabling smart batteries for portable devices. In: Proceedings of Global Forum on Mobile Handsets (London).

Hsieh, G.-C., Chen, L.-R., Huang, K.-S., 2001. Fuzzy-controlled Li-ion battery charge system with active state-of-charge controller. IEEE Trans. Ind. Electron. 48 (3), 585–593.

Huet, F., 1998. A review of impeadance measurements for determination of the state-of-charge or state-of-health of secondary batteries. J. Power Sources 70, 59–69.

Husseni, A., Kutkut, N., Batarseh, I., 2011. A hysteresis model for a lithium battery cell with improved transient response. In: Proceedings of 26th IEEE APEC 2011 Conference, pp. 1790–1794.

Israelsohn, J., 2001. Battery management included. EDN, 65–74.

Jossen, A., 2006. Fundamentals of battery dynamics. J. Power Sources 154, 530–538.

Karden, E., Buller, S., Doncker, R.W., 2000. A method for measurement and interpretation of impedance spectra for industrial batteries. J. Power Sources 85, 72–78.

Karfthoefer, B., 2005. Measure battery capacity precisely in medical design. Power Electron. Technol., 30–38.

Kassem, M., Delacourt, C., 2013. Postmortem analysis of calendar-aged graphite/LiFePO4 cells. J. Power Sources 235, 159–171.

Kassem, M., Bernard, J., Revel, R., Pélissier, S., Duclaud, F., Delacourt, C., 2012. Calendar aging of a graphite/LiFePO4 cell. J. Power Sources 208, 296–305.

Khun, E., Forgez, C., Lagonotte, P., Friedrich, G., 2006. Modeling NiMH battery using Cauer and Foster structures. J. Power Sources 158, 1490–1497.

Kularatna, N., 2008. Digital and Analogue Instrumentation. IET, London, pp. 113–162. Chapter 4.

Kutluay, K., Cadirici, Y., Ozkazanc, Y.S., Cadirici, I., 2005. A new on line state of charge estimation and monitoring system for sealed lead-acid batteries in telecommunication power supplies. IEEE Trans. Ind. Electron. 52 (5), 1315–1327.

Lai, J.S., Rose, M.F., 1992. High energy density double layer capacitors for energy storage applications. IEEE AES Mag. 7 (4), 14–19.

Landfords, J., Simonsson, D., Sokirko, A., 1995. Mathematical modelling of a lead/acid cell with immobilized electrolyte. J. Power Sources 55, 217–230.

Langnan, P.E., 2000. VRLA battery impedance analysis: cell evaluation via changes in the slope of the impedance curve. Power Qual. Assur., 24–29.

Leijen, P., Steyn-Ross, D., Kularatna, N., 2018. Use of effective-capacitance variation as a measure of state-of-health in a series connected automotive battery pack. IEEE Transactions on Vehicular Technology. 6 (3), March, pp. 1961–1968.

Li, J., Mazzola, M., Gafford, J., Younan, N., 2012. A new parameter estimation algorithm for an electrical analogue battery model. In: Proceedings of IEEE APEC 2012 Conference, pp. 427–433.

Li, S.E., Wang, B., Peng, H., Hu, X., 2014. An electrochemistry-based impedance model for lithium-ion batteries. J. Power Sources 258, 9–18.

Lin, F.J., Huang, M.S., Yeh, P.Y., Tsai, H.N., Kuan, C.H., 2012. DSP-based probabilistic fuzzy neural network control for Li-ion battery charger. IEEE Trans. Power Electron. 27 (8), 3782–3794.

Lin, H.-T., Liang, T.-J., Chen, S.-M., 2013. Estimation of battery state of health using probabilistic neural network. IEEE Trans. Ind. Inform. 9 (2), 679–685.

Lindahl, P.A., Cornachione, a., Shaw, S.R., 2012. A time-domain least squares approach to electrochemical impedance spectroscopy. IEEE Trans. Instrum. Meas. 61 (12), 3303–3311.

Marcicki, J., Canova, M., Conlisk, A.T., Rizzoni, G., 2013. Design and parametrization analysis of a reduced-order electrochemical model of graphite/LiFePO4 cells for SOC/SOH estimation. J. Power Sources 237, 310–324.

Mauracher, P., Karden, E., 1997. Dynamic modeling of lead/acid batteries using impedance spectroscopy for parameter identification. J. Power Sources 67, 69–84.

Micea, M.V., Ungurean, L., Carstoiu, G.N., Groza, V., 2011. Online state of health assessment for battery management systems. IEEE Trans. Instrum. Meas. 60 (6), 1997–2006.

Moura, S.J., Stein, J.F., Fathy, H.K., 2013. Battery Health conscious power management in plug-in hybrid electrical vehicles via electrochemical modeling and stochastic control. IEEE Trans. Control Syst. Technol. 21 (3), 679–694.

Nelson, P., Bloom, I., Amine, K., Henriksen, G., 2002. Design modeling of lithium-ion battery performance. J. Power Sources 110, 437–444.

Orazem, M.E., Tribollet, B., 2008. Electrochemical Impedance Spectroscopy. John Wiley, New Jersey. 523 pages.

Pan, Y.H., Srinivasan, V., Wang, C.Y., 2002. An experimental and modeling study of isothermal charge/discharge behavior of commercial Ni-MH cells. J. Power Sources 112, 298–306.
Pascoe, P.E., Anbuky, A.H., 2002. The behavior of the coup de fouet of valve-regulated lead-acid batteries. J. Power Sources 111, 304–319.
Pascoe, P.E., Anbuky, A.H., 2005. Standby power system VRLA battery reserve life estimation scheme. IEEE Trans. Energy Conv. 20 (4), 887–895.
Pattipati, B., Sankavaram, C., Pattipati, K.R., 2011. System identification and estimation framework for pivotal automotive battery management system characteristics. IEEE Trans. Syst. Man. Cybern. 41 (6), 869–884.
Phillips Semiconductors, 2003. The I2C Bus and How to Use It. Document # 98-8080-575-01.
Plett, G.L., 2004a. Extended Kalman filtering for battery management systems of LiPB-based HEV battery packs: part 1. Background. J. Power Sources 134, 252–261.
Plett, G.L., 2004b. Extended Kalman filtering for battery management systems of LiPB-based HEV battery packs: part 2. Modeling and identification. J. Power Sources 134, 262–276.
Plett, G.L., 2004c. Extended Kalman filtering for battery management systems of LiPB-based HEV battery packs: part 3. Sate and parameter estimation. J. Power Sources 134, 277–292.
Rakhmatov, D., Vrudhula, S., Wallach, D.A., 2003. A model for a battery lifetime analysis for organizing applications on a pocket computer. IEEE Trans. VLSI Syst. 11 (6), 1019–1030.
Ramadass, P., Haran, B., Gomadam, P.M., White, R., Popov, N., 2004. Development of first principles capacity fade model for Li-ion cells. J. Electrochem. Soc. 151, A196–A203.
Randles, J.E.B., 1947. Kinetics of rapid electrode reactions. Discuss. Faraday Soc. 1, 11–19.
Rao, R., Vrudhala, S., Rakhmatov, D.N., 2003. Battery modeling for energy aware system design. IEEE Comput. 36 (12), 77–87.
Renhart, W., Magele, c., Schweighofer, B., 2008. FEM based thermal analysis of NiMH batteries for hybrid vehicles. IEEE Tans. Magnet. 44 (6), 802–805.
Rong, P., Pedram, M., 2003. An analytical model for predicting the remaining battery capacity of Li-ion batteries. In: Proceedings of IEEE Design, Automation and Test Conference, Europe, pp. 1148–1149.
Rong, P., Pedram, M., 2006a. An analytical model for predicting the remaining battery capacity of Li-ion batteries. IEEE Trans. VLSI Syst. 14 (5), 441–451.
Rong, P., Pedram, M., 2006b. An analytical model for predicting the remaining capacity of lithium-ion batteries. IEEE Trans. VLSI Syst. 14 (5), 441–451.
Ruetschi, P., 2004. Aging mechanisms and service life of lead acid batteries. J. Power Sources 127, 33–44.
Sacarisen, P.S., Parvereshi, J., 1995. Lead acid fast charge controller with improved battery management techniques. In: South Conference, March 1995.
Schwartz, P., 1995. Battery management. In: Portable by Design Conference, pp. 525–547.
Schweighofer, B., Raab, K.M., Brasseur, G., 2003. Modeling of high power automotive batteries by the use of an automated test system. IEEE Trans. Instrum. Meas. 52 (4), 1087–1091.
Shahriari, M., Farrokhi, M., 2013. Online state-of-health estimation of VRLA batteries using state of charge. IEEE Trans. Ind. Electron. 60 (1), 191–202.
Shen, W.X., Chau, K.T., Chan, C.C., Lo, E.W.C., 2005. Neural network based residual capacity indicator for nickel metal hydride batteries in electric vehicles. IEEE Trans. Veh. Technol. 54 (4), 1705–1712.
Singh, P., Fennie, C., Reisner, D., 2004. Fuzzy logic modeling of state-of-charge and available capacity of nickel/metal hydride batteries. J. Power Sources 136, 322–333.
Stevanatto, L.C., Brusamarello, V.J., Tairov, S., 2014. Parameter identification and analysis of uncertainty measurements of lead-acid batteries. IEEE Trans. Instrum. Meas. 63 (4), 761–768.

Stiaszny, B., Ziegler, J.C., Krauz, E.E., Schmidt, J.P., Ivers-Tiffée, E., 2014. Electrchemical characterization and post-mortem analysis of aged $LiMnO_4Li(Ni_{0.5}Mn_{0.3}Co_{0.2})O_2$/graphite li-ion batteries, part 1. J. Power Sources 251, 439–450.

Swierczynski, M., Store, D.I., Stan, A., Teodorescu, R., Sauer, D.U., 2014. Selection and performance-degradation modeling of $LiMO_2/LI_4$ TI $_5O_{12}$ and Li FE PO_4/C battery cells as suitable energy storage systems for grid integration with wind power plants: an example for the primary frequency regulation service. IEEE Trans. Sustain. Energy 5 (1), 90–101.

Tan, X., Baras, J., 2005. Adaptive identification and control of hysteresis in smart materials. IEEE Trans. Automatic Control 50 (6), 827–839.

Tang, X., Zhang, X., Koch, B., Frisch, D., 2008. Modeling and estimation of nickel metal hydride battery hysteresis for SOC estimation. In: Proceedings of IEEE International Conference on Prognostics and Health Management, pp. 1–12.

Thele, M., Buller, S., Sauer, D.U., Doncker, R.W., Karden, E., 2005. Hybrid modeling of lead acid batteries in frequency and time domain. J. Power Sources 144, 461–466.

Unitrode Inc, 1997. Improved Charging Methods for Lead-Acid Batteries Using the UC 3906. Application Note U-104.

Urbin, M., Hinaje, M., Raë, S., Davat, B., Desperez, P., 2010. Energetical modeling of lithium-ion batteries including porosity effects. IEEE Trans. Energy Conv. 25 (3), 862–872.

US Department of Energy, 2010. Battery Test Manual for Plug_in Hybrid Electric Vehicles. Rev 2, INL/EXT-07-12536, 56 pages.

Valoen, L.O., Sunde, S., Tunold, R., 1997. An impedance model for electrode processes in metal hydride electrodes. J. Alloys Compd. 253 (254), 656–659.

Wang, J., Liu, P., Hicks-Garner, J., Sherman, E., Soukiazian, S., Verbrugge, M., Tataria, H., Musser, J., Finamore, P., 2011. Cycle-life model for graphite-LiFePO4 cells. J. Power Sources 196, 3942–3948.

Warburg, E., 1899. Über das Verhalten sogenannte unipolarisierbare Electroden gegen Wechselstrom. Ann. Phys. Chem. 67, 493–499.

Wright, R.B., Motloch, C.G., Belt, J.R., Christophersen, J.P., Ho, C.D., Richardson, R.A., Bloom, I., Jones, S.A., Battaglia, V.S., Henriksen, G.L., Unkelhaeuser, T., Ingersoll, D., Case, H.L., Rogers, S.A., Sutula, R.A., 2002. Calendar- and cycle-life studies of advanced technology development program generation 1 lithium-ion batteries. J. Power Sources 110, 445–470.

Wu, Q., Qiu, Q., Pedram, M., 2000. An interleaved dual-battery power supply for battery operated electronics. In: Proceedings: 2000 Conference, Asia and South Pacific Design Automation, pp. 387–390.

Wu, B., Mohammed, M., Brigham, D., Elder, R., White, R.E., 2001. A non-isothermal model of a nickel-metal hydride cell. J. Power Sources 101, 149–157.

Zhang, J., Ci, S., Sharif, H., Alahmad, M., 2010. An enhanced circuit-based model for single-cell battery. In: Proceedings of 25th IEEE APEC 2010 Conference, pp. 672–675.

Bibliography

Kalman, R.E., 1960. A new approach to linear filtering and prediction problems. Trans. ASME—J. Basic Eng., Ser. D 82, 35–45.

UNC CS 1960. The Seminal Kalman Filter Paper http://www.cs.unc.edu/~welch/kalman/kalmanPaper.html (Accessed 20.05.03).

Haykin, S., 1996. Adaptive Filter Theory, third ed. Prentice-Hall, Upper Saddle River, NJ.
Haykin, S., 2001. Kalman filters. In: Haykin, S. (Ed.), Kalman Filtering and Neural Networks. Wiley/Interscience, New York, pp. 1–22.
Burl, J., 1999. Linear Optimal Control: H2 and H1 Methods. Addison Wesley, Menlo Park, CA.
Kiehne, H.A., 2003. Battery Technology Handbook. Marcel Dekker Inc, New York.
Benchmarq Microelectronics Inc, Duracell Inc, Energizer Power Systems, Intel Corporation, et al., 1995a. Smart Battery System Specifications—System Management Bus Specifications. Revision 1.0,.
Benchmarq Microelectronics Inc, Duracell Inc, Energizer Power Systems, Intel Corporation, et al., 1995b. Smart Battery System Specifications—Smart Battery Data Specifications. Revision 1.0,.
Song, L., Evans, J.W., 2000. Electrochemical-thermal model of lithium polymer batteries. J. Electrochem. Soc. 147 (6), 2086–2095.

Recent developments of high-performance battery systems

5

5.1 Introduction

In Chapters 3 and 4, we have given a summary of mature rechargeable battery technologies and discussed the performance of rechargeable batteries from an engineer's viewpoint. With the major development of Li-ion batteries (LIB) and Sony Corporation commercializing them in 1990, the growth of portable products accelerated and by the turn of the 20th century, electric cars started appearing in the society. In addition, the demand for fossil fuel reduction, with reduced CO_2 emissions during the last two decades renewable energy systems kept growing rapidly, and this fuelled a demand for long-life battery banks, as well as alternatives to batteries.

This chapter provides an overview of newer battery chemistries including flow batteries, solid-state batteries (SSB), and recent advances of electrochemical storage systems.

5.2 Flow batteries for renewable energy systems

A flow battery is a rechargeable battery where the energy is stored in one or more electroactive species dissolved into liquid electrolytes. The electrolytes are stored externally in tanks and pumped through electrochemical cells, which convert chemical energy directly to electricity and vice versa, on demand. Flow batteries can be fitted to a wide range of stationary applications. Originally developed by NASA in the early 1970s as electrochemical energy storage systems for long-term space flights, flow batteries are now receiving attention for storing energy for durations of hours or days. Flow batteries have the advantages of low-cost devices, modularity, easy transportability, high efficiency, and can be deployed at a large scale (Badwal et al., 2014). The modularity and scalability of these devices means they can easily span the kW to MW range. As a result, their main development at present is focused on stand-alone remote area power systems or grid energy storage/support in combination with renewable energy generation.

With the renewable energy transition, supported by major initiatives such as the European Green Deal (The European Green Deal, 2019), large-scale energy storage systems with very long lifetime are required and the flow batteries is one of the solutions getting accepted by the energy companies, etc. The global energy transition scenario implies large-scale considerations when defining a solution. Lithium ion batteries (LIBs) are ubiquitous in our society and dominate the energy storage market powering portable devices, EVs and even smart grid facilities. In 2019, 8.8 GWh of LIB capacities were installed for stationary energy storage vs 0.25 GWh of redox flow batteries (RFBs) (Sánchez-Díez et al., 2021). However, due to its high

Energy Storage Devices for Renewable Energy-Based Systems. https://doi.org/10.1016/B978-0-12-820778-9.00012-7

maintenance cost and safety limitations, in addition to the limited availability of lithium, note the interest in developing alternatives for LIBs to efficiently store energy. Large-scale grid storage requires long life low-cost batteries, considering cyclability, calendar life, and round-trip efficiency. Installation and maintenance costs are still the main barriers for the penetration of storage on the grid. Thus, clear targets have been set in the European Sustainable Energy Technologies (SET) Plan, for stationary energy storage in terms of cost (0.05/kW/h/cycle) and durability (10,000 cycles and 20 years lifetime) for 2030 (Sánchez-Díez et al., 2021).

5.2.1 Principles of RFBs

Fig. 5.1 depicts the operation of a generic RFB. Two electrolyte tanks are coupled to two porous electrodes via respective pumps, pumping the electrolyte through the electrodes. Two current collectors are mounted onto electrodes to carry current through the external circuits. Two electrodes are separated by a porous membrane, which acts as a separator between the two electrodes. The energy capacity is a function of the electrolyte volume and the power is a function of the surface area of the electrodes.

Negative and positive electrolytes flow through porous electrodes, separated by the porous membrane, which allows protons to pass through it for the electron transfer process. During the exchange of charge a current flows over the electrodes, which can be used by a battery-powered device. During discharge the electrodes are continually supplied with the dissolved active masses from the tanks; once they are converted, the resulting product is removed to the tank. Different redox pairs are used in RFBs and their summary is provided in Badwal et al. (2014). Single-cell voltage values for different systems could vary from 1.0 to 2.4 V in general.

RFBs have emerged as relevant candidates to address the sustainable energy generation. Their unique capability to decouple power and energy based on their particular architecture results in advantages such as: flexible modular design and operation, excellent scalability, moderate maintenance costs, and long-life cycling. Thus, the system consists of three main components: energy storage tanks, stack of electrochemical cells, and the flow system.

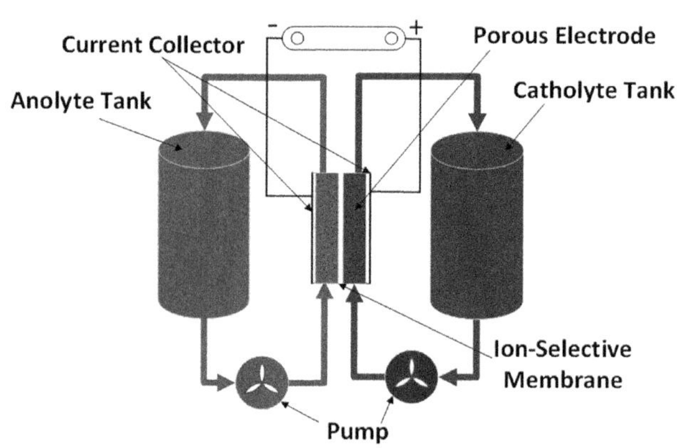

Fig. 5.1 Simplified concept of the redox flow batteries.

The energy storage proceeds as follows: (i) active species are contained in the tanks as a solution with a certain energy density, (ii) the solution, defined as the electrolyte, is pumped into the stack, where the electrochemical conversion takes place and collected back in the tanks. The size of the stack defines the power of the system while the amount of electrolyte stored in the tanks states the total energy. High round trip efficiency (RTE), depth of discharge (DoD), fast responsiveness, and negligible environmental impact (e.g., aqueous RFBs) are other key features for the technology's successful deployment. Power and energy density limitations in comparison to other technologies such as LIBs are generally overcome by the more cost-effective scalability.

5.2.2 Vanadium redox flow batteries

The most successful deployment of RFBs are the vanadium redox flow batteries (VRFB), which reached its commercialization in the 1980s (Sánchez-Díez et al., 2021). Fig. 5.2 depicts the case of a VRFB with its two electronic subsystems, namely power management system (PMS), which is a DC/AC bidirectional converter and the battery management system (BMS), which acts as the system supervisor for the

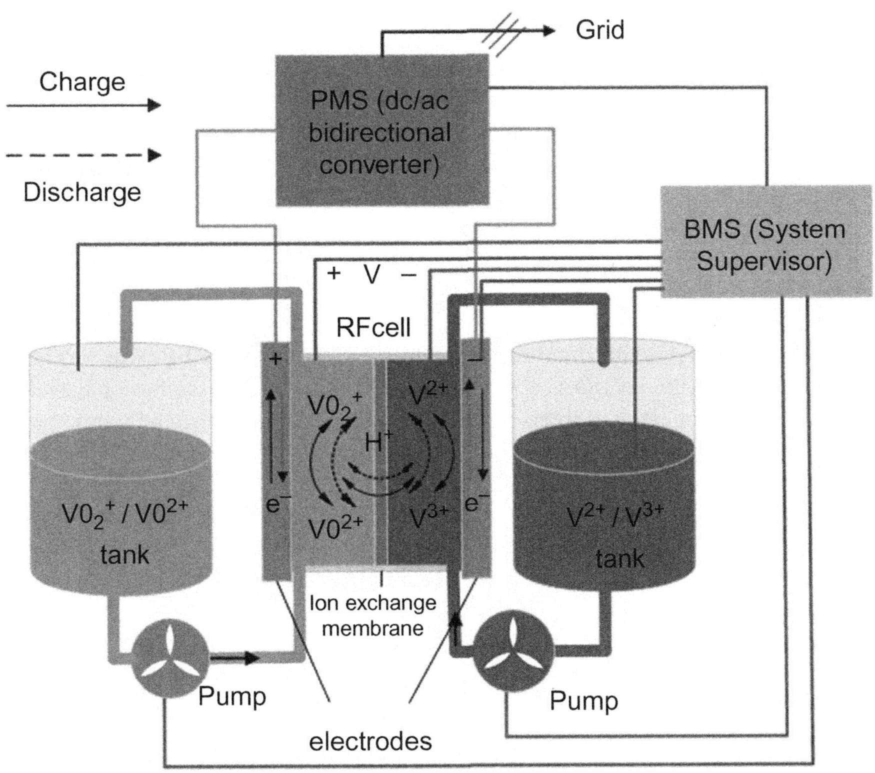

Fig. 5.2 kW class vanadium redox flow battery—Single-cell case.
Source Sánchez-Díez, E., et al., 2021. Redox flow batteries: status and perspective towards sustainable stationary energy storage, J. Power Sources. 481.

electrochemical system. Electroactive species in the two tanks of vanadium compounds are pumped into the ion-exchange membrane (redox flow cell), which works together with the associated electrodes.

As per Sánchez-Díez et al. (2021) there are over 25 worldwide companies who are active in developing VRFB systems suitable for renewable energy systems. Among the largest deployments are the Minami Hayakita Substation in Japan built by Sumitomo Electric Inc. and at Fraunhofer ICT in Germany. As at 2020, Rongke Power of China is building the largest VRFB systems in the order of 200–800 MW ratings. Essential electro-chemistry details of VRFBs are found in Sánchez-Díez et al. (2021). Single-cell voltages of VRFBs vary from 1.1 to 1.6 V depending on the chemistry of the system.

Due to the fast electrochemical kinetics, the response time is short, in the order of a millisecond, if the electrodes are kept full of electrolytes and pumps in standby promptly take over. This allows VRFBs to respond immediately to surge power demand from the grid, coping with power quality grid services such as sag compensation and frequency regulation (Sánchez-Díez et al., 2021). The electrolytes which remain inside the electrodes in standby also allow over-currents of 150% and more in the first few milliseconds.

5.2.3 Alternatives to standard RFBs—Hybrid RFBs

Alternatively to the standard RFBs, systems comprising of at least one solid electroactive material that is deposited or stripped within the stack have been widely explored (e.g., Zn-based RFBs). Those, so called hybrid RFBs, differ from the previous type in their capability to decouple power and energy, which is limited by the solid electrode (Fig. 5.3). Fig. 5.3 is a kW class hybrid flow battery (in single-cell form) based on a zinc-bromine system. More details are found in Sánchez-Díez et al. (2021).

RFBs that store electroactive material only in the outer tanks and flown through the cell in its reduced and oxidized states are called pure flow RFBs.

Thus, both pure flow RFBs and their hybrid counterparts have been successfully deployed, where VRFB and zinc-bromine redox flow batteries (ZBFBs) can be clearly defined as state-of-the-art (SoA) for the technology. Nevertheless, those have still a long way to go to meet the targets defined by energy institutions, and a new bunch of RFB systems is irrupting to oust VRFBs and show up as real alternatives to reach the market.

There are several more RFB systems currently under development, such as (i) aqueous inorganic pure flow batteries, (ii) aqueous organic redox flow batteries, (iii) pure flow membrane-less, and (iv) RFBs based on insoluble solid active materials. A summary is found in Sánchez-Díez et al. (2021).

5.3 Solid-state batteries

A solid-state battery is a technology that uses solid electrodes and a solid electrolyte, instead of the liquid or polymer gel electrolytes found in lithium-ion or lithium polymer batteries (Sánchez-Díez et al., 2021; Reisch, 2017). Materials proposed for use as

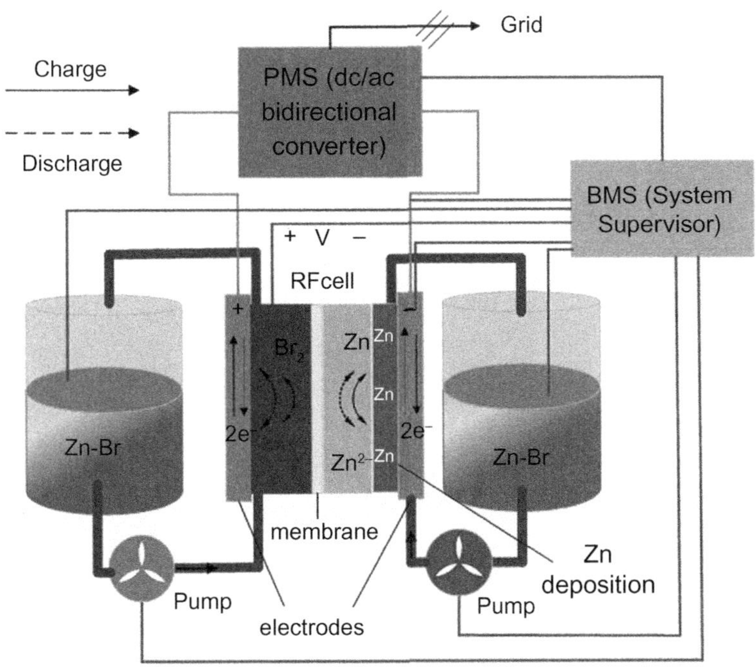

Fig. 5.3 kW class hybrid flow battery based on Zn-Br chemistry in single-cell form—Solid electrode based on zinc (Zn) deposition is the key difference in these compared to pure flow RFBs.

solid electrolytes in solid-state batteries include ceramics (e.g., oxides, sulfides, phosphates), and solid polymers. Solid-state batteries have found use in pacemakers, RFIDs, and wearable devices. They are potentially safer, with higher energy densities, but come at a much higher cost. Challenges to widespread adoption include energy and power density, durability, material costs, sensitivity, and stability. Subject of solid-state ionics, which deals with the study of ionic conductors (solid electrolytes) and ionic-electronic mixed conductors, is the subject which covers the electrochemistry in these futuristic batteries. Fig. 5.4 depicts the power density versus energy density for different solid-state ionic systems used for energy storage and conversion.

Solid-state batteries are being extensively studied and researched with a view to solving the well-known problems of LIBs such as the flammability, life cycle issues and storage, etc. Conventional batteries such as LIBs usually consist of a liquid electrolyte, which helps transport Li ions to and from the cathode and anode. However, this increases the chance of leakage of the electrolyte if any holes are present; this is one of the main drawbacks of the conventional LIB. Another problem inherent in the liquid electrolyte battery is the formation of dendrites of Li, which makes it prone to explosion (Kim et al., 2015; Zheng et al., 2018). In order to surmount these problems, a solid electrolyte can be positioned between the electrodes. This is the principle underlying the solid-state battery (Kim et al., 2015). The advantage of this type

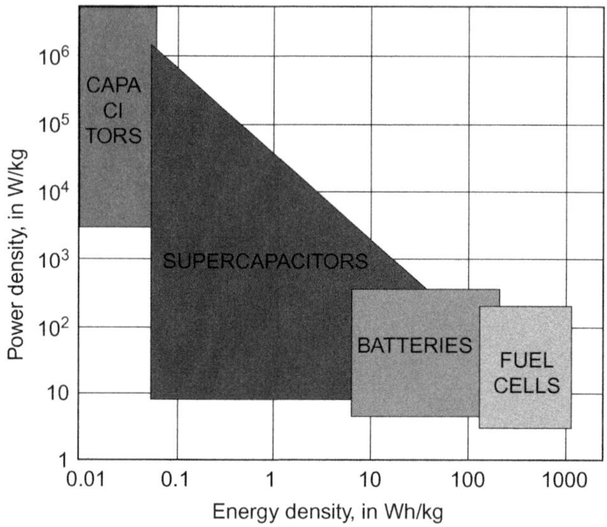

Fig. 5.4 Power density versus energy density for different solid-state ionic systems used in energy storage and conversion.

of battery is a reduction in the net weight and volume of the battery, greater energy output, and easy transfer of Li ions, which affords better efficiency (Kim et al., 2015). Solid-state batteries also exhibit some advantages over the other commonly known energy storage devices, such as capacitors. The advantages lie in the very small self-discharge of the solid-state batteries, minimal wear and tear, and yield of a more uniform output voltage. In recent decades, SSBs, especially solid-state lithium ion batteries, have been widely used (Zheng et al., 2018). Ideally, a solid-state electrolyte should have high cation conductivity, with good mechanical properties and good chemical stability that cannot be easily reduced by the metal itself. Moreover, owing to rapidly growing microelectronics and integrated optoelectronics circuits, there is an increasing demand for new, lightweight batteries with high-cycle life and high energy density. In a commercial lithium ion battery, the liquid electrolyte carries the risk of explosion and fire; moreover, the separators and packaging limit the size of the batteries.

All these factors have contributed to the development of all SSBs (Kim et al., 2015). However, solid-state batteries also present challenges, such as their relatively lower power density, high ionic resistance at room temperatures, and manufacturing cost. For SSBs, typical solid oxide electrolytes that have been developed so far consist of NASICON type, perovskite-type, garnet type, amorphous/glass type, and few others (Zheng et al., 2018). More details are available in Kim et al. (2015), Zheng et al. (2018), and Rahardian et al. (2019).

Fig. 5.5A compares the ion and electron transport of LIB and solid-state battery (SSB) and Fig. 5.5B depicts the charge and discharge process of lithium-based SSB with electrode materials.

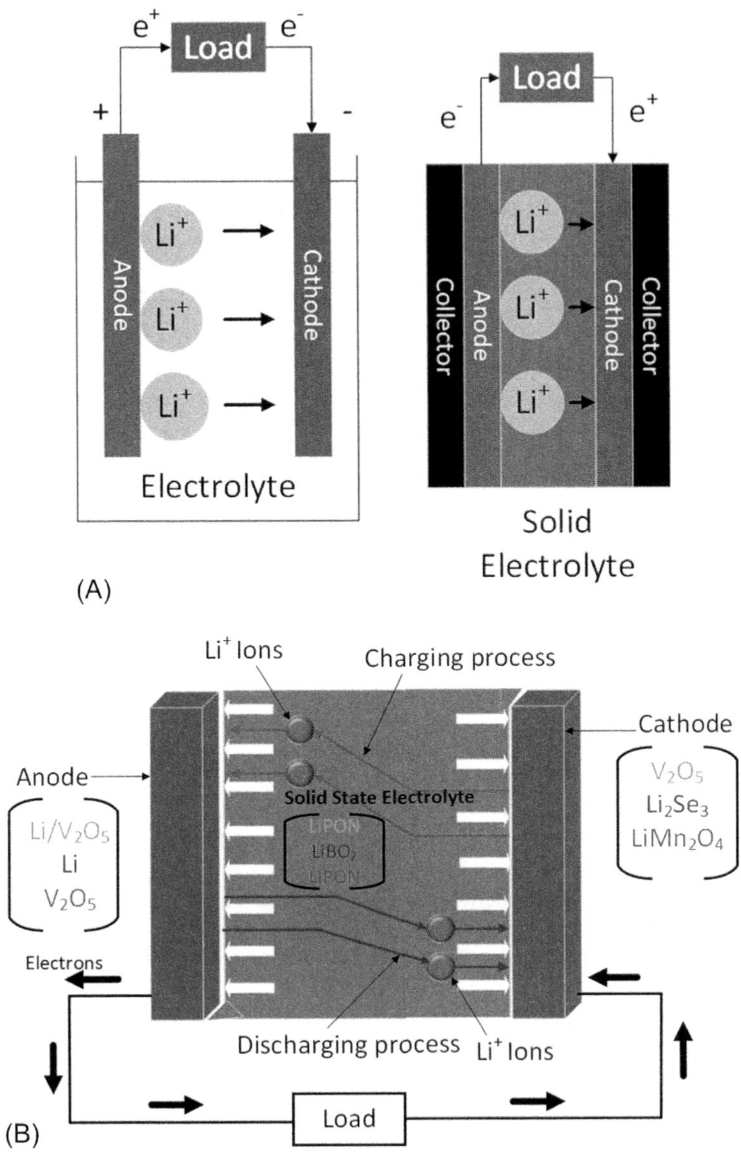

Fig. 5.5 Solid-state battery (SSB) concepts (A) ion and electron transport in LIB and SSB comparison (B) charge-discharge processes of Lithium-based SSB.
Source Kim, J.G., et al., 2015. A review of lithium and non-lithium based solid state batteries. J. Power Sources 282, 299–322.

5.4 More recent advances of traditional rechargeable batteries

Development of rechargeable battery technologies are mainly influenced by the automobile industry and the portable electronic products. While the newer internal combustion engine (ICE) based cars carry start-stop systems to save fuel in city driving conditions, which require better 12 V batteries with long life cycles, electric vehicles demand high energy LIBs or other types which could provide long range and long life. For the case of ICE-based cars, with advanced versions of 12 V batteries reaching the market, battery testing for aging and characterization has become an important subject (Koolmeyer and Jahns, 2019).

In LIB area many new electrolytes and active materials are reaching the industry from the research labs, and an overview of these developments as at latter part of 2020 is available in Armand et al. (2020).

References

Armand, M., et al., 2020. Lithium-ion batteries—current state of the art and anticipated developments. J. Power Sources 479, 228708.

Badwal, S.P.S., Giddey, S.S., Munnings, C., Bhatt, A.I., Hollenkamp, A.F., 2014. Emerging electrochemical energy conversion and storage technologies. Front. Chem. 2, 1–28.

Kim, J.G., et al., 2015. A review of lithium and non-lithium based solid state batteries. J. Power Sources 282, 299–322.

Koolmeyer, P.J., Jahns, T.M., 2019. Aging and performance comparison of absorbed glass matte, enhanced flooded, PbC, NiZn, and $LiFePO_4$ 12V start stop vehicle batteries. J. Power Sources 441, 227139.

Rahardian, S., et al., 2019. Review of solid-state battery technology progress. In: Proc of IEEE-ICEVT.

Reisch, M.S., 2017. Solid-state batteries inch their way toward commercialization. Chem. Eng. News 95 (46), 19–21.

Sánchez-Díez, E., et al., 2021. Redox flow batteries: status and perspective towards sustainable stationary energy storage. J. Power Sources 481.

The European Green Deal, 2019. COM 2019 (640), 11.12.2019.

Zheng, F., et al., 2018. Review of solid electrolytes for all-solid-state lithium-ion batteries. J. Power Sources 389, 198–213.

Capacitors as energy storage devices: Simple basics to current commercial families

6

6.1 Capacitor fundamentals

A capacitor is a device that stores electrical charge and releases them when it is required by the circuit. The arrangement shown in Fig. 6.1 illustrates the simplest configuration of a capacitor: two parallel conducting plates separated by a dielectric material (an electrical insulator) where parallel plates have an area A and separation d.

The charge on either one of the parallel conductors in the capacitor is zero in the uncharged state. When an external DC voltage, V, is applied to the capacitor electrodes, electric charges build up on the capacitor plates with electrons producing a positive charge ($+Q$) on one plate and an equal and opposite negative charge ($-Q$) on the other plate. Current, I, flow continues until the potential difference between the two electrodes becomes equal to the supply voltage, in a fully charged capacitor.

The ratio Q/V is defined as the capacitance

$$C = \frac{Q}{V} \tag{6.1}$$

with capacitance units of farad (F). For 1 F of capacitance, 1 C of charge is stored on the plates, when 1 V of external voltage is applied.

The behavior of a capacitor is described by the equation

$$i(t) = \frac{\mathrm{d}q}{\mathrm{d}t} = C\frac{\mathrm{d}v}{\mathrm{d}t} \tag{6.2}$$

where t is the time, i is the current through the capacitor at time t, and v is the voltage across the capacitor leads at time t. Integrating Eq. (6.2), capacitor voltage can be obtained as

$$v(t) = \frac{1}{C}\int_{o}^{T} i(t)dt + v(0) \tag{6.3}$$

where $v(0)$ is the initial voltage of the capacitor at the beginning of the charging process.

As shown in Fig. 6.1, the simplest form of a capacitor consists of two conducting plates of area, A, which are parallel to each other and separated by distance d by a dielectric material such as ceramic, polymer, or aluminum oxide. In such a

Energy Storage Devices for Renewable Energy-Based Systems. https://doi.org/10.1016/B978-0-12-820778-9.00003-6

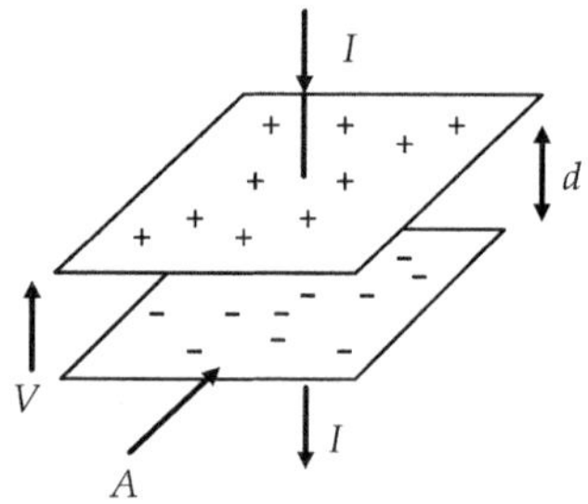

Fig. 6.1 Construction of a parallel-plate capacitor.

parallel-plate capacitor, the electric energy is stored statically in the electric field between two electrodes.

The capacitance formula for parallel-plate capacitors is given by

$$C = \frac{\epsilon A}{d} \tag{6.4}$$

Thus capacitance increases with electrode area (A) and with permittivity ϵ of the dielectric material, but decreases with the electrode separation d.

The most common capacitors fall within magnitudes normally specified in pF, nF, or μF. Conventional capacitors yield capacitance values in the range of 5 pF to 1 F with a voltage range of 10 to over 400 V. The trend in the industry is achieving larger capacitances. Capacitors with capacitance values greater than 1 F are placed in the special category called supercapacitors (Kim, 2003), which are discussed in detail in Chapter 7.

6.1.1 Capacitor charging

A typical charging circuit of a capacitor is shown in Fig. 6.2 using a constant DC voltage V_{in} and a series resistance R. This resistor represents the equivalent series resistance of the capacitor, resistance of external leads and connections, and any deliberately introduced resistance. Assume that the initial voltage of the capacitor, $V(0)$, is zero.

Using Kirchoff's voltage law,

$$V_{in} = i(t)R + \frac{1}{C}\int_o^T i(t)dt \tag{6.5}$$

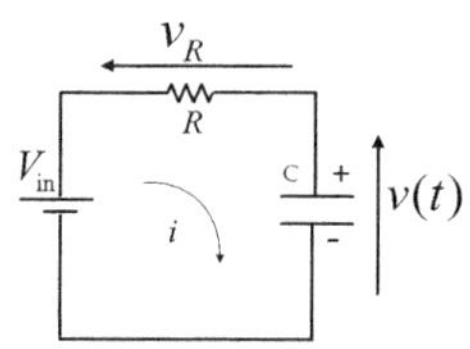

Fig. 6.2 Charging a capacitor from a voltage source.

Solving Eq. (6.5) for current we obtain

$$i(t) = \frac{V_{\text{in}}}{R} \exp\left[\frac{-t}{RC}\right] \tag{6.6}$$

Using Eq. (6.6), the capacitor voltage at a given time during the charging process is

$$v(t) = V_{\text{in}}\left[1 - \exp\left[\frac{-t}{RC}\right]\right] \tag{6.7}$$

Fig. 6.3A shows the voltage and current behavior of the capacitor during the charging process.

The rate of charge of the capacitor depends on the product of R and C. This is usually called the time constant, τ,

$$\tau = RC \tag{6.8}$$

Usually, it takes more than 5 time constants for a capacitor to fully charge or discharge.

6.1.2 Capacitor discharging

If the DC supply is replaced with a short circuit, the capacitor will discharge through the load resistance R. When discharging, the current behaves the same way as for charging, but flows in the opposite direction. Voltage across the capacitor will decay exponentially to zero. Equations for both voltage and current discharge are given by

$$v(t) = V_f \exp\left[\frac{-t}{RC}\right] \tag{6.9}$$

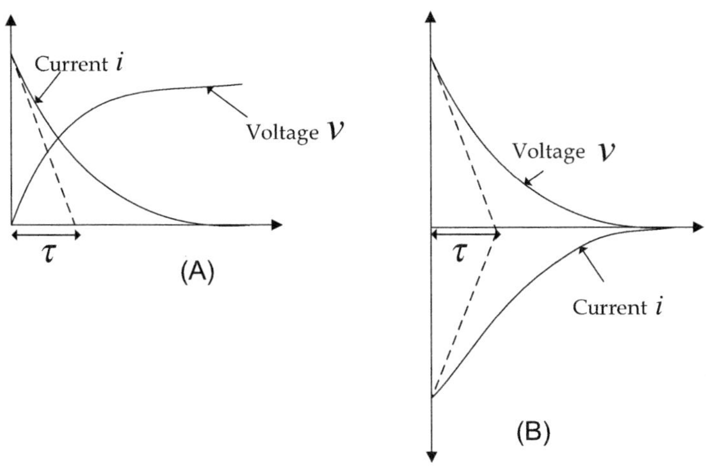

Fig. 6.3 Capacitor voltage-current behavior: (A) charging process and (B) discharging process.

$$i(t) = \frac{V_f}{R} \exp\left[\frac{-t}{RC}\right] \tag{6.10}$$

where V_f is the voltage at the start of capacitor discharge (see Fig. 6.3(B)).

6.1.3 Capacitor energy storage

A capacitor can store electrical energy and release it again. A capacitor acts as an energy reservoir that blocks the direct flow of current with DC voltage and allows the flow of current with AC or pulsating voltage depending on its capacitance and the given frequency. So the capacitor can assume a different role depending on the circuit:

- In a DC circuit it is used as an energy storage device.
- In an AC circuit it is used as a frequency-dependent resistor.

The total energy stored in a capacitor, E, is 0.5 times the product of the capacitance, C, and the square of the voltage, V.

$$E = \frac{1}{2} CV^2 \tag{6.11}$$

According to the law of conservation of charge, if two precharged capacitors are connected together as shown in Fig. 6.4, the total charge on the parallel combination is equal to the sum of the original charges on the capacitors.

The two capacitors shown in Fig. 6.4, C_1 and C_2, are charged to voltages V_1 and V_2, respectively. When the switch is closed, an impulse of current flows, and the charge is redistributed between the two capacitors. The total charge at the beginning of the parallel combination of the two capacitors is given by

$$Q_T = C_1 V_1 + C_2 V_2 \tag{6.12}$$

This charge is distributed between the two capacitors, so that the total charge across the two capacitors remains the same.

$$Q_T = C_1 V + C_2 V = (C_1 + C_2) V \tag{6.13}$$

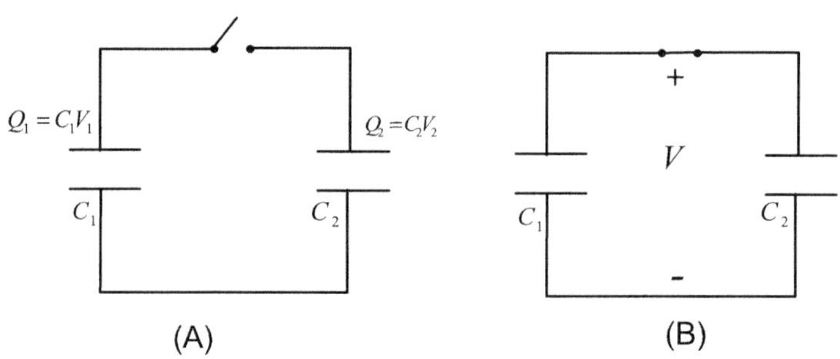

Fig. 6.4 Charge distribution between capacitors.

where V is the new voltage across the parallel combination.

$$V = \frac{C_1 V_1 + C_2 V_2}{C_1 + C_2} \tag{6.14}$$

The total energy loss E_L due to paralleling is given by

$$E_L = \left[\frac{1}{2}C_1 V_1^2 + \frac{1}{2}C_2 V_2^2\right] - \frac{1}{2}(C_1 + C_2)V^2 \tag{6.15}$$

which gives

$$E_L = \frac{1}{2}\frac{C_1 C_2}{C_1 + C_2}(V_1 - V_2)^2 \tag{6.16}$$

If C_2 is very much smaller than C_1, then relationship in Eq. (6.16) can be simplified as

$$E_L \approx \frac{1}{2}C_2(V_1 - V_2)^2 \tag{6.17}$$

6.2 Capacitor characteristics

6.2.1 Capacitors and Ragone plot

Energy storage devices (ESDs) are characterized by the energy and the available power for a load (Christen and Carlen, 2000). The Ragone plot is a chart used for the comparison of the performance of various energy-storing devices where the values of energy density (in Wh/kg) are plotted vs power density (in W/kg) [15]. ESDs such as batteries, capacitors, supercapacitors, flywheels, magnetic energy storage devices, etc., are located in characteristic regions in the power/energy plane in the Ragone plot. Both axes are usually presented in logarithmic scale, which allows the comparison of the performance of very different devices. As shown in Fig. 6.5, this plot compares the performances of various energy storage devices such as batteries, supercapacitors, and conventional capacitors (Dispennette, 2005; Sarjeant and Staffier, 1996). The choice of the ESD depends particularly on the power density required by the application.

The sloping lines (isotherms of characteristic time constant) on the Ragone plot indicate the relative time taken to get the charge in or out of the device (Sarjeant and Staffier, 1996). At one extreme, capacitor can pump power in and out within microseconds, whereas at the other extreme batteries cells that have a poor dynamic performance take a few hours to generate and deliver their energy.

Portable devices have a wide range of energy storage needs. Some portable devices require high power, high energy for hours of runtime, whereas others require only a few seconds of operation before recharging (Dispennette, 2005). As illustrated in the Ragone plot, batteries have high energy densities but low power densities making

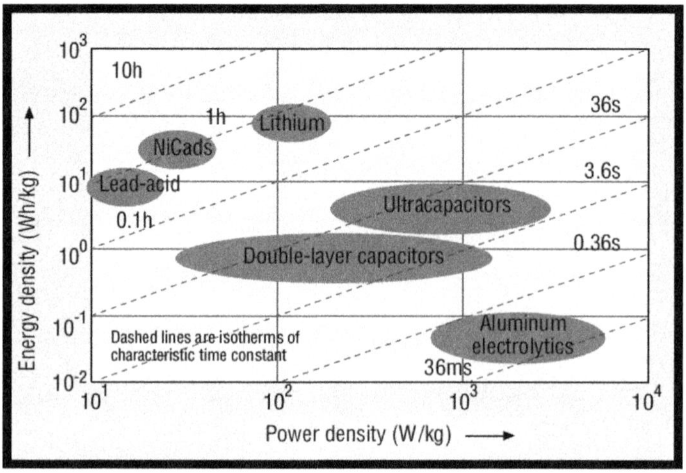

Fig. 6.5 The Ragone plot showing energy density vs power density (Sarjeant and Staffier, 1996).

them suitable for applications that need longer runtime. Batteries suffer from several drawbacks, such as limited lifetime and expensive maintenance. In contrast, capacitors have rather high-power densities but low energy densities with much longer lifetimes and cycle life (Sarjeant and Staffier, 1996).

6.2.2 Capacitor equivalent circuit models

In practical circumstances, the actual capacitor model is more complex as shown in Fig. 6.6A. Fig. 6.6B shows a simplified version of this equivalent circuit model.

R_{ac} and R_{dc} represent the equivalent resistance due to AC dielectric losses and leakage resistance due to DC dielectric losses, respectively (Kularatna, 2008). In certain circuits such as precision integrators or sample/hold circuits, the dielectric absorption of the capacitor is an important parameter. Dielectric absorption is a property in which the dielectric material does not become polarized instantly. This causes a memory effect and is represented by R_d and C_d in Fig. 6.6A. Dielectric absorption property prevents a capacitor from totally discharging, even when short circuited for a short period of time. In such a situation, if the capacitor is charged and then discharged, and left open, it will recover some of its charge which will gradually reappear as a small amount of DC voltage at the terminals of the capacitor due to the parasitic effect of R_d and C_d. Fig. 6.7 shows the effect of dielectric absorption when a capacitor is fully charged and then discharged by a short circuit (Kularatna, 2008). ESR and ESL indicate the equivalent series resistance and equivalent series inductance, respectively, which are important design parameters. More details of ESR and ESL are discussed next.

Fig. 6.6B shows the most commonly used equivalent circuit model when the capacitor is used at higher frequencies or in high-voltage situations. R_{leak} shows the insulation resistance of the dielectric in relation to the residual leakage current.

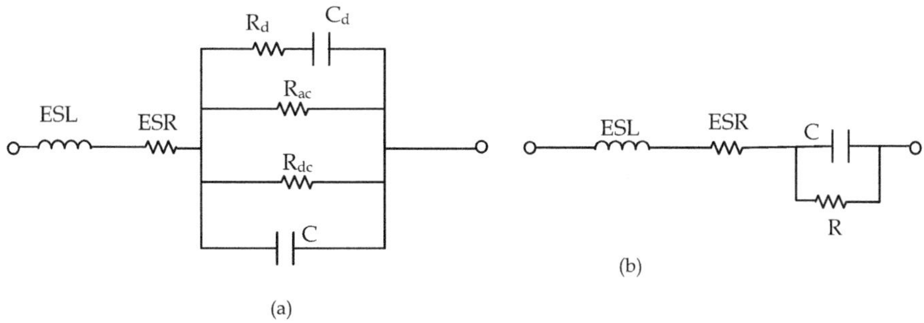

Fig. 6.6 Capacitor models: (A) a comprehensive equivalent circuit and (B) a simplified version.

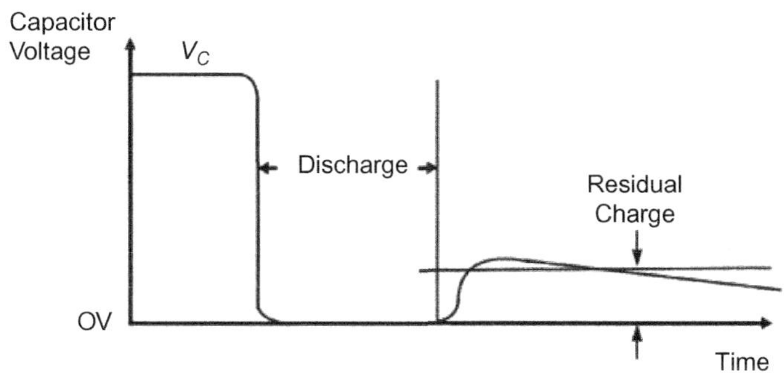

Fig. 6.7 Effect of dielectric absorption.

The capacitor self-discharge takes place through this resistance. ESR denotes the effective ohmic resistance existing between the connectors of the capacitor. It combines all the ohmic elements such connecting wires, contact resistances, and the electrodes. ESL specifies the self-inductance of a capacitor, which is determined by the supply leads, wires, and connecting parts and the respective package of the capacitor (ABC of Capacitors, n.d.).

R_{ac} and R_{dc}, two parameters are important design parameters when the capacitor is used at higher frequencies or in high-voltage situations. R_{ac} and R_{dc} represent the equivalent resistance due to AC dielectric losses and leakage resistance due to DC dielectric losses, respectively (Kularatna, 2008). In certain circuits such as precision integrators or sample/hold circuits, the dielectric absorption of the capacitor is an important parameter. Dielectric absorption is a property, in which the dielectric material does not become polarized instantly. This causes a memory effect and is represented by R_d and C_d in Fig. 6.6 A. Dielectric absorption property prevents a capacitor from totally discharging, even when short-circuited for a short period of time (http://en.wikipedia.org/wiki/Dielectric_absorption). In such a situation, if the capacitor is charged and then discharged, and left open, it will recover some of its charge which will gradually reappear as a small amount of DC voltage at the terminals of the capacitor due to the parasitic effect of R_d and C_d. Fig. 6.7 shows the effect of dielectric absorption when a capacitor is fully charged and then discharged by a short circuit (Kularatna, 2008; http://en.wikipedia.org/wiki/Dielectric_absorption).

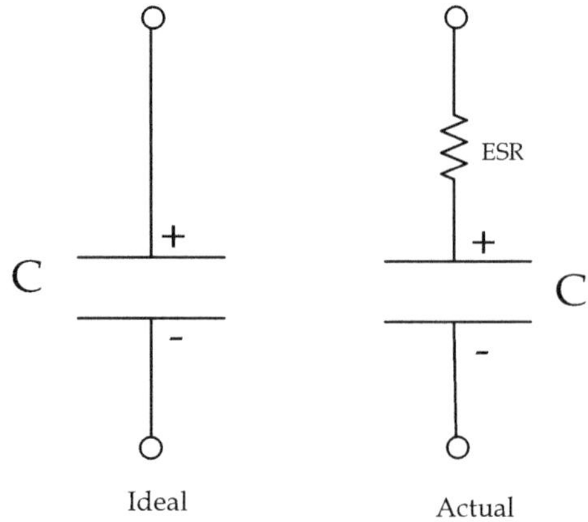

Fig. 6.8 (A) Ideal capacitor and (B) simplest first-order model.

In ideal conditions, the capacitor is simply represented as the capacitance, *C*, as shown in Fig. 6.8A. However, a simplified first-order model of the capacitor that has capacitance and equivalent series resistance (ESR) as shown in Fig. 6.8B is used in most DC applications.

6.2.3 Capacitor terminology

(a) Rated capacitance—the capacitance value usually measured at room temperature in the range 20–25°C.
(b) Rated voltage—the maximum continuous voltage that may be applied at rated temperature to ensure full life. As a result of construction, a capacitor can be safely charged to a defined rated voltage, which is also known as nominal voltage, operating voltage, and working voltage.
(c) DC leakage current—the dielectrics used in the capacitor construction are not perfect insulators and result in a small current "leaking" through the dielectric.
(d) ESR (equivalent series resistance)—the combined effect of all the resistive elements that form the resistive path in series with the capacitor, measured in ohms.
(e) Temperature range—the safe minimum and maximum ambient temperatures under which the capacitor may operate.
(f) DF (dissipation factor)—a measure of a capacitor's loss. It is the ratio of ESR/Xc. Dissipated energy typically turns into heat. The higher the DF, the greater the loss and heat generated.

6.3 Capacitor application scope

Some typical applications of a capacitor in the electronics industry are as follows (Holystone, n.d.):

- Store electrical energy as an energy buffer: Capacitors store electrical energy when disconnected from its charging circuit, so it can be used like a temporary battery. Capacitors are used as a charging unit and the energy is stored and released when required by the circuit.

- DC-blocking capacitor: When fully charged, the capacitor blocks the flow of DC current between two parts of a circuit.
- Bypass capacitor: The reactance of a capacitor decreases as the frequency increases. In certain applications, this property is utilized by placing a capacitor in parallel with other components to bypass a specified range of frequencies.
- Capacitor as a filter: In filter circuits, such as, low-pass, high-pass, and band-pass filters, capacitors are used as the main filter elements.
- Coupling capacitor: A capacitor to pass AC signal, which allows it to couple sections of an electronic circuit that requires DC isolation.
- Decoupling capacitor: A capacitor placed in parallel with the relevant DC power rail, applied to a small group of integrated circuits to minimize the logic noise and disturbances due to transients and the noise occurring on the power rail.
- Snubber capacitor: A snubber capacitor is used to limit the high-voltage transient across the circuit, which is frequently used in power electronic circuits.
- Power factor correction capacitors: In industrial applications where large inductive loads, such as motors, are used, these capacitors are used to counteract the effect of the inductive load to make load look more resistive or at unity power factor, improving efficiency.
- Motor starters: Some single-phase motors require a start winding that develops a magnetic field that is out of phase with the main winding to start the motor's rotation.

In real-time applications, capacitors do not show their ideal characteristics due to the effect of factors such as parasitic resistance, inductance, and capacitance variations over temperature and voltage. Choosing the wrong capacitor can lead to unpredictable circuit behavior, circuit instability, excessive power dissipation, or noise (Hewitt et al., 2016). A capacitor has a rated capacitance, and a tolerance, which is expressed as a percentage decrease or increase of the rated capacitance. The standard capacitance tolerance is $\pm 20\%$. In general, capacitors can have a tolerance in the range of -20% to as high as $+80\%$ (Sarjeant and Staffier, 1996; Kulkarni et al., 2012; H. A. Inc, n.d.).

The capacitance of a capacitor varies with temperature as the changes in temperature affect the dielectric properties. Furthermore, this variation depends on a small degree on the rated voltage and capacitor size. The normal working range for most capacitors is $-30°C$ to $+125°C$ with nominal voltage ratings (Sarjeant and Staffier, 1996; Kulkarni et al., 2012). ESR of a capacitor also increases with temperature.

The effective capacitance decreases as frequency increases. Impedance of an ideal capacitor is expected to drop with the inverse of the frequency, based on the relationship given by Eq. (6.18). However, the impedance of a nonideal capacitor model takes a dip at self-resonance, f_0 and then begins to increase with frequency as illustrated in Fig. 6.9. The inductive component (ESL) takes over at frequencies greater than f_0 (Kim, 2003; Gupta et al., 2018).

$$X_c = \frac{1}{2\pi f C} \tag{6.18}$$

The dielectric material of the capacitor is not an ideal insulator, as it results in the flow of very small current through the dielectric due to the influence of the powerful electric

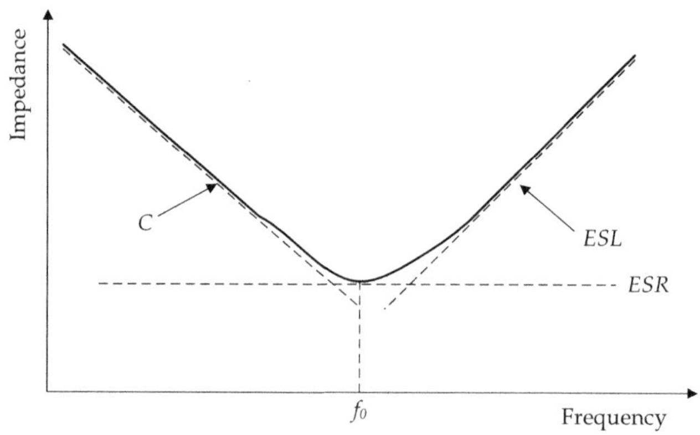

Fig. 6.9 Frequency dependency of a nonideal capacitor.

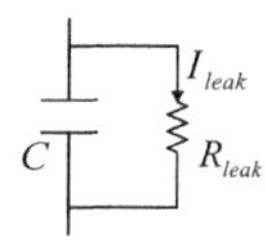

Fig. 6.10 Leakage model of a capacitor.

fields built up by the charge on the plates (Kulkarni et al., 2012). This small DC current flow, in the range of nanoamperes, is called the leakage current. However, electrolytic capacitors may have very high leakage currents typically in the order of about 5–20 μA (Kulkarni et al., 2012). Leakage can be represented as equivalent to a resistor in parallel with the capacitor as shown in Fig. 6.10. Furthermore, the capacitor leakage current also depends on the charging period, applied voltage, and temperature.

6.4 Capacitor types

As discussed previously, the capacitor is one of the most critical passive components used in electronic circuits to provide smoothing, buffering, filtering, bypassing, DC blocking, etc. There are a wide variety different types of capacitors, based on the capacitor construction and the technology used. The most commonly used capacitors are as follows (Sarjeant and Staffier, 1996; Ho et al., 2010; Sarjeant et al., 1998):

- film capacitors
- ceramic capacitors
- electrolytic capacitors

6.4.1 Film capacitors

The distinction between two commonly available types of film capacitors; polymeric and mica (or paper) is the use of synthetic vs natural film-like material (Ho et al., 2010). This technology provides a readily scalable capacitor technology from nano-joules to hundreds of kilojoules per unit. Film capacitors are polarity independent, general-purpose capacitors that are used in applications requiring tight capacitance tolerances and very low leakage currents. They have very low ESR and ESL values and much more stable electrical parameters. Film capacitors have relatively large sizes and weights, but have much higher surge or pulse load capabilities (Dispennette, 2005). In general, film capacitors are typically used in high-voltage applications (1–100 kV) which include moderate to large capacitance value for both DC and AC applications involving high-power electronics pulse-duty circuits, high-frequency filtering, continuous AC operation like lighting ballast and high-frequency inverters, solid-state switch snubbers, SCR commutation circuits, power factor correction, and fractional to high-power motor start and run (Sarjeant and Staffier, 1996).

6.4.2 Ceramic capacitors

Ceramic capacitors are also fixed-value capacitors that are suited for both AC and DC applications as they are nonpolarized capacitors (Sarjeant and Staffier, 1996). Electrical behavior of the capacitor is defined by the composition of the ceramic material.

This technology provides moderate energy density, high-power density, has formulation-dependent losses, is available in very small (pF) to moderate size (10–100 pF; 5–500 Vdc) capacitors including surface mount configurations, and are used primarily in low-voltage DC control circuits. Typical applications are blocking, buffering, bypass, coupling, low frequency filtering, tuning, and timing for frequencies up into the gigahertz regime (Sarjeant and Staffier, 1996; Ho et al., 2010).

The applications of the ceramic capacitors are divided into two stability classes (Kularatna, 2008):

Class 1 ceramic capacitors have high stability and low losses for resonant circuit application.
Class 2 ceramic capacitors have high volumetric efficiency for buffer, bypass, and coupling applications.

6.4.3 Electrolytic capacitors

Electrolytic capacitors are available in several types as aluminum, tantalum, and niobium versions (Ho et al., 2010). The internal structure of an aluminum electrolytic capacitor consists of two aluminum foils, which are separated by a porous material such as paper which is impregnated with an electrolyte as shown in Fig. 6.11. The aluminum foil which forms the anode of the capacitor will have its surface chemically highly etched to increase its surface area, and therefore its capacitance. This foil is anodized to produce a layer of aluminum oxide on its surface and this forms the insulation layer between the capacitor plates and must be sufficiently thick to withstand the rated voltage of the capacitor. The electrolyte is the cathode of the capacitor system, while a second aluminum foil which is lightly etched provides a means of electrically connecting to the liquid electrolyte. The use of a liquid to form the cathode of the system ensures that a good contact occurs between it and the highly etched anode, as the liquid can conform to the surface profile of the aluminum surface. The terminals of the capacitor are connected to the aluminum foils and the entire structure is rolled up and packaged into an aluminum can (Hewitt et al., 2016). Aluminum electrolytic

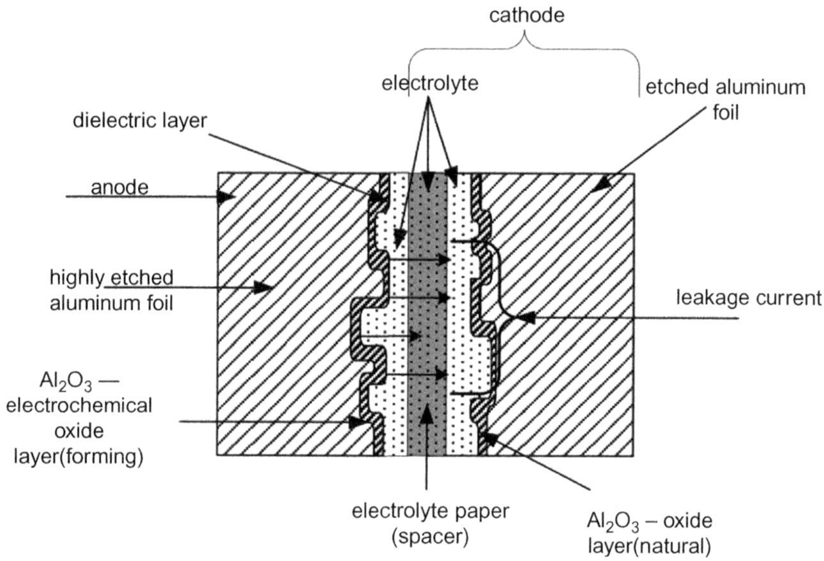

Fig. 6.11 Internal structure of an aluminum electrolytic capacitor (Kulkarni et al., 2012).

capacitors are available in the range of less than 1 μF to 1 F with working voltages up to several hundred DC volts, resulting in more capacitance and energy storage per unit volume (Sarjeant and Staffier, 1996; Ho et al., 2010). They are capable of providing a high ripple current capability together with a high reliability.

The distinction among different varieties of aluminum, tantalum, and niobium electrolytic capacitors is the presence of the type of liquid impregnate. This technology provides moderate energy and power density, but has relatively high power losses due to high ESR, and is polarity dependent, and therefore primarily used in DC circuits. Typical applications are moderate to large capacitors in the range of 1–100,000 μF at up to 600 V, mainly for DC applications involving filtering, rectified circuits, some pulsing circuits such as strobe lights and SCR commutation circuits, and fractional power motor starting (Sarjeant and Staffier, 1996).

Tantalum capacitors are generally available in low-voltage versions up to several hundred microfarads. Because of their smaller size for a given capacitance and lower impedance at high frequencies, they are popular in miniature applications (Kularatna, 2008). This capacitor category has a lower energy density and is produced to tighter tolerances than aluminum electrolytics (Kularatna, 2008).

Niobium capacitor was first introduced as cheaper alternative to tantalum capacitors. But, the material suppliers could not make a capacitor-grade niobium powder that matched the capacitance/voltage (CV) values required to compete with tantalum. Limitations of niobium capacitors include lower volumetric efficiency (CV in a given capacitor size), lower voltage range, and higher DC leakage compared to tantalum capacitors (Holystone, n.d.).

Each capacitor type is suitable for some applications but not adequate for others. A summary of essential features that is provided in Table 6.1 (Ho et al., 2010).

Table 6.1 Characteristics of various capacitor types.

	Aluminum	**Tantalum**	**Niobium**	**Ceramic**	**Film**
Dielectric	Aluminum oxide	Tantalum pentoxide	Niobium pentoxide	Based on barium titanate, etc.	Polyester, polypropylene, etc.
Shape	Screw terminal type, snap-in type, lead terminal type, chip type	Chip type, dip type	Chip type	Chip type, dip type	Dip type, case type for SMD
Advantages	Cheap, small size, and large capacity	Small size and comparatively large capacitance, semipermanent service life	Small size and comparatively large capacitance, semipermanent service life	Small-size, no polarity	Good characteristics, can be made for low- to high-voltage applications, high reliability
Disadv.	Short service life in hot environment, large tolerance, polarized	Can be used with some voltage leeway, polarity	Can be used with some voltage leeway, polarity	Large changes in capacitance due to changes in temperature and DC voltage	Large outside dimensions

Fig. 6.12 A sample of some commercially available capacitors.

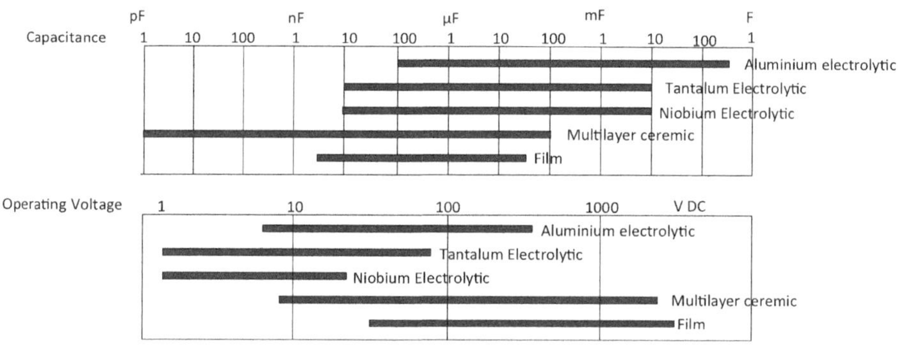

Fig. 6.13 Ranges of capacitance and operating voltage of various capacitors.

Fig. 6.12 shows some common capacitor types. Fig. 6.13 depicts a summary of ranges of capacitance and rated voltage of various types of capacitors (H. A. Inc, n.d.).

6.5 Capacitor aging, lifetime, and reliability

When talking about capacitor reliability, the various possible stages of its typical life cycle such as field aging, failure modes, expected service life, etc., are need to be considered. Capacitor failures could be due to manufacturing and design defects, capacitor field aging, external factors such as operating temperature, voltage, current, humidity, and mechanical stress (Gupta et al., 2018). Capacitor field aging results in the end of useful life of a product.

Chemical reactions are one of the mechanisms that lead to capacitor field aging. Combinations of temperature and chemical contaminates such as oxygen, moisture, and halogens lead to the deterioration of the dielectric materials. Electrolyte evaporation is a prominent deterioration mechanism which is caused by high temperatures within the capacitor core in electrolytic capacitors. This deterioration of the dielectric materials reduces the voltage withstand capability. In the case of metallized film capacitors, self-healing or localized dielectric breakdown due to overvoltage is the main aging mechanism (Emerson Network Power, n.d.).

Dielectric materials are considered as insulators which do not conduct current when voltage is applied across the insulation. However, insulating materials conduct a little current when voltage is applied, which is called a leakage current. Even though the leakage current is really small, when it flows in a very small channel, it leads to a rise in temperature causing heat and material electron interactions. Overtime, both the internal chemical reactions and the leakage current together lead to a reduction in the dielectric voltage withstand capability. The effects of these degradations of the aging process show slowly decreasing capacitance and the slowly increasing resistance values of the capacitor over the time (Hewitt et al., 2016; Emerson Network Power, n.d.).

This decrease in capacitance and increase in the resistance is linked to damage taking place inside of the capacitor. When the damage to the capacitor has been sustained, the probability for the capacitor to fail increases. The health of a capacitor is often explained using ESR and capacitance (Hewitt et al., 2016; Kulkarni et al., 2012). Capacitor industry guidelines which define the end of capacitor service life are based on a decrease in the capacitance and an increase in the series resistance. Both electrolytic and film capacitor technologies age by losing capacitance and developing higher internal resistance as provided in Table 6.2. If this process is allowed to continue, a point will be reached where the capacitor will fail into a short circuit and will lose the ability to withstand the rated voltage (Emerson Network Power, n.d.).

The capacitor aging process also generates gas which increases the internal gas pressure in both electrolytic and film capacitor constructions. The polymeric film capacitor has a pressure interrupter built into the case and when the pressure reaches a preset value, the interrupter opens inside of the capacitor case and disconnects the capacitor from the circuitry. When the pressure interrupter operates, the capacitor case expands upward and is very visible indicating that the capacitor has reached end of life

Table 6.2 Capacitor end of life parameters based on the technology (Emerson Network Power, n.d.).

	Film technology	**Aluminum electrolytic technology**
Max loss of capacitance	−5% to −10%	−15% to −20%
Max increase in ESR	+100% to +150%	+200% to +300%

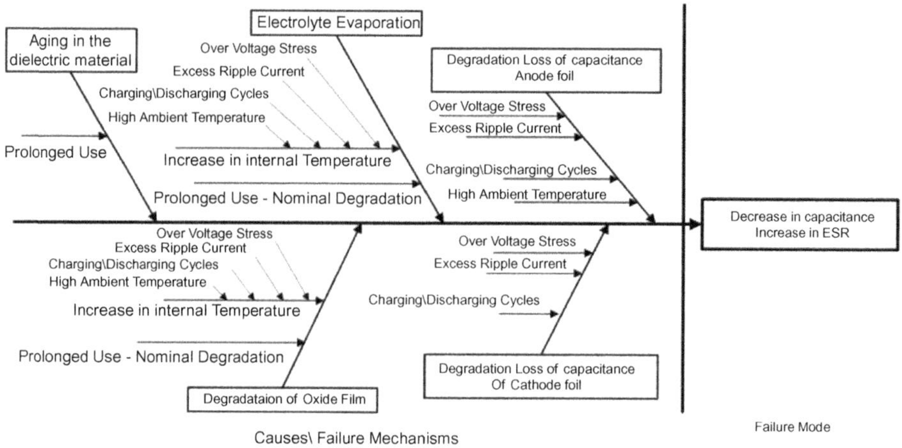

Fig. 6.14 Failure modes for aluminum electrolytic capacitors (Kulkarni et al., 2012).

and is no longer in the circuit. For aluminum electrolytic capacitors, the buildup in pressure pushes up a rubber bung at a preset pressure. This does not disconnect the capacitor from the circuit but it provides a visible sign when the capacitor is reaching end of life.

The fishbone diagram in Fig. 6.14 shows a summary of the most common set of failure modes for aluminum electrolytic capacitors (Kulkarni et al., 2012). This diagram identifies the relationship between root causes and failure modes observed in electrolytic capacitors. The primary reason for degradation in aluminum electrolytic capacitors is due to vaporization of electrolyte and degradation of electrolyte efficiency due to ion exchange during the charge/discharge cycles, which in turn leads to a drift in the equivalent series resistance and the capacitance.

References

ABC of Capacitors n.d.

Christen, T., Carlen, M.W., 2000. Theory of ragone plots. J. Power Sources, 210–216.

Dispennette, J., 2005. Ultracapacitors bring portability to power. Power Electron. Technol.

Emerson Network Power n.d. "Capacitors Age and Capacitors Have an End of Life," Emerson Network Power.

Gupta, A., Yadav, P., DeVoto, D., Major, J., 2018. A review of degradation behavior and modeling of capacitors. In: Presented at the 2018 International Technical Conference and Exhibition on Packaging and Integration of Electronic and Photonic Microsystems (InterPACK 2018), San Francisco, California, August.

H. A. Inc. n.d. "Basics of Capacitors." (Accessed).

Hewitt, D.A., Green, J.E., Davidson, J.N., Foster, M.P., Stone, D.A., 2016. Observation of electrolytic capacitor ageing behaviour for the purpose of prognostics. In: Presented at the 42nd Annual Conference of the IEEE Industrial Electronics Society, Florence, Italy, October.

Ho, J., Jow, T.R., Boggs, S., 2010. Historical introduction to capacitor technology. IEEE Electr. Insul. Mag. 26, 20–25.

Holystone. n.d. "Capacitor Applications." www.holystonecaps.com (Accessed).

Kim, Y., 2003. Ultracapacitor technology powers electronics circuits. Power Electon. Technol.
Kularatna, N., 2008. Electronic Circuit Design: From Concept to Implementation. CRC Press.
Kulkarni, C.S., Celaya, J.e.R., Biswas, G., Goebel, K., 2012. Accelerated aging experiments for capacitor health monitoring and prognostics. In: Presented at the IEEE AUTOTESTCON Proceedings, Anaheim, CA.
Sarjeant, W.J., Staffier, D.T., 1996. A report on capacitors. In: Presented at the Proceedings of Applied Power Electronics Conference. APEC '96, San Jose, CA, USA.
Sarjeant, W.J., Zirnheld, J., MacDougall, F.W., 1998. Capacitors. IEEE Trans. Plasma Sci. 26 (5), 1368–1392.

Electrical double-layer capacitors

7

Jayathu Fernando

7.1 Introduction

Chapter 6 provided a practical overview of common capacitor families used in electronic circuits and a revision of simple concepts related to capacitors. In general, common capacitor families do not have devices beyond several hundred thousand microfarads, but these devices come with a wide range of DC voltage ratings from 5 V to several kilovolts. In general, these devices are used in well-known circuit applications such as DC blocking, filtering, temporary energy storage, and building tuned circuits.

However, when we consider energy storage requirements similar to batteries, a relatively newer family of capacitors called electrochemical (EC) capacitors has entered the market from about the late 1980s, and within the last decade many manufacturers have introduced many variations in very high capacitance, low-voltage devices. These EC devices, variously referred to by manufacturers as supercapacitors, ultracapacitors, or electrochemical double-layer capacitors (EDLCs), store charge in an electrical double layer at the interface between a high surface area electrode and an electrolyte. Current commercial devices are relatively low-voltage single-cell devices with DC voltage ratings up to about 4 V or series-connected modules with voltage ratings up to and over 100 V DC. As we have indicated in Chapter 1, supercapacitors are very much capable of providing short-duration power bursts, and they are having life cycles of half a million to several millions, compared to batteries. As a general guideline, supercapacitors are over 10 times more capable in terms of specific power (W/kg) rating and about 10 times smaller compared to rechargeable batteries in terms of gravimetric energy density (Wh/kg) rating.

This chapter provides a general discussion on EDLC principles, modeling, equivalent circuits, device properties, and characterization. While this chapter provides an overview of historical developments up to 2015, concentrating mainly on symmetrical EDLCs, where both electrodes are nearly identical, Chapter 8 is an overview of more recent developments on hybrid types and pseudo-capacitors.

7.2 Historical background

Hermann von Helmholtz, a German physicist, first described the concept of double-layer capacitance in 1853. General Electric Company in 1957 first patented (Rightmire, 1966) the electrochemical capacitor based on the double-layer

Energy Storage Devices for Renewable Energy-Based Systems. https://doi.org/10.1016/B978-0-12-820778-9.00004-8

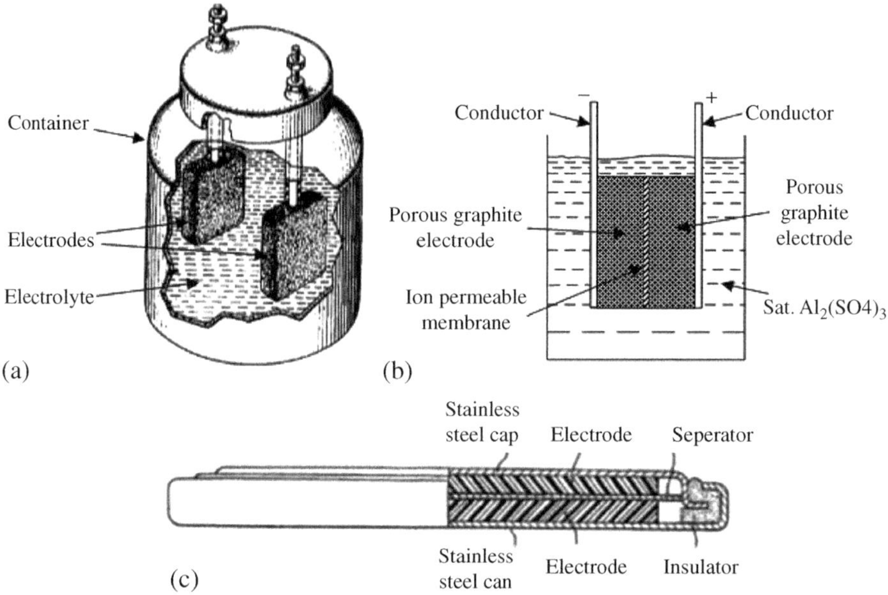

Fig. 7.1 Early supercapacitors (A) from the 1957 patent, (B) from the 1966 patent, and (C) early button cell supercapacitor.
Source: (A) Becker, H.I., 1957. Low Voltage Electrolytic Capacitor. US Patent 2800616, July 23; (B) Rightmire, R.A., 1966. Electrical Energy Storage Apparatus. US Patent 3288641, November 29; (C) Endo, M., Takeda, T., Kim, Y.J., Koshiba, K., Ishii, K., 2001. High power electric double layer capacitor (EDLC's); from operating principle to pore size control in advanced activated carbons. Carbon Lett. 1(3 and 4), 117–128.

capacitance structure (Fig. 7.1A), and this capacitor consisted of porous carbon electrodes with a double-layer capacitance. Commercial double-layer capacitors originated in the Standard Oil of Ohio (SOHIO) Research Center (Cleveland, United States) in 1961 and 1962 and a patent was granted to Rightmire titled "electrical energy storage apparatus" (Becker, 1957). Fig. 7.1B depicts the concept. Development work continued and the first advertising brochure was released in 1969. SOHIO also utilized the double-layer capacitance of high surface area carbon materials, but in a nonaqueous solvent containing a dissolved tetra-alkyl-ammonium salt electrolyte. At this time, SOHIO acknowledged that the *double layer* at the interface behaves like a capacitor of relatively high specific capacity. SOHIO went on to patent a disc-shaped capacitor in 1970 utilizing a carbon paste soaked in an electrolyte. Due to lack of sales, in 1971, SOHIO halted their development efforts and later licensed the technology to Nippon Electric Company (NEC) who developed and marketed commercial double-layer products. NEC of Japan licensed the technology from SOHIO and introduced the first EC products to the marketplace as memory backup devices.

Another group, Carborundum Company of Niagara Falls, New York, (formerly SOHIO), started selling double-layer capacitor products under the name MAXCAP and these devices are still sold by a company named Kanthal Globar (Kanthal Globar, n.d.). Several series with capacitance values from about 0.047 to 1 F with ESR values from fractional ohms to a few ohms, high temperature range from −25°C to 85°C, small diameter or low height packages are offered. Product brochures describing the MAXCAP products show a similar construction to the NEC capacitors and are another source of early products demonstrating the success of the double-layer effect (Endo et al., 2001). These early devices, due to the high ESR compared to modern-day supercapacitors, are only useful for limited applications such as memory backup or similar applications where a high ESR could be tolerated at low load currents.

During the 1980s, Matsushita Electric Industrial Co., Japan, now known as Panasonic, patented methods for producing double-layer capacitors containing improved electrodes (Endo et al., 2001). One, made from activated carbon (AC) fibers woven into a fabric, was the initial basis for the line of electric double-layer capacitors called Gold capacitors. Fig. 7.1C shows the cross-sectional view of a practical button cell using an AC fiber as a polarizable electrode (Endo et al., 2001). During the last decade, the market of electric double-layer capacitor has been drastically developing due to the increasing demand from devices such as mobile phones and laptop computers, and larger systems such as electric vehicles (EVs). Moreover, these environmentally friendly devices (compared to batteries) have become attractive for clean energy technology such as renewable sources with the increased demand for environmentally friendly solutions.

7.3 Electrical double-layer effect and device construction

Supercapacitors are based on the same simple principles applicable to electrostatic capacitors where larger area plates and shorter distance between plates could give a higher effective capacitance. In supercapacitors, the electrical double layer formed next to a large-area electrode and an electrolyte is effectively used, and hence these devices are technically called electric double-layer capacitors (EDLCs).

At this stage, it is worth summarizing the difference between electrochemical (EC) cells and electrochemical capacitors. In EC capacitors, actual positive and negative electrostatic charges physically reside at some two-dimensional plates set. In an EC cell, charges are stored indirectly and can be admitted or withdrawn from the electrodes based on Faradaic processes to transfer charges across the interface, involving oxidation and vice versa. Two types of processes, therefore, are distinguished as non-Faradaic processes and Faradaic processes in capacitors and batteries, respectively.

7.3.1 Electric double-layer effect

Fig. 7.2 depicts the basic idea of an EDLC, where two electrodes are inserted into an electrolyte which can act as an EDLC. As shown in Fig. 7.2A, when a power source is applied to the electrodes made up of a large surface area such as AC, at the collectors attached to electrodes, positive and negative charge layers built up with oppositely charged layers are formed within the electrolyte, next to the electrode surfaces. The use of a porous carbon, such as AC, is motivated by its high electrical conductivity, low cost, and high chemical stability with a large specific surface area (SSA) (or large surface area/mm^2). The large surface area is due to the distribution of pores, creating large areas as shown in Fig. 7.2B, for the case of each electrode-electrolyte surface in Fig. 7.2A.

Unlike batteries where energy storage is achieved via reduction and oxidation (redox) reactions creating electron transfers between chemical species, EDLCs are based on the charge separation occurring at an electrode-electrolyte interface. Fig. 7.3 depicts the simplified case of an EDLC construction to explain the behavior of a porous electrode. In a given particle of the porous material, different pore sizes are mixed and in each of these randomly distributed pores, an electrical double layer is formed, giving the simple case of a capacitor plate with a very high surface area.

Fig. 7.4 indicates an expanded view of the cross section of the case of a porous electrode, where the electrode surface is composed of multiple diameter pores with different pore size distribution (PSD). Depending on the diameter of the pores, it is categorized into subsets such as macropore, mesopore, and micropore as in

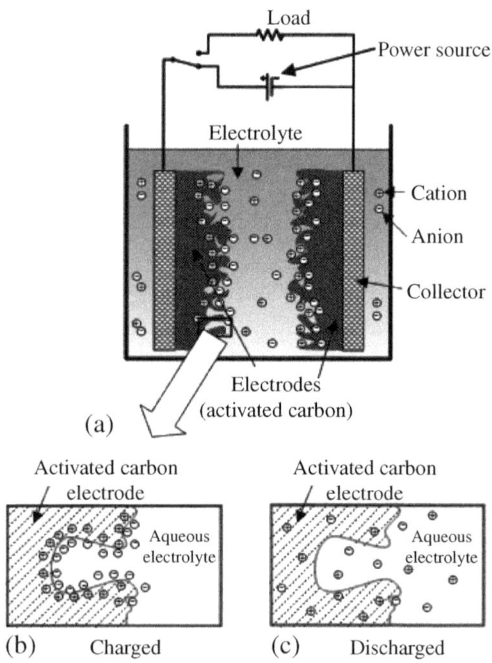

Fig. 7.2 Idea of an EDLC capacitor formed by the double layers at each electrode. (A) Conceptual electrochemical capacitor, (B) charge conditions, and (C) enlarged view of porous carbon under discharged condition.

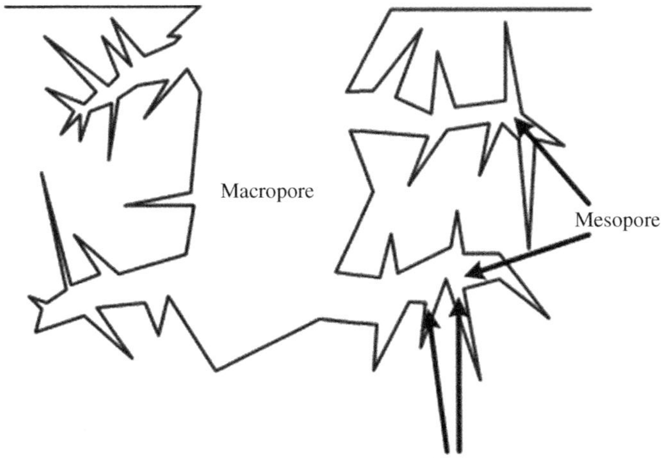

Fig. 7.3 A detailed view of the double layer showing macro, meso, and microlevel pores. Adapted from Itagaki, M., Suzuki, S., Shitanda, I., Watanabe, K., Nakazawa, H., 2007. Impedance analysis of electrical double layer capacitor with transmission line model. J. Power Sources 164, 415–424.

Fig. 7.3. This indicates that a porous electrode like AC can have a very a large effective surface area, to create a large capacitor at each electrode. In order to mathematically predict the behavior of the porous electrode, Levie (1963) developed analytical models based on circular cylindrical surfaces of fixed radius, which lead to the transmission line models (TLMs) usable for porous electrode behavior. Fig. 7.4A depicts this approximation in a pictorial manner, as adapted from this early publication. More details on this can be found in Itagaki et al. (2007) and Levie (1963). In order to achieve high energy density and power density in commercial devices, electrode materials must possess a large SSA available for ion electroadsorption, PSD optimized for the desired combination of energy and power characteristics, good electrical conductivity, and wettability by the electrolyte utilized in the device construction. The SSA and PSD are often considered to be two key factors controlling the energy and power density of EDLCs (Wei and Yushin, 2012). To improve the state-of-the-art devices, various porous carbon materials, such as ACs, carbide-derived carbons (CDCs), zeolite-templated carbons (ZTCs), carbon nanotubes (CNTs), carbon onions, carbon aerogels, and graphene, have been studied extensively as electrodes in EDLCs (Wei and Yushin, 2012). Out of these, though CDCs and ZTCs offer tight pore size control, high SSA, high conductivity, etc., these techniques are yet to get commercialized (Wei and Yushin, 2012).

7.3.1.1 Device construction

A practical DLC consists of two electrodes attached to current collectors, which are separated by a separator soaked in electrolyte, as shown in Fig. 7.5A. By selecting a large SSA-type material, such as AC, the capacitance per gram of material is

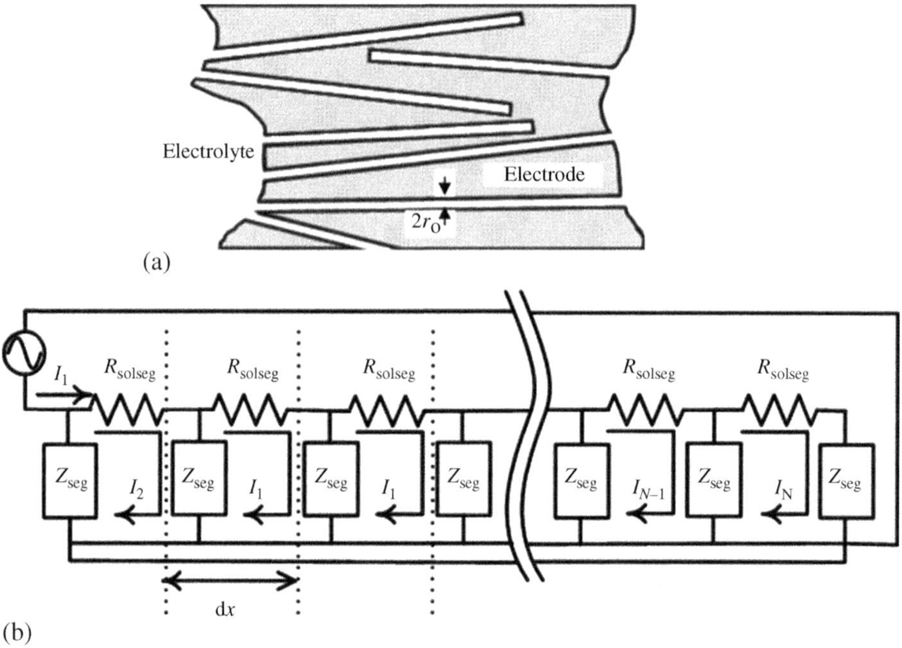

Fig. 7.4 Simplification of porous structure providing a large specific surface area (SSA). (A) Simplified case of a porous electrode by Levie (1963) with circular cylinders of identical diameter for developing a transmission line model (TLM) and (B) a transmission line model for a porous electrode.
Adapted from Itagaki, M., Suzuki, S., Shitanda, I., Watanabe, K., Nakazawa, H., 2007. Impedance analysis of electrical double layer capacitor with transmission line model. J. Power Sources 164, 415–424.

increased. The inner surface of each electrode is not a smooth surface, but is rather padded with activated porous carbon, resulting in a surface area that is about 100,000 times as large as the surface area of an ordinary electrostatic capacitor (Bakhourn, 2009). Fig. 7.5B depicts the actual construction of a cylindrical supercapacitor such as from Honda (Itagaki et al., 2007).

Electrode materials

Electrode material is the key for high capacitance and other important properties of electrochemical capacitors. High surface area, based on the porosity of the electrode, is the basic criteria to achieve, among other properties. Three different types of supercapacitors are possible based on electrode material, namely, (i) carbon/carbon, (ii) metal oxide, and (iii) conducting polymers. Carbon/carbon-type devices with an organic electrolyte are quite common.

Activated carbon While there are several electrode material options as already discussed, AC is the most commonly used type of porous carbons in supercapacitors. While there are recorded uses of AC in Egypt as far back as 1550 BC, industrial production of ACs in the United States was started in 1913 (Wei and Yushin, 2012).

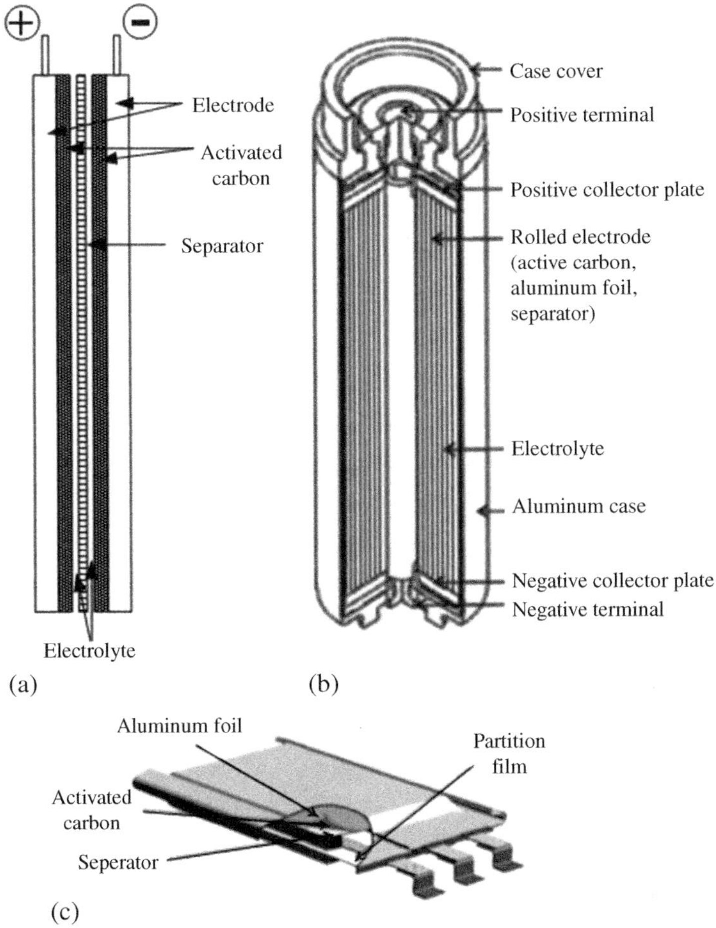

Fig. 7.5 Practical construction of a supercapacitor. (A) Basic idea, (B) cross section of a practical device by Honda (Itagaki et al., 2007), and (C) cross section of a practical device by Murata (http://www.murata.eu/downloads/DMF-DMT_TechnicalGuide_MEED.pdf).

At present, nearly all commercial EDLCs utilize a high surface area AC due to their well-developed manufacturing technologies, easy production in large quantities, relatively low cost, and great cycle stability. Petroleum coke, pitch, and coal used to be the most common precursors for commercial AC production, but the decreasing availability of fossil fuels, the growing global energy demand, and an increased awareness of the environmental impacts of fossil fuel combustion led to AC production from sustainable and renewable resources, such as nutshells, wood, starch, sucrose, cellulose, corn grain, banana fiber, coffee grounds, sugarcane bagasse, and others. The costs of some of the raw materials and the carbon yields of pyrolysis from these carbon sources are presented in Table 7.1 as per Wei and Yushin (2012). It is likely that renewable natural sources in Table 7.1 will likely dominate the future AC production.

Table 7.1 Raw material costs and carbon yield from pyrolysis of natural products used for activated carbon (AC).

Raw material	US$ cost (kg)	Carbon yield (percentage by weight)
Petroleum coke	1.4	90
Charcoal	1.2	90
Lignite	0.75	50
Coconut shell	0.25	30
Wood	0.8	25
Potato starch	1.0	45
Sucrose	0.25	<45
Cellulose	0.65	<45
Corn grain	0.25	<45
Banana fiber	4	<45

Source: Wei, L., Yushin, G., 2012. Nanostructured activated carbons from natural precursors for electrical double layer capacitors. Nano Energy 1, 552–565.

There are different types of ACs, with different SSA values based on the processing of AC materials. SSA values could vary from about 1200 to 2300 m^2/g as with resistivity values varying from 0.4 to 3 Ωcm for a range of ACs from the same manufacturer depending on the processes used for each case (Gamby et al., 2001). This performance variation depends on the PSD in each case and other properties of the material in general. Over the last quarter century, carbon has proved to be a very useful and lower cost electrode material due to its (a) chemical stability, (b) easy processability, (c) well-established physical and chemical methods for activation, (d) operational temperature range, and (e) usability without any binding material (Frackowiak and Béguin, 2001). As per studies in the mid-1990s, while petroleum coke could provide SSA in the range of 2600–3100 m^2/g, coconut shell-based AC provides SSA in the range of 950–2100 m^2/g (Morimoto et al., 1996). Up to 2015, AC was the most commonly used commercial electrode material for EDLCs. Raw material and activation methods are two of the most important factors which may influence the final structure and cost of ACs. For example, coal variations such as hypercoal, which is an ashless version obtained by mild thermal extraction of coal (hypercoal process), have very much acceptable properties for AC. Zhao et al. (2014) discusses a process using KOH activation of hypercoal to achieve SSA of 2440 m^2/g with specific capacitance (capacitance per gram of material) of around 44 F/g.

Most of the current research into electrode material places a heavy emphasis on different forms of ACs, metal oxides, and conducting polymers.

Graphene Since carbon materials can meet the requirements of EDLC electrodes, various carbonaceous materials such as ACs (in different forms), aerogels, xerogels, nanotubes, nanofibers, mesocarbon microbeads, expanded graphite, and ordered mesoporous carbons have been studied extensively as electrode materials (Du et al., 2010). Although ACs always have high SSA values, due to PSD, compared to the size of ions in relation to pore sizes, all the available surface area is not effectively usable for EDLCs. This is simply because EDL is formed by the ions coming in direct contact with the electrode surface. If any micropores are too small to provide contact with ions, EDL is not formed to give high capacitance (Du et al., 2010).

Graphene is a two-dimensional carbon plane with one atomic thickness. Since the paper on the freestanding graphene was published by Geim and coworkers (Novoselov et al., 2004), the preparation, structure, and property of graphene have attracted great attention owing to its particularly useful properties such as mechanical hardness, electrical conductivity, and so on. These properties suggest wide applications, including field effect transistors, adsorbents, lithium-ion batteries, and supercapacitors. Du et al. (2010) discuss the possibility of mass-producing grapheme nanosheets as electrode material for EDLCs.

Graphene-based EDLCs with ultrahigh-performance EDLCs were developed around 2010, to demonstrate the possibility of efficient 120 Hz line frequency component filtering, which is not well achieved with most commercially available SCs (Miller et al., 2011). As discussed by Miller et al. (2011), due to time constants of values around 1 s in commercially available SCs, they are not too efficient in filtering twice the line frequency components in AC to DC converters. Fig. 7.6A depicts the case of phase angle plots for various capacitors with various time constants including the case of an ideal capacitor. Fig. 7.6B depicts the phase angle performance of a graphene-based EDLC research (Miller et al., 2011) compared with a commercially available 350 F EDLC. While a 45 degrees phase angle for a commercial ELDC occurs around 0.15 Hz, this 175 F grapheme nanosheet capacitor provides an EDLC with a 45 degrees phase angle occurring at 15 kHz. More details on the subject have been provided by Miller et al. (2011), and basics related to this subject are in Appendix A.

Carbon nanotubes In general, the ESR of an EDLC results mainly from the electrical contact resistance in electrodes, the resistance in the solution bulk phase, and the resistance to ion migration in carbon micropores. In order to reduce the ESR, CNTs could be used since they are known to have high electrical conductivity and chemical stability. These two features render CNTs an ideal additive for electrodes to improve the conductivity. For this reason, there is an ongoing research to explore CNTs as a suitable electrode-enhancing material in EDLCs (Huang et al., 2008). In this work, Huang et al. (2008) developed a technique to grow CNTs on a polyacrylonitrile (PAN)-based AC fiber, in an attempt to enhance the electronic conductivity between the constituting fiber threads. For EDLCs assembled with this carbon fiber, the CNT-grafting modification is shown to significantly enhance the conduction of both electrons and electrolyte ions during double-layer formation. In this experiment, PAN-based AC fiber was used in the form of a woven cloth with a thickness ranging from 0.4 to 0.6 mm, with an SSA of 1200 m^2/g and pore volume of 0.59 cm^3/g. By using nickel as a catalyst for the growth of CNTs on carbon cloth, CNT-grafted AC electrodes were achieved. Fig. 7.7 depicts the comparison of this CNT-grafted fiber and the bare fiber based on Nyquist plots.

Recently, miniaturized EDLCs have been suggested for powering microelectromechanical systems, because of their dimensions and energy density.

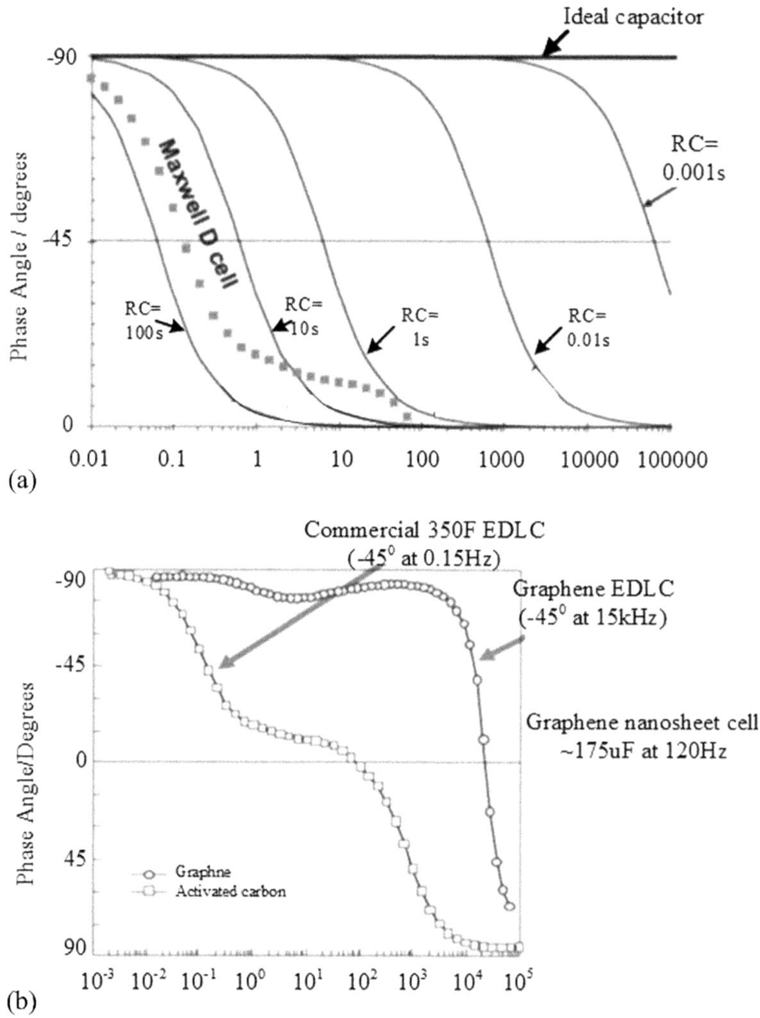

Fig. 7.6 Phase angle plots of capacitors and AC line frequency filtering capability of graphene-based EDLCs. (A) Phase angle of a series *RC* circuit and (B) Comparison of phase angle versus frequency of commercial EDLC of 350 F and graphene-based EDLC.
Source: Miller, J.R., Outlaw, R.A., Holloway, B.C., 2011. Graphene electric double layer capacitor with ultra high performance. Electrochim. Acta 56, 10443–10449.

Liu et al. (2011) and Sung et al. (2006) have provided details of microcapacitors suitable for MEMs and flexible microsupercapacitors, and study by Liu et al. (2011) is based on CNTs. Fig. 7.8 shows the study performed by Sung et al. (2006). This all-polymeric cell can have any shape and it does show resistance to change of capacitance due to bending, etc. Zheng et al. (2014) indicate more recent works related to porous graphene and AC composites with high packing density and large surface area for SC electrode material.

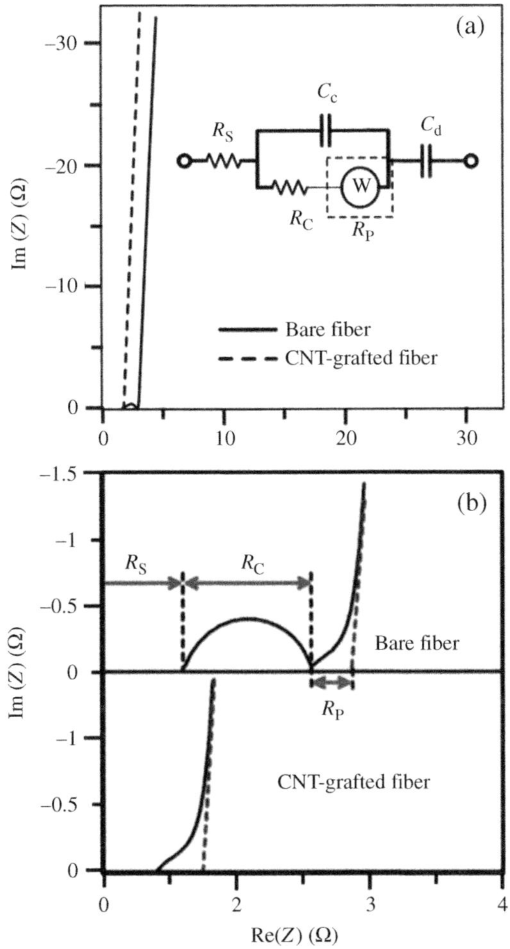

Fig. 7.7 Nyquist plot-based comparison of CNT-grafted fiber and bare fiber from 10 mHz to 100 kHz. (A) Equivalent circuit showing resistive components R_s, ion migration resistance; R_c, electrical contact resistance; R_p, distributed pore resistance. (B) Magnified view of the high frequency region of the Nyquist plot.
Source: Huang, C., Chuang, C., Ting, J., Teng, H., 2008. Significantly enhanced charge conduction in electric double layer capacitors using carbon nanotube-grafted activated carbon electrodes. J. Power Sources 183, 406–410.

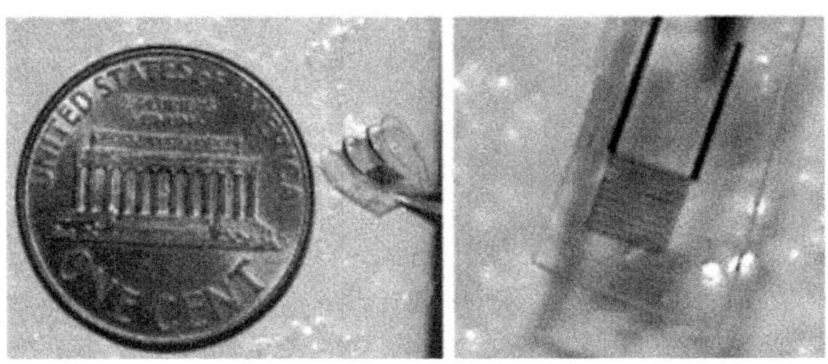

Fig. 7.8 Flexible microsupercapacitor.
Source: Sung, J., Kim, S., Jeong, S., Kim, E., Lee, K., 2006. Flexible micro-supercapacitors. J. Power Sources 162, 1467–1470.

Recently, some companies such as Skeleton Technologies have announced that they will be using a material patented as "curved graphene" in their new supercapacitor families.

Metal oxides Transition metal oxides such as oxides of Fe, Cu, Ni, Mn, Cu, and RuO_2 for electrode materials offer rich redox reactions such as in electrochemical cells providing high specific capacitance values for supercapacitors. In these situations, the EDL effect is combined with a Faradaic process like in batteries, and this additional effect is called *pseudocapacitance*, which will be discussed later. To date, it is well established that ruthenium oxide (RuO_2) is a prominent material exhibiting high specific capacitance and excellent reversibility. However, this metal oxide is expensive and toxic, and the search for environmentally friendly metal oxides is ongoing. Nowadays, transition metal oxides such as CuO are researched for supercapacitors (Dubal et al., 2013). CuO-based supercapacitors have recently attracted significant attention due to promising capacitance, environmental friendliness, good electrochemical performance, and lower cost (Dubal et al., 2013). More details have been provided by Dubal et al. (2013).

Another new approach using TiO_2 nanodots on the walls of multiwalled carbon nanotubes (MWNTs) to develop a TiO_2/MWNT nanocomposite for electrodes has been discussed by Sankapal et al. (2014). This seems to be a further development of many past efforts to fabricate different metal oxide/MWNT composites. MnO_2/MWNT, SnO_2/MWNT, NiO/MWNT, and V_2O_5/MWNT are other composites developed by researchers worldwide (Zhao et al., 2014). In these efforts, pseudocapacitance and electrostatic capacitance effects are combined to get an overall capacitance of high value. New interest on TiO_2 is mainly due to its higher stability, high abundance, better safety, and environmental friendliness. In the binder-free approach of TiO_2/MWNT, capacitance yield is reported to be 329 F/g and the scalability for larger area for industrial fabrication looks feasible (Sankapal et al., 2014).

Conducting polymers Electronically conducting polymers (ECPs) are promising materials for the realization of high-performance supercapacitors. Conducting polymers such as polypyrrole, polyaniline, polythiophene, polyindole, etc. are typical prototype materials for supercapacitors based pseudocapacitance. In particular, conducting polymers are especially suitable in view of their ease of fabrication and flexibility (Ramya et al., 2013). The ability to store charges in the case of conducting polymers arises from a doping process.

ECPs are characterized by high specific capacitances since their charging processes pertain to the whole polymer mass and not only to the surface, as in the case of double-layer ACs, and by high conductivities in the charged states. Furthermore, their charge-discharge processes are generally fast. These features help to develop devices with low ESR and high specific energy and power. The most promising polymer supercapacitor configuration in terms of charge storage capacity and cell voltage able to outperform the double-layer carbon supercapacitors is the n/p type, i.e., the configuration in which an n-doped polymer and a p-doped one are used as negative and positive electrodes, respectively (Mastragostino et al., 2002). However, this is not an easy-to-realize configuration and the difficulties are mainly related to the n-doping process. The charge-discharge processes of polymer-based supercapacitors are Faradaic and those of the n/p devices take place in a cell potential range settled by the n- and p-doping-undoping process potentials of the polymer electrodes. Thiophene-based polymer materials are used in these. Specific capacitance values of 95–220 F/g were achieved for p-doping and 80–165 F/g were achieved for n-doping as detailed by Mastragostino et al. (2002).

Composites of CNTs and conducting polymers can be prepared via chemical synthesis, electrochemical deposition on preformed CNT electrodes, or by electrochemical codeposition. These composites could combine the large pseudocapacitance of the conducting polymers with the fast charging/discharging double-layer capacitance and excellent mechanical properties of the CNTs. More details are available in the review paper by Peng et al. (2008).

Electrolytes

Different kinds of electrolytes have been used in EDLCs with AC electrodes: (i) organic (most common solution of tetraethylammonium tetrafluoroborate (TEATFB) salt either in anhydrous acetonitrile (AN) or propylene carbonate (PC) solvents), (ii) aqueous (solution of acids, bases, and salts), and (iii) ionic liquid (IL). Aqueous electrolytes are typically stable in symmetric EDLCs to 0.6–1.0 V, organic electrolytes to 2.2–2.9 V, and ILs to 2.6–4.0 V.

Organic electrolytes Organic electrolyte-based EDLCs offer a life cycle in excess of 5,000,000 and a moderately high operational voltage. They can operate in the temperature window from −30°C to +50°C (with an acetonitrile solvent) and are used in the majority of commercial devices with AC electrodes. However, they seem to suffer from high flammability and potential explosion risk because organic solvents exhibit high vapor pressure. Most practical devices today are based on organic electrolytes. Morimoto et al. (1996) have provided some details on early coin-cell type capacitors with PC solutions for memory backup applications. These early coin-cell devices were achieving capacitance values up to 0.5 F with ESR values from 10 to 20 Ω. Based on the same approach, with stacking layers of basic anode-cathode pairs, power capacitors with a capacitance of 4700 F were also achieved. Acetonitrile (AN) is widely used as a solvent for electrolytes in most commercial supercapacitors, with a dissociating salt such as tetraethylammonium tetrafluoroborate ($Et_4N^+BF_4$). The main advantage of this combination is low viscosity leading to high conductivity, but unfortunately it is too volatile and flammable. Organic electrolytes provide a greater potential window when compared to aqueous electrolytes, but low thermal stability, flammability, and toxicity limit their use in high-performance supercapacitors.

Aqueous electrolytes The advantages of using aqueous electrolytes include their very low-cost and high ionic conductivity. Serious disadvantages, however, include their low maximum applied voltage and the corrosion of EDLC electrodes observed at higher temperatures and voltages (particularly for acid-based electrolytes, such as H_2SO_4 solutions), which limits the lifetime and life cycle of the EDLCs. These have the potential for bulk storage capability.

Ionic liquids ILs are room-temperature molten salts, which are entirely composed of cations and anions, the nature of which influence the chemical/electrochemical and physical properties. In particular, the cation limits the negative potential window and the anion affects the positive potential window as well as the melting point, which in turn affects the working temperature range. ILs typically show a very low vapor pressure, high thermal stability, wide electrochemical window, and good conductivity at temperatures above 60°C, which make them suitable for supercapacitors required for applications such as in EVs (Arbizzani et al., 2007). Due to problems with common solvents used in organic electrolytes such as AN (which releases toxic elements released when burned) or γ-butyrolactone (which is more viscous, heavier, and presenting a less separating effect) room-temperature ILs are currently researched as alternatives for organic electrolytes (Abdallah et al., 2012).

The high-voltage stability of ILs makes IL-based EDLCs of very high energy density. In addition, ILs are nonflammable, nonvolatile, and nontoxic, which are important attributes for many mobile applications, including their potential use in hybrid engines (Wei and Yushin, 2012). Serious shortcomings of ILs are their high (often prohibitively high) cost and commonly low ionic mobility at room temperature. More details are available in Wei and Yushin (2012) and the citations therein. Supercapacitors based on reduced graphene oxide as electrodes and ILs as electrolyte were reported by Paulo et al. (2014) with electrochemical stability of 3 V with a specific capacitance of 71 F/g. Carbon-carbon (AC in both electrodes) electrodes with $PYR_{14}TFSI$ as an IL-specific capacitance value of 60 F/g was reported in 2007 with a maximum potential voltage of 4.5 V at 60°C, without any deterioration of ESR up to 40,000 cycles (Balducci et al., 2007). More recent work combines graphene nanosheets (GNSs) with IL to achieve specific capacitances in the range of 114 F/g (Kim and Kim, 2014).

Summary of research directions: Electrodes and electrolytes

Table 7.2 depicts a summary of research achievements as per citations in the chapter, comparing the specific capacitance, maximum voltage, and cycle life based on the work of worldwide electrochemists. It is important to note that specific capacitance depends on scan rate and information provided in the table shows the highest capacitance achieved in the specific work.

While early research efforts were concentrated toward increasing the properties of carbon materials by surface modification (Fang et al., 2006; Prabaharan et al., 2006), voltage capability toward 4 V (Prabaharan and Michael, 2004), and the use of high SSA material like C-Cloth (Niu et al., 2006), on another front application of CNTs, aerogel, xerogel, and nanoporous carbon materials, etc. (Chang et al., 2013; Du and Pan, 2006; Fang and Binder, 2006; Nasibi et al., 2013) were investigated. After 2010, graphene, graphene oxide, and GNSs combined with metal oxides were investigated (Gund et al., 2013; Kim and Kim, 2014; Zhang et al., 2014) for enhanced performance and using pseudocapacitance effects.

Overall research into supercapacitors is presently divided into two main areas that are based primarily on their mode of energy storage, namely: (i) the redox supercapacitor (pseudocapacitor) and (ii) the electrochemical double-layer capacitor.

Table 7.2 Summary of research achievements in SCs as per research work during 2004–14.

Electrodes	Electrolyte	Specific surface area (SSA) m^2/g	Specific capacitance (F/g)	Maximum voltage (Volts)	Number of test cycles performed measured	Reference source
Carbon-carbon	IL ($PYR_{14}TFSI$)	1428	90	3.5	40,000	Balducci et al. (2007)
SFG6-SFG6	LiPF6/EC + DMC		14.5	4.5	1000	Prabaharan and Michael (2004)
C-Cloth	H_2SO_4	2500	230–485	1.15		Niu et al. (2006)
MnO_2 NPs-Fe_3O_4 NPs	K_2SO_4 aq.		21.5	1.8	5000	Aiguo et al. (2009a)
GNS/IL	KOH		114		3000	Kim and Kim (2014)
AC	KOH		265	1	10,000	Xianzhong et al. (2012)
AC	$LiSO_4$		210	1.6	10,000	Xianzhong et al. (2012)

Manufacturing aspects of AC-based devices While the manufacture of supercapacitors aims at long life cycles of over a million charge-discharge cycles, the cost of manufacturing is one of the important business criteria. While a lot of new electrode materials and electrolytes are researched, still the industry is inclined to use cheap materials (like AC) for electrodes with organic electrolyte based on acetonitrile. A standard setup has emerged over the last decade: aluminum contact foils (coated on both sides with AC (Pandolfo and Hollenkamp, 2006)), an organic electrolyte, and a separator (e.g., paper). We here focus on tetraethylammonium tetrafluoroborate (TEA + TFB −) in acetonitrile (AN), which has the advantage of high conductivity even at low temperatures. The electrodes are not only composed of AC but also contain a polymeric binder in order to allow facile coating of the aluminum foils, and a conductive additive, usually carbon black, to enhance electrical interconnections between AC particles. EDLCs can show a short "burn-in" phase where their electrical characteristics change considerably; however, this is very fast (hours to days). It is clear that the ultimate performance of carbon-based supercapacitors will be closely linked to the physical and chemical characteristics of carbon electrodes. A review on AC properties and their role in EDLCs is provided by Pandolfo and Hollenkamp (2006). Aging of EDLCs has been discussed by Bittner et al. (2012). The impact of pore structure on series resistance has been discussed by Celzard et al. (2002). It is found that microporous carbon helps develop high capacitance slow charge-discharge, while mesoporous carbon helps high rate devices in organic electrolytes (Wang et al., 2007).

In practical devices, separator material characteristics influence the high- and low-frequency series and parallel resistance, low- and high-frequency capacitance, time constant, energy, and power density. PSD, the thickness of the separator, and SSA are important parameters for device performance, and research findings have been provided by Tõnurist et al. (2009, 2013).

7.4 Pseudocapacitance and pseudocapacitors

In electrochemical double-layer capacitance discussed so far, actual positive charges due to electron deficiency on one plate and negative charges due to excess electrons on the other plate are separated by a large-area material to give very high capacitance with long life cycles. In comparison to batteries where redox reactions (or Faradaic processes) occur at electrode surfaces, these devices are similar to electrostatic capacitors, where a dielectric separator is used in between plates. As stated by Burke (2000), the capacitance, $\frac{dQ}{dV}$, in this case is relatively constant and independent of voltage.

In parallel with the development of DLCs, another type of large capacitors termed as *pseudocapacitors* was developed by Conway et al. (1997) where Faradaic currents from reactions and the resultant charge transfer give rise to a *pseudocapacitance*. In general, this capacitance cannot be measured, but is utilizable as a capacitance and the value is voltage dependent. However, there are a number of metal oxides-based pseudocapacitive materials with constant capacitance with changing voltage (Ruiz et al., 2013).

Pseudocapacitance arises in cases where Faradaic charge-transfer processes that lead to passage of extents of charge, q, that, for thermodynamic reasons, depend on potential, unlike ideal Nemstian processes, where a singular, (ideally) constant electrode potential, e.g., for a metal/metal-ion equilibrium, arises, independent of the extent of the reaction. Examples of processes which are associated with pseudocapacitance are as follows:

(i) two-dimensional depositions (so-called underpotential deposition) of adatom arrays on electrode surfaces, e.g., H or Cu on Pt, Pb on Au, Bi on Au, Bi on Ag, H on Rh or Pt;
(ii) redox processes in liquid or solid solutions where the electrode potential is a function of the log of the ratio of reductant converted to oxidant (or vice versa) in a redox system; and
(iii) where, in some cases, chemisorption of anions at electrode interfaces takes place with potential-dependent partial Faradaic transfer of an electronic charge related to the so-called electrosorption valency. In general, there can be coupling between non-Faradaic double-layer charging and Faradaic surface processes (Conway et al., 1997). Electrochemistry related to these devices was provided by Conway et al. (1997) and Conway (1990). In general, while EDLCs based on non-Faradaic processes can provide specific capacitances from 20 to 50 μF/cm^2 (which is effectively giving large capacitance devices due to the use of large SSA materials), pseudocapacitance-based devices with Faradaic processes can offer values in the range of 200–500 μF/cm^2 and for multistate overlapping processes around 2000 μF/cm^2 (Conway and Pell, 2003). These devices lead to the development of hybrid devices.

Chapter 8 provides a representative summary of new commercial versions of hybrid devices.

7.5 Hybridization of electrochemical capacitors and rechargeable batteries

Rechargeable batteries are typically characterized by low specific power and high specific energy, while the opposite applies for electrochemical capacitors. Similarly, a myriad of different applications have different demands in terms of the power spectrum to be provided by the power source. It is therefore clear that a single system cannot fulfill the requirements of the different applications. High-energy high-power devices that are able to fulfill the requirements of more applications are highly desirable. In this respect, the goal of having high energy and high power in the same systems has been pursued in the last decades with different approaches. One of these is the hybridization of electrochemical capacitors with rechargeable batteries, with a view to combining the high specific power of the electrochemical capacitor with the high specific energy of the battery.

ECs are characterized by high specific power typically above 10kW/kg but low specific energy values of less than 10Wh/kg. Rechargeable batteries such as Li-ions today provide specific energies above 250Wh/kg for industrial products (Cericola et al., 2011). However, they are characterized by low specific power such as 1kW/kg. In an electrochemical cell, the whole volume of the active material is used for charge storage, and therefore the specific energy of a battery is much higher than that of EDLCs (Cericola et al., 2011). However, the rate of discharge in a battery is typically limited by the solid-state diffusion of Li in the bulk of the active materials, which is inherently slow, and therefore the specific power of batteries is significantly lower than in the case of EDLCs.

Both the lithium-ion battery and the EDLCs have similar technical aspects. Both current practical systems utilize an aprotic organic electrolyte, and the charging processes of the respective electrodes occur in similar potential ranges. These are important considerations for the practical hybridization of the two systems.

The hybridization of electrochemical capacitors and rechargeable batteries can, in principle, be performed according to different approaches. Readily available batteries and electrochemical capacitors can be externally hardwire connected, in serial or in parallel.

By analogy, the same approaches can be proposed at the internal level within one device, leading to the internal serial or the internal parallel hybrid. Between these four possible approaches, only two were widely exploited in the available scientific literature (Cericola et al., 2011).

The hybridization of electrochemical capacitors and rechargeable batteries has been proposed by several authors in the last decades (Cericola and Kötz, 2012). Many different approaches utilizing many different materials have been discussed but there is no unambiguous definition of an electrochemical hybrid energy storage device. A simple sorting of possible electrochemical capacitor-rechargeable battery hybrid systems is proposed in Fig. 7.9. The hardwired connection of a readily available electrochemical capacitor and a readily available rechargeable battery is defined as an "external hybrid." Such a connection can be arranged in a serial way where the

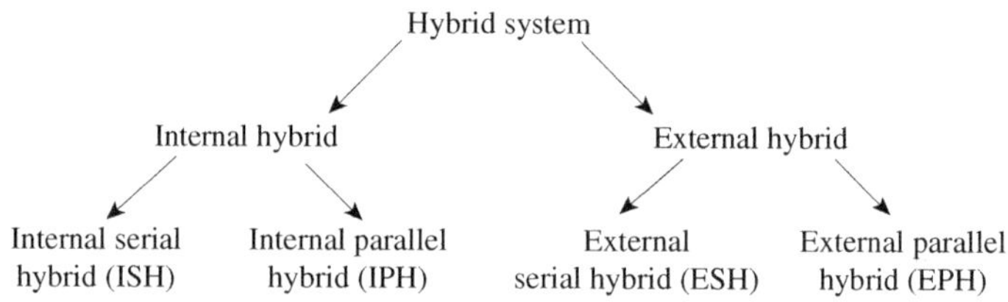

Fig. 7.9 Hybridization approaches in EDLCs and rechargeable batteries (Cericola and Kötz, 2012).

electrochemical capacitor is electrically connected in series with the battery resulting in an "external serial hybrid" device, or they can be connected in a parallel arrangement giving an "external parallel hybrid" device.

A similar way of hybridization may be imaged at the internal level. In this case, the "internal hybrid" devices are developed by hybridization on the electrode level. Therefore, a device based on a battery electrode and an electrochemical capacitor electrode will be defined as an "internal serial hybrid." These kinds of systems were firstly named as *asymmetric electrochemical capacitors*, shortened to *asymmetric capacitors*, and introduced by Razumov et al. (2001). The term *hybrid capacitor* was introduced and trademarked by Evans Capacitor Company to describe an electrolytic capacitor using a pseudocapacitive negative electrode (Razumov et al., 2001). The name "lithium-ion capacitor," sometimes abbreviated as "lithium capacitor," was introduced to describe a device using a graphite negative electrode and an AC-based positive electrode in combination with an Li+-containing electrolyte. Nowadays, however, these and other names are often loosely used in the scientific literature to describe various internal hybrids. Fig. 7.10 depicts the serial and parallel possibilities for two devices. The benefits of hybridization can be discussed by examining the two circuits in Fig. 7.10. The two devices in the circuits idealize either an actual battery or electrochemical capacitor, or one of their electrodes. In Fig. 7.10A, devices 1 and 2 are connected in series and in Fig. 7.10B, the two devices are connected in parallel. Both devices can be described by a certain capacitance (C), resistance (R), and mass (m). The ratio r is defined as the ratio between the specific capacitance of device 2 over the specific capacitance of device 1, where $r = \frac{(C_2/m_2)}{(C_1/m_1)}$. In this example, device 1 has a specific capacitance that is always smaller than or equal to device 2. Both circuits for a given set of capacitance and resistance values can be simplified to an equivalent RC circuit where energy and power are derived as $E = \frac{1}{2} C_{\text{tot}} V^2$ and $P = \frac{V^2}{4R_{\text{tot}}}$, respectively, where C_{tot} and R_{tot} are the total capacitance and total resistance of the equivalent circuit, and V is the voltage applied to the equivalent circuit. The so calculated energy E and power P are then normalized to the mass of the two devices in the corresponding circuits in Fig. 7.10A or B. Fig. 7.10 depicts the variation of specific power and specific energy versus the mass ratio, m_r, where $m_r = \frac{m_1}{m_1 + m_2}$.

Cericola et al. (2011) describe a case of hybrid devices attempted using the electrode material used for Li-ion batteries such as LTO and LMO (discussed in Chapter 3), and Cericola and Kötz (2012) provide an excellent summary on the

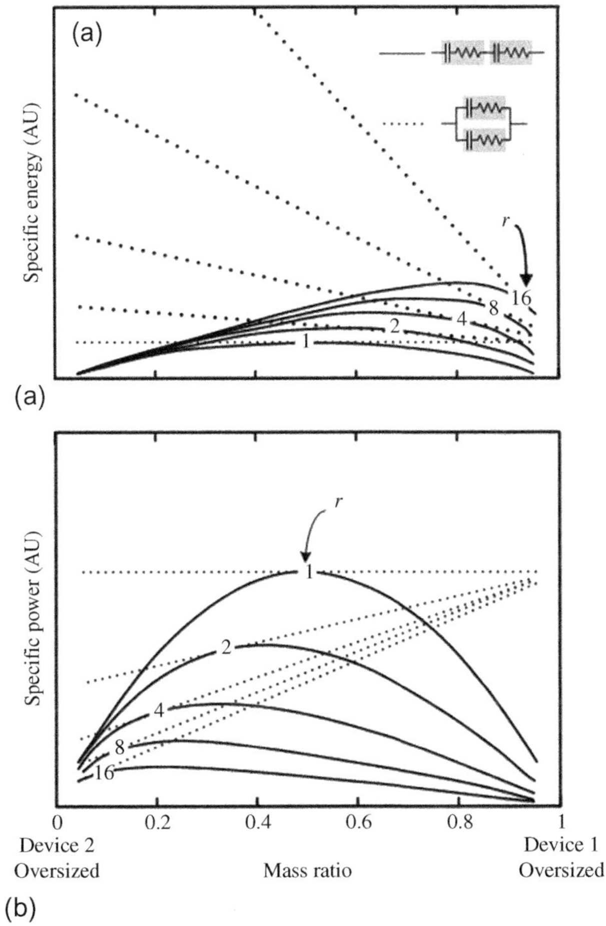

Fig. 7.10 Specific energy and specific power of a hybrid device. (A) Specific energy versus mass ratio and (B) specific power versus mass ratio.
Source: Cericola, D., Kötz, R., 2012. Hybridization of rechargeable batteries and electrochemical capacitors: principles and limits. Electrochim. Acta 72, 1–17.

fundamentals of hybrid devices together with a performance comparison of various electrode materials such as MnO_2, Ni based, Pb based, variety of metal oxides, and Li insertion materials, based on research papers published up to 2011. As per the concluding remarks of Cericola and Kötz (2012), the following provides a summary of a complex subject addressed by several studies:

(i) An external parallel approach is beneficial with respect to the battery (alone) in systems where pulse-type charge/discharge cycles occur at high current.

(ii) The use of additional electronic circuitry could enhance the performance of an external direct parallel system.

(iii) Internal serial hybridization allows to achieve devices with specific energy greater than that of an EC capacitor, usually with twice the capacity. A battery electrode performs better than a complete battery with respect to life cycle and the power capability due to lower SOC achieved.

(iv) The replacement of a capacitor electrode with a battery electrode typically allows the cell voltage to increase (this has allowed practical devices over 4 V into the commercial market). Electrolyte depletion is an issue.
(v) Internal parallel hybridization is an attractive approach to come up with devices with moderately high specific energy and power. Devices are highly tunable.

Sikha and Popov (2004), Pmar et al. (2010), and Catherino et al. (2006) provide a detailed discussion on external-type battery-capacitor hybrid systems and their applicability in cases like EVs, etc. Pasquier et al. (2004) and Michael and Prabaharan (2004) provide some details on early work related to Li electrodes combined with EDLCs for higher voltage capability.

7.6 Modeling and equivalent circuits

The discussions so far provide a background to the development of supercapacitors and the electrochemistry-based fundamentals applicable to EDLCs. The application of these devices requires simplified models suitable for design calculations, an estimation of their limitations in a given application, and most importantly the calendar and cycle lifetimes. In Chapter 3, we have discussed the electrochemistry-based concepts leading to simplified models of rechargeable battery chemistries. Almost all those concepts apply to EDLCs also, except for the case that Faradaic reactions or redox reactions are not dominating in commonly used commercial devices. One exception is pseudocapacitors, where limited redox reactions are used to achieve high specific capacitance values. Zhang and Zhao (2009), Sharma and Bhatti (2010), and Sevilla and Mokaya (2014) provide an overview of the current state of the art in supercapacitors, which indicates the domination of carbon-based materials in commercial devices.

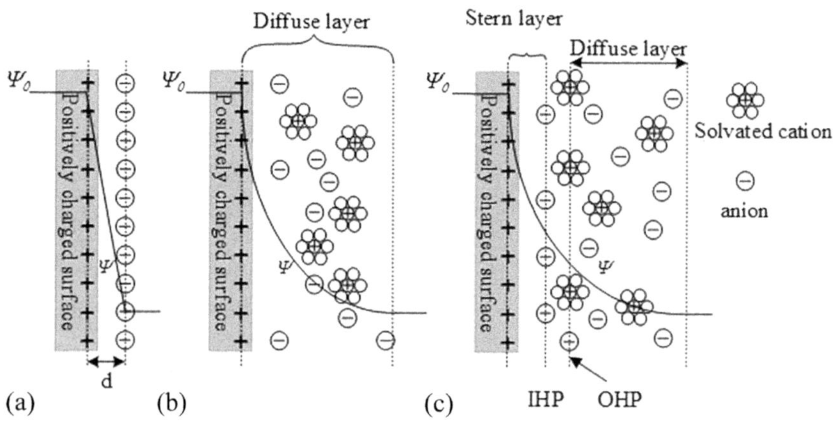

Fig. 7.11 Physical models for the electric double layer at a positively charged surface. (A) The Helmholtz model, (B) the Gouy-Chapman model, and (C) the Stern model.

Conventional capacitors store little by way of energy, due to the limited charge storage areas and geometric constraints of the separation distance between the two charged plates. However, supercapacitors based on the EDL mechanism can store significantly more because of the large interfacial area and the atomic range of charge separation distances. As schematically illustrated in Fig. 7.11A, the concept of EDL was first described and modeled by von Helmholtz in the 19th century, when he investigated the distribution of opposite charges at the interface of colloidal particles (Sharma and Bhatti, 2010). As shown in Fig. 7.11A, the Helmholtz double-layer model states that two electrodes of opposite charge form at the electrode/electrolyte interface and are separated by atomic distances. This double layer can be thought of as a molecular capacitor, where one plate is represented by the charges in the metal and the other plate by the ions at a minimum distance in the solution. The distance is of the order of only 0.5 nm, so the capacitance values are enormous. This simple model will be valid only for rather high concentration electrolyte solutions. In this simple model, the double layer is depleted from charges.

In more dilute solutions, the transition will not be so abrupt and the thickness of the double layer will increase, as shown in Fig. 7.11B. The thickness is related to the distance from the metal surface at which the ions can escape to the bulk by thermal motion. In that case, the counter-ion atmosphere will be more like the ionic atmosphere around an individual ion, and this is commonly referred to as the *diffuse electric layer*. In order to account for this situation, the simple Helmholtz EDL model was further modified by Gouy and Chapman (Sharma and Bhatti, 2010). In the combined theory of Gouy and Chapman, the exchange of counter-ions between the double layer and the bulk solution due to thermal motion is taken into account. Both coulombic forces and thermal motion hence influence the equilibrium distribution of counter-ions in their model for the diffuse double layer as shown in Fig. 7.11B. However, the Gouy-Chapman model leads to an overestimation of EDL capacitance. A discussion on the assumptions for their theory has been provided by Grimnes and Martinsen (2008).

Later in 1923, Stern combined these two models to explicitly recognize the two regions of ion distribution—the inner region called the compact layer or the Stern layer and the diffuse layer or the Gouy-Chapman layer (Grimnes and Martinsen, 2008; Sevilla and Mokaya, 2014; Sharma and Bhatti, 2010). Fig. 7.11C depicts the scenario. The inner Helmholtz plane and outer Helmholtz plane are used to distinguish the two types of adsorbed ions. The capacitance in the EDL, C_{dl}, can be treated as a combination of capacitances from two regions, the Stern type of compact double-layer capacitance, C_{H}, and the diffusion layer capacitance, C_{diff}. Thus, a simple relationship can be expressed as

$$\frac{1}{C_{\text{dl}}} = \frac{1}{C_{\text{H}}} + \frac{1}{C_{\text{diff}}} \tag{7.1}$$

In a practical EDLC, the electrode is a porous material with a high SSA, and the EDL behavior at the porous electrode is much more complex than that of an infinite planar electrode assumed in models of Fig. 7.11. This behavior is much affected by a number of parameters such as the mass transfer path, space constraints inside the pores, ohmic resistances associated with the electrolyte, and the wetting behavior of the porous electrode. Fig. 7.12 schematically depicts the behavior of a practical EDLC composed of the two current collectors and electrolyte absorbed into the porous electrode

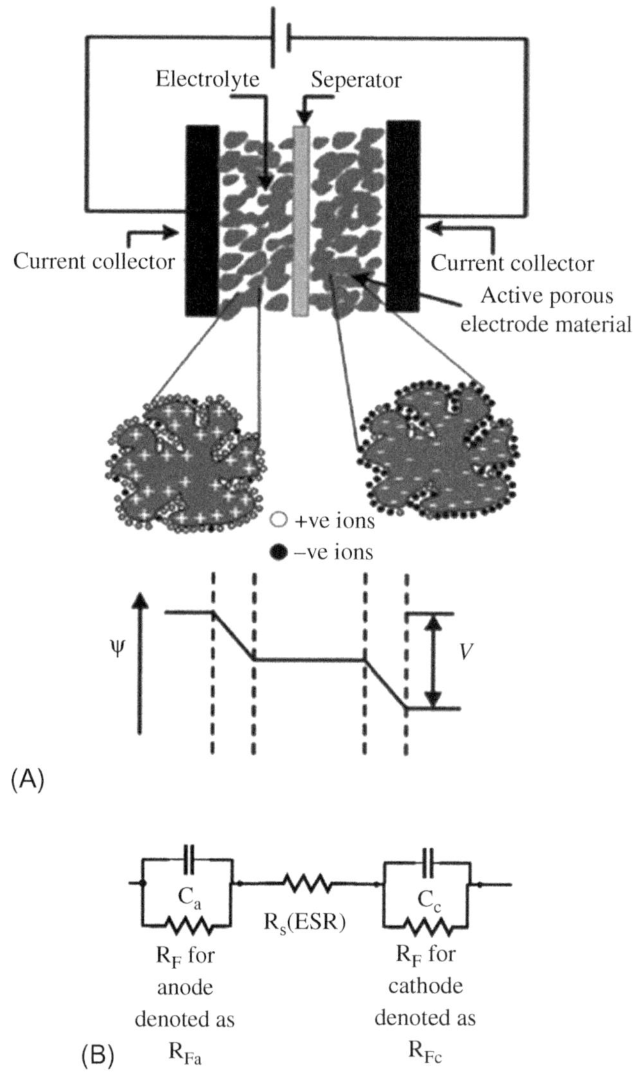

Fig. 7.12 Schematic representation of an EDLC based on porous electrode materials and the two separate capacitors formed at each electrode. (a) simplified physical layout and electro-chemistry aspects (b) simplified electrical equivalent circuit with ESR and leakage resistances

Adapted from Zhang, L.L., Zhao, X.S., 2009. Carbon-based materials as supercapacitor electrodes. Chem. Soc. Rev. 38, 2520–2531.

material and the separator. The capacitance estimation is generally assumed to follow that of a parallel-plate capacitor with value C given by

$$C = \frac{\varepsilon_r \varepsilon_0}{d} A \tag{7.2}$$

where ε_r is the dielectric constant of the electrolyte, ε_0 is the permittivity of the vacuum, A is the surface area available to electrolyte ions, and d is the effective thickness of the EDL, which is also called the Debye length (Grimnes and Martinsen, 2008). Based on Eq. (7.2), there should be a linear relationship between specific capacitance and SSA. However, experimental results discussed by Sharma and Bhatti (2010) indicate that this assumption is not accurate, and these new experimental findings cannot be fully interpreted by the EDL theory.

As shown in Fig. 7.12, with two electrodes combined in an EDLC, we are achieving an overall capacitance of C_T based on the anode and cathode capacitances C_a and C_c, respectively, which gives rise to the relationship,

$$\frac{1}{C_T} = \frac{1}{C_a} + \frac{1}{C_c} \tag{7.3}$$

The performance of commercially available supercapacitors is mainly evaluated based on six important criteria: (i) power density substantially greater than batteries with acceptably high energy densities (usually $>10\,\text{Wh/kg}$), (ii) very high life cycle (more than 100 times that of batteries), (iii) fast charge/discharge process within seconds, (iv) low self-discharge, (v) safe operation, and (vi) low cost. As depicted in the simplified RC equivalent circuit of Fig. 7.12, with the equivalent circuits of the two electrodes shown separately, equivalent series resistance (R_s), and the terminal voltage, determine the power density. Anode and cathode leakage resistances R_{Fa} and R_{Fc}, respectively, account for the self-discharge. To have the ability to fast charge transfer (charge/discharge), the $R_s C_T$ time constant of the device should be as small as possible. Since the capacitance of an SC is enormous, the R_s value should be extremely small. However, the current commercial SCs have ESR values in the range of $0.1\,\text{m}\Omega$ to a few $\text{m}\Omega$s.

7.6.1 Different types of equivalent circuits

Depending on the application, one can use different types of equivalent circuits with various complexities. The simplest version of ESR and the capacitance in series assumes a constant capacitance (independent of the voltage) and the ESR also to be constant. However, experimental behavior and measurements indicate that this is only a very first approximation, and this ignores the leakage. For this simplification, the R_{Fa} and R_{Fc} shown in Fig 7.12(B) is ignored and uses an ESR value and the total capacitance C_T as in Eq. (7.3).

It is important to understand that the simplified but independent electrode-based equivalent circuit is a major simplification of the overall theoretical model, which could be used to describe the porous electrodes and its behavior based on a theoretical

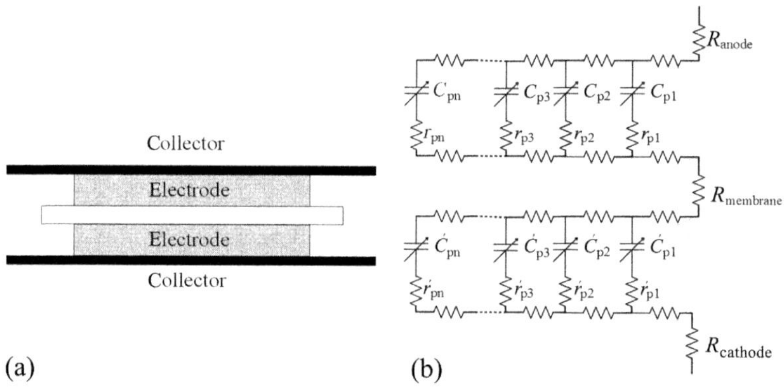

Fig. 7.13 Supercapacitor packaging principle and detailed equivalent circuit. (A) Packaging principle and (B) theoretical model.
Source: Belhachemi, F., Rael, S., Davat, B., 2000. A physical based model of power electric double layer capacitor. IEEE, 3069–3075.

model is shown in Fig. 7.13. Fig. 7.13A shows the simplified arrangement of two current collectors, two electrodes, and the porous insulting membrane, and Fig. 7.13B shows the final distributed model of the complete system. This is related to various models described in Fig. 7.11.

The different resistances that appear in Fig. 7.13 depend on numerous parameters such as (a) electrode material resistivity, (b) pore sizes, (c) membrane porosity, and (d) packaging technology (impregnation and collector material resistivity). A more detailed discussion has been provided by Belhachemi et al. (2000). As discussed in Fig. 7.4B, this is directly related to the TLM of the porous electrode. However, this is far too complex to handle when designing the circuits using SCs. Then we can simply classify this into different models such as two *RC* section-based or three *RC* section-based models for more accurate calculations. The literature indicates many different complex models based on including pore models and self-discharge-included models. Fig. 7.14 indicates a few popular and commonly discussed models as per discussions by Belhachemi et al. (2000), Yasser et al. (2008), and Lisheng and Crow (2008).

A practical observation on supercapacitors is that almost all important parameters such as ESR, capacitance, and leakage are dependent on voltage, current, and temperature in general. Fig. 7.15 depicts the dependence of ESR and capacitance on current and voltage.

Given the above situation, there are many publications which discuss different aspects of these dependencies as well as the differences in capacitor values based on the measurement method as well as the state of charge (SOC). Yang et al. (2010), Kurzweil et al. (2005), Spyker and Nelms (2000), and Nelms et al. (1999) provide a useful discussion on these aspects.

Useful practical models for design purposes are the two or three branch series model and the voltage-dependent ladder model. In general, Fig. 7.16 shows a general model that can be simplified into many different versions that we have discussed, and how duration of charge and discharge affects the mode (Lajnef et al., 2007).

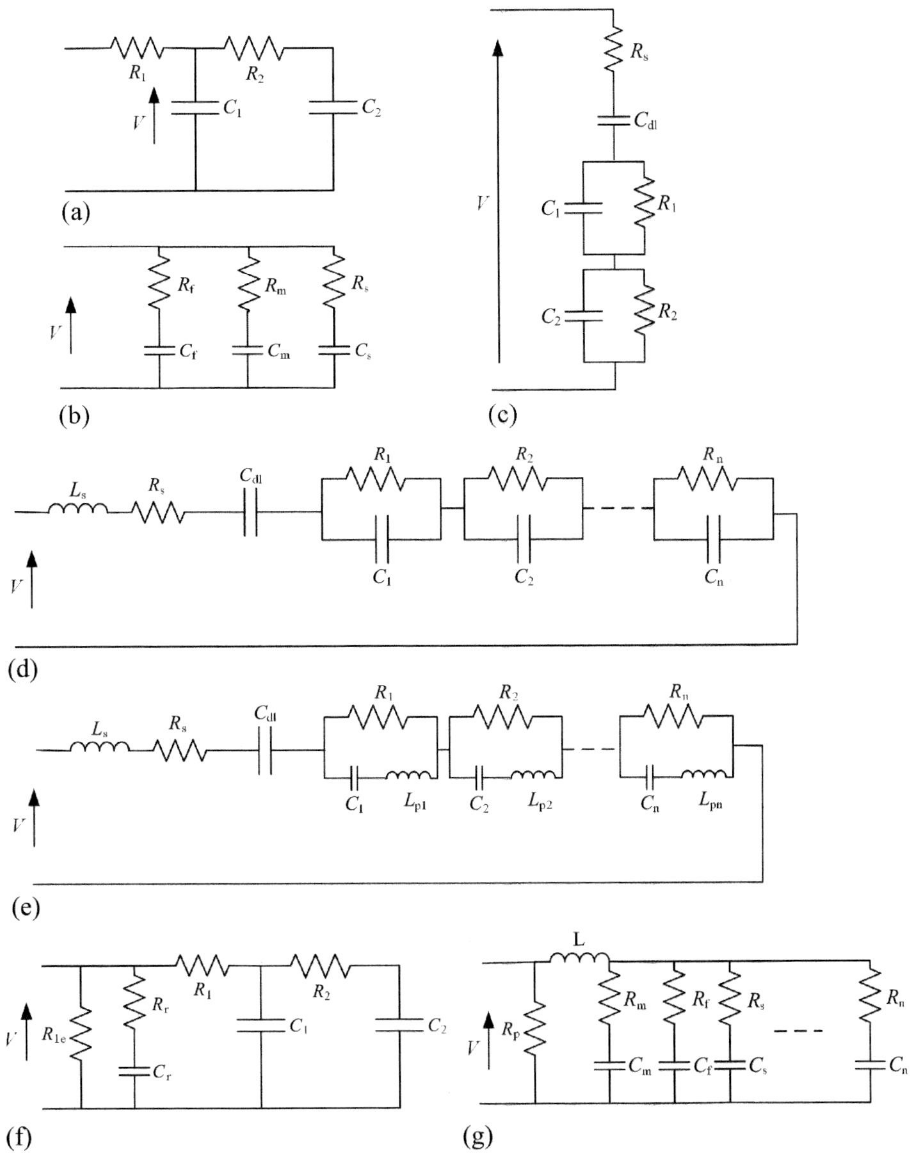

Fig. 7.14 Different models for supercapacitors. (A) Two branch model, (B) three branch parallel model, (C) series-parallel model, (D) impedance pore model, (E) complex pore model with pseudo inductance, (F) self-discharge and diffusion controlled model, and (G) general *RC* network model.

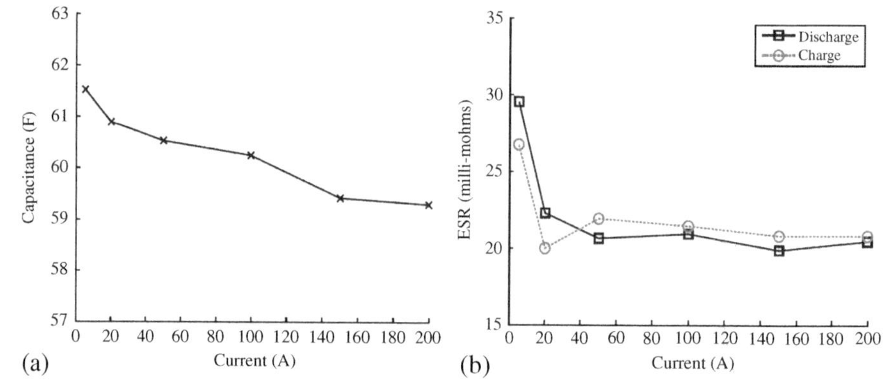

Fig. 7.15 Capacitance and ESR variation of a commercial supercapacitor family. (A) Capacitance versus current and (B) ESR versus current.
Source: Yang, F., Lnguang, L., Yuping, Y., He, Y., 2010. Characterization, analysis and modeling of an ultracapacitor. In: Proceedings of EVS25, November pp. 1–12.

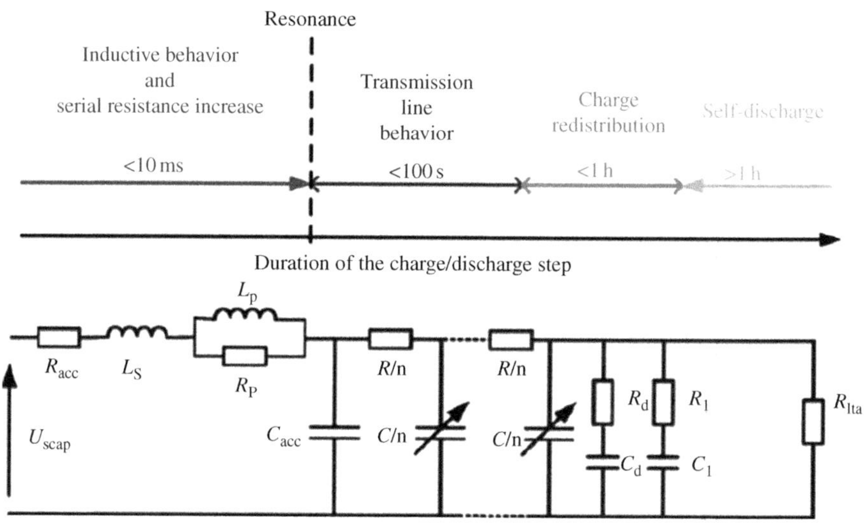

Fig. 7.16 A generalized model for a supercapacitor and its operational durations and states.
Source: Lajnef, W., Vinassa, J.-M., Briat, O., Azzopardi, S., Woirgard, E., 2007. Characterization methods and modelling of ultracapacitors for use as peak power sources. J. Power Sources 168, 553–560.

7.7 Testing of devices and characterization

Characterization methodologies applied to supercapacitors are not standardized and it is difficult to compare the laboratory test results with those of manufacturers. At present, there are not many international standards, except one set of IEC standards (IEC-40/1378/DIN IEC 62391-1, 2004; IEC-40/1379/DIN IEC 62391-2, 2004). Most supercapacitor parameters are dependent on parameters such as ambient temperature, current intensity, frequency, and voltage. Therefore, reliable and reproducible experimental results are required. This section describes some of the techniques that are used to estimate the parameters of supercapacitors. Some methods are mostly used by electrochemists in research laboratories, while some are used in industrial environments.

7.7.1 Charge-discharge method

This is one of the common methods used for capacitance and ESR measurement. In this process, an SC is discharged or charged at a constant current, and a solid-state load is used for this test. In this test, the capacitor is simplified as a single *RC* time constant device, and voltage variation for constant current charge or discharge of the device is monitored. Fig. 7.17A depicts this where the voltage across the supercapacitor is initially maintained at its rated voltage, V_R for 1 h, and then the device is discharged at a constant current, *I*. The capacitance, *C*, determination is estimated during constant current discharge between the voltage values from 80% to 40% of the rated voltage V_R according to the linear relationship $C = I \cdot \Delta t_c / \Delta V_c$ as shown in Fig. 7.17A. The dotted line in Fig. 7.17 shows the linear discharge, while the actual behavior is nonlinear as shown by the continuous trace.

Fig. 7.17B shows the 200 A constant current charge and discharge curves for a 3000 F supercapacitor. With reference to the graphs in Fig. 7.17B, the ESR measurement is obtained using the sudden change of voltage (ΔV) at the start of the charge $\Delta V_{C,L}$, at the end of the charge $\Delta V_{C,H}$, at the start of the discharge $\Delta V_{D,H}$, or at the end of the discharge $\Delta V_{D,L}$. For a constant charging or discharge current of *I*, the ESR value is given by $\Delta V/I$.

In the literature, it is reported that these values obtained by charge-discharge tests are quite dependent on the current values used in the test. Fig. 7.15 indicates this as adapted from Lajnef et al. (2007).

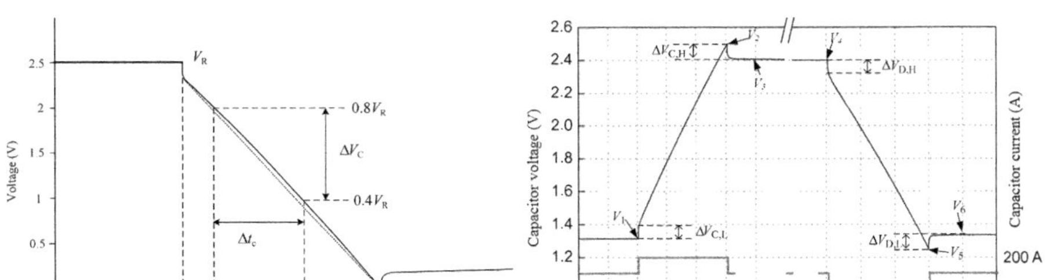

Fig. 7.17 Charge-discharge method. (A) Concept and (B) an actual measurement on a 3000 F capacitor.
Adapted from Lajnef, W., Vinassa, J.-M., Briat, O., Azzopardi, S., Woirgard, E., 2007. Characterization methods and modelling of ultracapacitors for use as peak power sources. J. Power Sources 168, 553–560.

Fig. 7.18 A solid-state load which can be used as a constant power device.

7.7.2 Constant power tests

Constant power tests are commonly used for the characterization of electric storage devices in general. In these tests, a solid-state load such as that shown in Fig. 7.18 could be used by setting its operation to constant power mode. For supercapacitors, the experimental procedure consists in discharging the element with constant power between two voltage limits of V_{nom} and $V_{nom}/2$ in order to measure the specific energy that this element can supply as a function of discharge power. As detailed by Pell and Conway (1996), the corresponding results can be graphically reported in the well-known Ragone diagram (Christen and Carlen, 2000) where the x-axis and the y-axis correspond, respectively, to the specific power (kW/kg) and to the specific energy (Wh/kg).

Fig. 7.19 shows the Ragone plot obtained for a 2600 F/2.5 V cell and with a specific power that has been limited by the maximum current capability of the tester at 800 A. Then the experimental points have been interpolated and extrapolated according to

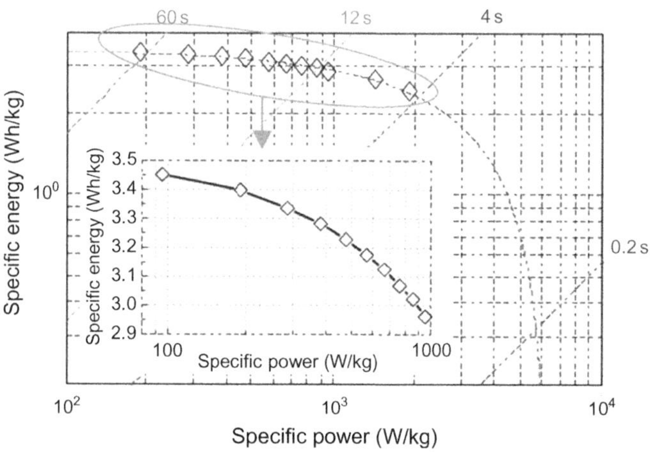

Fig. 7.19 Ragone plot for a 2600 F capacitor.

Eq. (7.4), where E is the energy, P is the power, R is the series resistance, and C is the rated capacitance (Christen and Carlen, 2000). U_0 is the initial voltage across the capacitor.

$$E_c(p) = \frac{C}{2}\left(RP \ln\left(\frac{RP}{U_0^2}\right) + U_0^2 - RP\right) \tag{7.4}$$

7.7.3 Impedance spectroscopy

Electrochemical impedance spectroscopy (EIS), summarized in Chapter 3, is often used in the characterization of electrochemical behavior of energy supercapacitor storage devices (Karden et al., 2002; Qu and Shi, 1998). To characterize a supercapacitor, the sweep in frequency must be done for various voltage levels and also for different temperatures. EIS allows the study of the influence of frequency on the electrode series resistance and on specific capacitance. A summary of the test process is given below.

The supercapacitor is polarized with a DC voltage. A small voltage ripple, typically 10 mV, is superimposed on the DC component. The ripple frequency is swept between 1 mHz and 10 kHz. The measurement of the current amplitude and phase with respect to the injected voltage permits the determination of the real and imaginary parts of the impedance as a function of the frequency (Buller et al., 2002). The measurements are usually performed in a controlled climatic chamber. Fig. 7.20 shows the Nyquist plot (complex-plane representation) of measured impedance spectra of a 1400 F/2.5 V capacitor.

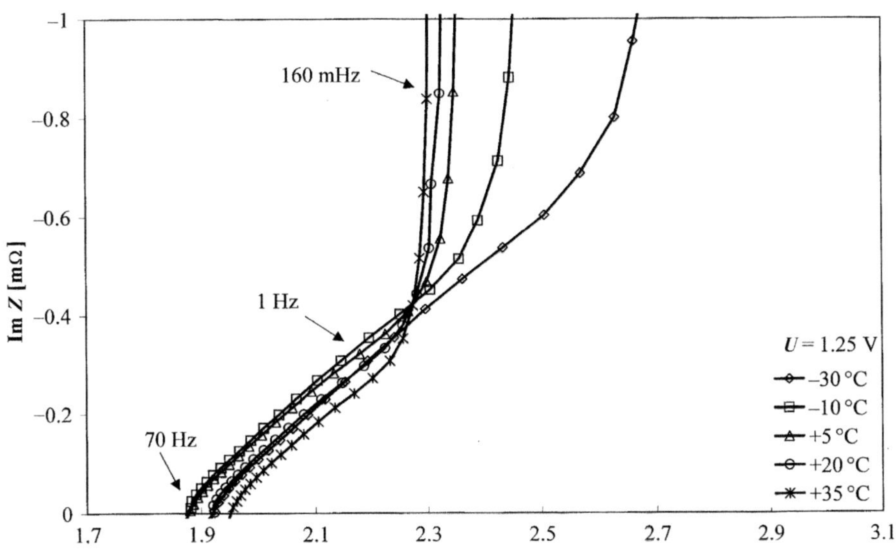

Fig. 7.20 Complex-plane representation of measured impedance spectra of a 1400 F/2.5 V supercapacitor.
Reproduced with permission from IEEE Buller, S., Karden, E., Kok, D., De Doncker, R.K., 2002. Modeling the dynamic behavior of supercapacitors using impedance spectroscopy. IEEE Trans. Ind. Appl. 38(6), 1622–1626.

7.7.4 Cyclic voltammetry

In this technique, the device under test is subjected to voltage rising at a fixed rate of sweep such as about 10 mV/s, and the corresponding device current is measured. The plot obtained by this test is quite frequently used in the process of device development stages. The graph obtained is called a *voltammogram.*

This test is useful to validate the capacitive behavior, to investigate the symmetry between the charge and discharge process, and to determine the potential limits of the device. It consists of the linear voltage sweep between two limit values, V_1 and V_2. If the ultracapacitor is modeled with a constant capacitor C in series with an internal resistor R_s, the current is expressed by Lajnef et al. (2007).

$$I(t) = C\frac{dv}{dt}\left[1 - \exp\left(-\frac{t}{CR}\right)\right]$$

In this equation, there are two terms. The first is permanent $I(t) = C\,dv/dt$, while the second is transient and disappears after a few time constants. When the capacitance depends on voltage and assumes very low serial resistance, the current is proportional to the capacitance by $I = C_{\text{diff}}\frac{dv}{dt}$, given that the voltage drop due to the resistor can be neglected.

Fig. 7.21 illustrates the voltammogram which is obtained by recording the current versus voltage with a sweep rate of 10 m/Vs. This figure shows that the current is constant between 0 and 0.3 V, i.e., the differential capacitance is constant within this voltage range. Then the current increases as well as the differential capacitance. In the 1.25–2.5 V range, the linear fit of the differential capacitance, which has been

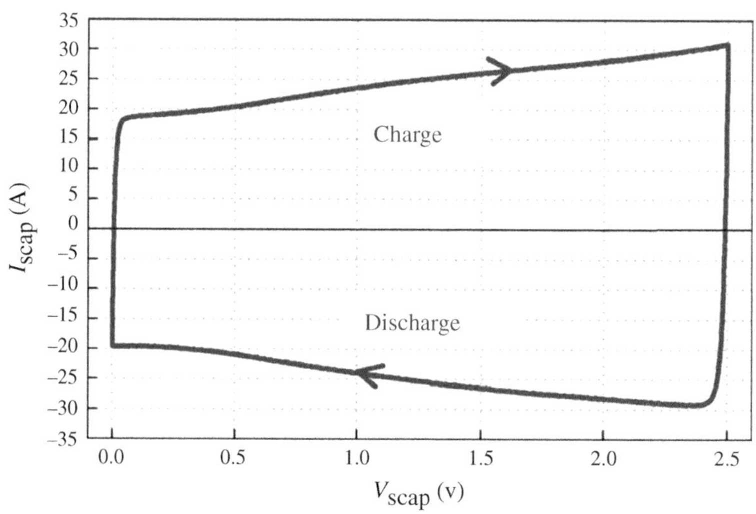

Fig. 7.21 Voltammogram measured with a sweep rate of 10 mV/s on a 2600 F/2.5 V cell. *Source*: Lajnef, W., Vinassa, J.-M., Briat, O., Azzopardi, S., Woirgard, E., 2007. Characterization methods and modelling of ultracapacitors for use as peak power sources. J. Power Sources 168, 553–560.

determined from constant current tests, seems to be right. Also, the evolution shape of the current versus voltage seems to be symmetric between charge and discharge that corresponds to a reversible behavior, i.e., the differential capacitance in charge is the same as in discharge.

Fig. 7.22 shows the comparison of different parameter extraction techniques (Kurzweil et al., 2005). The molecules at the electrochemical interface between the electrode and electrolyte show a nonlinear response to the electric field. This figure indicates that in a supercapacitor, capacitive and Faradaic charges could coexist. This charge, divided by the potential range from which it is transferred, constitutes a pseudocapacitance since it is not associated with a purely capacitive phenomenon. Transient techniques, which generate fast voltage changes, reflect the fast charge/discharge processes with "frozen" diffusion at the outer electrode surface. At slow-voltage changes, the slow electrode reactions in the inner electrode surface (strongly depending on voltage) additionally contribute to the measured capacitance values. In Fig. 7.23, SOC varies from full discharge to charge.

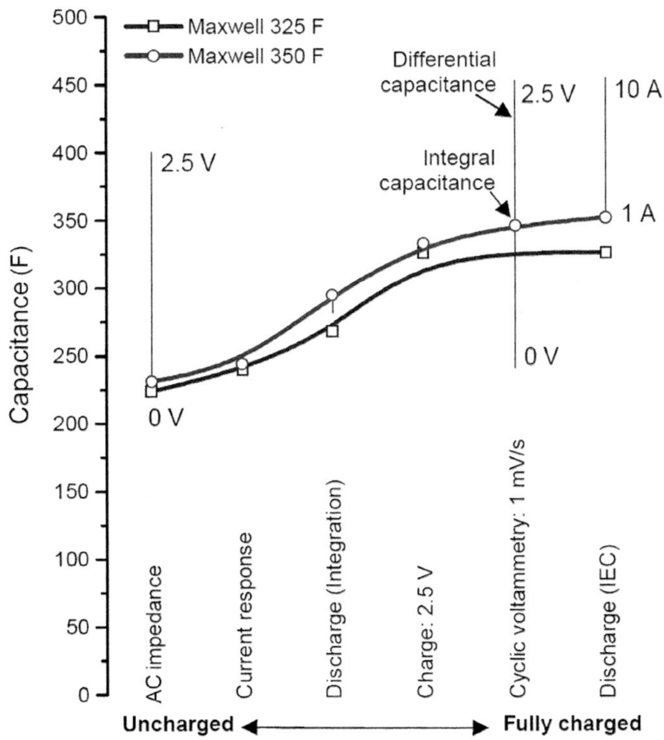

Fig. 7.22 Capacitance of Maxwell capacitors (BCAP0350) determined by different measuring techniques, creating a different state of charge of the capacitor.
Adapted from Kurzweil, P., Frenzel, B., Gallay, R., 2005. Capacitance characterization methods and ageing behaviour of supercapacitors. In: The 15th International Seminar on Double Layer Capacitors.

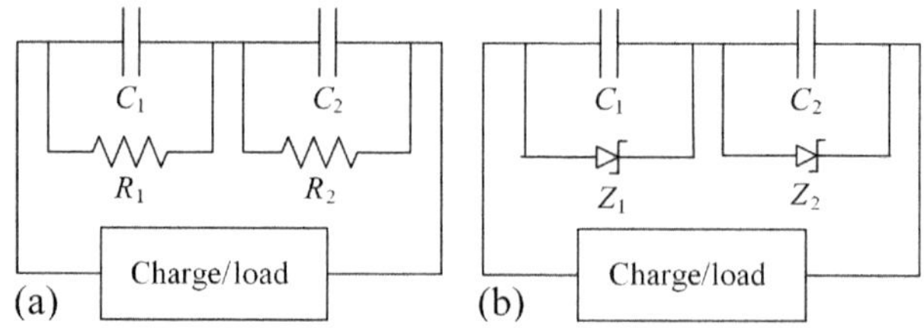

Fig. 7.23 Passive resistance balancing. (A) Resistor based and (B) zener diode based (Sharma and Bhatti, 2010).

7.8 Modules and voltage balancing

Supercapacitor cells generally have a very low operation voltage of between 2.5 and 3.0 V. However, most high-power applications require considerably higher voltages. For example, a car battery pack situation may mean either 12 or 42 V, which means that we need anything from about 5 to 20 elements in series to meet the nominal voltage value. In order to operate at higher application voltages, supercapacitors are connected in series and parallel configurations to form a module, to meet the DC voltage, energy capability, and the total ESR value, which determines the maximum power capability of the bank.

Supercapacitors coming out of mass scale production lines are not exactly identical, and hence the series-connected supercapacitor stacks develop unequal voltage distributions (Aiguo et al., 2009a,b; Chunhe et al., 2009; Gai Xiao et al., 2008; Rengui et al., 2008). The situation could become worse, because the device capacitance is also varying with different DC bias voltages. Voltage distribution in a series stack of supercapacitors is initially a function of capacitance. After the stack has been held at voltage for a period of time, voltage distribution becomes a function of internal parallel resistance (leakage current). For example, consider a stack of 20 capacitors initially charged to 50 V. If the cells have identical capacitance, the voltage should divide evenly so that each capacitor charges up to 2.5 V. If the cells have any variation in capacitance, individual cell voltages will vary based on capacitance. The cells with greater capacitance will be charged to lower voltages, and the cells with smaller capacitance will be charged to higher voltages. This is because each cell conducts the same charging current, and the voltage is a function of current and capacitance.

Those undesired unequal capacitances could attribute to several factors as follows (ELNA, n.d.; Ness Capacitor Co., Ltd, n.d.):

- the manufacturing tolerances of 5%–10%,
- the temperature gradient in the system, and
- the cell aging.

In this case, some supercapacitors will have an overvalue for their voltage, with an associated decrease of their lifetime. For the other supercapacitors, the maximum voltage will be under the voltage limit; the energy storage in supercapacitors will not be at its maximum level. A capacitor with a smaller capacitance needs a shorter charging time and thus reaches its maximum voltage faster than a larger capacitor. To avoid damaging the smaller capacitor, voltage equalization circuits have to be employed. Manufacturers of large devices supply SC modules similar to battery packs, with different types of cell-balancing circuits used inside the assembly.

7.8.1 Types of balancing

There are two common approaches to cell balancing in SC modules. They are (i) passive balancing and (ii) active balancing. Passive balancing circuits use passive components and diodes, etc. in general and they can be classified into two subgroups such as resistance balancing and zener diode balancing.

7.8.1.1 Passive balancing techniques

Resistance balancing

In order to prevent the adverse effects of unequalized charging of supercapacitors, an individual supercapacitor needs to be maintained at an equalized charge level. This can be achieved by regulating the voltage of individual cells. A simple approach to equalize the supercapacitor stacks is to use a bypass resistor in parallel as shown in Fig. 7.23. The amount of current drawn by the resistors is proportional to the cell voltage, which results in more current being diverted to the resistor as the cell voltage decreases. This will tend to reduce the voltage differences between the different cells within the stack since higher voltage cells will be further discharged by the resistors in parallel. One drawback of this approach is that the recovered energy is being converted into additional losses in the resistors. In addition, the amount of current drawn by the resistors is not regulated. As a result, the cell voltages are not fully regulated. The principle of equalization can be explained thus. As indicated in Fig. 7.23, once the capacitors are fully charged, two resistors will share the input charger voltage by the resistor divider action combined with the leakage currents of the capacitors. This will help to distribute the total stack voltage evenly across the entire series of capacitors. The voltage equalization is closely achieved by connecting the resistor in parallel across the capacitors, but great power dissipation is expected in the resistor.

Zener diode balancing

In order to better regulate the current drawn by the shunt elements, diode circuitry can be used. The use of a zener diode is one such approach. Because of the breakdown voltage of the device, it is used as a voltage regulator when placed in parallel across a load to be regulated. Fig. 7.23B indicates this concept where two zeners are connected across the two capacitors, C_1 and C_2. As soon as a capacitor C_1 reaches its maximum voltage, if the voltage is almost the breakdown voltage of the diode Z_1, it starts conducting and deviates the charging current. A similar situation works with the C_2-Z_2 pair also. In this scheme, the amount of lost energy is minimized since the shunt circuitry is only active when the cell voltage exceeds the preset breakdown level. However, this approach also leads to increased losses and, furthermore, suffers from the temperature dependency of the Zener diodes.

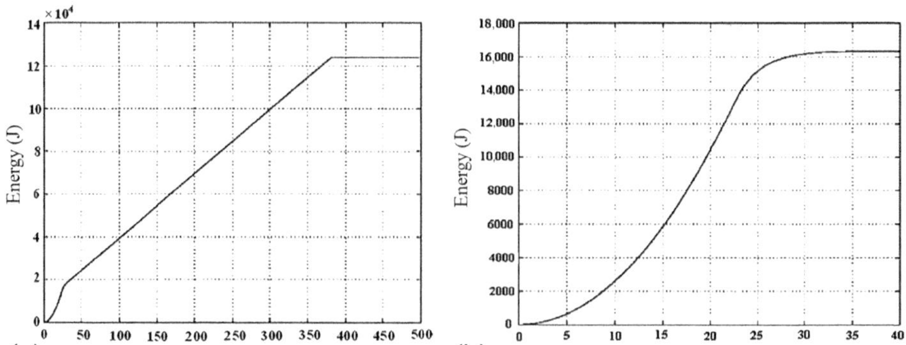

Fig. 7.24 (A) Simulated energy loss while discharging an EDLC bank balanced with resistors (Sharma and Bhatti, 2010). (B) Simulated energy loss while discharging a bank balanced with zener diodes (Sharma and Bhatti, 2010).

Simulated energy losses for voltage balancing with resistors and zener diodes are compared in Fig. 7.24A and B. In both cases, four 1000 F and one 800 F devices were connected in series and charged to 12.5 V. Energy stored at the end of the charging process amounted to 15 kJ. In the case of the voltage-balancing resistors (each 0.1 Ω), 120 kJ was required to fully charge the EDLCs, resulting in an efficiency of 12.5%. When zener diodes were used, only 16.3 kJ was required, resulting in an efficiency of 92% (Barrade et al., 2000). Charging efficiency is therefore much improved by using zener diodes for voltage balancing instead of resistors.

7.8.1.2 Active balancing

In the passive balancing scheme, the recovery energy is dissipated into energy dissipating resistive current shunts. As a result, the dissipated equalization schemes are best suited for low power applications or low current charge/discharge rates. This will tend to minimize the lost energy. In applications such as EVs, the current charge/

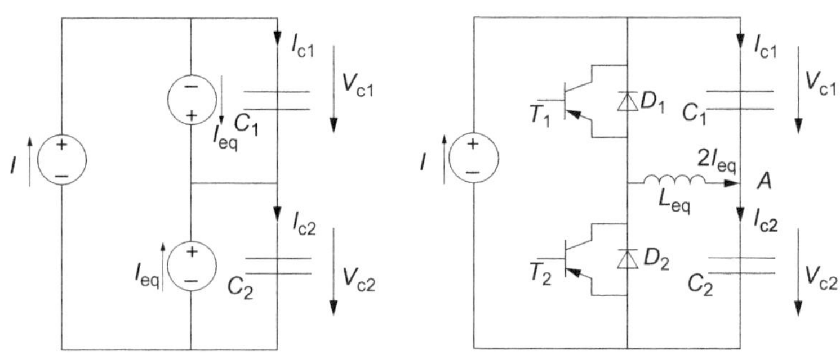

Fig. 7.25 Active balancing. (A) Concept and (B) buck-boost topology for active voltage balancing (Sharma and Bhatti, 2010).

discharge rates are pretty high (10%–100%) and the charge/discharge times are quite short. As a result, the charge equalization currents could be of the same order of magnitude. Therefore, in order to minimize lost energy and optimize the cell's performances, nondissipative equalization techniques (active balancing scheme) need to be used.

The general philosophy behind an active voltage-sharing device is shown in Fig. 7.25A (Barrade et al., 2000). The main charging current I is augmented by the equalizing currents I_{eq}. The size and direction of the equalizing currents are controlled according to the local voltages present across the corresponding device. The equalizing current sources can be realized by the buck-boost topology as shown in Fig. 7.25B. If the voltage U_{c1} is detected as being significantly greater than voltage U_{c2}, transistor T_1 will be switched at a certain frequency to generate a positive equalizing current, $2I_{eq}$. If U_{c2} is greater than U_{c1}, then T_2 will be switched to generate a negative equalizing current. This process continues until the voltages are balanced.

References

Abdallah, T., Lemordant, D., Claude-Montigny, B., 2012. Are room temperature ionic liquids able to improve the safety of supercapacitors organic electrolytes without degrading the performances? J. Power Sources 201, 353–359.

Aiguo, X., Shaojun, X., Xiaobao, L., 2009a. Dynamic voltage equalization for series-connected ultracapacitors in EV/HEV applications. IEEE Trans. Veh. Technol. 58 (8), 3981–3987.

Aiguo, X., Xiaobao, L., Shaojun, X., 2009b. Research on dynamic voltage equalization circuit for series connected ultracapacitors. In: IEEE International Conference on Industrial Technology, 2009 (ICIT '09).

Arbizzani, C., Beninati, S., Lazzari, M., Soavi, F., Mastragostino, M., 2007. Electrode materials for ionic liquid-based supercapacitors. J. Power Sources 174, 648–652.

Bakhourn, E.G., 2009. New mega-farad ultracapacitors. IEEE Trans. Ultrason. Ferroelectr. Freq. Control 56 (1), 14–21.

Balducci, A., Dugasa, R., Taberna, P.L., Simona, P., Plée, D., Mastragostino, M., Passerini, S., 2007. High temperature carbon-carbon supercapacitor using ionic liquid as electrolyte. J. Power Sources 165, 922–927.

Barrade, P., Pittet, S., Rufer, A., 2000. Energy storage system using a series connection of supercapacitors, with an active device for equalizing the voltages. In: International Power Electronics Conference, Tokyo, Japan.

Becker, H.I., 1957. Low Voltage Electrolytic Capacitor. US Patent 2800616, July 23.

Belhachemi, F., Rael, S., Davat, B., 2000. A physical based model of power electric double layer capacitor. In: IEEE, pp. 3069–3075.

Bittner, A.M., Zhu, M., Yang, Y., Waibel, H.F., Konuma, M., Starke, U., Weber, C.J., 2012. Aging of electrochemical capacitors. J. Power Sources 203, 262–273.

Buller, S., Karden, E., Kok, D., De Doncker, R.K., 2002. Modeling the dynamic behavior of supercapacitors using impedance spectroscopy. IEEE Trans. Ind. Appl. 38 (6), 1622–1626.

Burke, A., 2000. Ultracapacitors: why, how and where is the technology. J. Power Sources 91, 37–50.

Catherino, H.A., Burgel, J.F., Shi, P.L., Rusek, A., Xiulin, Z., 2006. Hybrid power supplies: a capacitor assisted battery. J. Power Sources 162, 965–970.

Celzard, A., Collas, F., Mareché, J.F., Furdin, G., Rey, I., 2002. Porous electrodes-based double layer capacitors: pore structure versus series resistance. J. Power Sources 108, 153–162.

Cericola, D., Kötz, R., 2012. Hybridization of rechargeable batteries and electrochemical capacitors: principles and limits. Electrochim. Acta 72, 1–17.

Cericola, D., Novák, P., Wokaum, A., Kötz, R., 2011. Hybridization of electrochemical capacitors and rechargeable batteries: an experimental analysis of the different possible approaches utilizing activated carbon, $Li_4Ti_5O_{12}$ and $LiMn_2O_4$. J. Power Sources 196, 10305–10313.

Chang, Y., Wu, C., Wu, P., 2013. Synthesis of large surface area carbon xerogels for electrochemical double layer capacitors. J. Power Sources 223, 147–154.

Christen, T., Carlen, M.W., 2000. Theory of Ragone plots. J. Power Sources 91, 210–216.

Chunhe, C., et al., 2009. Research of supercapacitor voltage equalization strategy on rubber-tyred gantry crane energy saving system. In: Asia-Pacific Power and Energy Engineering Conference, 2009 (APPEEC '09).

Conway, B.E., 1990. Transition from "supercapacitor to battery" behavior in electrochemical energy storage. In: Proceedings of Power Sources Symposium, pp. 319–327.

Conway, B.E., Pell, W.G., 2003. Double layer and pseudocapacitance types of electrochemical capacitors and their applications to the development of hybrid devices. J. Solid State Electrochem. 7, 637–644.

Conway, B.E., Birss, V., Wojtowicz, J., 1997. The role and utilization of pseudocapacitance for energy storage by supercapacitors. J. Power Sources 66, 1–14.

Du, C., Pan, N., 2006. Supercapacitors using carbon nanotubes films by electrophoretic deposition. J. Power Sources 160, 1487–1494.

Du, X., Song, H., Chen, X., 2010. Graphene nanosheets as electrode material for electric double layer capacitors. Electrochim. Acta 55, 4812–4819.

Dubal, D.P., Gund, G.S., Lokhande, C.D., Holze, R., 2013. Cuo cauliflowers for supercapacitor application: novel potentiodynamic deposition. Mater. Res. Bull. 48, 923–928.

ELNA. http://www.elnaerica.com/company.htm.

Endo, M., Takeda, T., Kim, Y.J., Koshiba, K., Ishii, K., 2001. High power electric double layer capacitor (EDLC's); from operating principle to pore size control in advanced activated carbons. Carbon Lett. 1 (3-4), 117–128.

Fang, B., Binder, L., 2006. A modified activated carbon aerogel for high-energy storage in electric double layer capacitors. J. Power Sources 163, 616–622.

Fang, B., Wei, Y.-Z., Kumagai, M., 2006. Modified carbon materials for high-rate EDLCs application. J. Power Sources 155, 487–491.

Frackowiak, E., Béguin, F., 2001. Carbon materials for the electrochemical storage of energy in capacitors. Carbon 39, 937–950.

Gai Xiao, D., et al., 2008. Analysis on equalization circuit topology and system architecture for series-connected ultra-capacitor. In: IEEE Vehicle Power and Propulsion Conference, 2008 (VPPC '08).

Gamby, J., Taberna, P.L., Simon, P., Fauvarque, J.F., Chesneau, M., 2001. Studies and characterizations of various activated carbons used for carbon/carbon supercapacitors. J. Power Sources 101, 109–116.

Grimnes, S., Martinsen, Ø.G., 2008. Bioimpedance and Bioelectricity Basics, second ed. Elsevier-Academic Press. 471 p.

Gund, G.S., Dubal, D.P., Patil, B.H., Shinde, S.S., Lokhande, C.D., 2013. Enhanced activity of chemically synthesized hybrid graphene oxide/Mn_3O_4 composite for high performance capacitors. Electrochim. Acta 92, 205–215.

Huang, C., Chuang, C., Ting, J., Teng, H., 2008. Significantly enhanced charge conduction in electric double layer capacitors using carbon nanotube-grafted activated carbon electrodes. J. Power Sources 183, 406–410.

IEC-40/1378/DIN IEC 62391-1, June 2004. Fixed Electric Double Layer Capacitors for Use in Electronic Equipment. Part I: Generic Specification.

IEC-40/1379/DIN IEC 62391-2, June 2004. Fixed Electric Double Layer Capacitors for Use in Electronic Equipment. Part II: Sectional Specification: Electric Double Layer Capacitors for Power Application.

Itagaki, M., Suzuki, S., Shitanda, I., Watanabe, K., Nakazawa, H., 2007. Impedance analysis of electrical double layer capacitor with transmission line model. J. Power Sources 164, 415–424.

Kanthal Globar. n.d. Maxcap[fi] Double Layer Capacitors-Product Information & Application Data, Document M-2004A-09/07.

Karden, E., Buller, S., De Doncker, R.K., 2002. A frequency domain approach to dynamical modeling of electrochemical power sources. Electrochim. Acta 47, 2347–2356.

Kim, J., Kim, S., 2014. Preparation and electrochemical property of ionic liquid-attached graphene nanosheets for an application of supercapacitor electrode. Electrochim. Acta 119, 11–15.

Kurzweil, P., Frenzel, B., Gallay, R., 2005. Capacitance characterization methods and ageing behaviour of supercapacitors. In: The 15th International Seminar on Double Layer Capacitors.

Lajnef, W., Vinassa, J.-M., Briat, O., Azzopardi, S., Woirgard, E., 2007. Characterization methods and modelling of ultracapacitors for use as peak power sources. J. Power Sources 168, 553–560.

Levie, R.D., 1963. On porous electrodes in electrolyte solutions. Electrochim. Acta 8, 751–780.

Lisheng, S., Crow, M.L., 2008. Comparison of ultracapacitor electric circuit models. In: IEEE, 20–24, pp. 1–6.

Liu, C., Tsai, D., Chung, W., Li, K., Lee, K., Huang, Y., 2011. Electrochemical micro- capacitors of patterned electrodes loaded with manganese oxide and carbon nanotubes. J. Power Sources 196, 5761–5768.

Mastragostino, M., Arbizzani, C., Soavi, F., 2002. Conducting polymers as electrode materials in supercapacitors. Solid State Ionics 148, 493–498.

Michael, M.S., Prabaharan, S.R.S., 2004. High voltage electrochemical double layer capacitors using conductive carbon as additives. J. Power Sources 136, 250–256.

Miller, J.R., Outlaw, R.A., Holloway, B.C., 2011. Graphene electric double layer capacitor with ultra high performance. Electrochim. Acta 56, 10443–10449.

Morimoto, T., Hiratsuka, K., Sanada, Y., Kurihara, K., 1996. Electric double-layer capacitor using organic electrolyte. J. Power Sources 60, 239–247.

Nasibi, M., Golozar, M.A., Rashed, G., 2013. Nanoporous carbon black particles as an electrode material for electrochemical double layer capacitors. Mater. Lett. 91, 323–325.

Nelms, R.M., et al., 1999. A comparison of two equivalent circuits for double-layer capacitors. In: Fourteenth Annual Applied Power Electronics Conference and Exposition (APEC '99): vol. 2, pp. 692–698.

Ness Capacitor Co., Ltd. http://www.nesscap.com/prod/ba3.htm.

Niu, J., Pell, W.G., Conway, B.E., 2006. Requirements for performance characterization of C double-layer supercapacitors: applications to a high specific-area C-cloth material. J. Power Sources 156, 725–740.

Novoselov, K.S., Geim, A.K., Morozov, S.V., Jiang, D., Zhang, Y., Dubonos, S.V., Grigorieva, I.V., Firsov, A.A., 2004. Electric field effect in atomically thin carbon films. Science 306, 666–669.

Pandolfo, A.G., Hollenkamp, A.F., 2006. Carbon properties and their role in supercapacitors. J. Power Sources 157, 11–27.

Pasquier, A.D., Plitz, I., Gural, J., Badway, F., Amatucci, G.G., 2004. Power-ion battery: bridging the gap between Li-ion and supercapacitor chemistries. J. Power Sources 136, 160–170.

Paulo, C.J., Rodrigo, T., Lavall, L., Silva, G., 2014. Supercapacitors based on modified graphene electrodes with poly(ionic liquid). J. Power Sources 256, 264–273.

Pell, W.G., Conway, B.E., 1996. Quantitative modeling of factors determining Ragone plots for batteries and electrochemical capacitors. J. Power Sources 63, 255–266.

Peng, C., Zhang, S., Jewell, D., Chen, G., 2008. Carbon nanotube and conducting polymer composites for supercapacitors. Proc. Natl. Acad. Sci. U. S. A. 18, 777–778.

Pmar, N., Mierlo, J.V., Verburugge, B., Bossche, P.V.D., 2010. Power and life enhancement of battery—electrical double layer capacitor for hybrid electric and charge-depleting plug-in vehicle applications. Electrochim. Acta 55, 7524–7531.

Prabaharan, S.R.S., Michael, M.S., 2004. High voltage electrochemical double layer capacitors using conductive carbons as additives. J. Power Sources 136, 250–256.

Prabaharan, S.R.S., Vimala, R., Zainal, Z., 2006. Nanostructured mesoporous carbon as electrodes for supercapacitors. J. Power Sources 161, 730–736.

Qu, D., Shi, H., 1998. Studies of activated carbons used in double-layer capacitors. J. Power Sources 74, 99–107.

Ramya, R., Sivasubramanian, R., Sangaranarayanan, M.V., 2013. Conducting polymers-based electrochemical supercapacitors—progress and prospectus. Electrochim. Acta 101, 109–129.

Razumov, S.N., Klementov, A., Litvinenko, S., Beliakov, A., 2001. US Patent No. 6222723.

Rengui, L., et al., 2008. A new topology of switched capacitor circuit for the balance system of ultra-capacitor stacks. In: IEEE Vehicle Power and Propulsion Conference, 2008 (VPPC '08).

Rightmire, R.A., 1966. Electrical Energy Storage Apparatus. US Patent 3288641, November 29.

Ruiz, V., Roldán, S., Villar, I., Blanco, C., Santamariá, R., 2013. Voltage dependence of carbon based supercapacitors for pseudocapacitance quantification. Electrochim. Acta 95, 225–229.

Sankapal, B.R., Gajare, H.B., Dubal, D.P., Gore, R.B., Salunkhe, R.R., Ahn, H., 2014. Presenting highest supercapacitance for TiO_2/MWNTs nanocomposites: novel method. Chem. Eng. J. 247, 103–110.

Sevilla, M., Mokaya, R., 2014. Energy storage applications of activated carbons: supercapacitors and hydrogen storage. Energy Environ. Sci. 7, 1250–1280.

Sharma, P., Bhatti, T.S., 2010. A review on electrochemical double-layer capacitors. Energy Convers. Manag. 51, 2901–2912.

Sikha, G., Popov, B., 2004. Performance optimization of a battery-capacitor hybrid system. J. Power Sources 134, 130–138.

Spyker, R.L., Nelms, R.M., 2000. Classical equivalent circuit parameters for a double-layer capacitor. IEEE Tran. Aerosp. Electron. Syst. 36 (3), 829–836.

Sung, J., Kim, S., Jeong, S., Kim, E., Lee, K., 2006. Flexible micro-supercapacitors. J. Power Sources 162, 1467–1470.

Tõnurist, K., Jänes, A., Thomberg, T., Kurig, H., Lust, E., 2009. Influence of mesoporous separator properties on the parameters of electrical double-layer capacitor single cells. J. Electrochem. Soc. 156, A334–A342.

Tõnurist, K., Thomberg, T., Jänes, A., Romann, T., Sammelselg, V., Lust, E., 2013. Influence of separator properties on electrochemical performance of electrical double-layer capacitors. J. Electroanal. Chem. 689, 8–20.

Wang, L., Morishita, T., Toyopda, M., Inagaki, M., 2007. Asymmetric electric double layer capacitors using carbon electrodes with different pore size distributions. Electrochim. Acta 53, 882–886.

Wei, L., Yushin, G., 2012. Nanostructured activated carbons from natural precursors for electrical double layer capacitors. Nano Energy 1, 552–565.

Xianzhong, S., Xiong, Z., Haitao, Z., Dacheng, Z., Yanwei, M., 2012. A comparative study of activated carbon-based symmetric supercapacitors in Li_2SO_4 and KOH aqueous electrolytes. J. Solid State Electrochem. 16, 2597–2603.

Yang, F., Lnguang, L., Yuping, Y., He, Y., 2010. Characterization, analysis and modeling of an ultracapacitor. In: Proceedings of EVS25, November, pp. 1–12.

Yasser, D., Venet, P., Gualous, H., Rojat, G., 2008. Electrical, frequency and thermal measurement and modeling of supercapacitor performance. In: Proceedings 3rd ESSSCAP, 2008. [CD-ROM].

Zhang, L.L., Zhao, X.S., 2009. Carbon-based materials as supercapacitor electrodes. Chem. Soc. Rev. 38, 2520–2531.

Zhang, Q., Xu, C., Lu, B., 2014. Super-long life supercapacitors based on the construction of Nifoam/graphene/Co_3S_4 composite film hybrid electrodes. Electrochim. Acta 132, 180–185.

Zhao, X., Huang, S., Cao, J., Xi, S., Wei, X., Kamamoto, J., Takarada, T., 2014. KOH activation of hypercoal to develop activated carbons for electric double layer capacitors. J. Anal. Appl. Pyrolysis 105, 116–121.

Zheng, C., Zhou, X., Cao, H., Wang, G., Liu, Z., 2014. Synthesis of porous graphene/activated carbon composite with high packing density and large surface area for supercapacitor electrode material. J. Power Sources 258, 290–296.

New developments of larger supercapacitors: Symmetrical devices, hybrid types, and battery-capacitors

8

8.1 Introduction

In Chapter 7 we have given an overview of developments related to electrical double layer capacitors, starting from the historic developments from 1853 and recent developments up to 2014. According to IDTechEx reports (Harrop, 2020), by 2041 the supercapacitors business could be reaching 10% of the batteries business. With new technologies and materials such as graphene and carbon nanotubes adopted in manufacturing supercapacitors, there will be a large number of worldwide manufacturers for different forms of supercapacitors. This chapter is aimed at presenting an engineer's viewpoint of new developments related to larger commercial devices and supercapacitor modules within the last 5–6 years.

8.2 Supercapacitor modules

Supercapacitor adoption in automobiles is one of the main drivers for the new supercapacitor modules and larger single-cell devices. Within the last few years, facilitating the start-stop function (ECO drive) capability of the cars was mainly based on supercapacitor modules (SCMs). Fig. 8.1 depicts the case of an SCM in a traditional internal combustion engine-based car, which reduces the burden on the 12-V battery due to frequent engine starting required in facilitating the start-stop function. In a car without a start-stop function, battery is used to crank the engine only 3–5 times a day, but the start-stop function could increase that number by 10 times and stress the battery and reduce its lifetime in the car.

In addition, SCMs are frequently used in hybrid electric vehicles (HEVs) and electric vehicles (EVs) to help in achieving high power output, when required, without stressing the battery pack used. Higher power density of the supercapacitors, due to their low equivalent series resistance (ESR), which is useful at the time of acceleration comes into play here. Fig. 8.2 depicts different types of SCMs produced by different manufacturers. Fig. 8.2A shows a 6-V capable module made by connecting in series two 3.0-V capable single cells. Such modules are frequently used in memory backup circuits and other electronic circuit modules, where short-term fluctuations of the respective DC supply is not tolerated by the circuit. Fig. 8.2B shows SCMs used in 12-V battery-based automotive systems, where they help the start-stop function.

Energy Storage Devices for Renewable Energy-Based Systems. https://doi.org/10.1016/B978-0-12-820778-9.00015-2

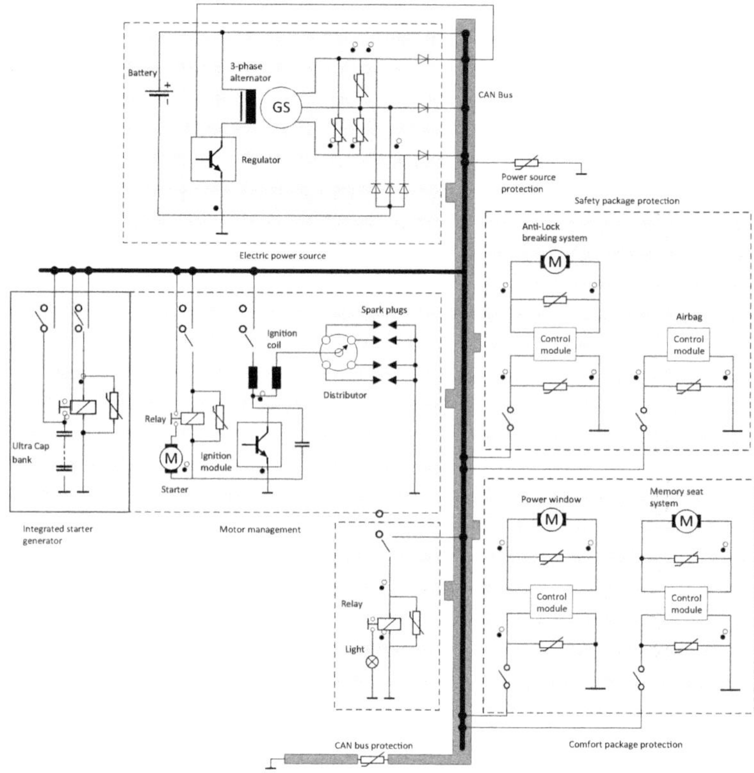

Fig. 8.1 Use of a supercapacitor module (an ultra-capacitor bank) in a representative automotive electrical system.
Source: Kularatna, N., Steyn-Ross, A., Fernado, J., James, S., 2019. Design of Transient Protection Systems: Including Supercapacitor Based Design Approaches for Surge Protectors. Elsevier, 263 pp.; Courtesy of EPCOS AG: From TDK EPCOS, 2018. SIOV Metal Oxide Varistors, App Note January 2018, 17 pp.

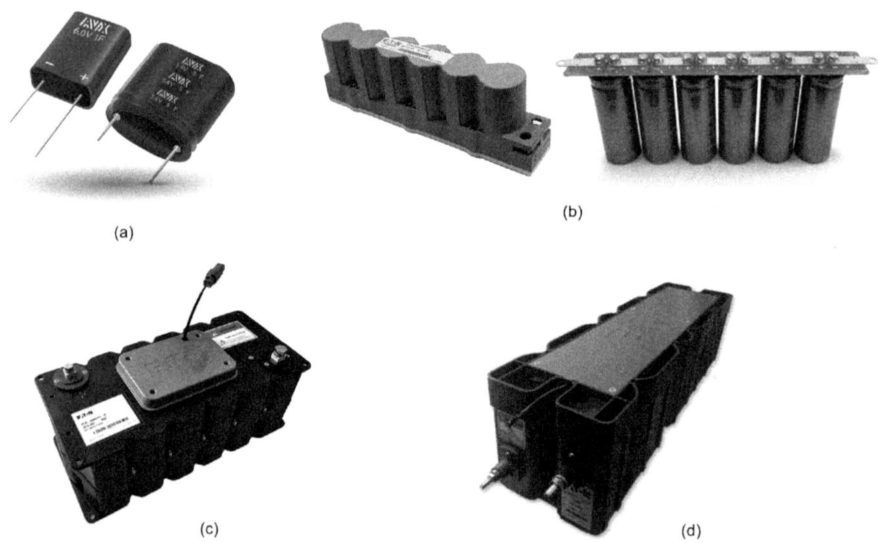

Fig. 8.2 Different types of SCMs (A) 6-V capable module, (B) 12–16-V range modules for 12-V battery-based cars, (C) 48-V module, and (D) 62-V module.

Fig. 8.2C shows a 48-V module used in 42-V battery-based automotive systems, and Fig. 8.2D shows a high-voltage module of 62-V capability. In most of these systems, balancing circuits are used to keep all cells at equal voltage levels. These modules can be in the range of few Wh to over 100 Wh in general with much longer life times than traditional flooded lead-acid, absorbed glass matt-type lead-acid or Li-ion battery packs.

Given the complex issues of rechargeable batteries such as short cycle life and recycling issues (Harrop and Collins, 2020; Holland and Jiao, 2020), newer supercapacitor technologies are gaining momentum in the component manufacturing sector.

The automobile sector's take up of supercapacitors for their high power density is demonstrated by the most recent introduction of Lamborghini's Sián model, which utilizes supercapacitor energy storage for rapid acceleration (Slovick, 2019). While most hybrid cars use electric motors to downsize the accompanying gasoline engine, in this case Lamborghini combines an electric motor with a massive V12 gasoline engine. It is rated at 785 hp (577 kW) at 8500 rpm—the highest output ever from a Lamborghini. Combined with the additional 34 hp from the electric motor of the hybrid system, the Sián delivers a total of 819 hp (602 kW), enabling it to reach a top speed of over 350 km/h (217.5 mph). A 48-V e-motor delivering 34 hp is incorporated into the gearbox to provide immediate response and improved performance. This is claimed to be first application where a low-voltage hybrid system has a direct connection between the electric motor and wheels, so that the e-motor can take on the strain during low-speed maneuvers such as reversing and parking. Rather than a lithium-ion (Li-ion) battery, the Sián employs a supercapacitor array, which Lamborghini boasts is three times more powerful than a battery of equivalent weight. The company originally pioneered supercapacitor technology in its Lamborghini Aventador to provide added cranking power for efficient restarting of the 12-cylinder-engine vehicle.

While Li-ion batteries can store around 20 times more energy for their weight than supercapacitors, the battery's flaw is that it takes time to discharge this energy due to resistance within the device. The fast movement of charge in supercapacitors is due to very low equivalent series resistance (ESR). In effect, this means they can be charged and discharged at high specific current values (A/kg) that may be over 100 times that of batteries, without damaging the unit (Kouchachvili et al., 2018). As developed for the Sián, the SCM (supercapacitor array) is said to be three times lighter than a battery producing the same power.

The Lamborghini Sián incorporates a newly developed regenerative braking system, a technology that recaptures the kinetic energy lost as heat when a car decelerates and turns it into electricity. Using the essentially free energy from the car's brakes stored in the supercapacitor system, the electrical system helps get the car moving in tandem with the V12. The energy stored provides an instantly available power boost, allowing the driver to draw immediately on increased torque when accelerating away, up to 130 km/h (81 mph) when the e-motor automatically disconnects. As a result, it is more than 10% faster than a car without this system.

In both hybrid electric vehicles (HEV) and electric vehicles (EV), SCMs could be combined for achieving better overall performance in terms of power to energy ratio, cycle life, and the overall efficiency achieved in energy conversion (Kouchachvili et al., 2018; Mastragostino et al., 2007).

8.3 Recent advances in supercapacitor technologies and commercial devices

Over the last 10 years, with a lot of new electrode and electrolyte materials developed in research labs getting adapted by the supercapacitor manufacturers, many different versions of supercapacitor families have entered the market. As indicated by the Ragone plot in Fig. 8.3, supercapacitors are gradually trying to reach the energy density of lead-acid batteries while providing higher life cycle capability than all rechargeable battery chemistries.

In order to have a discussion on the commercially available supercapacitor families from an engineer's viewpoint, we should first compare general capacitors with supercapacitor families in terms of canister sizes, typically used in packaging electrolytic or film capacitors. In the lower row of Fig. 8.4, we have shown different sizes of traditional electrolytic capacitors of different capacitance values with their $1/2\,CV^2$ value-based energy storage capability. In the upper row of Fig. 8.4, we have placed a set of low-cost supercapacitors by various manufacturers where internal construction is based on identical active carbon electrodes. Given this comparison, it is easy to say that the supercapacitors have almost one million times larger capacitance for the same volume of the canister. However, their maximum DC voltage rating is less than 3.0 V per cell

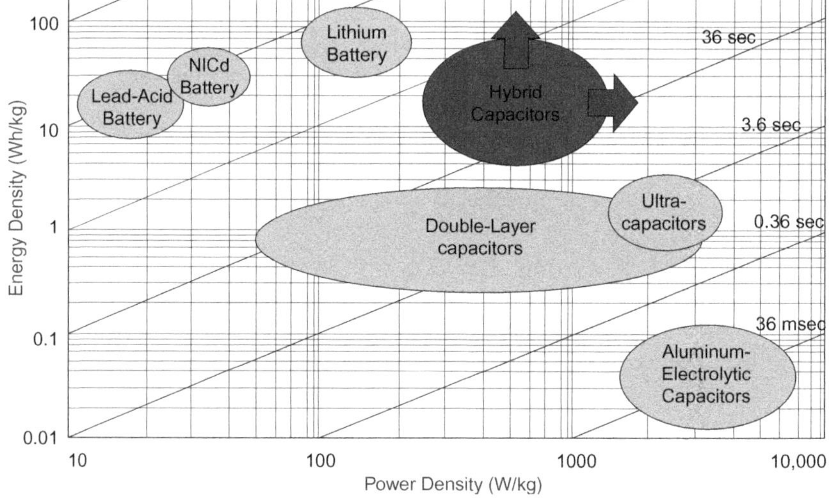

Fig. 8.3 Ragone plot for energy storage systems indicating more recent achievements. *Source*: US Defense Logistics Agency.

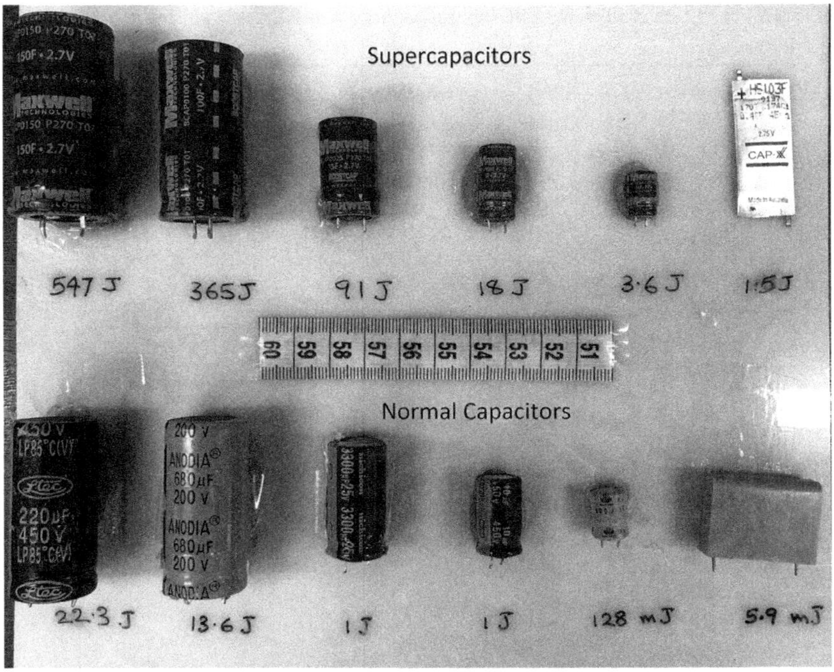

Fig. 8.4 Comparison of symmetrical double-layer capacitors (low-cost devices) with electrolytic capacitors of the similar canister sizes.

for the case of symmetrical double-layer capacitors. From the data in Fig. 8.4, we can simply see that despite their lower DC voltage ratings, they could store almost one- to two-order larger energy.

A general question asked by the circuit designers while considering supercapacitors is the possible price increase compared to electrolytic capacitors or film types. Table 8.1 prepared by VINATech (United Kingdom) provides an answer indicating that the supercapacitors below 100 F (with 3.0-V DC rating) are comparable with the prices of electrolytic capacitors of the range 100–500 μF.

After 2005, supercapacitor manufacturers were able to commercialize newer types of electrode combinations, borrowing the experiences from Li-ion battery technologies creating asymmetrical types now commercially known as hybrid supercapacitors. Fig. 8.5 compares the internal concept of these hybrid types where one of the electrodes are made of lithium titanate or similar electrode materials creating an internal structure of two series capacitors of unequal value. Fig. 8.5A is a symmetrical electrical double-layer capacitor with two identical activated carbon or similar electrodes. Fig. 8.5B is an asymmetric supercapacitor where one electrode is based on activated carbon and other is similar to an electrode in a lithium battery, creating a series pair of unequal capacitors on either side of the electrolyte layer. In general, this newer construction achieves almost 2–3 times higher capacitance for the same canister volume. However, this capacitance gain comes with a penalty of slightly lower cycle life.

Table 8.1 The price comparison of supercapacitors with similar canister size electrolytic capacitors.

Canister size	**3-V ELDC (F)**	**Price per 1000 pieces (US$)**	**Electrolytic (μF)**	**Price per 1000 pieces (US$)**
8 × 13	1	0.4	100, 25 V	0.18
10 × 20	5	0.53	470, 35 V	0.24
10 × 30	10	0.6	1000, 25 V	0.4
16 × 25	25	1.6	4700, 10 V	0.88
18 × 40	50	1.8	470, 50 V	1.7

Courtesy of VINATech UK office; Source: Kularatna, N., Jayananda, D., 2020. Supercapacitor based long time constant circuits: a unique design opportunity for new power electronic circuit topologies, IEEE Ind. Electron. Mag., 40–56.

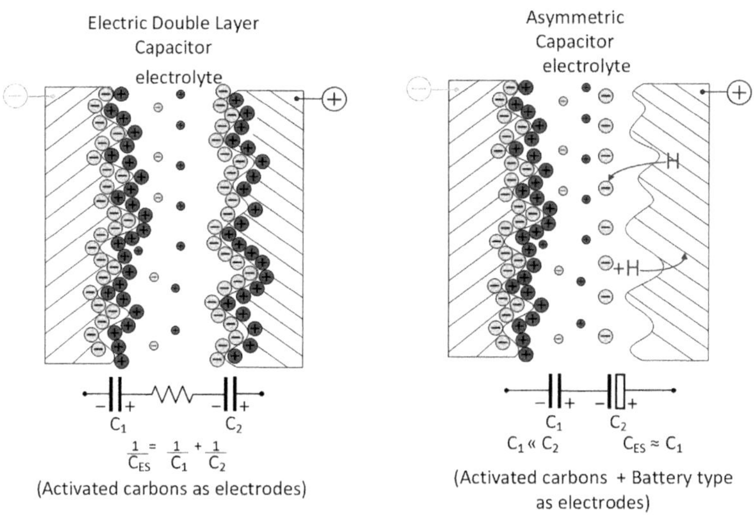

Fig. 8.5 Comparison of electrode structure in symmetrical double-layer supercapacitors and hybrid devices (A) electric double-layer capacitor with symmetrical activated carbon electrodes (B) asymmetric supercapacitor with nonidentical electrodes.

In a relatively newer construction based on pseudo-capacitance, a newer family known as battery caps or capa-batteries have entered the market with almost 10–25 times larger capacitance for the same canister volume. Fig. 8.6 is a photograph of a 3000 F, 3.0-V symmetrical double-layer capacitor (on the left-hand side) compared with a 7500 F, 2.8-V hybrid device (in the middle) and a 40,000 F 2.7 V capa-battery made by Smawha Electric, Korea.

As we see from Table 8.2, all these supercapacitor families have the volumetric power density specifications, which are significantly higher than the lead-acid or Li-ion batteries. This is due to very low ESR values achieved by the supercapacitor devices. As shown in Fig. 8.7, another disadvantage of batteries in general is their

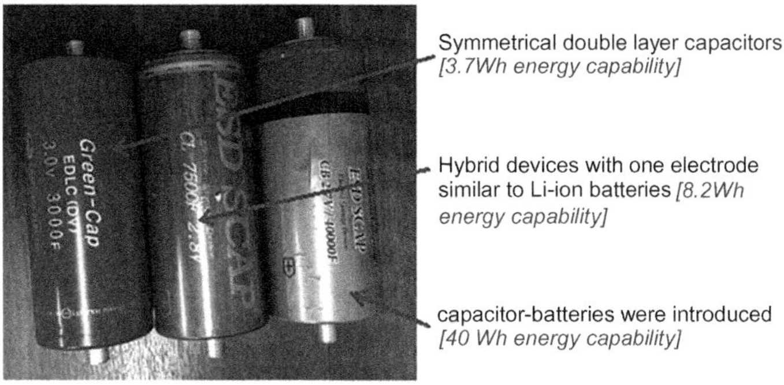

Source : Samwha Electric

Fig. 8.6 Photographic comparison of large supercapacitors of three different kinds with similar canister size.
Adapted from Kularatna, N., Jayananda, D., 2020. Supercapacitor based long time constant circuits: a unique design opportunity for new power electronic circuit topologies, IEEE Ind. Electron. Mag. 40–56.

Table 8.2 Compares the essential specifications of these supercapacitor families with lead-acid and Li-ion batteries.

Parameter	EDLCs	Hybrid SCs	Capa-batteries	Lead-acid	L_1-ion
Energy density (Wh/L)	5–8	10–14	50–120	50–125	250–670
Power density (W/L)	8000	2500–4,000	1600–3200	25–100	375–1,750
Cycle life (cycles)	1,000,000	40,000–50,000	15,000–20,000	500–2000	1000–1200
Rated voltage (V)	2.7	2.7	2.7	2	3–3.6
Capacitance (F)	1–3000	200–7500	1000–70,000	N/A	N/A

An interesting observation here is that capa-batteries manufactured by Samwha Electric, Korea, have almost reached the volumetric energy density of the lead-acid batteries (ESD-SCAP catalogue, 2018).

gradually increasing internal resistance with the percentage discharge of its energy. In Fig. 8.7, an Energizer 6-V battery is compared with a 3000 F, 2.7-V symmetrical EDLC since they have similar volume of package. In a battery, the internal resistance increases by one to two orders when the energy is discharged from fully charged condition to fully discharged condition. In HEVs and EVs this creates a major issue, and SCMs comes to the rescue. In these hybrid energy storage systems, when the automobile requires rapid acceleration, SCM comes effectively.

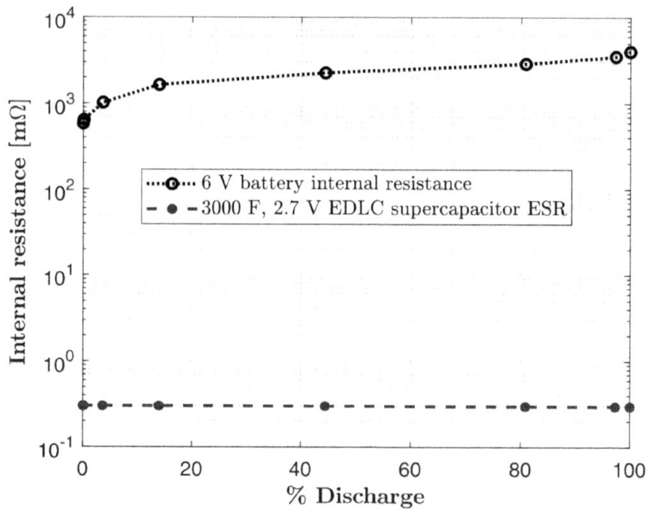

Fig. 8.7 Comparison of internal resistance variation with the percentage discharge—6-V energizer battery and a 3000F 2.7V EDLC from VINATech.

8.4 Comparison of discharge curves of different supercapacitor families

Given the case that significant differences in electro-chemistry applicable to these capacitor types exist, it is worth comparing the discharge curves of these three types. Fig. 8.8 compares the significant differences of discharge curves for the three types compared here, under constant resistance load conditions. While a symmetrical EDLC demonstrates a near exponential discharge curve, in the hybrid and capa-battery families curves become similar to rechargeable battery discharge curves, demonstrating their nonlinear nature of the capacitance.

Fig. 8.9 depicts the case of constant current discharge of a hybrid capacitor compared with a capa-battery at different C rates. In this case, we have estimated the Ah capability of 7500 and 40,000F single-cell devices (treating them like a battery) and estimated the corresponding C rate used for the discharge.

8.5 Future developments of larger supercapacitors

Given the rapid commercial developments over the last 5 years summarized above for larger cells, and the need for high-performance energy storage systems for transportation systems including cars, heavy vehicles, and future electric aircrafts new supercapacitor materials and formats will be an important subject topic. As per Harrop and Collins (2020), by IDTechEx summarizes, new electrolytes and

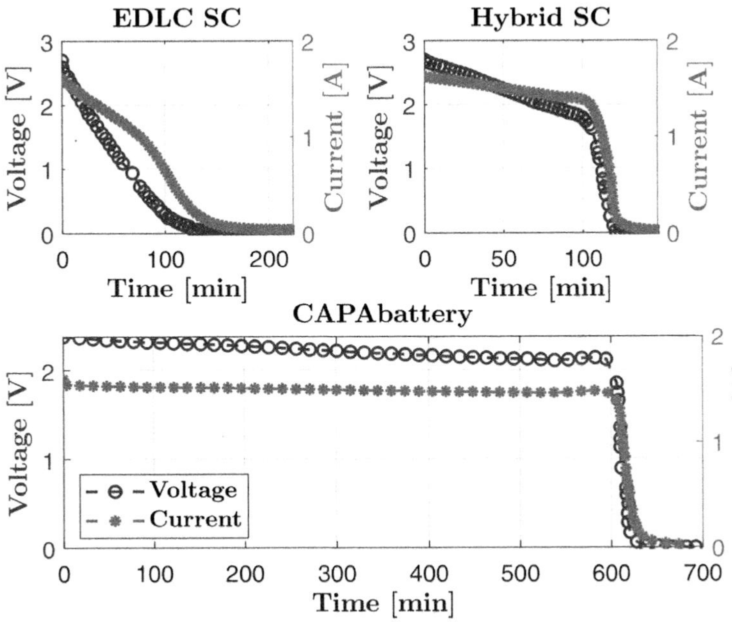

Fig. 8.8 Discharge curves for three SC technologies for constant resistance—3000 F, 2.7 V EDLC SC, 7000 F, 2.7 V hybrid SC and 40,000 F, 2.8 V capa-battery.

electrode materials will play a significant role in introducing new battery-like devices with high life cycle, energy density, and power density. Meanwhile some car companies are working on load-bearing supercapacitor materials, as well as tires and cables to work like supercapacitors. Flexible supercapacitors and micro-supercapacitors adaptable for the semiconductor fabrication area are another interesting research topic to develop. This is mainly due to well-known issues of the existing and developing rechargeable battery technologies, which are harder to eliminate fully.

Another device family getting into mature manufacturing stage is the lithium supercapacitors with DC voltages around 3.8 V and above and is worth a discussion. Fig. 8.10 depicts a few commercial examples of these 3.8-V Li-supercapacitors, and for these the manufacturers' recommendation is not to discharge them below about 2.5 V.

As per the experience of the authors team here, these new developments will ignite not only pure-supercapacitor-based or supercapacitor-dominated transportation systems, but also unique new families of supercapacitor-assisted converter techniques, where designers should look at these devices as one million times larger capacitors—creating new design opportunities for supercapacitor-based long-time constant circuits for renewable energy systems (Kularatna and Jayananda, 2020).

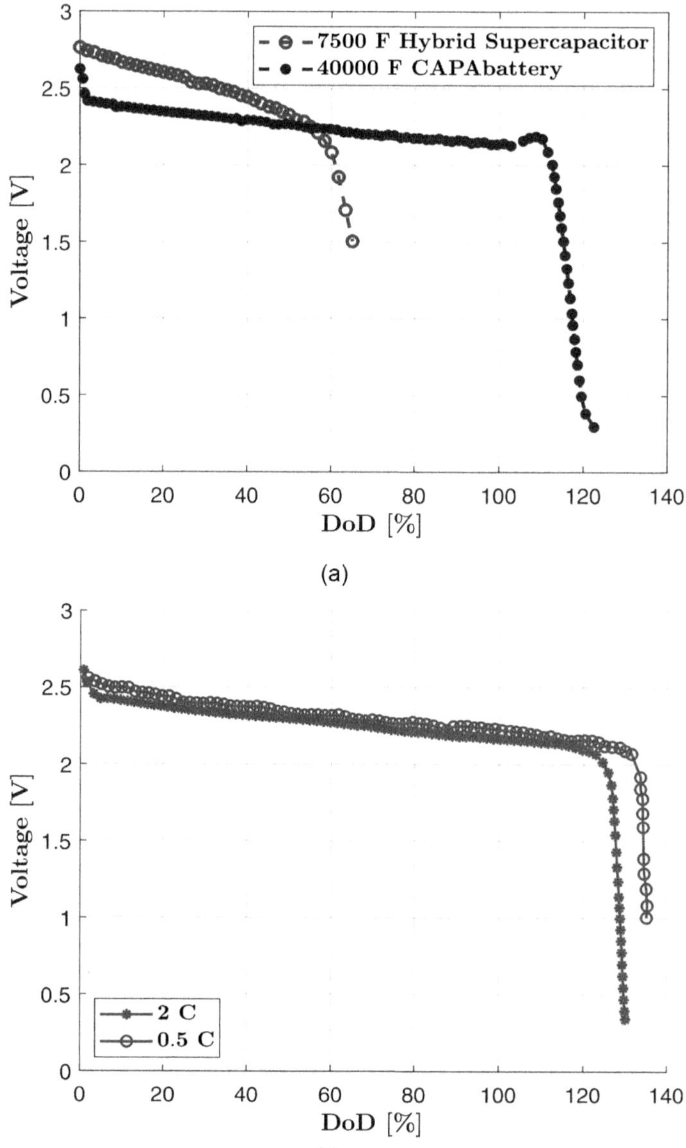

Fig. 8.9 Comparison of hybrid capacitors and capa-batteries at different constant current discharge rates.

Fig. 8.10 Lithium supercapacitor families raising the maximum operating voltage beyond 3.8 V.
Courtesy of VinaTech, United Kingdom.

References

ESD-SCAP catalogue, 2018. Samwha Electric Capacitor Co, Yongin-sl, South Korea.

Harrop, P., 2020. Supercapacitor Markets, Technology Roadmap, Opportunities 2021-2041. IDtechEx report.

Harrop, P., Collins, R., 2020. Supercapacitor Materials and Formats 2020–2040. IDtechEx report.

Holland, A., Jiao, N., 2020. Li-ion Battery Recycling: 2020-2040: Technologies and Processes, Markets, Value Chain, Players, Economics and Business Cases, Forecasts. IDtechEx Reports.

Kouchachvili, L., et al., 2018. Hybrid battery/supercapacitor energy storage system for the electric vehicles. J. Power Sources 374, 237–248.

Kularatna, N., Jayananda, D., 2020. Supercapacitor based long time constant circuits: a unique design opportunity for new power electronic circuit topologies. IEEE Ind. Electron. Mag., 14 (2), 40–56.

Mastragostino, M., et al., 2007. Strategies for high-performance supercapacitors for HEV. J. Power Sources 174, 89–93.

Slovick, M., 2019. Lamborghini Hybrid Uses Supercapacitors in Place of Batteries. Electronic design, 20 Sept.

Supercapacitor assisted (SCA) techniques and the supercapacitor-assisted loss management (SCALoM) concept

9

Thilanga Ariyarathna

9.1 Introduction

Until recently, capacitors in electrical circuits were in the range of few picofarads to about 100,000 μF. This in general forms RC circuits with time constants of less than 100 ms. Supercapacitors come in similar canister sizes to electrolytic or film capacitors and offer 5–7 order larger devices. This translates to the practical possibilities to build circuits with very large time constant circuits with seconds to minutes order vales. Given the preliminary discussion about capacitor basics in Chapter 6, in this chapter, we study the specific case of supercapacitor-based circuits with one-to-several-order larger time constants, developing the theoretical basis for a very different and unique set of new power electronic circuit topologies, known as supercapacitor-assisted (SCA) techniques. The chapter provides an overview of useful applications based on two basic extended design approaches using long-time constant *RC* circuits, namely, (i) utilizing a useful resistive load as the resistive component and (ii) slowing down the charging operation significantly by changing the capacitor to a supercapacitor. This chapter details the theoretical foundations for a set of unique SCA techniques, applicable to high-efficiency DC-DC converters, surge absorbers, rapid heat transfer, and renewable energy-based LED lighting, white goods, etc.

9.2 Typical capacitor charging and discharging process

As detailed in Chapter 6, basic capacitor charging and discharging process can be formulated as below. Special case of starting with an uncharged capacitor is considered here.

Voltage and current versus time curve for capacitor charging circuit is shown in Fig. 9.1. Voltage across the capacitor cannot be changed instantly and this starts at zero and gradually reaches the supply voltage. When we turn the switch on, initially the capacitor acts as short circuit and current is at its maximum value. This behavior is described by Eqs. (9.1), (9.2).

Energy Storage Devices for Renewable Energy-Based Systems. https://doi.org/10.1016/B978-0-12-820778-9.00008-5

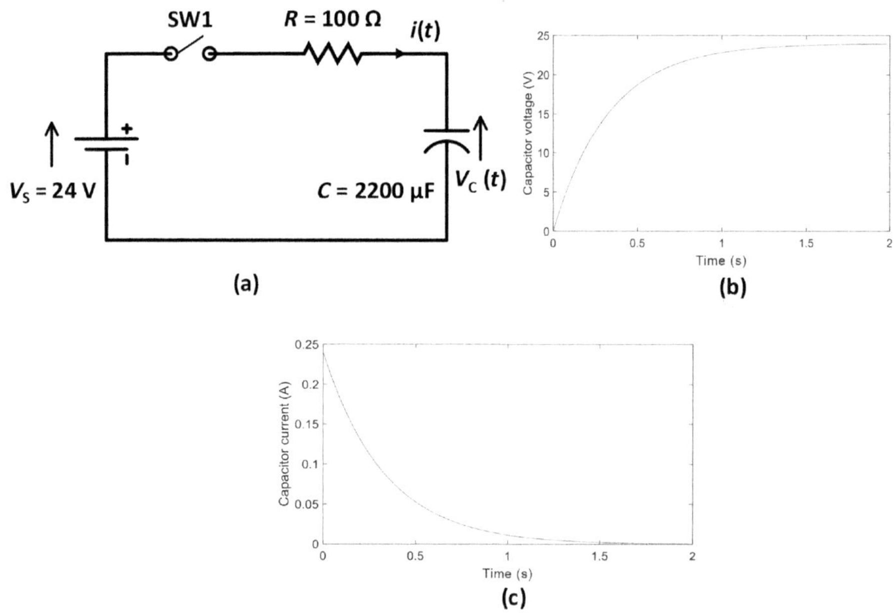

Fig. 9.1 *RC* charging loop: (A) circuit, (B) voltage curve, and (C) current curve.

Capacitor voltage $v(t)$ at time t is given by

$$v(t) = V_S\left(1 - e^{-t/RC}\right) \tag{9.1}$$

The current through the circuit at time t is given by

$$i(t) = \frac{V_S}{R}\left(e^{-t/RC}\right) \tag{9.2}$$

where the total loop resistance is given by R, capacitance of the capacitor is given by C, and V_S represents the supply voltage. Based on the circuit time constant given by $\tau = RC$, capacitor takes more than five time constants (5τ) approximately to reach its final voltage. Table 9.1 compares discharging time constant for a few different

Table 9.1 Approximate time constant τ for different capacitor resistor combinations.

Capacitor	ESR Ω	τ with ESR	Approximate τ with external $R = 10\Omega$
1µF 50 V	4.5	4.5 µs	14.5 µs
10µF 50 V	1.6	16 µs	116 µs
100µF 50 V	0.1	10 µs	1 ms
1F[a]	0.24	240 ms	10 s
10F[a]	0.025	250 ms	100 s
120F[a]	0.008	960 ms	1,200 s

[a] Manufactured by AVX

capacitors with and without a 10Ω load resistance. Effect of time constant τ on the capacitor charging curve is shown in Fig. 9.2. It is important to highlight here that when a capacitor is charged from an initial voltage value of zero, the capacitor stores an energy content of 0.5 CV^2 and the loop resistance wastes the same energy during the charging period. Therefore in a capacitor charging circuit (beginning with a zero initial voltage on the capacitor) 50% of the energy is wasted in the resistive parts of the loop, irrespective of the value of the loop resistance.

General capacitor discharging circuits are shown in Fig. 9.3 where an ideal short circuit is placed across its terminals (Fig. 9.3A) and with an external resistor of 10 Ω is placed in Fig 9.3B. Parasitic resistance of the charging path is considered negligible.

Now consider the capacitor discharge circuit as shown in Fig. 9.4. Initial voltage of the capacitor is considered as V_I then the voltage at time t, $V(t)$ is given by

$$v(t) = V_I\left(e^{-t/RC}\right) \tag{9.3}$$

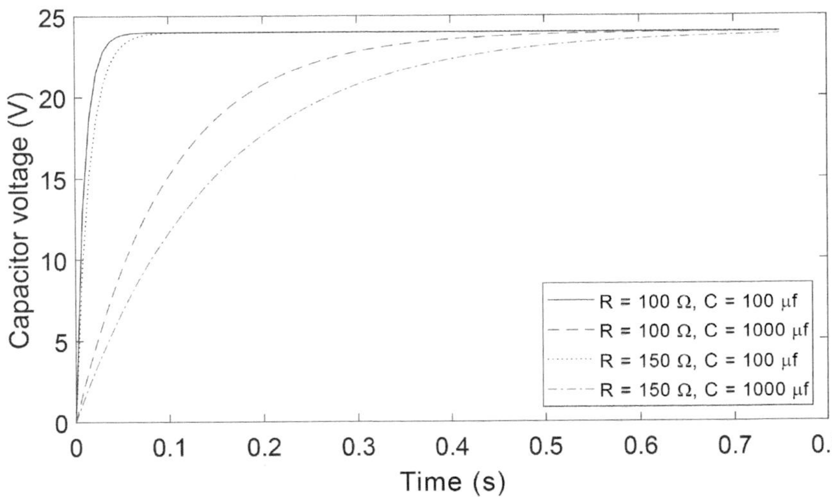

Fig. 9.2 Effect of τ on charging curve of a capacitor.

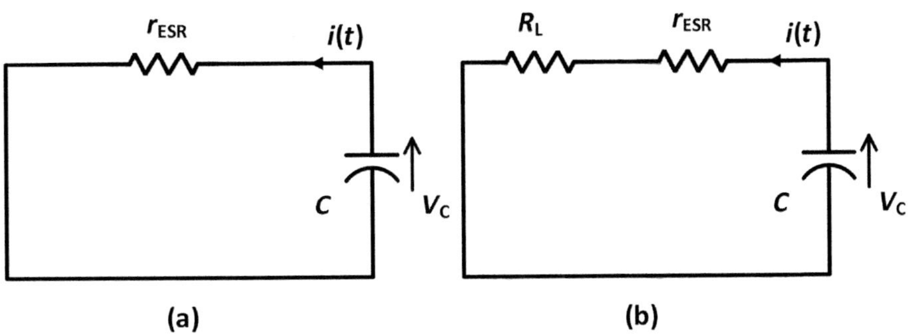

Fig. 9.3 Discharging a capacitor (A) to its own ESR (short circuited) and (B) to 10 Ω load resistance.

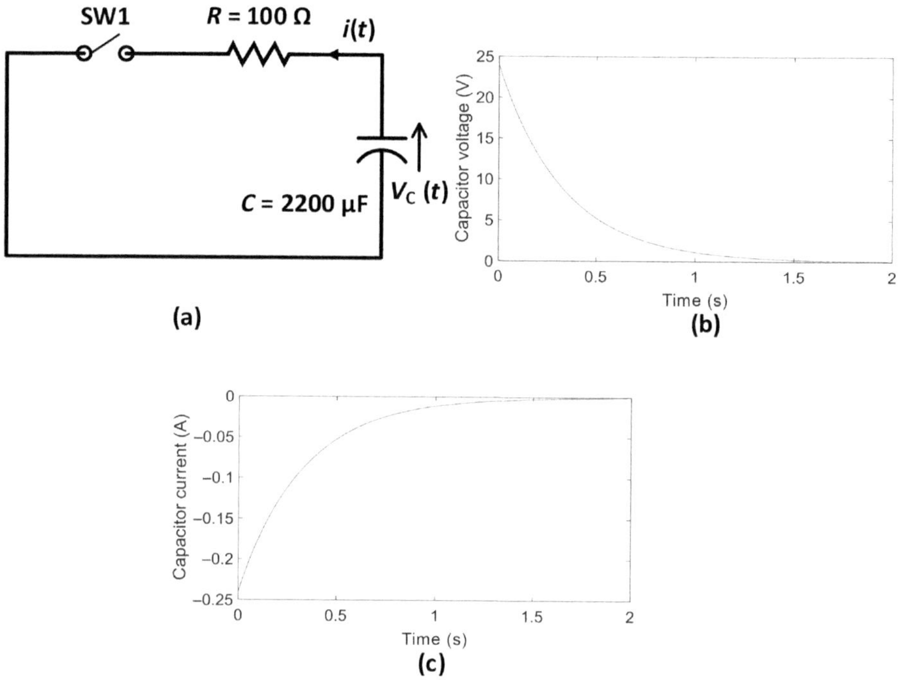

Fig. 9.4 *RC* discharging: (A) circuit, (B) voltage curve, and (C) current curve.

The current flow of the circuit $i(t)$ at time t is given by

$$i(t) = -\frac{V_1}{R}\left(e^{-t/RC}\right) \tag{9.4}$$

When a capacitor is charged from an initial value of V_1 to a final value of V_2 total energy accumulated by the capacitor is given by

$$E = \frac{1}{2}C\left(V_2^2 - V_1^2\right) \tag{9.5}$$

9.3 Generalized case of the RC circuit; with precharged capacitor connected to a DC source, which is higher than the rated DC voltage of the capacitor

The general capacitor charging efficiency is recapped in the previous section when the capacitor starts from a zero voltage condition. This section discusses the charging and discharging efficiency of a precharged capacitor. When a capacitor is precharged, charging efficiency of the charging loop could be higher, since the voltage drop across the loop resistance is lower. Effect of two different power supplies to charging curve is shown in Fig. 9.5. Here $t_2 < t_1$ which means a capacitor can reach given voltage (V_1) earlier under higher voltage power supply. In a practical situation, no capacitor should

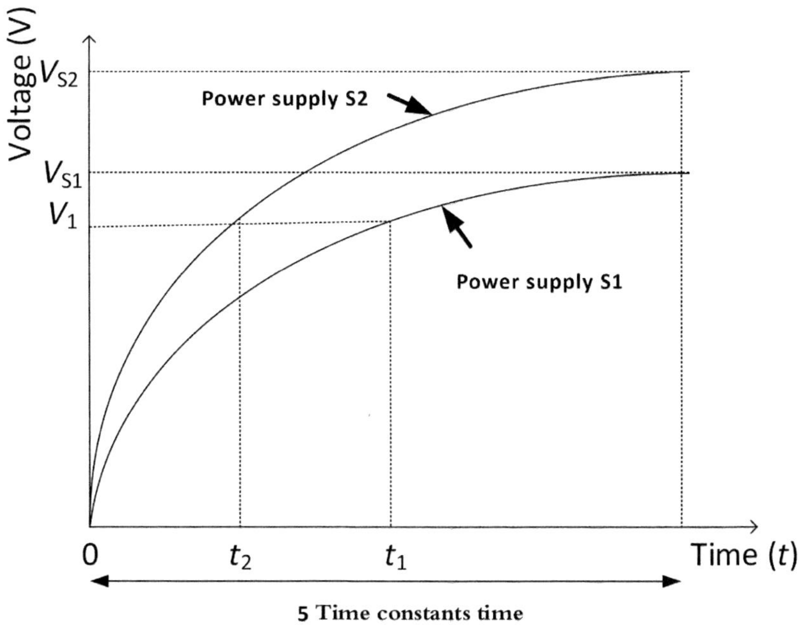

Fig. 9.5 Charging capacitor under different power supplies S1 and S2.

be charged beyond its rated DC voltage. In general for the safety of the component, we charge a capacitor to a lower value than its maximum DC voltage rating. (We later indicate this design maximum as the working voltage of the capacitor). Given this case, V_1 should be within the DC rating of the capacitor. This case is specially applicable to an RC circuit based on a supercapacitor (where supercapacitor's DC rating is in the order of few volts only) and the circuit time constant is significantly increased, allowing the use of a low speed switch to stop capacitor overcharging.

However, when we charge the capacitor with a higher voltage, charging efficiency is low due to higher voltage across the loop resistance. In order to analyze this general case, specifically applicable to a loop with a supercapacitor, we consider the case of a source voltage of m times the rated DC voltage of the supercapacitor, V_w ($m > 1$), and the capacitor is precharged to a voltage of k times V_W.

When a fully discharged capacitor bank charges from 0 to V_W from a DC supply of mV_W, the capacitor voltage at time t is

$$v(t) = mV_W\left(1 - e^{-t/RC}\right) \tag{9.6}$$

Following the curve in Fig. 9.6B, let t_1 and t_2 are the times required for $v(t)$ to reach kV_W ($0 < k \leq 1$) and V_W, respectively. Given a set of new practical circuit topologies discussed in later chapters, in this part, we analyze how does a capacitor charging circuit responds, when the power supply voltage (V_S) is higher than the rated voltage (V_W) of the capacitor, which carries some precharge represented by the voltage kV_W. We seek to determine how the efficiency of the charging process varies as a function of k (the initial voltage fraction) and m (the power supply overvoltage factor). In a

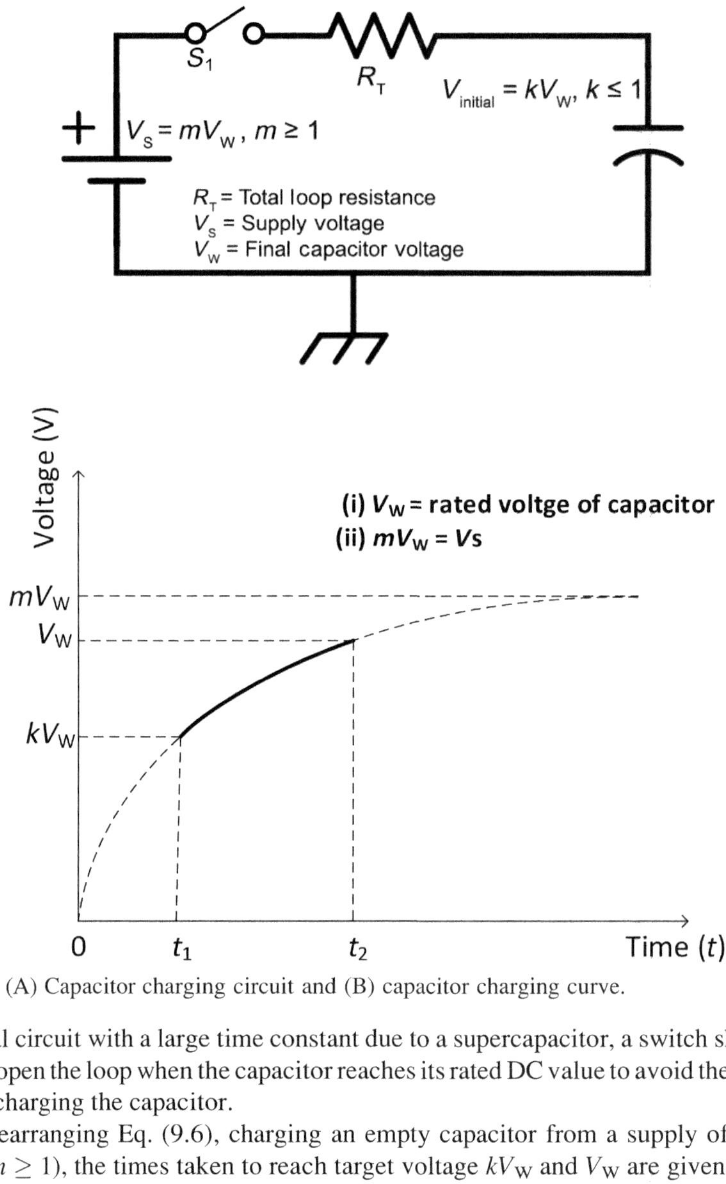

Fig. 9.6 (A) Capacitor charging circuit and (B) capacitor charging curve.

practical circuit with a large time constant due to a supercapacitor, a switch should be used to open the loop when the capacitor reaches its rated DC value to avoid the disaster of overcharging the capacitor.

By rearranging Eq. (9.6), charging an empty capacitor from a supply of voltage mV_W ($m \geq 1$), the times taken to reach target voltage kV_W and V_W are given, respectively, by the expressions

$$e^{-t_1/RC} = \frac{m-k}{m} \quad \text{and} \quad e^{-t_2/RC} = \frac{m-1}{m} \tag{9.7}$$

Instantaneous voltage across capacitor $V_C(t)$ at time t is given by

$$\begin{aligned} mV_W - V_C &= mV_W - mV_W(1 - e^{-t/RC}) \\ &= mV_W e^{-t/RC} \end{aligned}$$

The increment in stored energy between time t_1 and t_2 is given by

$$E_C(t_2,t_1) = \frac{1}{2}CV_W^2 - \frac{1}{2}C(kV_W)^2 = \frac{1}{2}CV_W^2(1-k^2) \tag{9.8}$$

while the energy loss in the resistor over this time interval is given by the integral

$$E_R(t_2,t_1) = \int_{t_1}^{t_2} i^2 R dt \tag{9.9}$$

with

$$i^2(t)R = (mV_W - v(t))^2/R = \frac{m^2V_W^2}{R}e^{-2t/RC}$$

Integrating Eq. (9.9) and substituting end-point expression (9.7) lead to

$$E_R(t_2,t_1) = \frac{1}{2}CV_W^2(k^2 - 2mk + 2m - 1)$$

giving an energy efficiency ratio of

$$\eta = \frac{E_C}{E_C + E_R} = \frac{1-k^2}{2m(1+k)} = \frac{1+k}{2m} \tag{9.10}$$

In the limit $k \to 0$ (no capacitor precharge) with $m \to 1$ (no power supply overvoltage factor), efficiency reaches the general case of $\eta \to \frac{1}{2}$, giving us the common case of capacitor charging discussed in many standard textbooks. Eq. (9.10) indicates us that charging efficiency is higher when the capacitor is not fully discharged.

If our goal is to minimize charging time, then we would set an over voltage factor of $m > 1$, but this implies higher current and *reduced* efficiency since $\eta \sim \frac{1}{m}$. Generalized case for variable k and m values is depicted in Fig. 9.7, showing that the best efficiency is reached when k and m are close to unity.

9.3.1 Applying the generalized theory in an AC-DC converter

The earlier analysis could be used to explain the case of an AC-DC converter based on a bridge rectifier and a capacitor, with a load of R_L connected as in Fig. 9.8A. Fig. 9.8C depicts the capacitor charging phase, where $2R_d$ represents the two diodes in the bridge rectifier, coming in series with the loop parasitic resistances represented by r_p. Smoothing capacitor is represented by C, with its equivalent series resistance (ESR) is represented by r_{ESR}.

Given the case of load R_L connected to the circuit, for the Thevenin's equivalent circuit across the capacitor terminals, R_T can be derived as

$$R_T = ((2R_d + r_P) \ll R_L) + r_{ESR}$$

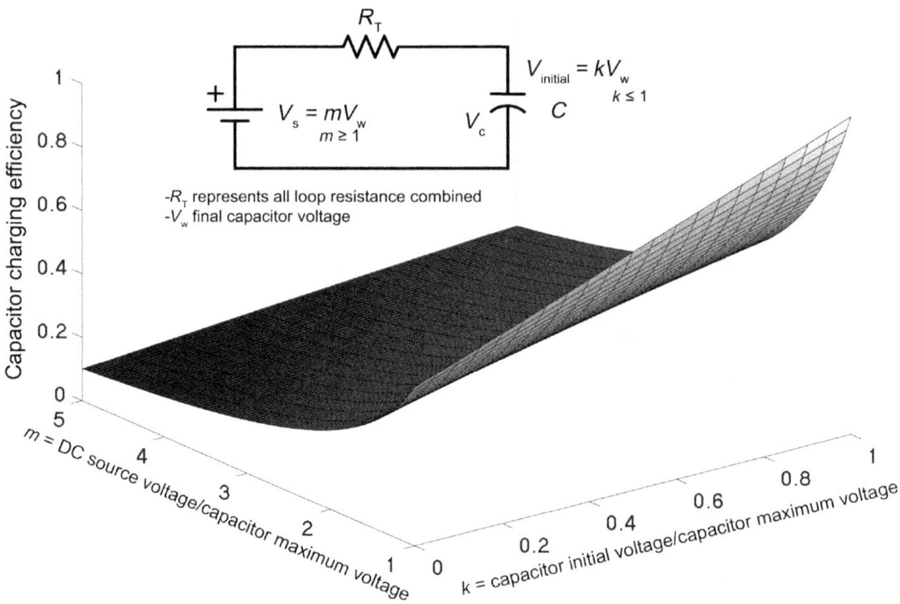

Fig. 9.7 Graphical representation of the circuit behavior in terms of efficiency versus m and k.

However, since in many practical rectifier circuits $(2R_d + r_p) \ll R_L$, Thevenin equivalent resistance (Fig. 9.9) can be reduced to

$$R_{total} = (2R_d + r_P) + r_{ESR}$$

This outcome is shown in Fig. 9.9B.

If the DC source voltage is less than the rated value of the smoothing capacitor (with the case of $m = 1$), capacitor voltage will fluctuate from V_P to kV_P as shown in Fig. 9.8E, with the assumption that diode voltage drop is negligible. Eq. (9.10) now reduces to

$$\eta = \frac{1+k}{2} \tag{9.11}$$

This outcome is visualized in Fig. 9.10. According the graph, efficiency of the operation improves as k getting closer to the unity. Precharge factor k can be related to ripple voltage $V_{r(p\text{-}p)}$ and input peak voltage V_P as

$$k = \frac{V_P - V_{r(p-p)}}{V_P} = 1 - \frac{V_{r(p-p)}}{V_P} \tag{9.12}$$

If this analysis is applied to a typical rectifier charging circuit with peak voltage 12 V (after rectification) and peak-to-peak ripple of 2 V, 91.65% theoretical efficiency can be achieved.

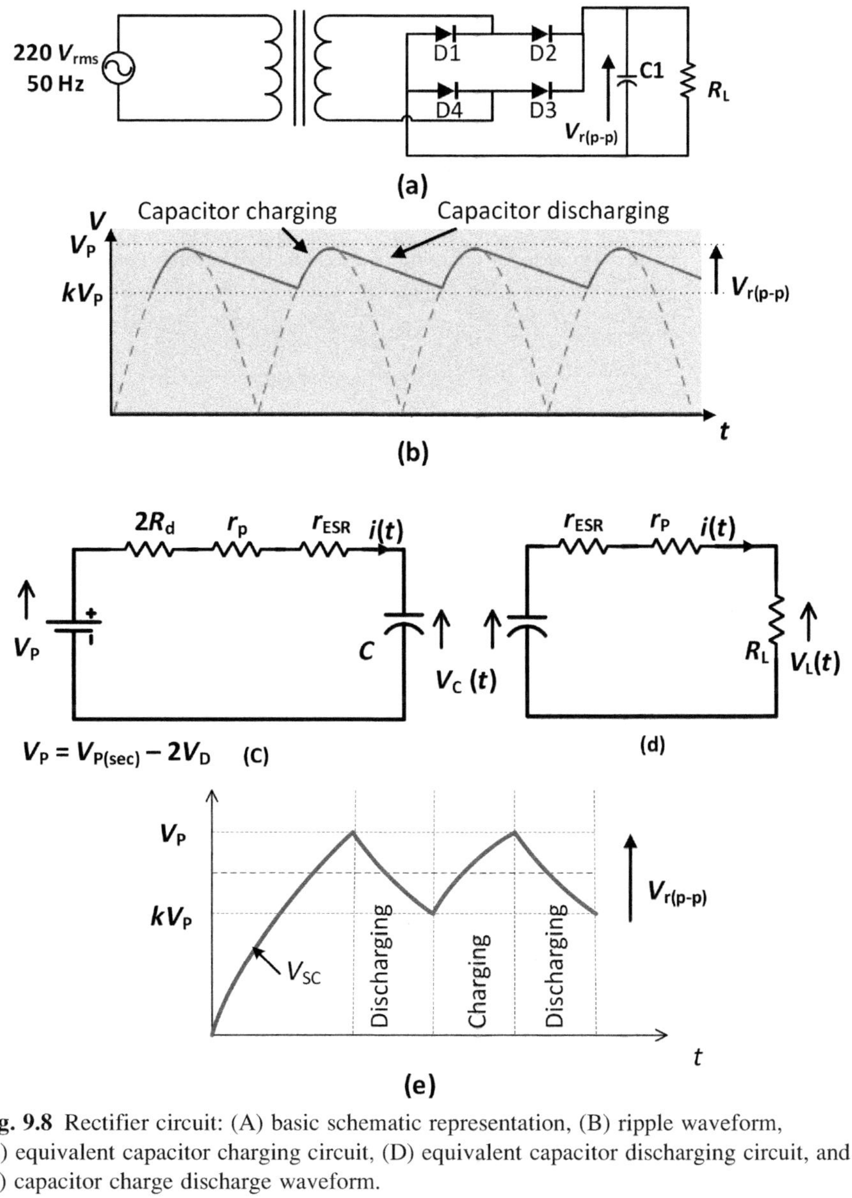

Fig. 9.8 Rectifier circuit: (A) basic schematic representation, (B) ripple waveform, (C) equivalent capacitor charging circuit, (D) equivalent capacitor discharging circuit, and (E) capacitor charge discharge waveform.

When a higher current is drawn from the circuit $V_{r(p-p)}$ increases and k deviate from the unity. This reduces the efficiency of the overall system. This energy will be dissipated in series resistance of the capacitor, on-resistance of the diode and path resistance of the circuit as heat.

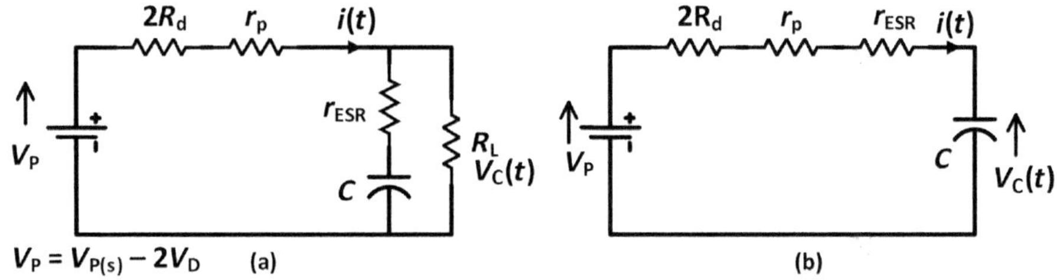

Fig. 9.9 Rectifier circuit: (A) capacitor charging circuit and (B) Thevenin equivalent circuit.

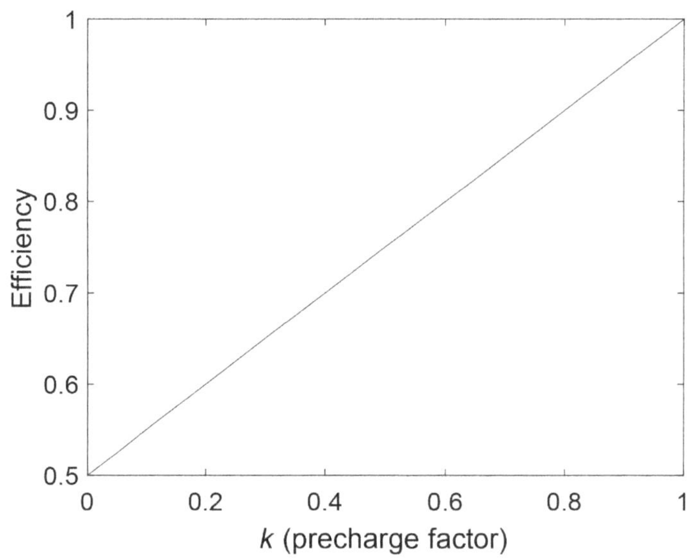

Fig. 9.10 Charging efficiency of smoothing capacitor in bridge rectifier circuit.

9.4 Analysis of discharging efficiency

Fig. 9.8D depicts the equivalent circuit for capacitor discharging to the load. ESR of the capacitor (R_{ESR}), path resistance (r_P), and the load (R_L) appear series to the capacitor. The total resistance (R_T) of the discharge circuit is given by

$$R_T = r_{ESR} + r_P + R_L$$

Initial current at the discharging phase is given by (capacitor is charged up to V_P)

$$I_I = \frac{V_P}{R_T}$$

Current at time t, $i(t)$ is given by the following equation:

$$i(t) = I_{\mathrm{I}}\left(e^{-t/R_{\mathrm{T}}C}\right)$$

However, as this is a series circuit where same current appear across each component, average current and voltage can be used for the calculations. Here, the term $I_{\mathrm{L(avg)}}$ represents average load current and $V_{\mathrm{L(avg)}}$ represents average voltage across the load over discharging time.

Where $E_{\mathrm{R_L}}$ represents energy consumed by the load, $E_{\mathrm{r_{ESR}}}$ represents energy dissipated in series resistance of the capacitor and $E_{\mathrm{r_P}}$ represents energy dissipated in the path resistance, efficiency η of the discharge circuit can be calculated as

$$\eta = \frac{E_{\mathrm{R_L}}}{E_{\mathrm{R_L}} + E_{\mathrm{r_{ESR}}} + E_{\mathrm{r_P}}}$$

$$\eta = \frac{R_{\mathrm{L}}}{R_{\mathrm{L}} + r_{\mathrm{P}} + r_{\mathrm{ESR}}}$$

Sometimes load is not purely linear and in such cases average resistance $R_{\mathrm{L(avg)}}$ can be calculated as

$$R_{\mathrm{L(avg)}} = \frac{V_{\mathrm{L(avg)}}}{I_{\mathrm{L(avg)}}}$$

Generally, the capacitor discharging phase is more efficient than the charging phase. Using capacitors with low ESR and maintaining load resistance at a higher value compared to ESR and the path resistance, efficiency of the discharge circuit can be enhanced. In other words, at low current, capacitor discharging process is more efficient.

According to the equivalent circuits of charging and discharging phases, load resistance has no major effect on the charging phase; however, it has a higher impact on the discharging phase. Because of this reason, the time constant of the discharge circuit is larger than the charging circuit and the shape of the ripple waveform can be easily explained. Efficiency of *RC* charging loop can be enhanced by introducing a useful resistance and this depends on the ratio between parasitic resistance and the useful resistance. This efficiency enhancement is shown in Fig. 9.11.

9.5 Supercapacitor as a lossless dropper

When capacitors and inductors are considered, capacitors have a clear volumetric advantage as an energy storage device. In order to store 100 J of energy, 50 H inductor is required when it is carrying 2 A current which is given by $\frac{1}{2}Li^2$, and this is an extremely large inductor. However, a supercapacitor rated at 100 F and 2.7 V, with a volume of only 17.1 ml, can store 364 J of energy when it is fully charged. Due to low ESR of supercapacitors, instant high currents can be drawn and it makes

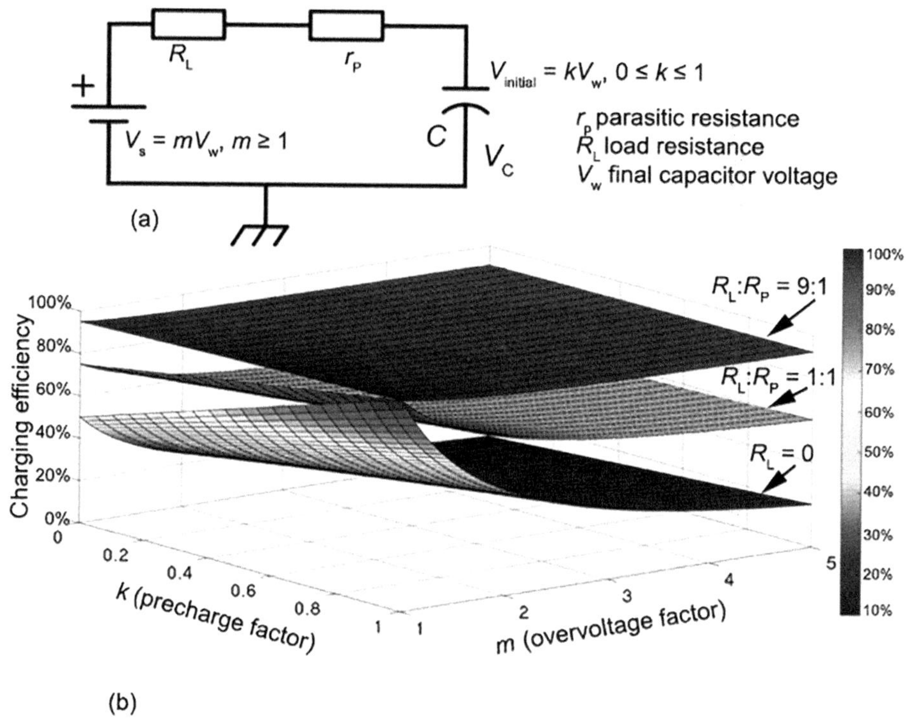

Fig. 9.11 Insertion of a useful resistance into the RC charging loop: (A) modified circuit with useful resistance R_L inserted and (B) graphical representation of the circuit behavior in terms of efficiency vs. m and k.

SCs are less dissipative compared to batteries. Time constant of commercial SCs could be very large and current can be considered as constant in a short-time span. Given the case of very low ESR of supercapacitors, they can be used as lossless voltage droppers in power converters (discussed in Chapter 10).

Voltage difference ΔV of a capacitor circuit can be explained by the following equation:

$$\Delta V = \frac{I_C \Delta t}{C} \tag{9.13}$$

where I_C represents the capacitor current and Δt is the time span. The previous simple relationship indicates that when the capacitor become larger, voltage change occurring at the capacitor terminals becomes smaller. In a practical case of a 100 F supercapacitor, which could be of the same physical size as a 100 μF electrolytic capacitor, voltage will be almost one million times smaller, for the same charging current.

This gave the practical base to develop the supercapacitor-assisted low-dropout regulator (SCALDO) technique discussed later, where we treat an ideal supercapacitor as a lossless voltage dropper.

9.6 The first useful application: SC as a lossless dropper in SCALDO

SCALDO is the first application of the loss circumventing technique discussed earlier. The team was able to develop a new power converter technique by introducing SC or an array of SCs as a lossless voltage dropper in the series path of a low-dropout regulator. The basic idea and the two phases of the simplest SCALDO converter are shown in Fig. 9.12. Key features of this converter can be highlighted as

- load receives the low-noise and high-current slew rate capable DC output of a linear regulator;
- switching frequency is extremely low;
- RFI/EMI issues are eliminated;
- output currents ranging from mA to over 100 A; and
- DC-UPS capability.

By using the SCALDO approach, end-to-end efficiency (ETEE) of 5–3.3 and 12–5 V converters can be improved by multiplication factors of 1.33 and 2, respectively. A linear 5–3.3 V converter with nominal 66% of ETEE can be theoretically increased to 99%, and 12–5 V linear converter 42% efficiency can be increased to 84% (Kankanamge and Kularatna, 2014). Efficiency improvement in popular linear regulators is shown in Table 9.2. Detailed discussion of SCALDO converter is in Chapter 10.

Qualitatively explained, a linear regulator (LDO—a low-dropout regulator is a special case of a linear regulator) is based on the case of a series transistor and feedback loop working in combination to give a constant output DC value when the input DC source and/or the load current fluctuates. In SCALDO, we place an extremely large capacitor (a supercapacitor) in series as in Fig. 9.12A which acts as a lossless dropper. In a typical 12−5 linear regulator we could have 7 V drop across the series transistor. However, if we combine a supercapacitor precharged to 6 V, and place in

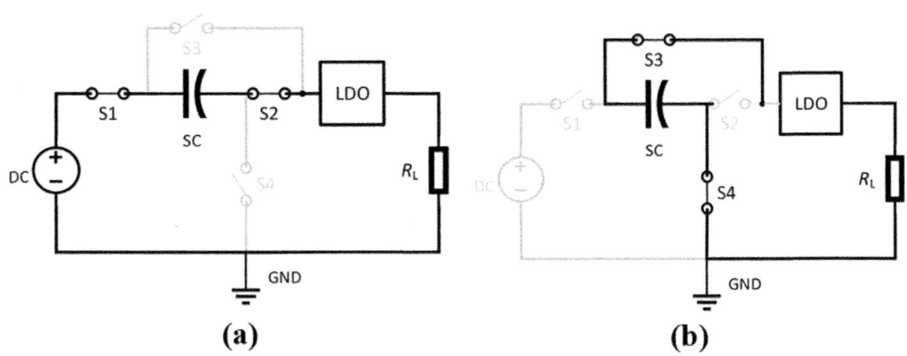

Fig. 9.12 The concept of supercapacitor energy storage and recovery: (A) minimizing the series element dissipation and (B) releasing stored energy.

Table 9.2 Comparison of common linear regulator configurations with SCALDO.

DC-DC converter configuration	5–1.5 V	12–5 V	5–3.3 V	325–48 V
Efficiency without SCALDO	30%	42%	66%	14.75%
Efficiency with SCALDO	90%	84%	88%	88.6%
No. of supercaps used (n)	2	1	3	5
No. of switches required (k) [$k = 3n + 1$]	7	4	10	16
Efficiency improvement factor	3	2	1.33	6
Remarks	High V_{in}/V_{out}	High V_{in}/V_{out}	Low V_{in}/V_{out}	Google 48 V architecture

series it shows only 6 V input to transistor. Now the linear regulator (LDO) works as a 6–5 V regulator with much higher efficiency. Excess energy accumulated during the charge phase is released back to LDO in discharge phase, keeping the input DC source disconnected.

9.7 Resistor-loss and capacitor energy capability in SCASA technique and issues of direct implementation

Supercapacitors are practically capable of delivering and receiving high currents. In general, no harm is made when they are charged with high-voltage sources up to their operating voltage. These properties of SCs are used to develop supercapacitor-assisted surge absorber (SCASA), which is another nontraditional application of SCs. This is a patented and commercialized SCA technique which significantly reduce component count of typical surge absorbers and the cost. In the capacitor charging circuit, power dissipated in the loop resister decreases as capacitor accumulates energy and reaches higher voltage since current across resistor becoming lower. 100 μF capacitor can accumulate 0.5 J energy when it is charged up to 100 V.

Given the earlier argument about very high energy storage capability of supercapacitors, when they are placed in a resistive charging loop, they create very large time-constant circuit. If we apply a high DC voltage to the circuit, which lasts only a very short time (such as in lightning induced voltage on power lines) of T_{step} in Fig. 9.13B capacitor will reach only very small DC value, while most energy in the step DC input will be wasted in the resistive part of the loop. This simple concept is innovatively extended into the SCASA technique, discussed in Chapter 10.

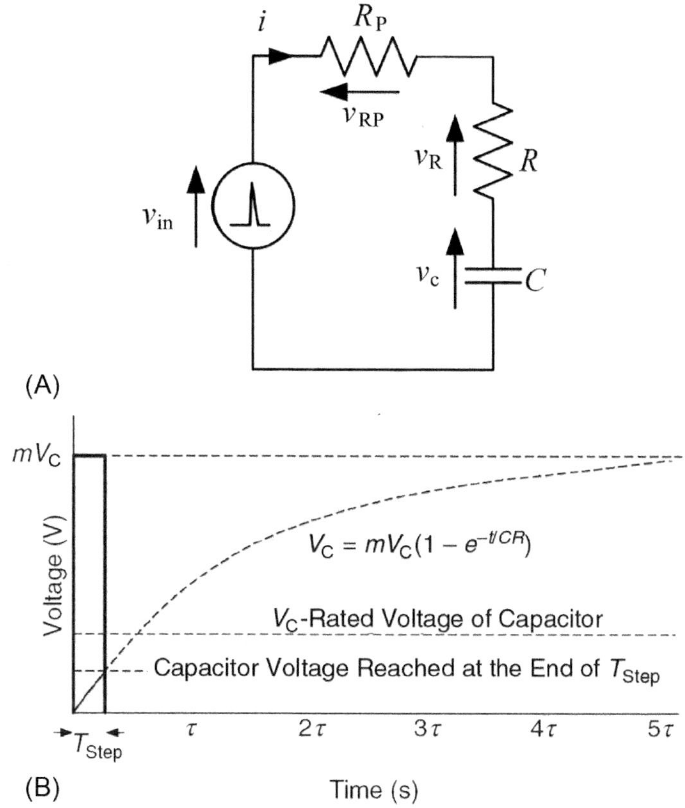

Fig. 9.13 Applying short duration pulse on supercapacitor: (A) equivalent *RC* circuit and (B) voltage rise in the capacitor.

Fig. 9.14 shows the cases of electrolytic caps and supercapacitors indicting why SCASA technique is feasible, despite the low DC voltage rating of supercapacitors (Fig. 9.14).

9.8 Basis of SCATMA technique

Given the very low ESR property of commercial supercapacitors, they are able to discharge very high current, compared to many rechargeable battery types with different chemistries. This property was effectively used to develop a solution to hot-water delay in domestic water faucets, due to long buried water pipes within a building supplying hot water from central source. SC-assisted temperature-modification apparatus (SCATMA) is an "instant water heating technique" based on prestored SC energy to overcome the delayed delivery of hot water in domestic water taps (Gurusinghe et al., 2017). This technique utilizes a precharged low-voltage SC bank with short-term

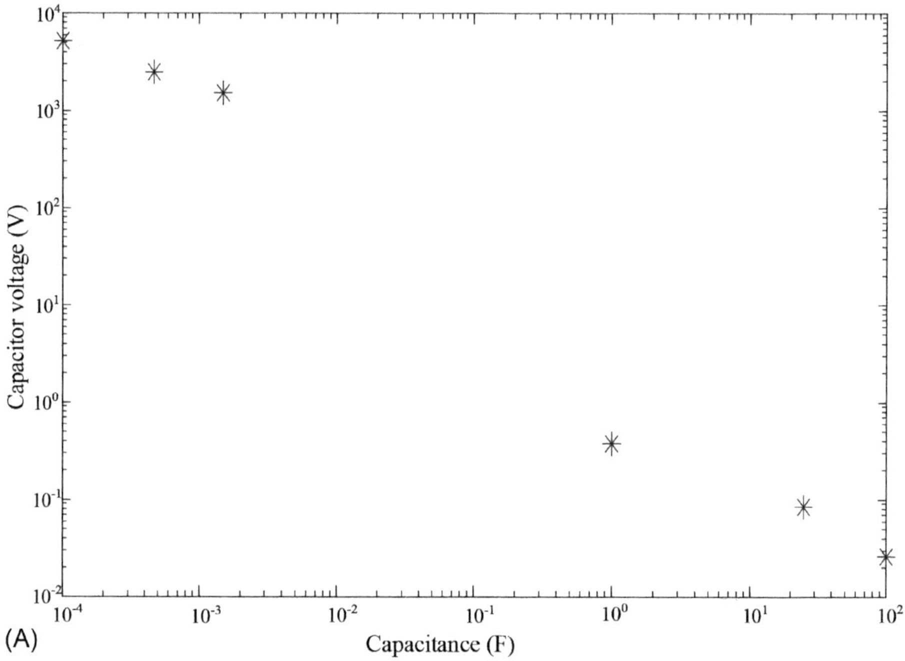

(A)

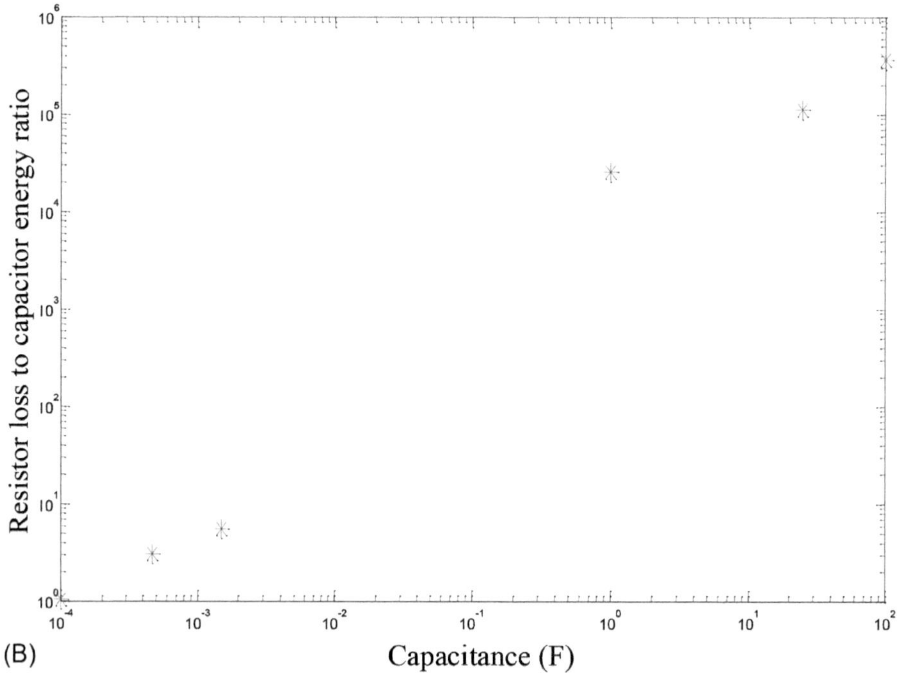

(B)

Fig. 9.14 (A) The voltage building up on different capacitors for a 6 kV transient surge and (B) comparison of energy lost and stored for electrolytic capacitors and SCs subjected to a 6 kV, 50 μs pulse.

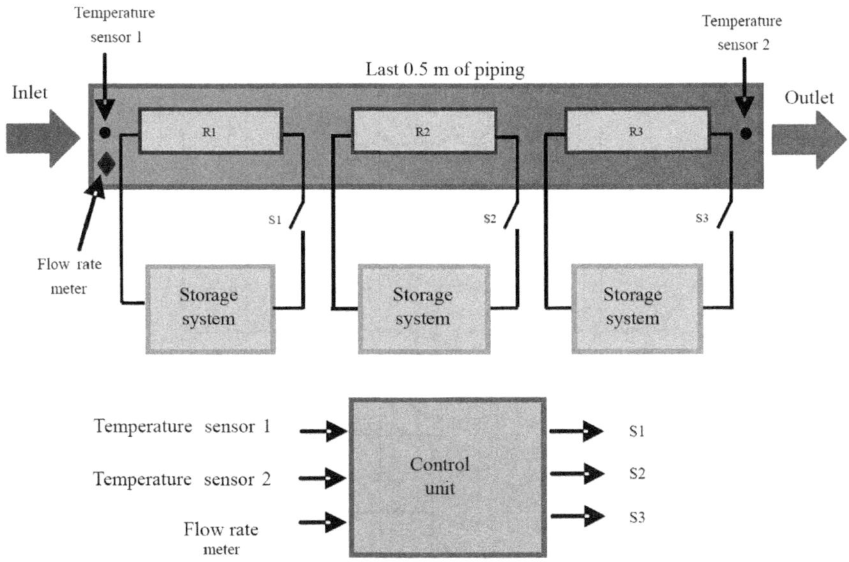

Fig. 9.15 SCATMA block diagram.

current delivery capability in the range of 600–2,400 A (Fig. 9.15). The system uses SC cells in the range of 1,200–5,000 F with storage capability of 0.6–3 Wh (Gurusinghe et al., 2017).

Energy-limited transient-mode fast supercapacitor charger topology (Gurusinghe et al., 2016a, 2017) is successfully used in the SCATMA technique for fast-charging supercapacitor bank. As shown in Fig. 9.16, in this topology, the capacitor bank is partitioned into two parts. The converter is directly fed with full-wave rectified AC line voltage without a smoothing capacitor. During the "ON" period as seen in Fig. 9.17, with the practical ability to absorb high charging currents, C_1 is directly fed by the current from the rectified AC source. The primary winding of the flyback transformer accumulates energy acting as an inductive load in the C_1 capacitor charging path. The terminal voltage of C_1 is monitored to avoid overcharging. When SW is turned "OFF" as shown in Fig. 9.18, diode D goes into conduction mode releasing the energy stored in the magnetic core into the second half of the SC bank C_2. Charging of C_2 is similar to the operation of a flyback converter where its resistive load at the output is replaced by C_2. This is also a useful practical instance of loss circumventing technique discussed in the chapter where transformer coupled with second SC bank is acting as the useful resistance.

The overall achievement of the topology is reducing the charging time of an SC bank by a significant factor. The converter does not require a bulky buffer capacitor at the unipolar input source derived from a rectified AC source (Gurusinghe et al., 2015, 2016a, b, 2017). A comprehensive discussion on this low-voltage rapid-heat delivery technique is included in Chapter 14.

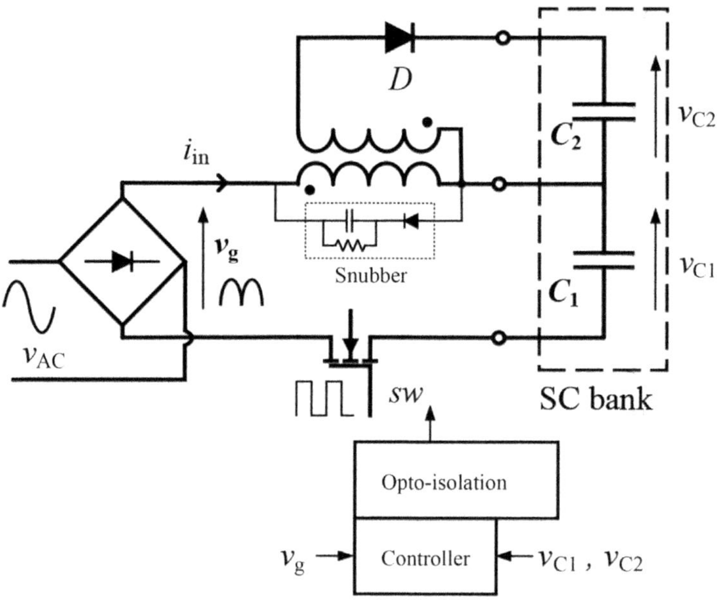

Fig. 9.16 Energy-limited transient-mode fast supercapacitor charger topology (Gurusinghe et al., 2017).

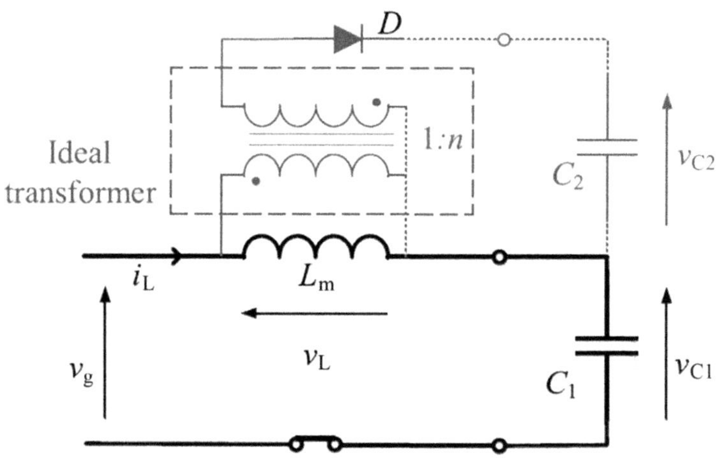

Fig. 9.17 On state of the MOSFET (Gurusinghe et al., 2017).

9.9 Renewable energy areas and SCALED basics

Loss circumventing principles in *RC* loop could be successfully applied in DC microgrids to light up 12 V LED lamps. In typical inverter-based renewable energy systems, ETEE efficiency is decided by the product of efficiency in individual

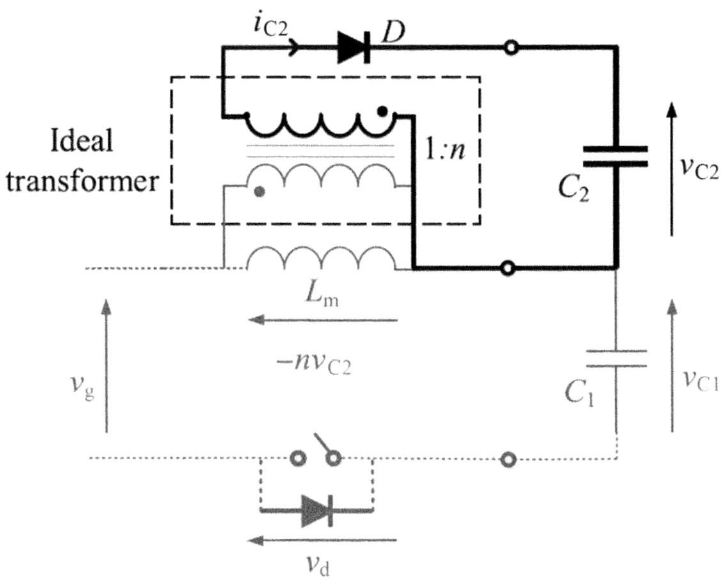

Fig. 9.18 Off state of the MOSFET (Gurusinghe et al., 2017).

converter stages such as DC-DC converter, inverter, and 230–12 V LED driver. However, commercially available 12 V LED lamps can operate over a wide range of positive or negative input voltage around their nominal voltage without affecting their brightness. Combining this useful property of 12 V LED lamps with a supercapacitor bank to act as an energy storage device for short-term renewable energy input fluctuations, supercapacitor-assisted LED (SCALED) was developed to light up LED lamps in DC microgrids efficiently (Fig. 9.19).

General benefits of SCALED technique (Kularatna and Jayanada, 2020) can be listed as follows:

- short-term energy storage providing DC-UPS capability;
- buffer unsafe voltages to power low-voltage LED lamps;
- higher ETEE;
- minimizing RFI/EMI issues; and
- tolerate wider DC-voltage fluctuations of the renewable source.

Detailed discussion of SCALED converter is available in Chapter 12.

9.10 Potential future applications

9.10.1 SC-assisted submodular inverter (SCASMI)

An inverter can be introduced to the capacitor charging loop as an useful resistance and SC-assisted submodular inverter (SCASMI) is developed based on this principle. Overall inverter is partitioned into smaller inverters. Half of the inverters are fed

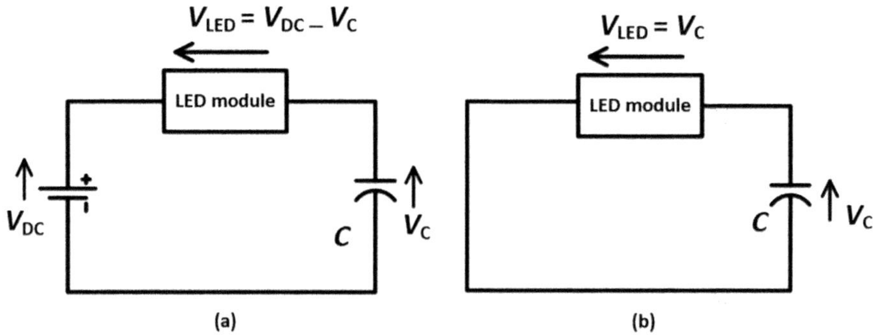

Fig. 9.19 Existing solar power-based LED lighting system.

by precharged SC modules, and the other other half is connected in series with SC modules (at the DC input side) and fed by the renewable source. Total output is determined by the sum of series connected small inverter voltages (Kularatna, 2014). Basic block diagram of this technique is shown in Fig. 9.20.

Fundamental concepts of this new approach can be summarized as

- dividing the overall inverter requirement into fractional inverters in terms of output power;
- feeding each fractional inverter by half of the nominal voltage of the DC source (such as the solar supply);
- using precharged SC banks to power half of the inverters;
- feeding the second half of the fractional inverters by an input DC source in series with the discharged SC banks; and
- ensuring the output voltage of the overall system is based on series-connected outputs of the fractional inverters.

At the time of completing this chapter, an on-going research project is active to develop an inverter system based on the SCASMI concept. One early byproduct of this research project, applying the SCA techniques and the overall SCALoM concept, is to show how a supercapacitor bank could be used with a commercially available inverter to extend its effective input voltage range (Gunawardane et al., 2021).

9.11 Conclusion

The discussion in this chapter on supercapacitor assisted (SCA) techniques is to demonstrate how a long-time constant formed by a supercapacitor and a resistive load opens up uniquely new avenues to develop new power conversion and protection techniques (discussed in Chapters 10–14). The overarching concept of supercapacitor-assisted loss management (SCALoM) can be considered as a unique new theoretical approach to developing long-time constant-based power converters and protection systems (including transient surge protectors and DC circuit breakers). This is a new multiuniversity research project on-going in New Zealand.

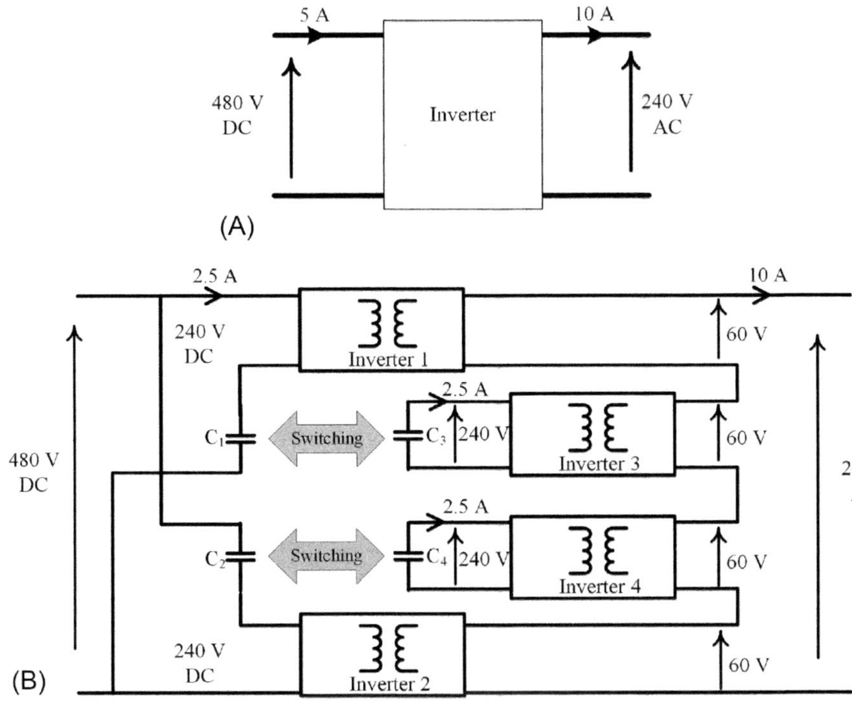

Fig. 9.20 The simplified concept of the SCASMI technique.

References

Gunawardane, K., Bandara, N., Subasinghage, K., Kularatna, N., 2021. Extending the input voltage range of solar PV Inverters with supercapacitor energy circulation. Electronics 10 (1), 88. https://doi.org/10.3390/electronics10010088.

Gurusinghe, N., Kularatna, N., Charleston, S.A., Fernando, J., Round, W.H., 2015. Hybridisation techniques for a supercapacitor-assisted temperature modification apparatus for inline water heating. In: IECON 2015—41st Annual Conference of the IEEE Industrial Electronics Society, November, pp. 002196–002200.

Gurusinghe, N., Kularatna, N., Round, W.H., Steyn-Ross, D.A., 2016a. Design approaches for fast supercapacitor chargers for applications like SCATMA, SRUPS. In: 2016 IEEE Applied Power Electronics Conference and Exposition (APEC), March, pp. 2479–2484.

Gurusinghe, N., Kularatna, N., Round, W.H., Steyn-Ross, D.A., 2016b. Single-stage fast supercapacitor charger with inherent power factor correction for energy-limited applications. In: IECON 2016—42nd Annual Conference of the IEEE Industrial Electronics Society, October, pp. 1412–1416.

Gurusinghe, N., Kularatna, N., Round, W.H., Steyn-Ross, D.A., 2017. Energy-limited transient-mode fast supercapacitor charger topology. IEEE Trans. Power Electron. 32 (2), 911–914.

Kankanamge, K., Kularatna, N., 2014. Improving the end-to-end efficiency of DC-DC converters based on a supercapacitor-assisted low-dropout regulator technique. IEEE Trans. Ind. Electron. 61 (1), 223–230. https://doi.org/10.1109/TIE.2013.2245613.

Kularatna, N., 2014. Dynamics and modeling of rechargeable batteries: what electrochemists? Work tells the electronic engineers. IEEE Power Electron. Mag. 1 (4), 23–33. https://doi.org/10.1109/MPEL.2014.2361264.

Kularatna, N., Jayananda, D., 2020. Supercapacitor-based long time-constant circuits: a unique design opportunity for new power electronic circuit topologies. IEEE Ind. Electron. Mag. 14 (2), 40–56. https://doi.org/10.1109/MIE.2019.2959199.

Supercapacitor as a lossless dropper in DC-DC converters—SCALDO technique

10

10.1 Introduction

Supercapacitors are typically used as energy storage devices (ESDs) and such applications are well documented. However, in state-of-the-art devices, with the capacitance values increased by a factor of nearly one million (compared to large electrolytic capacitors), a less significant fact is that the equivalent series resistance (ESR) values have dropped down to a range of 0.1 to about 100 mΩ. Given this important observation, when a supercapacitor of the value C is charged or discharged with a constant DC current of I_C we can expect an approximate voltage change across the terminals given by $\frac{I_C \Delta t}{C}$, where Δt is the time. When the capacitance value is very large, and the time duration is short we can expect only a very small voltage change across the terminals of the capacitor. Combining this effect with the milliohms order ESR, we can expect the device to have a near constant voltage drop across a terminal, together with a very low resistive drop. In other words, if we place a supercapacitor in series with an electronic circuit (supplied by a voltage source value of the maximum value of V_p), which typically consumes a finite current, the capacitor will take a very long time to block the path by developing a voltage difference across its terminals, which is equal to the difference between V_p and $V_{\text{in}}^{\text{min}}$ where $V_{\text{in}}^{\text{min}}$ is the minimum input voltage required to operate the circuit as expected. Fig. 10.1 depicts this scenario, where the supercapacitor acts as a lossless dropper in a practical circuit.

Next important situation here is to realize that the typical voltage drops across the ESR of the supercapacitor, could be much less than the typical DC voltage drop across a semiconductor power switch such as a metal oxide semiconductor field-effect transistor (MOSFET), insulated gate bipolar transistor (IGBT), or bipolar. State-of-the-art-power MOSFETs have typical DC on resistances ($R_{\text{DS(on)}}$) within the range of 1 to few 100 mΩs.

Combining the following two cases has given rise to a unique DC-DC converter topology, developed by the supercapacitor applications teams at the University of Waikato and Auckland University of Technology (i) voltage change across supercapacitor can be small due to charge/discharge currents over a short period and (ii) voltage drop across the ESR due to charge/discharge current could be much smaller than typical static DC losses across a power semiconductor switch (Kularatna and Fernando, 2009, 2011). This chapter provides a discussion on how a supercapacitor can be used as a lossless dropper in a linear DC-DC converter where end-to-end efficiency (ETEE) of a linear regulator based on low-dropout regulators (LDO) can be improved by a significant multiplication factor.

Energy Storage Devices for Renewable Energy-Based Systems. https://doi.org/10.1016/B978-0-12-820778-9.00007-3

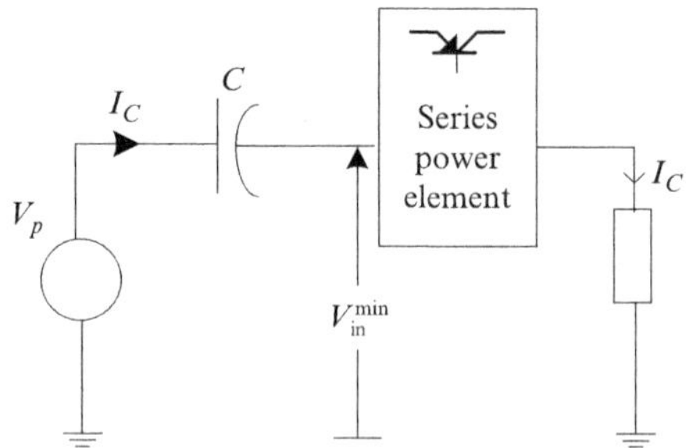

Fig. 10.1 The basic concept: supercapacitor as a lossless dropper in a practical circuit.

The chapter commences with a brief overview on DC power management and DC-DC converter topologies and directly proceeds to a designer viewpoint on the novel application of supercapacitor as a lossless dropper technique with some discussion on results achievable in typical DC-DC converters such as 12–5 V and 5–3.3 V cases.

10.2 DC-DC converters and DC power management

All electronic systems require a very stable DC source with constant DC value, minimum ripple and noise at the output, and fast response to changes of load current without a dip or an overshoot at the output rail. Another important requirement in modern electronic systems is energy efficiency in the converters, and longest possible battery run-time for battery-powered systems. Energy efficiency is a very hot topic among product designers where ETEE from AC power source to final DC rails is to be kept as high as possible, while the battery run time is optimized by many advanced techniques combining different types of DC-DC converters (Kularatna, 2011).

10.2.1 Commonly used DC-DC converters

In modern electronics, the following three basic approaches are available for DC-DC conversion:

1. linear regulators
2. switched-mode power supplies (SMPS)
3. switched capacitors (charge pumps)

In linear regulators, a power semiconductor device such as a power MOSFET, IGBT, or a bipolar power transistor is used as a voltage dropper in the series path between the unregulated DC input and the regulated output. A control loop compares a sample of regulated output voltage with a constant voltage DC reference source and feeds the control input of the power semiconductor switch to keep the output DC value at a constant value. Historically, the linear regulators were the first development wave for DC-DC converters in the 1960–70s, but these were very inefficient with typical efficiencies in the range of 30%–60%. However, they had very low-noise output with the fast transient response to fast changes of load current.

Switched-mode converters use inductors and capacitors as ESDs to convert one DC voltage to another with a theoretical efficiency of 100%. There are several commonly used converter topologies such as buck converter, boost converter, and buck-boost converter, and several related topologies. Kularatna (2011) provides a very good discussion on the subject and compares the techniques with linear regulators. All these topologies use power semiconductors as switches operating in the range of few tens of kHz to over 2 MHz.

Switched capacitor converters or charge pumps use capacitor-based switching techniques but they do not use bulky inductors or transformers which are typically used in switch-mode power supplies. These are low current capable converters, typically limited to few 100 mA only.

Table 10.1 compares the popular DC-DC converter techniques. More details have been provided by Walt Kester (http://www.analog.com/static/imported-files/tutorials/ptmsect3.pdf), Jovalusky (2005), Linear and Switching Voltage Regulator Handbook on Semiconductor (2002), Simpson (2012), Kester et al. (1998), and Palumbo and Pappalardo (2010).

10.2.2 LDOs as a special case of linear regulators

Since the subject of this chapter is based on the use of LDOs, which is a special case of linear regulators, let us start with a brief discussion on linear regulator designs. As shown in Fig. 10.2, the major components of a typical linear regulator include the pass element (series power semiconductor which acts as voltage dropper element in the load current path), the precision reference, the feedback network, and the error

Table 10.1 Comparison of popular DC-DC converter techniques.

Feature	Linear regulators	Charge pump converters	Switched-mode power supplies
Design complexity	Low	Moderate	Moderate to high
Cost	Low	Moderate	Moderate
Noise	Lowest	Low	Low to moderate
Efficiency	Low to moderate	Moderate to high	High
Thermal management	Poor to moderate	Good	Best
Output current capability	Moderate	Low	High
Requirement of magnetic parts	No	No	Yes
Limitations	Cannot step up	V_{in}/V_{out} ratio	Layout considerations

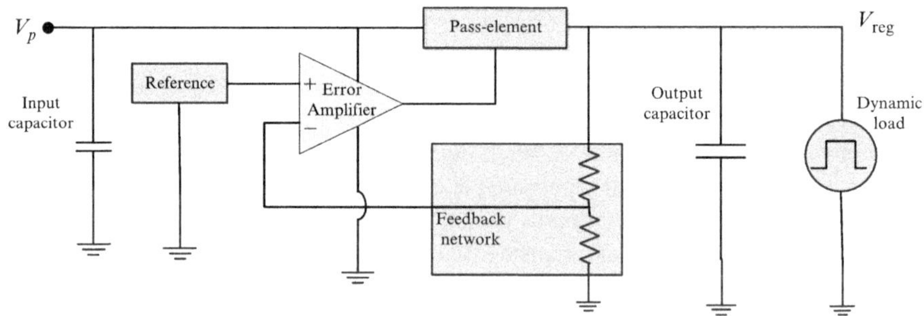

Fig. 10.2 Linear regulator block diagram.

amplifier. Chapters 1 and 2 of a book by Kularatna (2011) give a detailed account of the design aspects of DC power supplies based on linear regulators.

In general, when a linear regulator can be designed with a low DC voltage difference between the input and the output DC rails, it is called an LDO. Term low dropout comes from the fact that these single-chip silicon devices, very commonly used in portable products such as phones, cameras, and laptops, usually combined with switch-mode types, can operate down to a very low voltage difference between the input and output, which is called the dropout voltage at which it stops working as a useful DC-DC converter.

Fig. 10.3 illustrates the commercial utilization of LDO and switching regulator families with predictions for the future years as per market research reports from Databeans, indicating the total business potential in millions of dollars (lower trace-M), total unit sales in million units (middle trace-MU), and the average selling price (upper trace-ASP) of these components (Inouye et al., 2010).

Due to the difference in pass elements, in general, there are three basic types of linear regulator topologies:

1. standard (NPN Darlington) regulator,
2. quasi-LDO regulator, and
3. low-dropout or LDO regulator.

In a linear regulator, the minimum possible difference between the input and the output voltages while retaining the regulatory function is referred to as the dropout voltage. In the above topologies, the dropout voltage shows a remarkable difference across the three. A linear regulator that operates with the highest dropout voltage, dissipates the most internal power so has the lowest efficiency.

Usually in linear regulator configurations, most of the energy is dissipated across the pass element. Typical linear regulator construction does not have any energy storing mechanisms, so the power that is not delivered to the load is dissipated as heat within the regulator.

The power which is dissipated within the regulator (P_D) can be described as follows (Falin, 2006; Lee, 1999; Marasco, n.d.; Rincon-Mora, 2009; Texas Instruments, 2009; Zendzian, 1999):

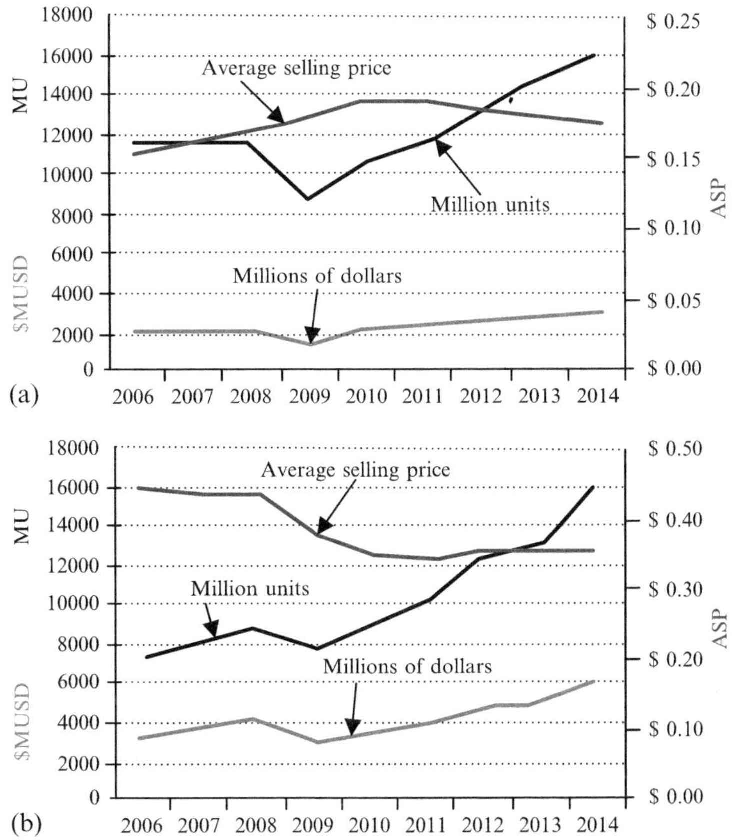

Fig. 10.3 Worldwide market forecast: (A) switching regulators; (B) LDO regulators (Inouye et al., 2010).

$$P_D = \left[V_p - V_{\text{reg}}\right] I_{\text{L}} + V_p I_q \tag{10.1}$$

where V_p, V_{reg}, I_{L}, and I_q are the input voltage, the regulated output voltage, the load current, and ground pin current (current drawn by the control circuits), respectively.

The first part of this relationship is the dissipation of the pass device and the second part is the power consumption of the controller portion of the circuit. Linear regulators generate excessive heat if the voltage difference between input and output voltage is high. But, LDOs have extremely low dropout voltages which enhance the efficiency.

The efficiency of a linear regulator can be approximated by

$$\eta = \frac{V_{\text{reg}} I_{\text{L}}}{V_p I_p} \approx \frac{V_{\text{reg}} I_{\text{L}}}{V_p \left(I_{\text{L}} + I_q\right)} \approx \frac{V_{\text{reg}}}{V_p \left[1 + \dfrac{I_q}{I_{\text{L}}}\right]} \tag{10.2}$$

As shown in Eq. (10.2), the efficiency of an LDO regulator is also limited by the ground-pin current of the circuit. To have a high-efficiency LDO, the ground-pin current should be minimized (http://www.analog.com/library/analogdialogue/archives/43-08/ldo.html; Texas Instruments, 2009). Modern LDOs have reasonably low I_q, and for simplicity, such a value can be neglected in efficiency calculations if I_q is very small compared to I_L (Man et al., 2007; Rincon-Mora, 1996).

Fig. 10.4 shows the regulator efficiency as a function of load current (I_L) with different I_q under a particular output voltage (V_{reg}). It shows that the use of very low I_L improves the regulator efficiency at small I_q, which often happens when systems are operating in standby mode to extend battery life.

10.2.3 Losses associated with switch-mode converter-based DC power supplies

An important advantage of switching regulators is their relatively high-power efficiency. Assuming the use of ideal energy storage elements such as capacitors and inductors with perfect lossless switches theoretically 100% efficiency can be achieved. But, in practical situations, inductors, capacitors, and transistor switches do not show their ideal characteristics and thus deviate much from the theoretical maximum efficiency. Therefore, in practice, the range of efficiencies could be in the range of 80%–90% for the DC-DC converter stage (Walt Kester, n.d.). If the AC-DC conversion stage is included as is required in a typical off-the-line power supply, ETEEs will be much lower. The following example illustrates this practical scenario.

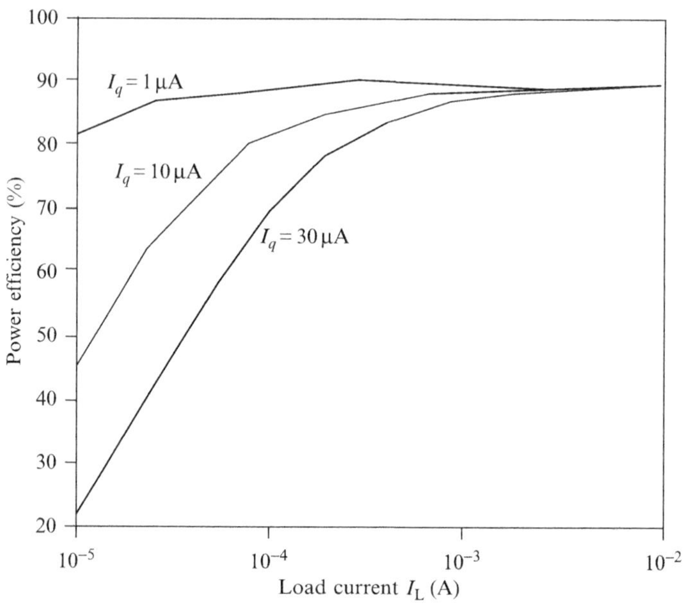

Fig. 10.4 Effect of ground-pin current on efficiency of an LDO (Man et al., 2007).

There can be various sources of loss within a typical SMPS, depending on its power level, topology, and circuit complexity. The approximate loss contribution range of each is shown in Fig. 10.5, as a percentage of the input power, for a typical supply that delivers 10 W of load power (Jovalusky, 2005; Kularatna, 2011).

In general, the major loss contributors in a switching power supply are

1. static losses in diodes;
2. transistor losses such as $I^2R_{DS(on)}$ in MOSFETs or losses due to saturation voltage of BJTs;
3. dynamic losses in switching transistors and diodes occurring at high frequency due to the charging and discharging of different parasitic capacitances;
4. losses due to charge/discharge currents through ESR (in most cases AC ESR comes into effect);
5. magnetic components losses; and
6. control circuits losses/PCB track losses.

Switched capacitor converters are capable of achieving efficiencies greater than 90%. However, these converters are not suitable to maintain high efficiency for a wide range of ratios of input to output voltages, unlike the case for switching and low-dropout regulators. The load current capability is limited by the size of the capacitors and the current-carrying capacity of the switches. In practice, typically switched-capacitor inverter chips are available with 10–500 mA load currents (Kester et al., 1998).

10.2.4 Power management approaches using combinations of switch modes and LDOs

Power supply designs for portable electronic products have become an exponentially growing segment due to the evolutional advancements in both switch-mode and linear voltage regulator designs. With sinking operating voltages and rising current demands, the designer must consider factors such as power wastage, long battery life,

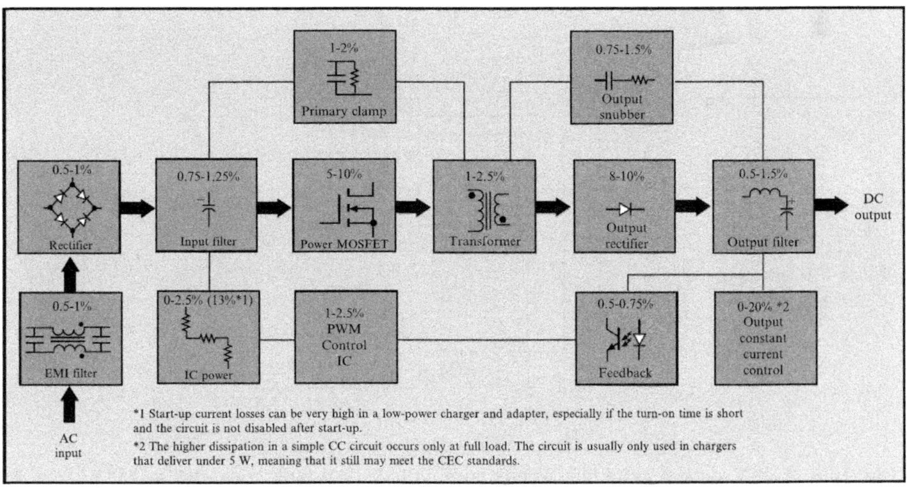

Fig. 10.5 Block diagram of the power-consuming circuits within a SMPS (Jovalusky, 2005).

small size, and low cost. These design parameters change and rearrange the design priorities for power supplies and make the choice between linear and switching mode power supplies ambiguous.

LDOs are frequently used in battery-powered portable products such as point of load (POL) power supplies where adaptation of switch-mode power supplies becomes difficult, primarily due to noise performance, slew rate requirements, and the complex nature of their output impedance (Linear regulators, 2001; Travis, 1998). An example of a POL approach in a portable product with a 48–8 V intermediate bus converter is given in Fig. 10.6 where all other voltages are derived from the 8 V bus. In this kind of situation, to minimize compound efficiency loss, which can be significant at times, the POL converters should offer efficiency close to or above 90% (Bull and Smith, 2003). The digital still camera (DSC) is shown in Fig. 10.7, indicating the combined use of switch modes and LDOs is another example. Fig. 10.8 depicts the part of power management solution offered for such a case from Texas Instruments TPS65010 IC. In TPS65010 IC, part of the circuitry allows configuring several switch-mode circuits, while the other part of the silicon is used to configure LDO circuits. More details have been provided by Day (2003) and (http://www.ti.com/product/tps65010). In some portable products where some parts of the load require very limited currents, in the orders of 10 to few 100 mA only, and with DC power rails higher than the battery voltage or the bulk switching power supply's output rail voltage, switched capacitor converters are also used (Palumbo and Pappalardo, 2010).

In practical products switchers SMPS and LDOs are combined for many practical reasons such as

1. noise level of SMPS and the need for RFI/EMI filters,
2. LDO output quality,
3. fast transient response of LDOs,
4. simplicity and small PCB footprint consumed by LDOs, and
5. inductors as a bulky component in switchers.

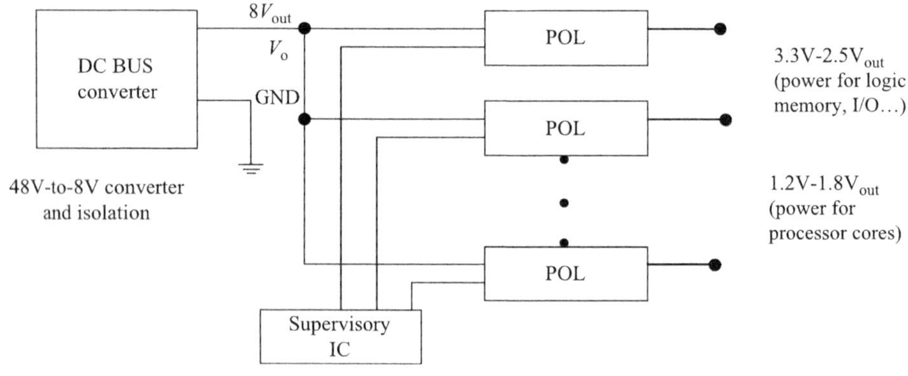

Fig. 10.6 An 8 V intermediate bus architecture with 48–8 V bulk converter and POL stages (Bull and Smith, 2003).

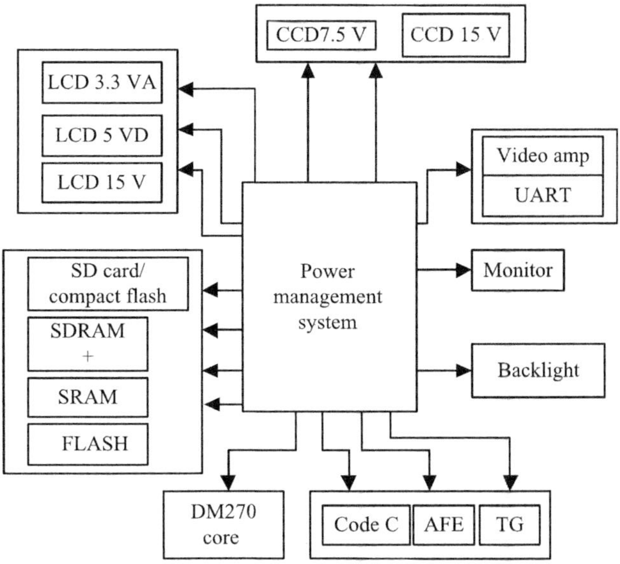

Fig. 10.7 An example of a digital still camera (DSC) architecture (Day, 2003).

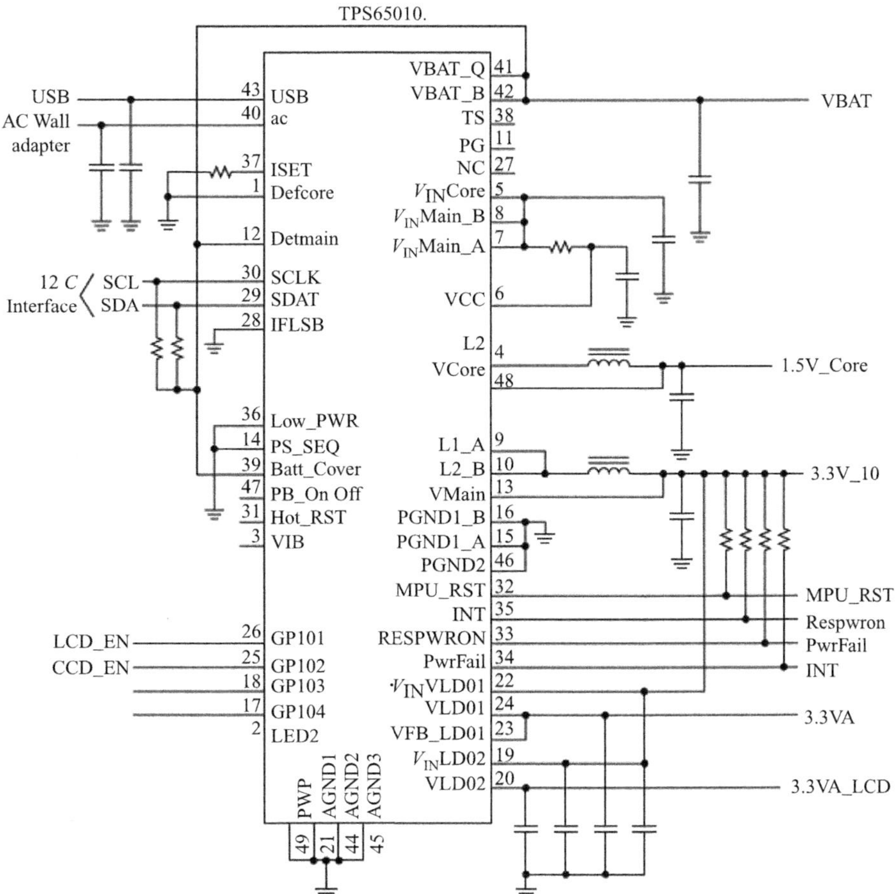

Fig. 10.8 Power management solution, Texas Instruments TPS65010 IC (Day, 2003).

10.3 Supercapacitor-assisted low-dropout regulator (SCALDO) technique

10.3.1 Essential fundamentals

So far our discussion summarizes the fact that in practical electronic systems, a combination of different DC-DC converter topologies is used to achieve the overall efficiency, noise minimization, transient response, and the packaging issues. If the control circuit consumes only a negligible amount of power, the best possible theoretical efficiency of a linear regulator is given by

$$\eta = \frac{V_{reg}}{V_p} \times 100\% \tag{10.3}$$

where V_{reg} and V_p are the regulated output and the input voltages, respectively.

Usually, in a linear regulator configuration, most of the energy is dissipated across the pass-element or the series power semiconductor. Introducing a voltage dropper in series with the pass element will make the dissipation across the pass element lower.

Placing a resistor in series with the pass element as shown in Fig. 10.9 would move some of the energy to the resistor, but this will not improve the overall ETEE.

For this purpose, the ideal voltage dropper element which can be used in the series path is an energy storing element, such as a capacitor or an inductor. If an inductor voltage is used as a voltage dropper, current through the inductor cannot be a constant. A capacitor is a matching candidate as a series dropper element, but in such a case if the capacitance is very small it will block the circuit quickly, stopping the current in a negligible period of time. As discussed in Section 10.1, an extremely large ideal capacitor can be logically used as a lossless dropper in a linear regulator in the series path. The fundamental concept of choosing a supercapacitor as the series dropper element is as follows.

The voltage change (dv) across a very large capacitance C such as a supercapacitor is very small when a finite charge (or discharge) current $i(t)$ flows during a finite time t

$$dv = \int_0^t \frac{1}{C} i(t)\, dt \tag{10.4}$$

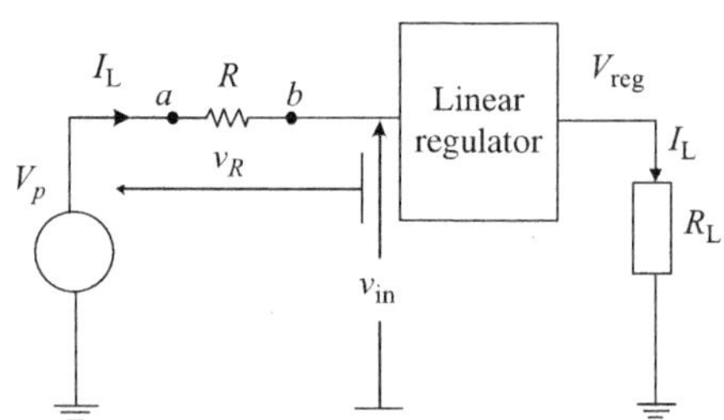

Fig. 10.9 An arrangement of placing a resistor in series with the pass element to reduce heat dissipation in the pass transistor.

A large capacitor, such as a supercapacitor (with negligible ESR value) with a DC voltage can be used as a voltage dropper in the series path, then the effect of the voltage change (dv) due to charging the capacitor for a finite amount of time, can be neglected in terms of the operational principles applicable to that particular circuit. For example, when a large capacitor is connected in series to a circuit for a finite time, the capacitor will not act as a blocking element to that circuit due to charging during that time. Based on this simple principle, using a single supercapacitor or an array of supercapacitors in the series path to store and release the energy, the supercapacitor-based efficiency improvement technique suitable for linear regulators was developed (Kularatna and Fernando, 2009, 2011).

10.3.2 Basic concept

Fig. 10.10 illustrates the concept: a supercapacitor C_{SC} is placed in series with the input of the LDO. Since the series capacitor is large (of the order of farads), it will pass current as it charges for a reasonable period of time. Based on this simple principle, a single supercapacitor can cyclically store and release energy. As per Fig. 10.10A in the first phase of the operation, the supercapacitor is charged from the unregulated power supply. When the value of LDO input voltage, V_{in} drops to the minimum allowable voltage V_{in}^{min} the operation is switched to the next phase. In this phase, disconnecting the unregulated power supply, the supercapacitor is connected in parallel with the input of the LDO as shown in Fig. 10.10B, to release the stored energy in the supercapacitor.

Assuming the initial voltage across the supercapacitor (C_{SC}) is $V_{SC}(0)$ after charging for a period t, the instantaneous voltage across the capacitor V_{SC} is given by

$$V_{SC} = V_{SC}(0) + \frac{I_L t}{C_{SC}} \tag{10.5}$$

where I_L is the load current entering the capacitor from the unregulated power supply. (We can assume that the load acts as a constant-current sink, and that the bypass current from the LDO to ground can be neglected.)

The unregulated source voltage V_p, equals the sum of the supercapacitor voltage and the LDO input voltage

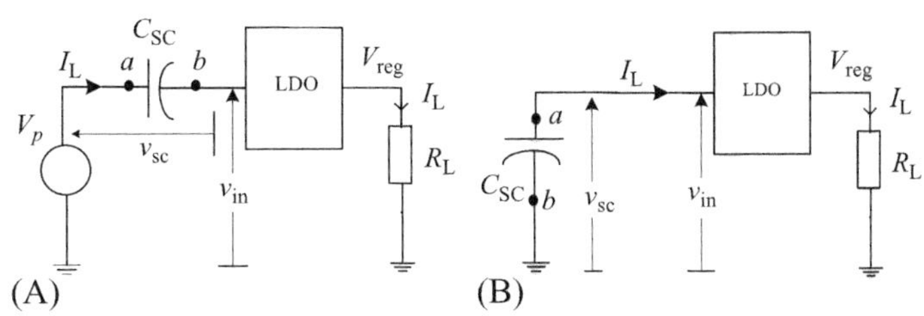

Fig. 10.10 The concept of supercapacitor energy storage and recovery: (A) minimizing the series element dissipation and (B) releasing stored energy.

$$V_p = V_{SC}(t) + V_{in}(t) \tag{10.6}$$

The supercapacitor charges until V_{in} reaches V_{in}^{min} while the voltage across the capacitor reaches $V_p - V_{in}^{min}$ at the end of the charging time. In order to discharge this supercapacitor at the next stage down to V_{in}^{min}, the criterion $V_p - V_{in}^{min} > V_{in}^{min}$ must be satisfied. This creates the condition $V_p > 2V_{in}^{min}$.

At the end of the charging cycle, the stored energy can be released as shown in Fig. 10.10B. Discharging continues until the LDO input voltage drops back to V_{in}^{min} satisfying the following equation:

$$V_p - 2V_{in}^{min} = \frac{I_L \Delta t}{C_{SC}} \tag{10.7}$$

The circuit draws power from the unregulated input only during half the time of its operating period: during the supercapacitor charging phase, the supercapacitor draws current from the unregulated supply, whereas during the discharging, the supercapacitor delivers power to the LDO keeping the unregulated power supply disconnected from the system.

Considering the charge balance of the supercapacitor during the charge-discharge cycle

$$I_L t_{ch} = I_L t_{dch} \tag{10.8}$$

where t_{ch} and t_{dch} are charge time and discharge time, respectively.

This gives

$$t_{ch} = t_{dch} \tag{10.9}$$

So that the average input current drawn from the unregulated power supply is

$$I_{avg} = \frac{I_L t_{ch}}{t_{ch} + t_{dch}} = \frac{I_L}{2} \tag{10.10}$$

Assuming that the capacitors and switching elements are ideal and neglecting the power consumed by the control circuits, the approximate end-to-end efficiency is given by

$$\eta = \frac{P_{out}}{P_{in}} = \frac{V_{reg} I_L}{V_p \frac{I_L}{2}} = \frac{2V_{reg}}{V_p} \tag{10.11}$$

The discussion above applies to a practical DC-DC converter such as 12–5 V converter, where the maximum ETEE possible is $\frac{5}{12} \times 100\%$ which is approximately 42%. In a practical implementation, a supercapacitor can be used with a DC voltage between 6.7 (12–5.3 V) and 5.3 V for the case with a commercial LDO chip with a minimum input voltage such as 5.3 V at the regulation. As we see from Eq. (10.11), maximum efficiency achievable is 84%. Practical values of 79%–81% were achieved in preliminary proof of concept projects (Kularatna and Fernando, 2009; Kularatna et al., 2010). More discussion on this subject will be found in Section 10.6. In general, the practical scenario is, the supercapacitor keeps charging above its average DC voltage within the cycle, in the series mode, and later the supercapacitor releases its charge until it reaches a minimum voltage required for the operation of the LDO, keeping the DC rail disconnected.

10.4 Generalized SCALDO concept

The simple case in the previous discussion requires only one SC and four low-speed switches. However, for general cases of step-down converters, we can extend the basic idea into a general case applicable to almost all cases of step-down converters.

10.4.1 Series to parallel SC array

As shown in Fig. 10.11 an array of n ideal series supercapacitors, C_{SC} each, is placed in series with the input of the LDO. This configuration is applicable to a SCALDO regulator when it satisfies the criterion $V_p > 2V_{in}^{min}$. Basically, this is the extended version of the single supercapacitor, $n = 1$ case. The number of supercapacitors, n, can be determined using the following formula:

$$n < \frac{V_p - V_{in}^{min}}{V_{in}^{min}} \tag{10.12}$$

As the equivalent capacitance of the series capacitor array is large (in the order of farads), it will pass current as it charges for a reasonable period of time. As per Fig. 10.11A in the first phase of the operation, the series capacitor array is charged until the value of the LDO input voltage, V_{in} is dropped down to its minimum possible input voltage, V_{in}^{min}.

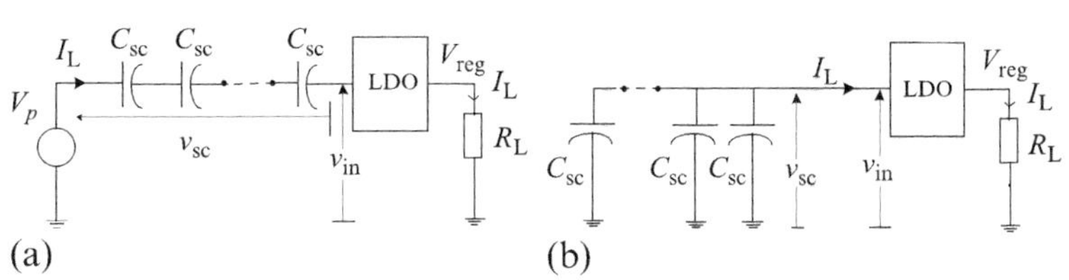

Fig. 10.11 The concept of SCALDO regulator for $V_p > 2V_{in}^{min}$: (A) charging n capacitors series minimizing the series element dissipation and (B) discharging n capacitors parallel releasing stored energy.

Assuming the initial voltage across a supercapacitor (C_{SC}) is $V_{SC}(0)$, after charging for a period t, the instantaneous voltage across the capacitor V_{SC} is given by

$$V_{SC} = V_{SC}(0) + \frac{I_L t}{C_{SC}} \tag{10.13}$$

where I_L is the load current entering each capacitor from the power supply during the charging process assuming the ground pin current is negligible.

The voltage across a single capacitor at the end of the charging is

$$V_{sc}^{max} = \frac{V_p - V_{in}^{min}}{n} \tag{10.14}$$

At the end of this phase, the array of supercapacitors is connected to the input of the LDO as an array of parallel capacitors as shown in Fig. 10.11B to release the stored energy. This discharging process continues until the LDO input voltage drops back to V_{in}^{min}.

The circuit draws power from the unregulated supply during the supercapacitor charging phase (t_{ch}), whereas during the discharge phase (t_{dch}) the supercapacitor delivers power to the LDO keeping the unregulated power supply disconnected from the system.

Considering the charge balance of a single capacitor during charge-discharge cycles

$$I_L t_{ch} = \frac{I_L}{n} t_{dch} \tag{10.15}$$

which gives

$$t_{dch} = n t_{ch} \tag{10.16}$$

The average input current taken from the unregulated power supply is

$$I_{avg} = \frac{I_L t_{ch} + 0 \times t_{dch}}{t_{ch} + t_{dch}} = \frac{I_L}{n+1} \tag{10.17}$$

Assuming that the capacitors and switching elements are ideal and neglecting the power consumed by control circuits, approximate ETEE for this configuration is given by

$$\eta = \frac{P_{out}}{P_{in}} = \frac{I_L V_{reg}}{V_p I_{avg}} = \frac{I_L V_{reg}}{V_p \frac{I_L}{n+1}} = (n+1)\frac{V_{reg}}{V_p} \tag{10.18}$$

10.4.2 Parallel to series SC array

As shown in Fig. 10.12A an array of n ideal parallel supercapacitors C_{SC} each, is placed in the input of the LDO. This is applicable to a SCALDO regulator when $V_p < 2V_{in}^{min}$. The number of supercapacitors, n, can be determined using the following formula:

$$n > \frac{V_{in}^{min}}{V_p - V_{in}^{min}} \tag{10.19}$$

As the result of the parallel capacitor array is large (in the order of farads), it will pass current as it charges for a reasonable period of time. As per Fig. 10.12A in the first phase of the operation, the parallel capacitor array is charged until the value of the LDO input voltage, V_{in} drops to V_{in}^{min}.

Assuming the initial voltage across a supercapacitor (C_{SC}) is $V_{SC}(0)$, after charging for a period t, the instantaneous voltage across each capacitor V_{SC} is given by

$$V_{SC} = V_{SC}(0) + \frac{I_L t}{nC_{SC}} \tag{10.20}$$

where I_L/n is the portion of load current entering each capacitor from the power supply during the charging process assuming the ground pin current is negligible.

The voltage across a single capacitor at the end of charging is

$$V_{sc}^{max} = V_p - V_{in}^{min} \tag{10.21}$$

As shown in Fig. 10.12B, in the next phase of operation, using an array of series capacitors, the stored energy is released until the LDO input voltage drops back to V_{in}^{min}.

Considering the charge balance of a single capacitor during charge discharge cycles

$$\frac{I_L}{n} t_{ch} = I_L t_{dch} \tag{10.22}$$

which gives

$$t_{dch} = \frac{t_{ch}}{n} \tag{10.23}$$

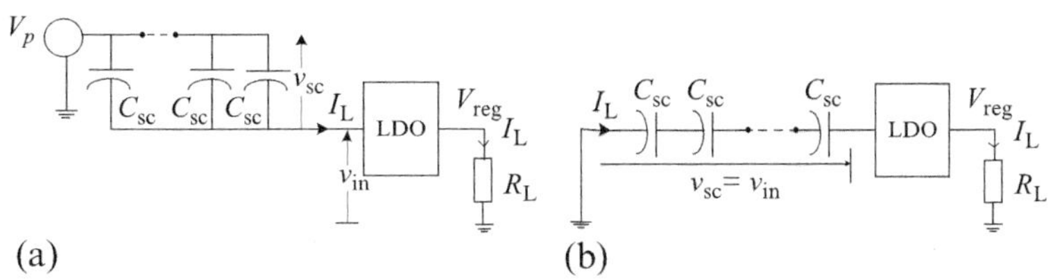

Fig. 10.12 The concept of SCALDO regulator for $V_p < 2V_{in}^{min}$: (A) charging n parallel capacitor array minimizing the series element dissipation and (B) discharging n capacitors series releasing stored energy.

Table 10.2 The parameters values for the three different cases of SCALDO topologies.

Parameter	$V_p > 2V_{in}^{min}$ **basic config (with single SC)**	$V_p > 2V_{in}^{min}$ **gen. config**	$V_p < 2V_{in}^{min}$ **gen. config**
n	1	$n < \frac{V_p - V_{in}^{min}}{V_{in}^{min}}$	$n > \frac{V_{in}^{min}}{V_p - V_{in}^{min}}$
k	4	$3n+1$	$3n+1$
V_{sc}^{max}	$V_p - V_{in}^{min} - I_L (2r + r_{sc})$	$\frac{1}{n}\left[V_p - V_{in}^{min} - I_L[(n+1)r + nr_{sc}]\right]$	$V_p - V_{in}^{min} - \frac{I_L}{n}(2r + r_{sc})$
t_{ch}	T_1	$\frac{T_n}{n}$	T_n
t_{dch}	T_1	T_n	$\frac{T_n}{n}$
η_r	2	$1+n$	$1+\frac{1}{n}$

n, number of supercapacitors; k, number of switches; V_{sc}^{max}, max voltage across the supercapacitor; t_{ch}, supercapacitor charge time, t_{dch}, supercapacitor discharge time; η_r, efficiency improvement factor.

Therefore, the average input current drawn from the unregulated power supply is

$$I_{avg} = \frac{I_L t_{ch} + 0 \times t_{dch}}{t_{ch} + t_{dch}} = \frac{I_L}{1 + \frac{1}{n}} \tag{10.24}$$

Similar to $V_p - 2V_{in}^{min}$ case, under ideal conditions, the approximate ETEE is given by

$$\eta = \frac{P_{out}}{P_{in}} = \frac{I_L V_{reg}}{V_p I_{avg}} = \frac{I_L V_{reg}}{V_p \frac{I_L}{1 + \frac{1}{n}}} = \left(1 + \frac{1}{n}\right) \frac{V_{reg}}{V_p} \tag{10.25}$$

Table 10.2 summarizes the characteristics of the basic configuration which satisfies $V_p > 2V_{in}^{min}$ criterion using a single supercapacitor and the two general configurations which satisfy $V_p > 2V_{in}^{min}$ and $V_p < 2V_{in}^{min}$ criteria using an array of supercapacitors. η_r the efficiency improvement factor is a benchmark indication of this technique, and for the two different general cases discussed in Section 10.4, these are $(1+n)$ and $(1+\frac{1}{n})$ for the two cases of $V_p > 2V_{in}^{min}$ and $V_p < 2V_{in}^{min}$, respectively. It is important to highlight here that the SCALDO technique is always a step-down converter, and it is not a variation of switched capacitor converters (Kankanamge and Kulatana, 2013)

10.5 Practical examples

The SCALDO technique can be easily configured for common DC-DC converters such as 12–5, 5–3.3, and 5–1.5 V, and the technique provides efficiency improvement factors of 2, 1.33, and 3, respectively, compared with linear converters with the same input-output combinations (Kularatna and Fernando, 2009; Kularatna et al., 2010, 2011a,b,c; Kankanamge and Kularatna, 2012). For example, in a 5–1.5 V SCALDO regulator, using thin profile supercapacitors in the range of fractional farads to a few farads, this translates to an approximate ETEE of nearly 90%. Table 10.3 indicates the efficiencies at each transfer stage, assuming that LDO is not abruptly disconnected. This table gives a clear overview of how this novel topology can assist with increasing the ETEE of a linear regulator, with a view to getting the best out of linear-regulator-based DC-DC converters. Practical implementations of these topologies will be discussed in the next chapter.

10.6 SCALDO implementation examples

10.6.1 12–5 V SCALDO regulator

Fig. 10.13 shows the schematic implementation of the 12–5 V SCALDO regulator which satisfies the criteria, $V_p > 2V_{in}^{min}$, in the form of its basic configuration, $n = 1$. In designing this prototype, an LDO of the type MCP1827 from Microchip Technology Inc., which has the specifications provided in Table 10.4, was used as the main low-dropout regulator. For high-efficiency LDO operation, an approximate LDO input voltage of 0.3 V above the regulated output voltage (5 V) was chosen, so that V_{in}^{min} was maintained at 5.3 V. As the power supply input voltage was around 12 V and the regulated output is 5 V, three supercapacitors of 4 F/2.5 V of the type PC series from Maxwell Technologies, were connected in series to maintain the required voltage level of the series dropper element. This resulted in 1.33 F/7.5 V resultant capacitor as the series capacitor.

Solid-state relays of the type PVN012 from International Rectifier were used as S_1, S_2, S_3, and S_4 switches. Switches S_1 and S_3 control supercapacitor charging, while S_2 and S_4 control the discharge phase. Solid-state relays were used in this proof-of-concept prototype to avoid complications arising from the body diode of common power MOSFETS. A PIC microcontroller of the type PIC16F684 was used as the controller to drive the photovoltaic switches. An independent 5 V regulator (LM7805) was used to power the microcontroller. The S_1, S_2, S_3, and S_4 switches were controlled through the RA0, RA5, RA1, and RA2 port pins of the microcontroller, respectively. The LDO input voltage was monitored through the RA4 port pin. The voltage level at this pin is converted to digital format using 10-bit analog to digital converter of the PIC which operates at 4 MHz oscillation frequency. The result is then compared with reference point (digital value of the minimum LDO working voltage, V_{in}^{min}) and the switching between the supercapacitor charge and discharge phases was made accordingly. The microcontroller firmware was developed to drive the switches based on algorithm shown in the flow chart in Fig. 10.14.

A switch-over period, t_{sw} was designed to accommodate any transition delays of the PVN012 switches. This creates a short time duration just before the circuit is transferred from the supercapacitor charge mode to the discharge mode. Time duration is set by the PIC microcontroller according to the switch delay. The time duration is

Table 10.3 Some useful parameters and their values for a selected converter requirements.

Configuration	End-to-end regulation requirement	LDO min. input voltage	Capacitor requirements				Approximate efficiency at different stages of operation in discharging			Efficiency improvement factor		Max theoretical efficiency of a standard regulator (%)
			Number of caps (n)	Voltage across a single cap in discharging (V)			@Max voltage (%)	@ Mid voltage (%)	@ Min voltage (%)			
				Max.	Mid	Min						
$V_p < 2V_{in(min)}$	5–3.3 V	3.5	3	1.50	1.40	1.20	73	78	92	$1+1/n$	1.33	66
$V_p > 2V_{in(min)}$	12–5 V	5.3	1	6.50	6.00	5.30	77	83	93	2	2	42
$V_p > 2V_{in(min)}$	5–1.5 V	1.6	2	1.70	1.65	1.60	88	91	94	$1+n$	3	30

Parameters: number of capacitors; capacitor voltages and approximate efficiencies of SCALDO regulators at transfer times.

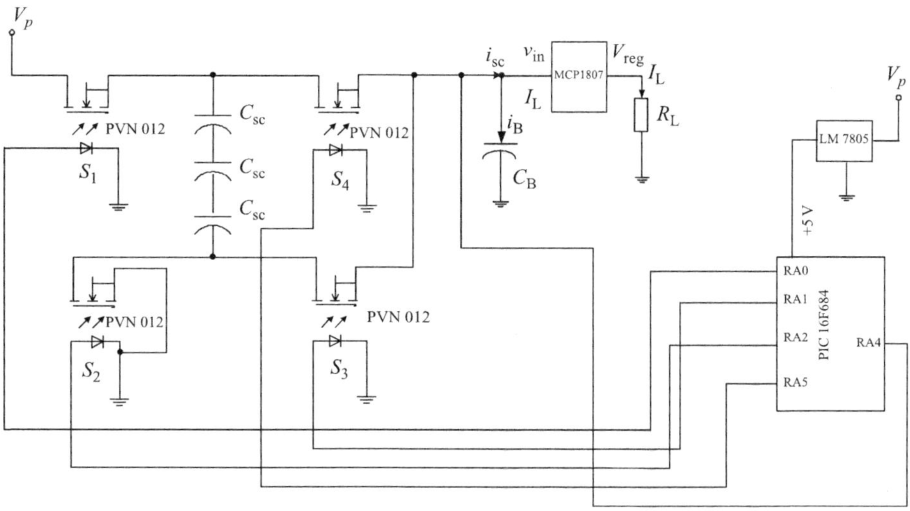

Fig. 10.13 The schematic representation of the 12–5 V SCALDO regulator designed using MCP1827 LDO, 4 F/2.5 V Maxwell supercapacitors, PVN012 Photovoltaic switches, PIC16F684 microcontroller, and LM7805 regulator.

Table 10.4 Important specifications of the LDOs used for the SCALDO prototypes.

Specification	MCP1827
Maximum output current	1.5 A
Input voltage range	2.3–6.0 V
Low-dropout voltage	330 mV at 1.5 A load
Adjustable output voltage range	0.8–5.0 V
Transient response	Fast response to load transients
Stability	Stable with 1.0-μF ceramic output capacitor

determined according to the datasheet of the switch and typically is about 300 ms. Fig. 10.15 shows the delay characteristics of the PVN012 switch.

To keep the linear regulator powered during the switch transitions, a buffer capacitor (C_B) with sufficient capacitance was connected in between the linear regulator input and the ground terminal. In t_{sw} time duration, the buffer capacitor is expected to discharge into the LDO input. The value of the buffer capacitor (C_B) can be determined as shown in Eq. (10.26) using the maximum load current capability of the LDO (I_L^{max}), switching time delays of a PVN012 switch (t_{sw}) and the minimum possible input voltage at the LDO input (V_{in}^{min}). A 1000-μF electrolytic capacitor was chosen as C_B for the 12–5 V SCALDO design.

$$C_B = \frac{I_L^{max} t_{sw}}{V_{in}^{min}} \tag{10.26}$$

Fig. 10.16 shows a prototype of 12–5 V SCALDO regulator.

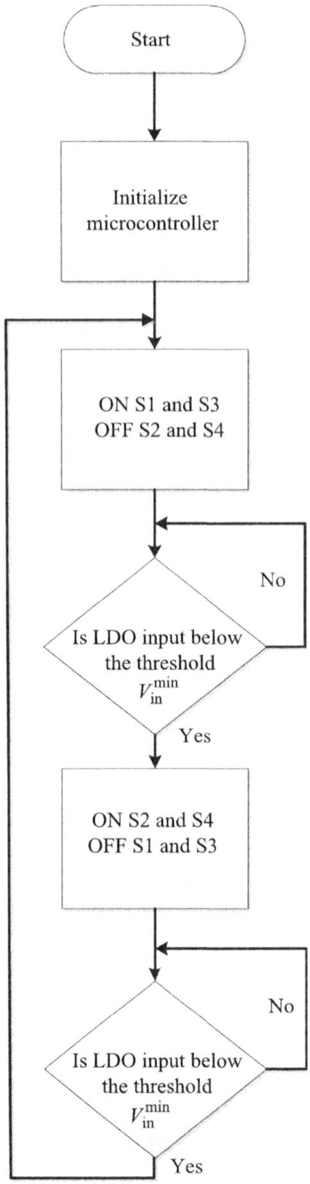

Fig. 10.14 The flowchart of the PIC algorithm.

10.6.1.1 Load regulation

Fig. 10.17A–D indicates the oscilloscope waveforms during the circuit operation for a load current of 200, 400, 600, and 800 mA, respectively. At these stages, the input power supply voltage was maintained at 12 V. As the load current increases, the switch operating frequency also increases. When the current flowing through the

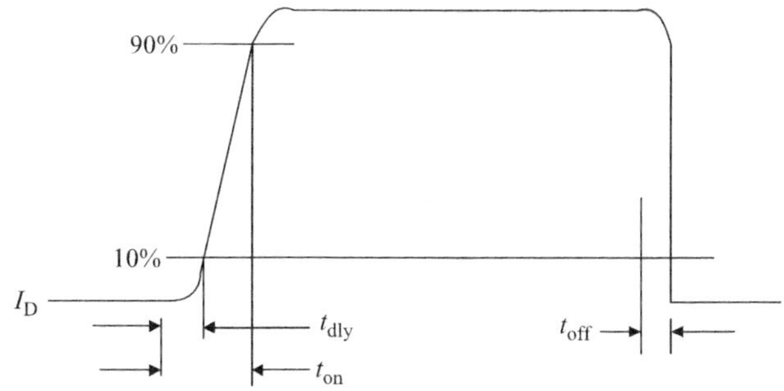

Fig. 10.15 Delay characteristics of PVN012 switches.

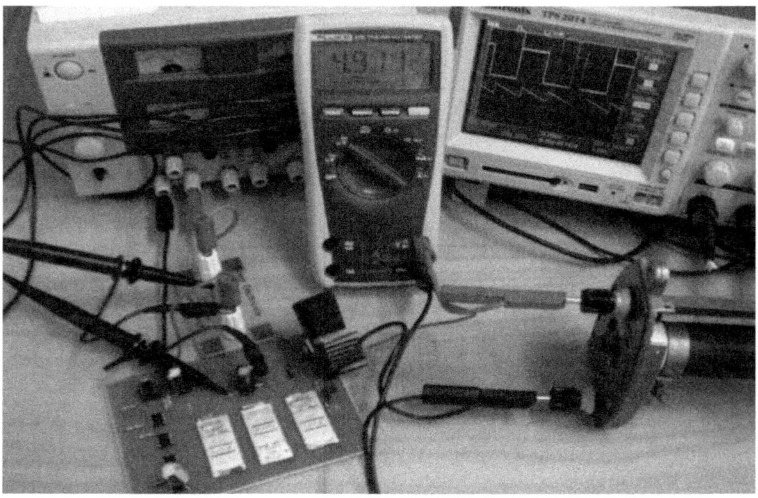

Fig. 10.16 The prototype design of the 12–5 V SCALDO regulator.

supercapacitor increases, the time taken to reach the threshold voltage level decreases. Therefore, both the charging and discharging cycles of the supercapacitor take a shorter time period. In Fig. 10.17 traces (1)–(4) represent the input voltage of the LDO (V_{in}), the voltage across one of the three supercapacitors (V_{SC}), the LDO output voltage (V_{reg}), and the input current drawn from the unregulated power supply (I_{in}), respectively. The trace (3), V_{SC}, clearly shows the supercapacitor charge-discharge cycles during the circuit operation. The trace (1) shows how the LDO input voltage changes during the charge-discharge cycles. The waveform of trace (4), I_{in} shows the current drawn from the unregulated power supply during the charge-discharge cycles. This clearly shows that the circuit draws a considerable current only in the charging phase of the supercapacitor. In the discharging phase, only a little current is drawn for the control circuits to function. This I_{in} was measured by connecting a series resistor of $0.1\,\Omega$ in series with the unregulated power supply and the input terminal of the circuit.

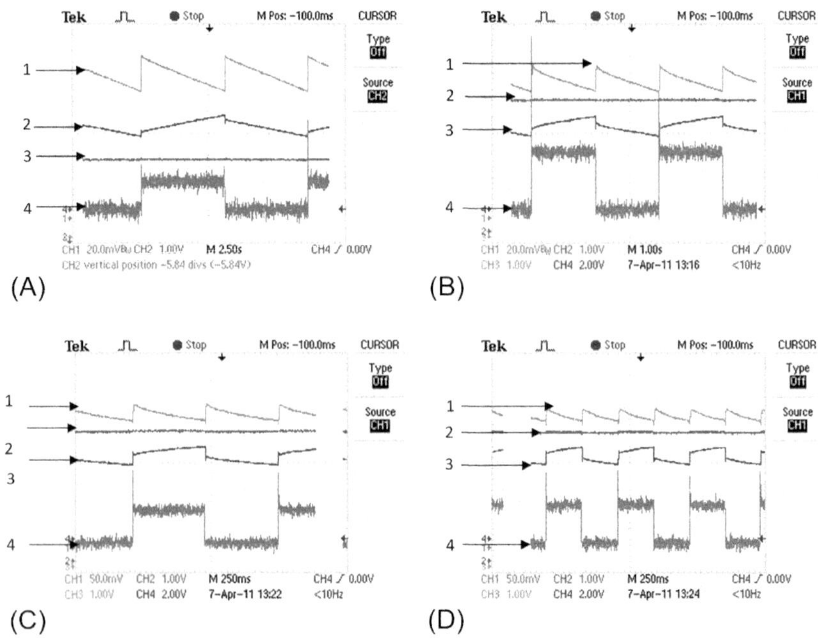

1 Input voltage of the LDO

2 Voltage across one of the three supercapacitors

3 LDO output voltage

4 Input current drawn from the unregulated power supply

Fig. 10.17 Oscilloscope waveforms of 12–5 V SCALDO regulator, for a load current of (A) 200 mA, (B) 500 mA, (C) 800 mA, and (D) 600 mA at an input voltage of 12 V.

In the SCALDO prototypes, the control circuit current was not set at a minimum value due to the time constraints and the limited budget. Therefore, in the calculations, the control circuit current is not considered.

In the following calculations, assume that V_p, I_{in}, V_{reg}, I_{L}, and I_c are the input voltage of the unregulated power supply, the current drawn from the unregulated power supply, the output voltage, the output current and the control circuit current of the SCALDO regulator, respectively. Neglecting the control circuit current, the average current drawn from the unregulated power supply during a supercapacitor charge-discharge cycle is

$$I_{\text{avg}} = \frac{[I_{\text{in}} - I_c]}{2} \tag{10.27}$$

The overall ETEE of SCALDO cycle can be estimated as

$$\eta = \frac{V_{\text{reg}} I_{\text{L}}}{V_p I_{\text{avg}}} 100\% \tag{10.28}$$

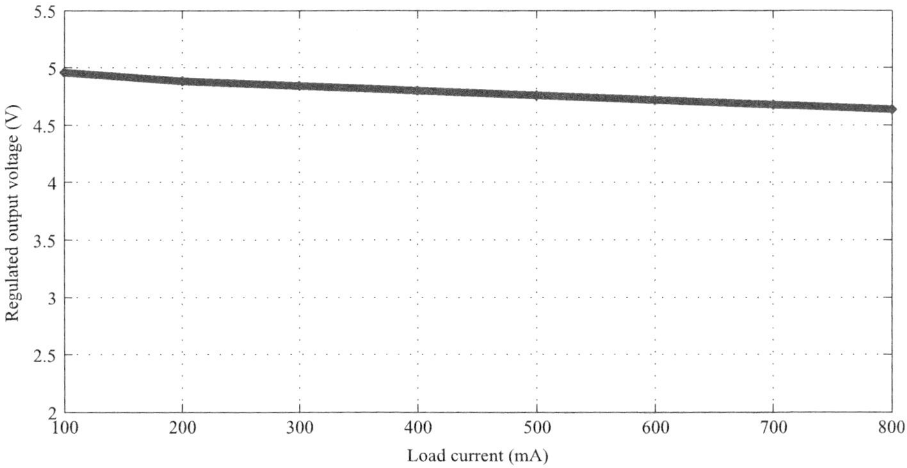

Fig. 10.18 The load regulation plot of the 12–5 V SCALDO regulator at an input voltage of 12 V.

The ETEE results of the 12–5 V SCALDO regulator which were estimated using Eqs. (10.27), (10.28) based on the experimental results. The circulation frequency increases from 0.05 to 4 Hz with the increase of the load current to 100–800 mA, respectively. Fig. 10.18 shows the load regulation plot based on the experimental results (Gunawardane, 2014) and load regulation is about 0.04%/mA.

10.6.1.2 Line regulation

When the input voltage increases, the operating frequency of the switches decreases. The reason is, when the difference between the unregulated input voltage and the threshold voltage level at the input of the LDO increases, the amount of energy is stored in the supercapacitor also increases. Therefore, both the charge and the discharge cycles of the supercapacitor take longer time periods, and this results in a lower operating frequency of the switches. Fig. 10.19A–D indicate the oscilloscope waveforms during the circuit operation, for an input voltage of 13.5, 13, 12, and 11.5 V, respectively. The load current is maintained at 250 mA at these stages. In Fig. 10.19 traces (1)–(4) represent, the input voltage of the LDO (V_{in}), the voltage across one of the three supercapacitors (V_{SC}), the LDO output voltage (V_{reg}), and the input current of the system (I_{in}), respectively. The trace (3), V_{SC}, clearly shows the circuit operation during the supercapacitor charge and discharge cycles. Trace (1) shows how the LDO input voltage changes during the charge and discharge cycles. The I_{in} trace clearly shows that the circuit draws a considerable current only in the charge phase of the supercapacitor and in the discharge phase only a little current is drawn for the control circuits to operate. Fig. 10.20 depicts the corresponding line regulation plot. As with load regulation calculations, the control circuit current was neglected in the calculations.

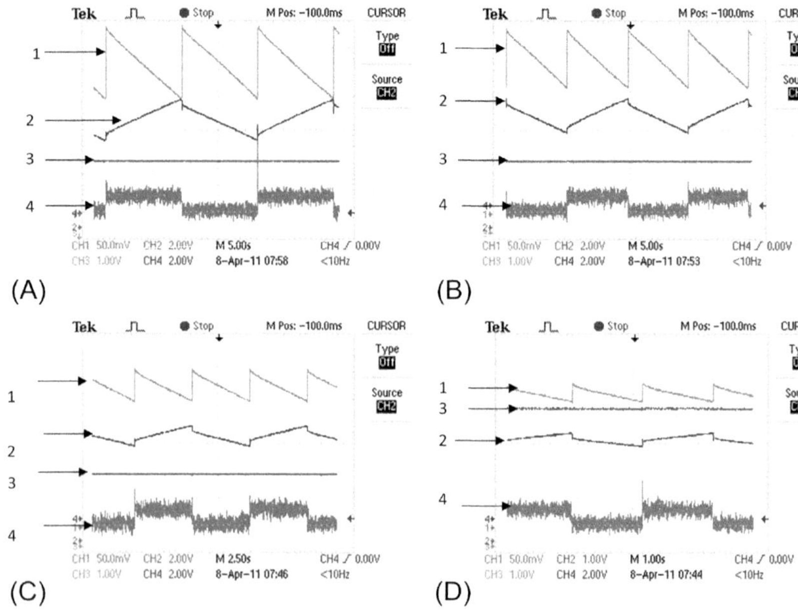

1	Input voltage of the LDO
2	Voltage across one of the three supercapacitors
3	LDO output voltage
4	Input current drawn from the unregulated power supply

Fig. 10.19 Oscilloscope waveforms of the 12–5 V SCALDO regulator for an input voltage of (A) 13.5 V, (B) 13 V, (C) 12 V, and (D) 11.5 V at a load current of 250 mA.

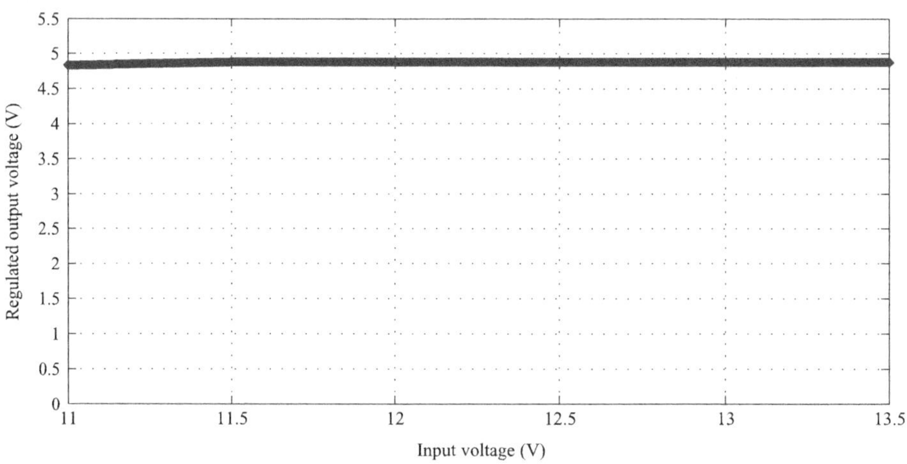

Fig. 10.20 The line regulation experimental results of 12–5 V SCALDO regulator at a load current of 250 mA.

10.6.1.3 End-to-end efficiency

According to the load regulation experimental results (Gunawardane, 2014), the ETEEs were calculated. The corresponding efficiency plotted against load current is shown in Fig. 10.21. Using the line regulation experimental data (Gunawardane, 2014), efficiency plotted against the input voltage was drawn as in Fig. 10.22. This figure shows a comparison of the theoretical performance of SCALDO technique (upper trace), practical performance of SCALDO technique (middle trace), and performance of standard linear regulator (lower trace) for a 12–5 V regulator. The prototype achieves overall ETEEs in the range of 70%–80%, compared to the maximum theoretical efficiency of 42% for a linear regulator used for the same input-output combination. The middle trace of Fig. 10.22 shows the improved efficiency in the new SCALDO implementation. The graphs indicate a 3%–5% difference between the theoretical and measured power efficiencies for the lossless-dropper supercapacitor technique. This discrepancy is due to fact that the theoretical curve ignores the losses in switch resistances and the capacitor ESR. It is useful to compare this case of only 3%–5% loss contribution due to ESR and switch resistances, with that of commercial switched-mode configurations where the losses are typically 10%–30% due to static and dynamic losses in diodes, high-frequency switching transistors, and other parasitic effects. This comparison demonstrates the clear advantage of the new supercapacitor-based lossless-dropper technique of the SCALDO regulator.

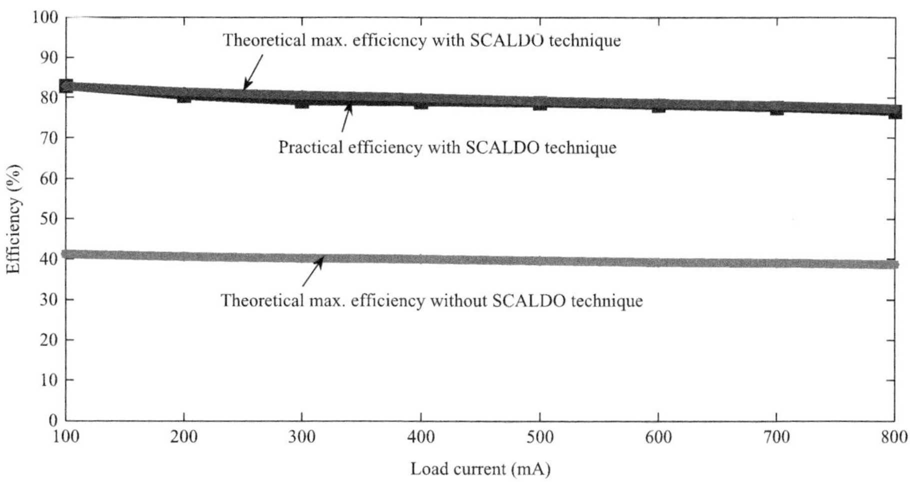

Fig. 10.21 Efficiency vs load current of the 12–5 V SCALDO regulator at an input voltage of 12 V.

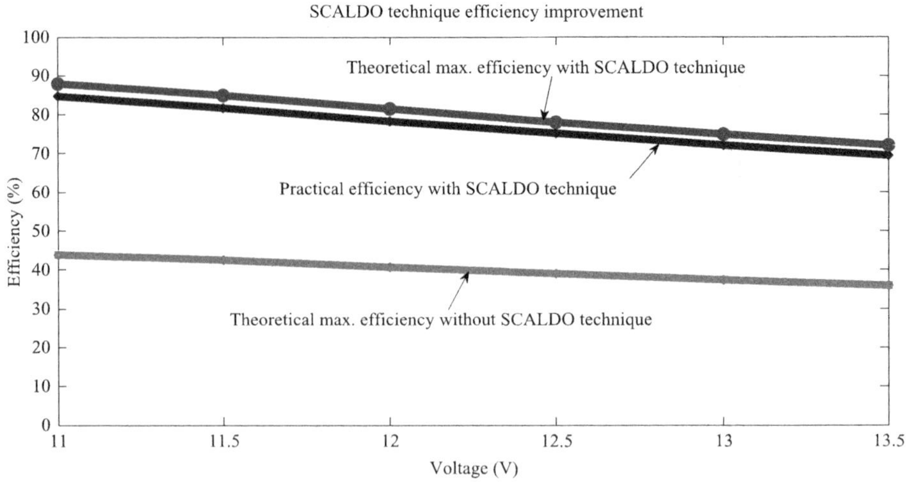

Fig. 10.22 Efficiency comparison for a 12–5 V regulator: Theoretical results with SCALDO technique (upper trace), practical results with SCALDO technique (middle trace), and basic 12–5 V linear regulator theoretical performance (lower trace) at a load current of 250 mA.

10.6.2 Load transient response

10.6.2.1 Output current slew rate of DC power supplies

A well-regulated power supply is critical for the proper operation and reliability of modern high-speed microprocessors. They require fast delivery of large currents in a smaller time, tight supply-voltage tolerance, and intelligent voltage programming. This voltage regulation must be as tight as ±5% and this tolerance is the resultant of DC set point accuracy, accuracy over temperature and input voltage variations, and transient response. If the voltage on the processor is too high or low or even a processor voltage is temporarily too high or too low can cause problems with the end operation of the processor. In achieving such rigorous voltage requirements, the total accuracy of the power supply including the accuracy of the output voltage in a steady-state condition and the transient response is vital (Glaser, 2011). The latest CPU voltage-regulator specifications from Intel and AMD call for load-current slew rates of 50–200 A/μs and peak currents of 60 A to more than 120 A (Gentchev, 2000).

There is an increasing demand for point-of-load (POL) converters that maintain good regulation in the presence of fast varying load transients (di/dt). A POL converter that is capable of handling transients up to 300 A/μs has recently been introduced. This level of performance presents new challenges to engineers during the converter design, and also during the test and verification stages of product development. The combination of high-frequency converter operation and fast load transients demands stringent design practices and a thorough understanding of every element of the design and test setup (Callanan, 2004).

In modern microprocessor-based systems, the operation at low voltages, GHz clock frequencies, high power, and the ability to survive fast current transients requires a new generation of voltage regulator modules (VRMs). Transient response is the most important VRM test and the most difficult to perform. A VRM must maintain regulation under high transition rate loading, so it is crucial to generate a fast load pulse or single transition to verify the voltage transient levels and transient response time under specified loading. As with traditional power supplies, different loading patterns can cause significant differences in transient response time and voltage level. Therefore, an electronic load that simulates fast load changes is necessary to verify the transient response of a VRM. In order to verify the transient response, measurements of the rise and fall times upon a step change in the load are necessary. Generally, this type of test requires a load that is able to produce a rise and fall time approximately five times faster than the power supply (Lee, 2001).

10.6.2.2 Load transient response of 12–5 V SCALDO regulator

The load transients of the SCALDO prototypes were measured using TEXIO PXL-151A DC electronic load.

Once the switch load current was set at the TEXIO DC load, it was connected to the SCALDO prototype and corresponding changes in the output voltage were observed from the Tektronix digital oscilloscope. For better observation of the variations, AC coupling mode of the oscilloscope was enabled. For a given slew rate in the rising and falling edges of the load, corresponding output voltage variations were recorded in terms of transient voltage ripple and transient time. The experimental setup used for the transient measurement is shown in Fig. 10.23. Typical current step of the TEXIO PXL-151A DC electronic load is shown in Fig. 10.24.

Fig. 10.25 shows the transient voltage changes for 12–5 V SCALDO regulator, where the trace (1) shows the current step, and the trace (2) shows the corresponding transient voltage change.

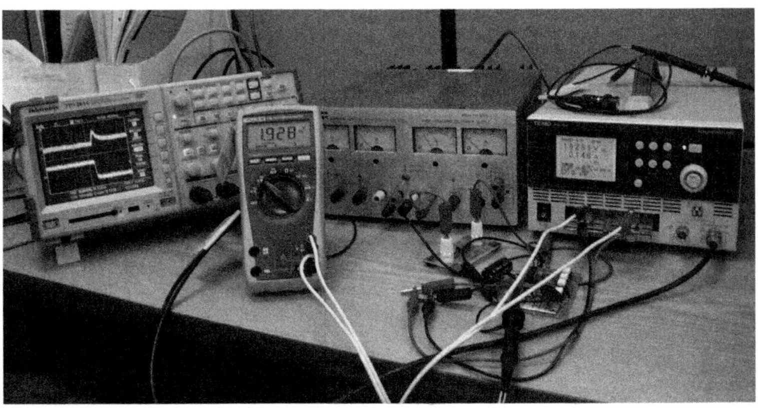

Fig. 10.23 Transient measurement set up of a SCALDO regulator using TEXIO DC electronic load.

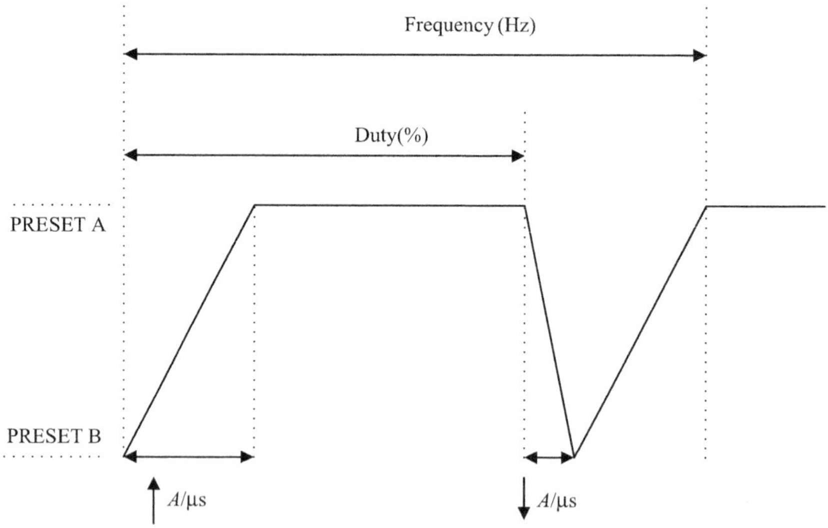

Fig. 10.24 Typical current step of the TEXIO PXL-151A DC electronic load.

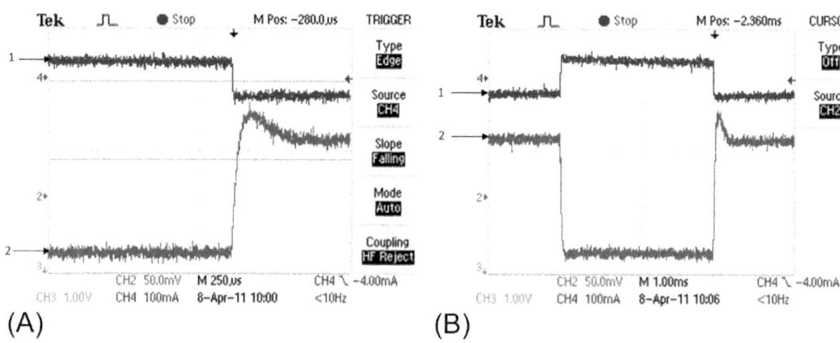

Fig. 10.25 Transients measurements of 12–5 V SCALDO: (A) falling edge; (B) falling/rising edge.

10.6.3 5.5–3.3 V SCALDO regulator

Fig. 10.26 block diagram shows the 5.5–3.3 V SCALDO regulator which is an implementation of $V_p < 2V_{in}^{min}$ criterion, $n = 3$ case. Fig. 10.27 shows the schematic representation of this SCALDO implementation. LDO LT1963-3.3 from Linear Technology which has the following specifications was used as the main low-dropout regulator in this prototype (Fig. 10.28):

- output current of 1.5 A,
- fixed output voltages at 3.3 V,

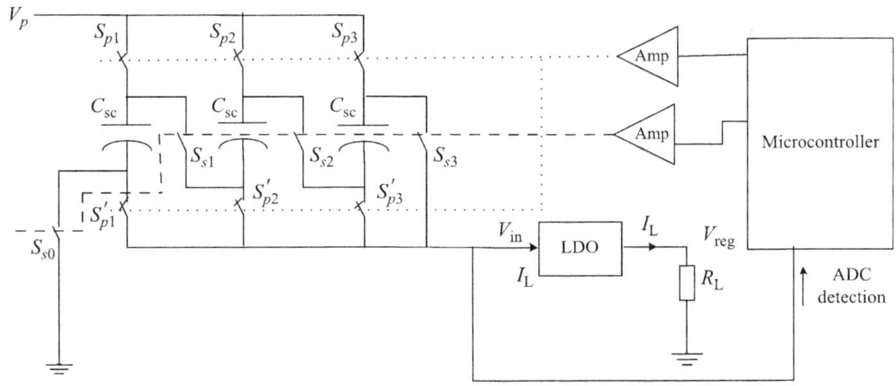

Fig. 10.26 Block diagram of the 5.5–3.3 V SCALDO regulator.

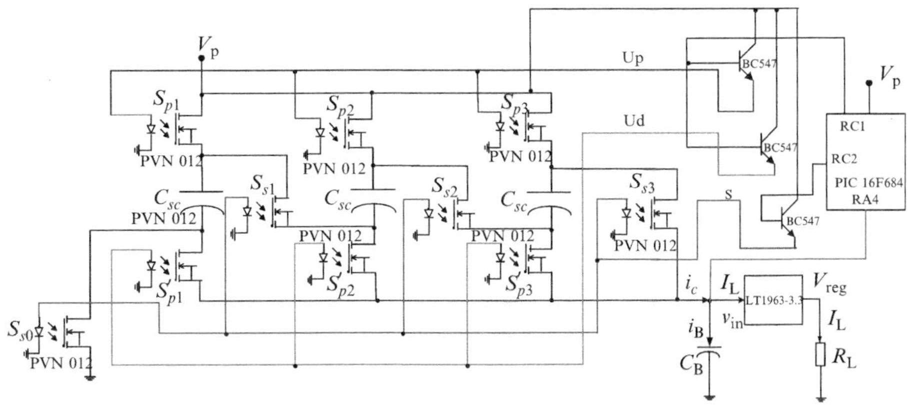

Fig. 10.27 Schematic representation of the 5.5–3.3 V SCALDO regulator designed using an LDO LT1963-3.3, three 1.2 F/2.5 V Maxwell supercapacitors, ten PVN012 Photovoltaic switches, a PIC16F684 microcontroller, two current amplifiers, and a LM7805 regulator.

Fig. 10.28 Prototype of the 5.5–3.3 V SCALDO regulator.

- dropout voltage of 340 mV,
- quiescent current of 1 mA, and
- optimized for fast transient response.

In order to maintain a better LDO operation, the LDO input voltage was kept at least 0.3 V above the required output voltage, so that, $V_{\text{in}}^{\text{min}}$ was maintained at 3.6 V. Originally, it was planned to design a 5.5–3.3 V regulator but due to switch resistance losses and supercapacitor ESR losses, the input voltage was maintained at 5.5 V to overcome the issues of series resistances which drop the $V_{\text{in}}^{\text{min}}$ to unacceptable level. Three supercapacitors of 1.2 F/2.5 V of the type Cap-XX, were used as the group of identical supercapacitors (C_{SC}). A PIC microcontroller of the type PIC16F684 was used as the controller to drive the photovoltaic switches. The LDO input voltage was monitored through the RA4 port pin of the PIC microcontroller. The switch set (6 nos) S_{p1}, S_{p2}, S_{p3}, S'_{p1}, S'_{p2}, S'_{p3} and the switch set (4 nos) S_{s0}, S_{s1}, S_{s2}, S_{s3} were controlled through the RA1 and RA2 port pins, respectively. In order to control several switches using a single control signal, two current amplifiers were used for the parallel and series set of switches, respectively. The three supercapacitors were kept in parallel using the switch set S_{p1}, S_{p2}, S_{p3}, S'_{p1}, S'_{p2}, S'_{p3}. This capacitor group was connected in series with the LDO input to keep the supercapacitors charging from the unregulated power supply, while maintaining the requirement for the LDO operation. During the charging process, when 3.6 V for the LDO was reached, the controller switched on switch group S_{s0}, S_{s1}, S_{s2}, S_{s3} and switched off the other group of switches to keep the three supercapacitors in series to release the stored energy. The stored energy was released until the 3.6 V condition for the LDO was met and then switched back to the charging configuration again. To keep the linear regulator powered during this transition, a buffer capacitor (C_{B}) with sufficient capacity was connected in between the linear regulator input and the ground terminal. Ten PVN012 photovoltaic relays were used as switches to keep the set of supercapacitors in charge and discharge configurations. The microcontroller firmware developed for this technique was based on the same algorithm as the $V_p > 2V_{\text{in}}^{\text{min}}$ configuration in the flow chart of Fig. 10.14. Additional details including the microcontroller firmware are provided by Gunawardane (2014).

10.6.3.1 Load regulation

Fig. 10.29A–D indicates the oscilloscope waveforms during the circuit operation for a load current of 200, 600, 1000, and 1300 mA, respectively. At these stages, input voltage was maintained at 5.5 V. As the load current increased, both the charge and the discharge cycles of the supercapacitor took a shorter time period to operate. In Fig. 10.29 traces (1)–(4) represent voltage across one of the three supercapacitors (V_{SC}), LDO output voltage (V_{reg}), the input current of the system (I_{in}), and input voltage of the LDO (V_{in}), respectively. Fig. 10.30 shows the load regulation graph.

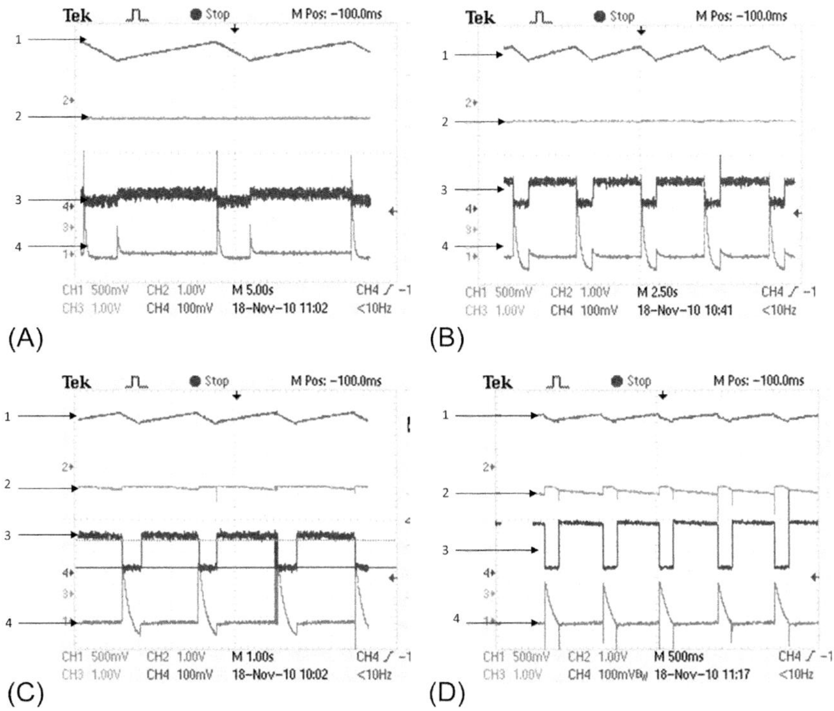

1 Voltage across the supercapacitor

2 LDO output voltage

3 Input current drawn from the unregulated power supply

4 Input voltage of the LDO

Fig. 10.29 Oscilloscope waveforms of 5.5–3.3 V SCALDO Regulator for a load current of (A) 200 mA, (B) 600 mA, (C) 1000 mA, and (D) 1300 mA at input voltage of 5.5 V.

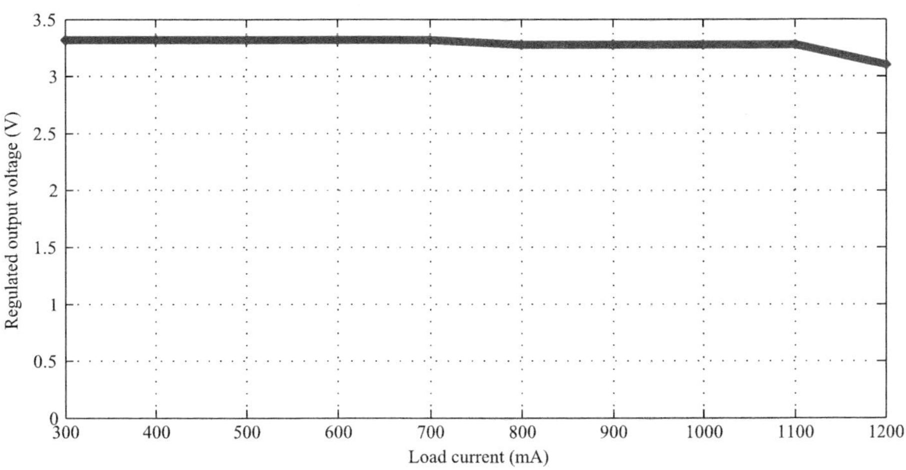

Fig. 10.30 Load regulation results of 5.5–3.3 V SCALDO regulator at input voltage of 5.5 V.

10.6.3.2 Line regulation

For 5.5–3.3 V SCALDO regulator, Fig. 10.31 depicts the corresponding line regulation plot based on the experimental results. The load current was fixed at 300 mA during these measurements.

10.6.3.3 End-to-end efficiency

Using the line regulation experimental data (Gunawardane, 2014), the ETEE plotted against the input voltage is shown in Fig. 10.32. This figure shows a comparison of the theoretical performance of SCALDO technique (upper trace), the practical

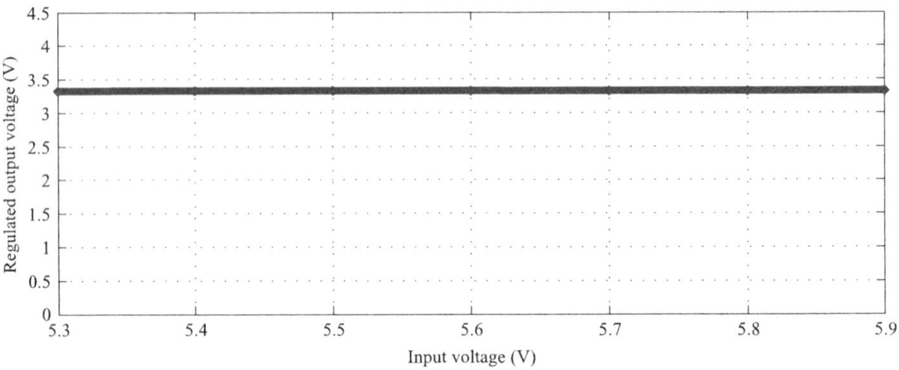

Fig. 10.31 Line regulation results of 5.5–3.3 V SCALDO regulator at a load current of 300 mA.

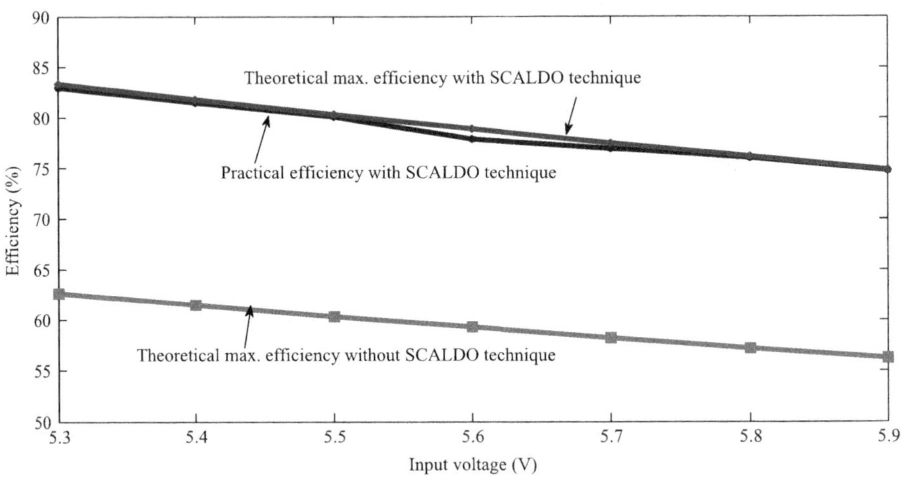

Fig. 10.32 Efficiency comparison for a 5.5–3.3 V SCALDO regulator: theoretical results with SCALDO technique (upper trace), practical results with SCALDO technique (middle trace), and basic 5.5–3.3 V linear regulator theoretical performance (lower trace).

performance of SCALDO technique (middle trace), and the performance of a standard linear regulator (lower trace) for a 5.5–3.3 V regulator. The prototype achieves overall ETEEs in the range of 75%–85%, compared to the maximum theoretical efficiency of 60% for a linear regulator used for the same input-output combination. The middle trace of Fig. 10.32 shows the improved efficiency in the new SCALDO implementation. The graphs indicate a 1%–2% difference between the theoretical and measured power efficiencies for the SCALDO technique. This discrepancy is due to fact that the theoretical curve ignores the losses in switch resistances and the capacitor ESR.

10.6.3.4 Transient response

Fig. 10.33 shows the oscilloscope figures of the transient measurements for 5.5–3.3 V SCALDO regulator. For a given slew rate, in the rising and falling edges of the load, output voltage variations were recorded in terms of transient voltage ripple and transient time (Gunawardane, 2014). Transient response is a complicated situation based

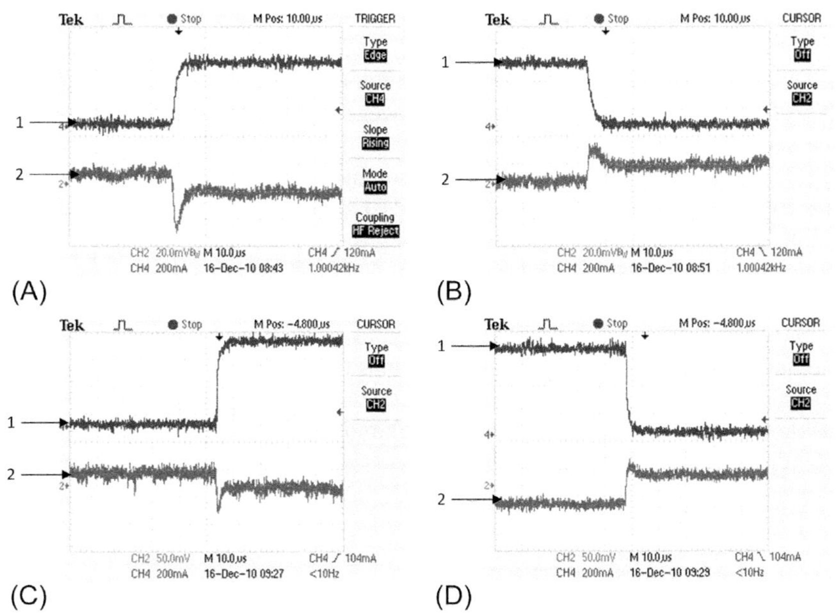

1 Transient load current
2 Change of output volatge

Fig. 10.33 Transients of 5.5–3.3 V regulator using SCALDO technique: (A) 1 A/μs 50–600 mA rising edge; (B) 1 A/μs 50–600 mA falling edge; (C) 20 A/μs rising edge 50–800 mA; and (D) 20 A/μs falling edge 50–800 mA.

on the output capacitor and the behavior of the LDO itself. In this SCLADO analysis, further detailed analysis on finer details was not carried out due to the time constraints of the project.

10.7 Wider applications of SCALDO technique

SCALDO technique is a unique way to design high-efficiency linear regulators without RFI/EMI issues. With commercial supercapacitors with capacitance values varying from 0.2 to over 500 F where the size of the SC is nearly similar to electrolytic capacitors, and their ESR is in the range of less than 100 mΩ (for smaller sizes) to about 1 mΩ, designer can use the technique to come up with different types of SCALDO implementations. Possible applications are as follows:

1. high current high-efficiency linear regulators,
2. desktop computer power supplies,
3. linear high-efficiency DC power supplies with DC UPS capability, and
4. VRMs.

Since modern supercapacitors can charge and discharge at very high currents such as 10 A to few 100 A in general, if suitable power switches are implemented, the technique can be useful in linear regulators with output current ratings up to few 100 A.

Technique can be extended to desktop computer power supplies, with the target of ETEE exceeding the current capability for desktop computers. This is another major possible application of the SCALDO technique.

Since SCs are available up to and beyond 50,000 F with extra-low ESR (less than 0.5 mΩ), the SCALDO technique can be simply extended to buffer a few-cycle loss of AC mains in an AC main-derived DC supply, by oversizing the supercaps. However, this requires additional circuit design extensions to take care of the precharging needs of the supercaps used. Application of the technique to develop fully linear VRMs is possible by modifying the base technique to minimize the number of power switches (Kularatna and Wickramasinghe, 2013).

10.7.1 RS-SCALDO technique

For high current capable LDO circuits, efforts should be taken to reduce the number of switches. One such improvement, applicable to the present SCALDO design has been identified. Using two identical LDOs, the number of switches can be reduced from $3n+1$ to $2n$. This reduced-switch SCALDO (RS-SCALDO) topology has been suggested as a solution for modern VRM requirements.

More theoretical and experimental details on this new modification are provided by Kularatna and Wickramasinghe (2013). As shown in Fig. 10.34, for the $V_p - 2V_{\text{in}}^{\text{min}}$ basic SCALDO design, using two identical LDOs, the number of switches can be reduced to two instead of the four switches used in the present design.

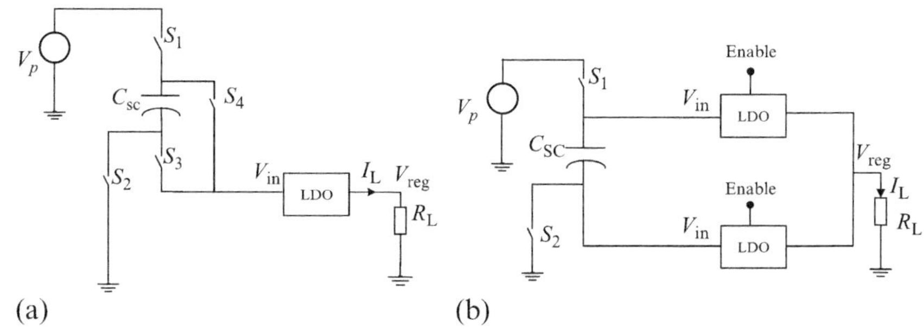

Fig. 10.34 (A) Basic SCALDO configuration with single LDO and four switches; (B) modified RS-SCALDO (reduced switches) configuration with two identical LDOs and two switches.

10.7.2 DO-SCALDO technique

Dual-output SCALDO (DO-SCALDO) topology is a novel improvement to provide dual polarity outputs from a single input DC voltage source maintaining the useful hallmarks of the original SCALDO approach. Chapter 11 provides a details of DO-SCALDO technique.

10.8 Comparison between SCALDO regulators and charge pumps

Supercapacitor-assisted low-dropout regulators (SCALDO) were proposed as an alternative design approach of DC-DC converters, where a very low-frequency supercapacitor circulation technique is combined with a commercial low-dropout regulator IC to achieve significantly high ETEE. In this technique, the supercapacitor is used as a lossless voltage dropper, and the energy reuse occurs at very low frequencies such as from a few to a few 100 Hz, eliminating any RFI/EMI issues. The efficiency advantage of this approach is very much closer to the efficiencies of practical switching regulators and also eliminates the utilization of bulky inductors.

It is important to note that the SCALDO technique does not use supercapacitors for any DC-DC conversion purpose, as in the case of switched-capacitor type DC-DC converters (charge pumps) such as voltage doublers, multipliers, and inverters. Charge pumps utilize microfarad-order capacitors to achieve a voltage conversion using switching frequencies in the order of hundreds of kHz. The SCALDO technique utilizes the supercapacitor as a voltage dropper only, and the very small ESR in modern supercapacitors allows the capacitor losses in the series path to be minimized, well approximating a lossless voltage dropper. This SCALDO technique can be extended to provide high output currents in the order of 5–50 A at DC rail voltages suitable for modern processors. Given the low leakage currents in new families of supercapacitors, usually within a range of 5 to about 50 μA (Mars, 2012; http://www.cap-xx.com/products/products.htm), compared to their high charge/discharge current capabilities in the range of 10 to several 100 A, leakage current-related losses are negligible. Also, the supercapacitor is the best candidate for this new application, as the ESR values (in the range of fractional milliohm to less than 100 mΩ) are less than or comparable $R_{DS(on)}$ values of low voltage MOSFETs. Given that the new technique works in the range of fractional hertz to few 100 Hz order frequencies, and with the new supercapacitor families having very low leakage currents, dynamic losses will be significantly lower than a charge pump. Compared to linear or switching regulators, the output voltages of charge pump circuits are usually not regulated and are designed to operate at a fixed switching frequency of few 100 kHz with a rated output load current and voltage. Because of practical limitations on the size of capacitors and switches, applications of charge pumps are limited mainly to low-and-medium power levels of several tens of milliwatts.

Table 10.5 A comparison summary of SCALDO technique and switched capacitor converters.

SCALDO technique	Switched-capacitor techniques
A modified version of a linear regulator with an enormously large capacitor in the series path as a lossless voltage dropper. The capacitor and the switches do not convert the voltage	Basically a high-frequency switching technique for voltage conversion
Always a step-down configuration	Practically used to step-up or invert a DC voltage
Enormously large capacitors (supercapacitors) are used	Capacitors used are in the range of a few nanofarads to a few tens of microfarads
Operation frequency is variable and depends on the load current. Switching operation is based on the case of a maximum/minimum voltage detected across the input of the LDO	Circuit design starts with a fixed oscillator supplying the switching frequency
Very low frequency (10 to few 100 Hz) switching used for energy storage and reuse in capacitors	Switching frequency is in the range of 10 to several 100 kHz
Load always sees the precise output of a linear/low drop out regulator	Load regulation is not precise and requires a another voltage regulator (linear/low drop out type) for precise output voltage
Capacitor never comes parallel to the input unregulated supply	In one part of the cycle capacitor comes in parallel to the unregulated supply
Technique is applicable to very large load currents (high current capable LDO is required)	Technique is suitable only for very low load currents
Negligible dynamic losses in switches	Significantly high dynamic losses in switches
Theoretically an efficiency multiplication factor is defined for a given configuration	Theoretically a voltage conversion factor applies to a given configuration

Table 10.5 provides capabilities and limitations of SCALDO technique comparing the practical performance with typical switched-capacitor converters.

Given the above simple summary, it is very clear to see that the SCALDO technique is quite different from the operation of charge pumps, for three primary reasons:

1. SCALDO technique uses a capacitor as a lossless voltage dropper, together with a linear low-dropout regulator for precise output regulation.
2. Extremely low switching frequency is variable, and it depends on the load current.
3. If an ultra LDO can be developed for the required output current, there is no limit to precise output regulation at high load currents. (This is due to the availability of single-cell supercapacitors (commercially available) which are in the range of fractional farads to a few thousand farads.)

References

Bull, C., Smith, C., 2003. Integrated Building Blocks for Dual-output Buck Converter. Power Electronic Technology.

Callanan, S., 2004. Testing High di/dt Converters. Artesyn Technologies. Available: www.powerelectronics.com.

Day, M., 2003. Integration Saves Time and Board Space Converter. Power Management Special Supplement. Power Electronics Technology, pp. 64–67. Available: www.powerelectronics.com.

Falin, J., 2006. A 3-A, 1.2-V_{OUT} linear regulator with 80% efficiency and P_{LOST} <1w Converter. Analog Appl. J., 10–13. Available: www.ti.com/aaj.

Gentchev, A., 2000. Designing High-Current, vrm-Compliant CPU Power Supplies. Analog Devices. Available: www.edn.com.

Glaser, C., 2011. Optimal Transient Response for Processor-Based Systems. Texas Instruments. Available: www.powerelectronics.com.

Gunawardane, K., 2014. Analysis on Supercapacitor Assisted Low Dropout Regulators. The University of Waikato. Ph.D. dissertation.

Inouye, S., Robles-Bruce, M., Scherer, M., 2010. Power Management—General Purpose Analog Service. Databeans Incorporated. Tech. Rep. Available: www.databeans.net.

Jovalusky, S., 2005. New Energy Stand ards Banish Linear Supplies. Power Electronics Technology, Power Integrations, San Jose, CA. Available: www.powerelectronics.com.

Kankanamge, K., Kularatna, N., 2012. Implementation aspects of a new linear regulator topology based on low frequency supercapacitor circulation. In: IEEE Applied Power Electronics Conference, USA.

Kankanamge, K., Kulatana, N., 2013. Supercapacitor assisted LDO (SCALDO) technique-an extra low frequency design approach to high efficiency DC-DC converters & how it compares with the classical switched capacitor converters. In: Proc of 28th Annual IEEE-APEC-2013, pp. 1979–1984.

Kester, W., Erisman, B., Thandi, G., 1998. Switched Capacitor Voltage Converters. Analog Electronics Technical Report.

Kularatna, N., 2011. DC Power Supplies Power Management and Surge Protection for Power Electronic Systems. CRC Press.

Kularatna, N., Fernando, J., 2009. A supercapacitor technique for efficiency improvement in linear regulators. In: IEEE 35th Annual Conference of Industrial Electronics, pp. 132–135.

Kularatna, N., Fernando, J., 2011. High Current Voltage Regulator. US Patent 9707 430 B2, March 15.

Kularatna, N., Wickramasinghe, J., 2013. Supercapacitor assisted low dropout regulators (SCALDO) with reduced switches: a new approach to high efficiency VRM designs. In: IEEE International Symposium on Industrial Electronics (ISIE), pp. 1–6.

Kularatna, N., Fernando, J., Kankanamge, K., Tilakaratna, L., 2010. Very low frequency supercapacitor techniques to improve the end-to-end efficiency of DC-DC converters based on commercial off the shelf LDOs. In: Proceeding of IEEE Industrial Electronics Conference, pp. 721–726.

Kularatna, N., Kankanamge, K., Fernando, J., 2011a. Supercaps improve LDO efficiency. Part 1: low noise linear supplies. Power Electron. Technol. Mag. April.

Kularatna, N., Kankanamge, K., Fernando, J., 2011b. Supercapacitor enhance LDO efficiency. Part 2: implementation. Power Electron. Technol. May.

Kularatna, N., Fernando, J., Kankanamge, K., Zhang, X., 2011c. A low frequency supercapacitor circulation technique to improve the efficiency of linear regulators based on LDO ICs. In: IEEE Applied Power Electronics Conference, Texas, USA, pp. 1161–1165.

Linear and Switching Voltage Regulator Handbook on Semiconductor, February, 2002. Available: http://onsemi.com.

Lee, B.S., 1999. Understanding the Terms and Def initions of LDO Voltage Regulators. Texas Instruments.

Lee, J., 2001. High Slew Rate Electronic Load Checks New Generation Voltage Regulator Modules. Power Electronics. Chroma ATE Inc., Irvine, CA.

Linear regulators in portable applications (2001, May). Maxim Integrated Products. [Online]. Available: http://www.maxim-ic.com/an751.

Man, T.Y., Mok, P.K.T., Chan, M., 2007. High slew-rate push-pull output amplifier for low-quiescent current low-dropout regulators with transient-response improvement. IEEE Trans. Circuits Syst. 54 (9), 755–759.

Marasco, K., How to Successfully Apply Low Dropout Regulators. Analog Devices. [Online], Available: http://www.analog.com/library/analogdialogue/archives/43-08/ldo.html. www.analog.com.

Mars, P., 2012. Coupling a supercapacitor with a small energy harvesting source. EDN Mag. [Online]. Available: http://www.edn.com/design/components-and-packaging/4374932/1/Coupling-a-supercapacitor-with-a-small-energy-harvesting-source.

Palumbo, G., Pappalardo, D., 2010. Charge pump circuits: an overview on design strategies and topologies. IEEE Circuits Syst. Mag., 31–45.

Rincon-Mora, G.A., 1996. Current Efficient, Low Voltage, Low Drop-out Regulators. Georgia Institute of Technology. Ph.D. dissertation.

Rincon-Mora, G.A., 2009. Analog IC Design With Low-Dropout Regulators. McGraw-Hill.

Simpson, C., 2012. Linear and Switching Voltage Regulator Fundamentals. National Semiconductor Application Note [Online], Available: http://www.ti.com/lit/an/snva558/snva558.pdf.

Texas Instruments, 2009. Technical Review of Low Dropout Regulator Operation and Performance. Tech Report, August.

Travis, B., 1998. Linear vs. Switching Supplies: Weighing All the Options. EDN, January 1, 40.

Walt Kester, B.E., Switching Regulators. [Online], Available: http://www.analog.com/static/imported-files/tutorials/ptmsect3.pdf.

Zendzian, D., 1999. A High Performance Linear Regulator for Low Dropout Applications. Unitrode Corporation. [Online], Available: http://www.ti.com.cn/cn/lit/an/slua072/slua072.pdf.

Extended applications of SCALDO technique

11

Kasun Subasinghage

11.1 Introduction

As discussed in Chapter 10, the SCALDO technique is based on a low headroom, high-efficiency linear regulator, preceded by a supercapacitor-based lossless voltage dropper. This combination effectively creates a high-efficiency DC-DC converter, where a supercapacitor maintains its charge balance by recirculating its excess energy back to the linear regulator at a very low speed in the range of few millihertz to few Hertz. This technique is versatile and useful over a large range of load currents from few milliamperes to several hundred amperes, which is mainly governed by the selected power semiconductors. This chapter provides a discussion on extended applications of the SCALDO technique.

The SCALDO regulator and its derivatives fall in the nonconventional family of DC-DC converter topologies as shown in Fig. 11.1 (Subasinghage, 2019). Apart from the basic SCALDO topology and the generalized topology, three further extensions of the SCALDO technique have been introduced: (a) reduced-switch (RS)-SCALDO regulator, (b) DC-UPS SCALDO regulator, and (c) dual-output (DO) SCALDO regulator.

In the DC-UPS SCALDO method, all the fundamental circuit components are kept the same while the size of the supercapacitor is increased to store more energy to keep a continuous regulation during a DC power source disconnection (Gunawardane et al., 2018). RS-SCALDO technique is developed to increase the current handling capability of the basic SCALDO technique targeting voltage regulator modules (VRMs) or similar high-current applications by reducing the number of switches in the converter (Wickramasinghe et al., 2017; Wickramasinghe, 2016), where the DO-SCALDO regulator is designed for high-current capable and split-rail dual-output DC-DC converter applications (Subasinghage, 2019; Subasinghage et al., 2019).

11.2 RS-SCALDO technique

Discussion in Chapter 10 indicates that a single-stage SCALDO regulator requires four low-speed switches. Since two switches come in series in each mode of the converter, they contribute to the static losses of the SCALDO converter. For high-current capable LDO regulator circuits, efforts should be taken to reduce the number of

Energy Storage Devices for Renewable Energy-Based Systems. https://doi.org/10.1016/B978-0-12-820778-9.00011-5

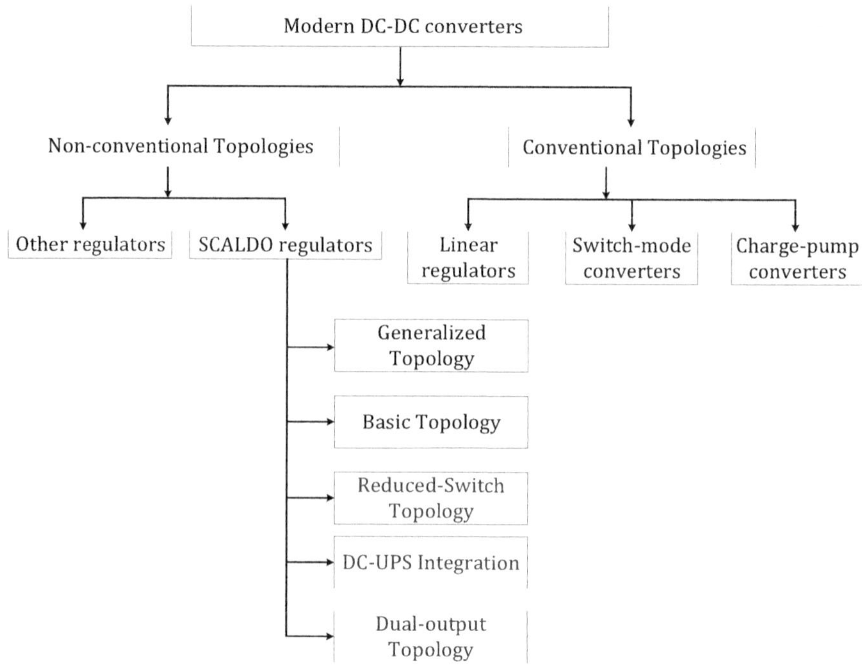

Fig. 11.1 SCALDO topology and its derivatives in the family of modern DC-DC converters.

switches. One such improvement, applicable to the present SCALDO design, has been identified as the reduced-switch SCALDO or RS-SCALDO topology. In this technique, the number of switches can be reduced from $(3n+1)$ to $2n$ (where n is the number of supercapacitors). For example, in the $V_p > 2V_{\min}$ basic SCALDO design where a single supercapacitor ($n = 1$) is used, the number of switches can be reduced to two instead of four switches as shown in Fig. 11.2 (Wickramasinghe et al., 2017; Wickramasinghe, 2016).

As with the basic SCALDO operation, the RS-SCALDO regulator also has two operating modes as displayed in Fig. 11.3, involving both charging and discharging modes. The SC is allowed to charge in series with the LDO(1) in the charging mode as depicted in Fig. 11.3A. At the same time, LDO(2) is disabled to avoid any passage of current through this regulator by turning off the gate or base of its series-pass device from the control circuit. The SC is kept in this arrangement until the input voltage of LDO(1) reaches its minimum working voltage. When the charging mode is completed, the SC is forced to discharge via LDO(2) as displayed in Fig. 11.3B. In this mode too, the short circuit path via the body diode of LDO(1) is blocked by a special internal arrangement of the LDO(1) and more implementation details are available in a study by Wickramasinghe et al. (2017). These two modes are continued by maintaining the charge balance of the SC. As the power source delivers energy only during the charging mode, it maintains the same end-to-end efficiency advantage of $(2V_o/V_p)$ similar to the fundamental SCALDO configuration.

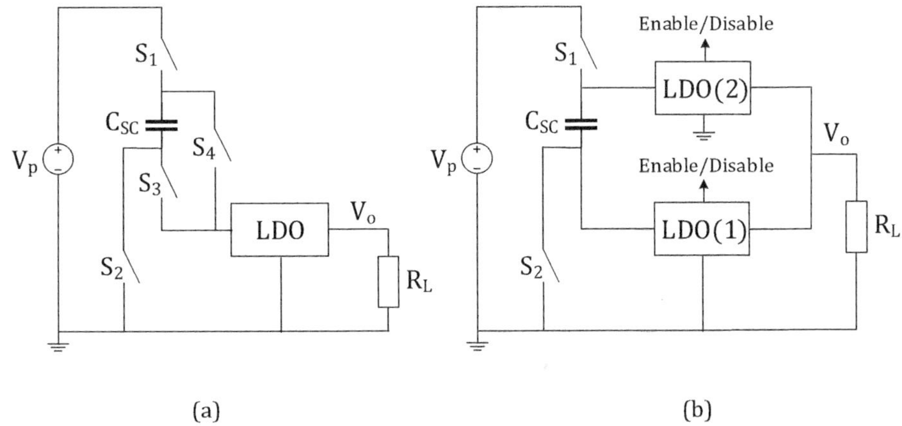

Fig. 11.2 Comparison of SCALDO and RS-SCALDO topologies: (A) basic SCALDO configuration with single LDO regulator and four switches and (B) modified RS-SCALDO (reduced switches) configuration with two identical LDOs and two switches.

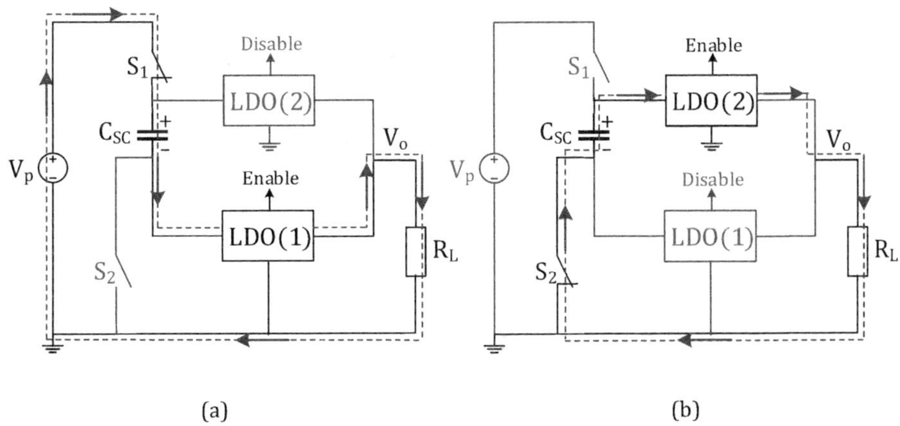

Fig. 11.3 The basic RS-SCALDO operating modes: (A) charging mode and (B) discharging mode.

11.3 DC-UPS SCALDO regulators

11.3.1 DC-UPS systems

If a power converter can have an energy storage capability to buffer any short-term or long-term failures of its input stage, it can keep the continuity of the output power. Uninterrupted power supply (UPS) systems typically use an energy storage system (ESS) within the system, so that the output stage sees a continuous flow of energy. These systems can be typically found as rechargeable batteries, and occasionally based on other ESS systems. With the supercapacitor families developing fast with higher energy density (Chapter 8), the use of supercapacitors for UPS systems is rapidly growing. One such example is PCS100 UPS-I sold by ABB which starts from 150kVA rating and a large array of series/ parallel connected supercapacitor banks is utilized as the ESS (2018).

The primary objective of a UPS system is to provide backup power against power source disconnections hence the critical loads can be gracefully shutdown or kept in operation until the source is reconnected (Hwang et al., 2011; da Silva et al., 2011). Additionally, they can provide power conditioning against voltage fluctuations and noise (Santos et al., 2016). While the AC-UPS systems are often used in consumer and industrial loads, an alternative approach is also used to safeguard DC loads such as computer systems, telecom systems, and data centers as all the electronics systems work with DC voltages (Fernandez et al., 2002; Kwon, 2012; Mondal and Keisling, 2012). This approach has many advantages over the AC-UPS systems due to its less complexity, the smaller number of conversion stages, low cost, and high efficiency. In today's market, DC-UPS systems use battery banks, fuel cells, capacitor banks, or combinations to store energy. Although the main goal of a DC-UPS system is to prevent power interruptions, there is a competition between manufacturers to introduce some additional features such as high efficiency, longer backup power, compactness, and cost. The recent development in this field has widened the application of DC-UPS systems in many directions including data centers (Mondal and Keisling, 2012; Pan et al., 2018).

11.3.2 Integration of DC-UPS feature in the SCALDO regulator

SCALDO regulator is a unique technique since it uses a supercapacitor as a near-lossless voltage dropper, and the secondary property of its supercapacitor to operate as an energy storage element for emergency usage. Interestingly, the size of the supercapacitor in this regulator can be increased to a level where the amount of stored energy in the supercapacitor is sufficiently large to handle a short duration DC line interruption. Since the supercapacitors are more compact in today's market with their common values range from 1 to 3000F (Subasinghage et al., 2019), this choice does not add bulkiness to the overall design. The following few paragraphs demonstrate how the supercapacitor of a SCALDO regulator can be oversized to achieve DC-UPS capability within the converter.

11.3.3 Operating principle

The hardware configuration and the primary switching method of the DC-UPS approach are similar to the basic SCALDO operation. Only the switching algorithm is changed to handle a line interruption (Gunawardane et al., 2018). Therefore, the DC-UPS integrated SCALDO regulator can offer reduced cost, lightweight, and less complexity together with high efficiency.

In the basic SCALDO regulator, the SC can be charged until the input voltage of the LDO regulator ($V_{in(LDO)}$) drops to its minimum working voltage or the V_{min} value. Conversely, in the DC-UPS approach, a new value for this minimum voltage ($V_{min(UPS)}$) is defined for the switching controller such that it is slightly higher than V_{min} of the LDO regulator as defined in Eq. (11.1). This extra voltage headroom enables the SC to store extra energy so that it can be discharged to the real minimum working voltage of the LDO regulator.

$$V_{\min(\mathrm{UPS})} > V_{\min} \tag{11.1}$$

According to the operation principles of the SCALDO regulator, the charging and discharging periods become shorter when the $V_{\min}$ value is increased. On the other hand, these periods become longer as the capacitance of the SC becomes larger. Therefore, to counteract the effect of using $V_{\min(\mathrm{UPS})}$ as per Eq. (11.1), an extra-large capacitance ($C_{\mathrm{sc(UPS)}}$) is used in this DC-UPS approach while maintaining the same low-frequency operation as with the basic SCALDO topology. The normal charge period ($t_{\mathrm{ch(nom)}}$) and discharge period ($t_{\mathrm{dis(nom)}}$) at the steady-state operation are identical as defined in Eq. (11.2).

$$t_{\mathrm{ch(nom)}} = t_{\mathrm{dis(nom)}} = C_{SC(\mathrm{UPS})} \frac{\left[V_p - 2V_{\min(\mathrm{UPS})} - 2I_{\mathrm{in(LDO)}}(2R_{SW} + R_{ESR})\right]}{I_{\mathrm{in(LDO)}}} \tag{11.2}$$

where R_{SW} is the resistance of a single switch, R_{ESR} is the ESR of the SC, V_p is the input source voltage, and $I_{\mathrm{in(LDO)}}$ is the input current of the LDO regulator when driving a constant load at the output.

The switching algorithm of the DC-UPS SCALDO regulator can be explained based on the operation cycle where the line interruption event occurs. If the interruption occurs in the charging mode, the SC is immediately switched into the discharging mode to release its stored energy back to the LDO regulator. As the line interruption can happen at any point of the normal charging cycle, it is proven that the LDO regulator is driven by the SC not greater than $t_{\mathrm{dis(UPS)}}$ according to Eq. (11.3). Due to the oversized SC and extra voltage headroom, the SCALDO regulator can be designed to keep $t_{\mathrm{dis(UPS)}}$ substantially large.

$$t_{\mathrm{dis(UPS)}} \le C_{SC(\mathrm{UPS})} \frac{\left[V_p - V_{\min(\mathrm{UPS})} - V_{\min} - 2I_{\mathrm{in(LDO)}}(2R_{SW} + R_{ESR})\right]}{I_{\mathrm{in(LDO)}}} \tag{11.3}$$

If the line interruption occurs during the discharging mode of the normal SCALDO operation, the same mode of operation can be further continued until the input voltage of the LDO regulator falls to $V_{\min}$. In this case too, if the DC line is disconnected when the input voltage of the LDO regulator is at $V_{\min(\mathrm{UPS})}$, the LDO can be driven by the stored energy of the SC for at least $t_{\mathrm{dis(UPS)}}$ as given by Eq. (11.4)

$$t_{\mathrm{dis(UPS)}} \ge C_{SC(\mathrm{UPS})} \frac{\left[V_{\min(UPS)} - V_{\min}\right]}{I_{\mathrm{in(LDO)}}} \tag{11.4}$$

This operating principle can be easily integrated into SCALDO regulator by modifying the algorithm of the microcontroller unit as shown in Fig. 11.4. In this diagram, the

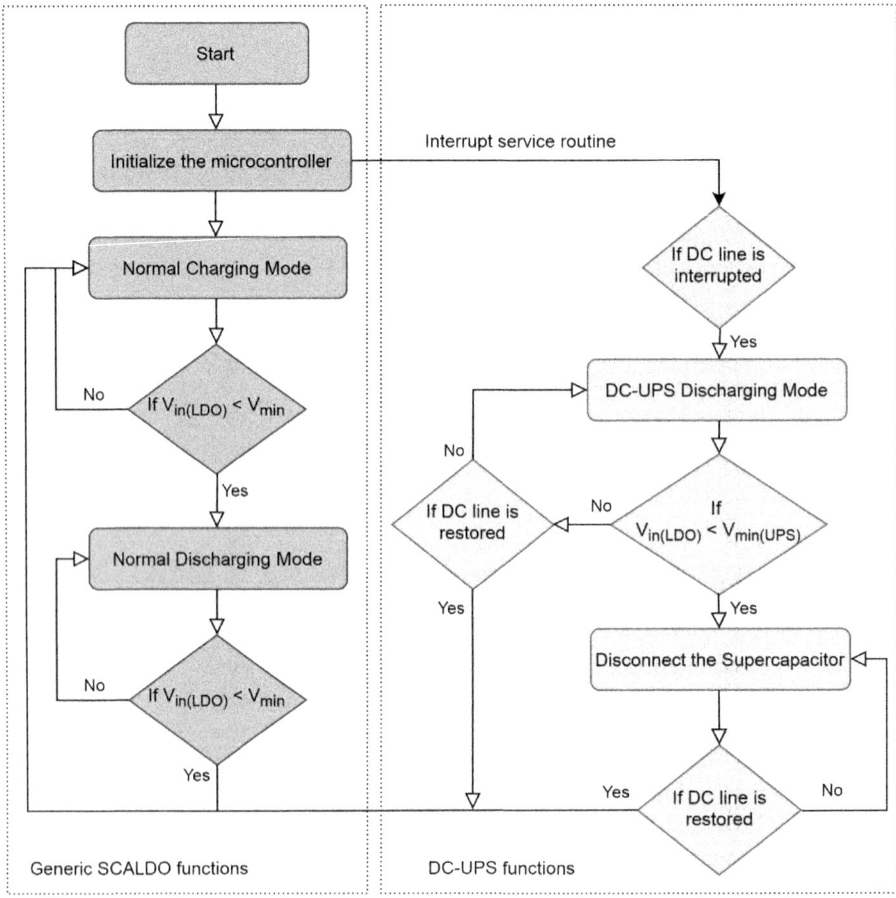

Fig. 11.4 Simplified DC-UPS algorithm.

blocks on the left-hand side depict the generic SCALDO algorithm (based on Chapter 10) where the blocks on the right-hand side show the modifications done to achieve the DC-UPS feature with interrupt service routine. Although this flowchart shows only the essential blocks of the DC-UPS algorithm, more implementation details are available in a study by Gunawardane et al. (2018).

Figs. 11.5 and 11.6 show the oscilloscope traces taken from a 12- to 5-V DC-UPS SCALDO regulator when the source voltage is 12 V. Since the real minimum working voltage of the LDO regulator in use is 5.4 V, the $V_{min(UPS)}$ value is defined as 5.7 V in the microcontroller firmware by keeping 300 mV voltage headroom.

The oscilloscope trace in Fig. 11.5A is recorded when the line disruption occurs at the nominal charging mode of the SCALDO operation when $C_{SC} = 3.3$F. This waveform is divided into four sections: A, B, C, and D. Sections A and D show the normal SCALDO operation where the input voltage of the LDO regulator changes from 5.7 to 6.1 V with periods of 2.2 s. At point b_1 where the input voltage of the LDO

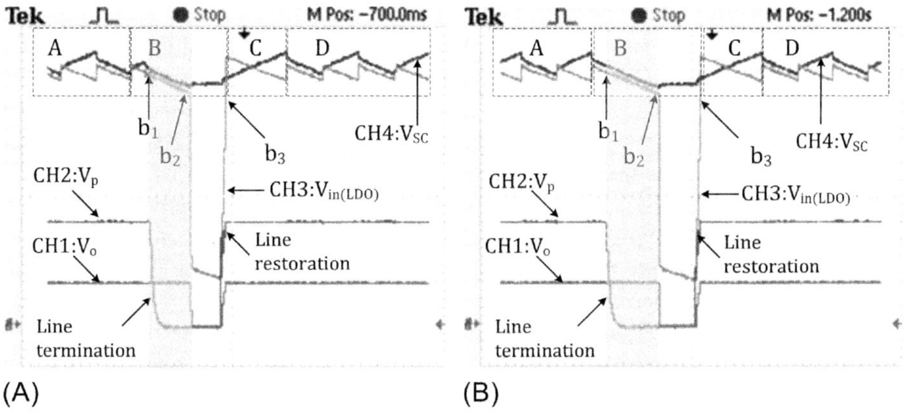

Fig. 11.5 Power source disruption event handling of the DC-UPS SCALDO regulator when $C_{SC} = 3.3\,F$: (A) charging phase and (B) discharging phase (oscilloscope settings: CH1-CH2 (DC-coupling: 5 V/div), CH3-CH4(DC-coupling: 1 V/div), time scale: 2.5 s/div).

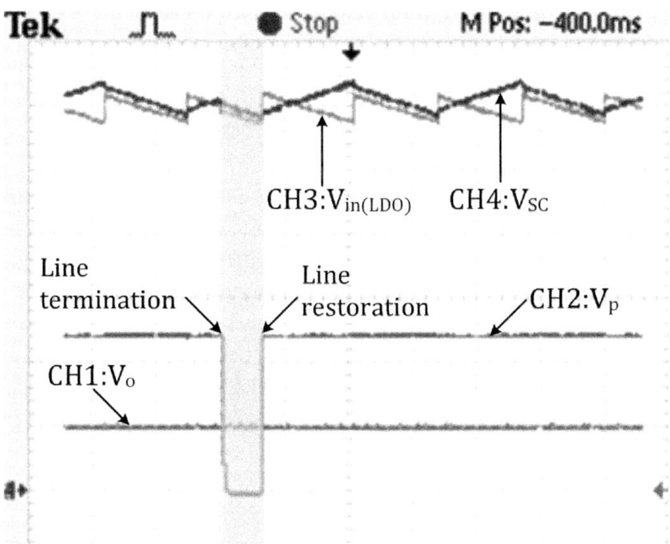

Fig. 11.6 Power source disruption event handling with continuous output regulation when $C_{SC} = 16.6\,F$ (oscilloscope settings: CH1-CH2(DC-coupling: 5 V/div), CH3-CH4(DC-coupling: 1 V/div), time scale: 10 s/div].

regulator is 6 V, the DC source is disconnected, and the SC is forced to the discharge mode. The regulation of the LDO is maintained for 2.5 s until its input voltage reaches 5.4 V. The shaded area in Fig. 11.5 shows the DC-UPS region of the SCALDO regulator. At this point (b_2), the regulation cannot be maintained further as the power source is not restored. Therefore, the SC is disconnected from the input of the

LDO regulator during the period from b_2 to b_3. When the DC source is restored at point b_3, the charging mode is initiated, and the normal SCALDO operation is begun.

When the line disruption occurs in a discharging mode as depicted in Fig. 11.5B, the SC is kept in the same mode until the input voltage of LDO regulator becomes V_{min}. This oscilloscope trace shows that the SC can power the LDO regulator for at least 3 s from where the DC source is disconnected. Since the DC power is not restored afterward, there is a dead time (period: b_2 to b_3) of the regulation. From point b_3, the normal SCALDO operation is started as the DC source is connected.

The dead time of the regulation observed in Fig. 11.5B can be further reduced by oversizing the supercapacitor based on the maximum power of the regulator and the maximum duration of energy backup. Compared to the previous two cases shown in Fig. 11.5, the oscilloscope trace in Fig. 11.6 shows an example where a longer (6 s) and continuous backup regulation is maintained when the SC is changed to 16.6 F.

In summary, the DC-UPS feature can be easily integrated into the SCALDO regulator by oversizing the supercapacitor and without the requirement of any additional hardware. With the availability of larger capacitances of the supercapacitors, the DC power outages can be handled for long periods. Therefore, this feature makes wider applicability of the SCALDO regulator in modern high-efficiency DC-DC converters for telecom devices, data centers, and computers.

11.4 Dual-polarity (DO-SCALDO) concept

Another versatile extension of the original SCALDO concept is its dual-output and split-rail version (Subasinghage, 2019; Subasinghage et al., 2019), which is also known as DO-SCALDO regulator. The same end-to-end efficiency (ETEE) advantage of the single-stage SCALDO regulator is achieved in this converter along with other useful characteristics such as low EMI/RFI, low noise, and high slew rate capability. Therefore, this new technique is not limited to noise-sensitive dual-output applications such as localized microprocessors, liquid crystal displays (LCDs), but also satisfies the requirement of high-current loads such as small-motor drivers and audio amplifiers.

11.4.1 Conceptual background

The DO-SCALDO regulator is based on the series connection of a positive-rail LDO regulator (LDO_P) and a negative-rail LDO regulator (LDO_N) as illustrated in Fig. 11.7. If the front-end SC switching network is not used, the sum of the ground pin current and the load current of LDO_P should be equal to the sum of the ground pin current and the load current of LDO_N satisfying the basic circuit laws. The flow of ground pin and load currents of the two LDO regulators is illustrated in dotted lines. Any unbalanced loading condition will terminate the regulation of the two LDOs. Clearly, the regulation is continued if and only if their output currents are equal as the ground pin currents of well-designed LDO regulators are very small.

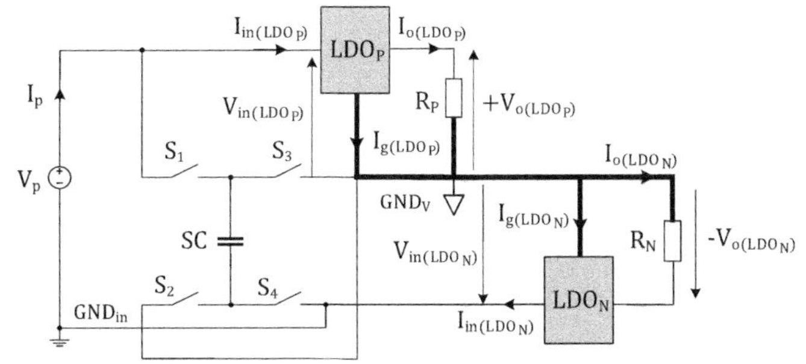

Fig. 11.7 The cascaded dual LDO regulators based on the split-rail DO-SCALDO approach.

In the DO-SCALDO concept with the use of a low-speed switching network, the difference of the load currents of two LDOs is handled while keeping the charge balance of the SC. In one phase of the circuit operation, the SC charges by allowing the differential current to pass through it, and in the next cycle, the stored energy of the SC is delivered back to the circuit. The LDO regulators keep their outputs regulated with reference to the virtual ground (GND_V) although their input voltages change with time due to the SC switching action.

The following design constraints should be satisfied to maintain the voltage regulation in the DO-SCALDO regulator:

$$V_{o(\mathrm{LDO_P})} = \left|V_{o(\mathrm{LDO_N})}\right| \tag{11.5}$$

$$V_{\min(\mathrm{LDO_P})} = \left|V_{\min(\mathrm{LDO_N})}\right| = V_{\min} \tag{11.6}$$

$$V_p > 2V_{\min} \tag{11.7}$$

$$V_{SC(\max)} \geq V_p - V_{\min} \tag{11.8}$$

where $V_{\mathrm{SC(max)}}$ is the maximum allowable operating voltage of the SC.

The DO-SCALDO method works based on four operating modes:

Mode 1: Initial charging process.
Mode 2: Output current of $\mathrm{LDO_P}$ is greater than the output current of $\mathrm{LDO_N}$ $I_{o(\mathrm{LDO_P})} > I_{o(\mathrm{LDO_N})}$.
Mode 3: Output current of $\mathrm{LDO_P}$ is lower than the output current of $\mathrm{LDO_N}$ $I_{o(\mathrm{LDO_P})} < I_{o(\mathrm{LDO_N})}$.
Mode 4: Two LDO regulators have equal output currents $I_{o(\mathrm{LDO_P})} = I_{o(\mathrm{LDO_N})}$.

11.4.2 Phase 1: Initial charging process

At the initial powering up of the DO-SCALDO regulator, the SC is precharged to a voltage slightly above the $V_{\min}$ value. The charging path is completed via R_C resistor, the switch S_4, and the additional switch S_5 as depicted in Fig. 11.8. Once the charging process is completed, the SC is switched to one of the inputs of the two LDO regulators, and the load current balancing operation is maintained.

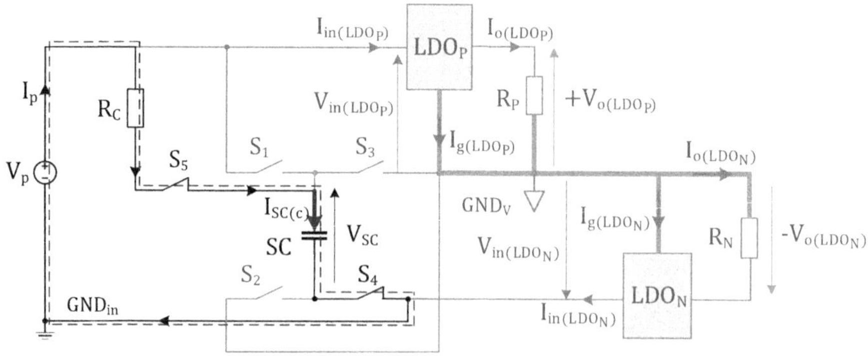

Fig. 11.8 Initial charging process.

11.4.3 Modes 2–4: Handling of differential load current

Fig. 11.9A and B shows the circuit operation when the $I_{o(\mathrm{LDO_P})} > I_{o(\mathrm{LDO_N})}$ condition is satisfied. The SC comes in parallel with $\mathrm{LDO_N}$ and the differential load current $I_{o(\mathrm{LDO_P})} - I_{o(\mathrm{LDO_N})}$ finds its way through the SC in the charging phase as depicted in Fig. 11.9A. This charging current is shown in the dashed lines. The source current (I_P) becomes approximately $I_{\mathrm{in(LDO_P)}}$ as the SC charges and the input voltage of $\mathrm{LDO_N}$ rises from $V_{\min}$ value. This increasing voltage of the SC forces the input voltage of $\mathrm{LDO_P}$ to descend from (V_P-$V_{\min}$) value and when the input voltage of $\mathrm{LDO_P}$ becomes $V_{\min}$ the charging process is terminated.

In the discharging phase as depicted in Fig. 11.9B, the SC and $\mathrm{LDO_P}$ are connected in parallel. As the SC discharges, the input voltage of $\mathrm{LDO_P}$ continues to reduce from (V_P-$V_{\min}$) value and the source current approximately becomes $I_{\mathrm{in(LDO_N)}}$. When the input voltage of $\mathrm{LDO_P}$ reaches $V_{\min}$, the discharging phase is completed, and the SC is switched back to the charging phase.

Conversely, when $I_{o(\mathrm{LDO_P})} < I_{o(\mathrm{LDO_N})}$ criterion is met (in Mode 3), the circuit operation becomes exactly opposite to that in Mode 2. The charging phase of the SC is in parallel to $\mathrm{LDO_P}$ and the discharge phase is in parallel to $\mathrm{LDO_N}$ as displayed in Fig. 11.9C and D.

When the two load currents are equal, the SC neither charges nor discharges and this circuit characteristic allows the SC to be connected in parallel to any of the inputs of two LDO regulators as shown in Fig. 11.9E and F. More details about the circuit parameters such as charging and discharging periods, source current and voltage differences are available in a study by Subasinghage (2019).

11.4.4 Efficiency and losses

The major power and energy losses of the DO-SCALDO regulator are summarized in Table 11.1.

Considering all the losses defined in Table 11.1, the theoretical maximum end-to-end efficiency ($\eta_{\max}$) of the DO-SCALDO regulator is defined by Eq. (11.9).

$$\eta_{\max} = \frac{V_{o(\mathrm{LDO_P})} + \left|V_{o(\mathrm{LDO_N})}\right|}{V_P} \tag{11.9}$$

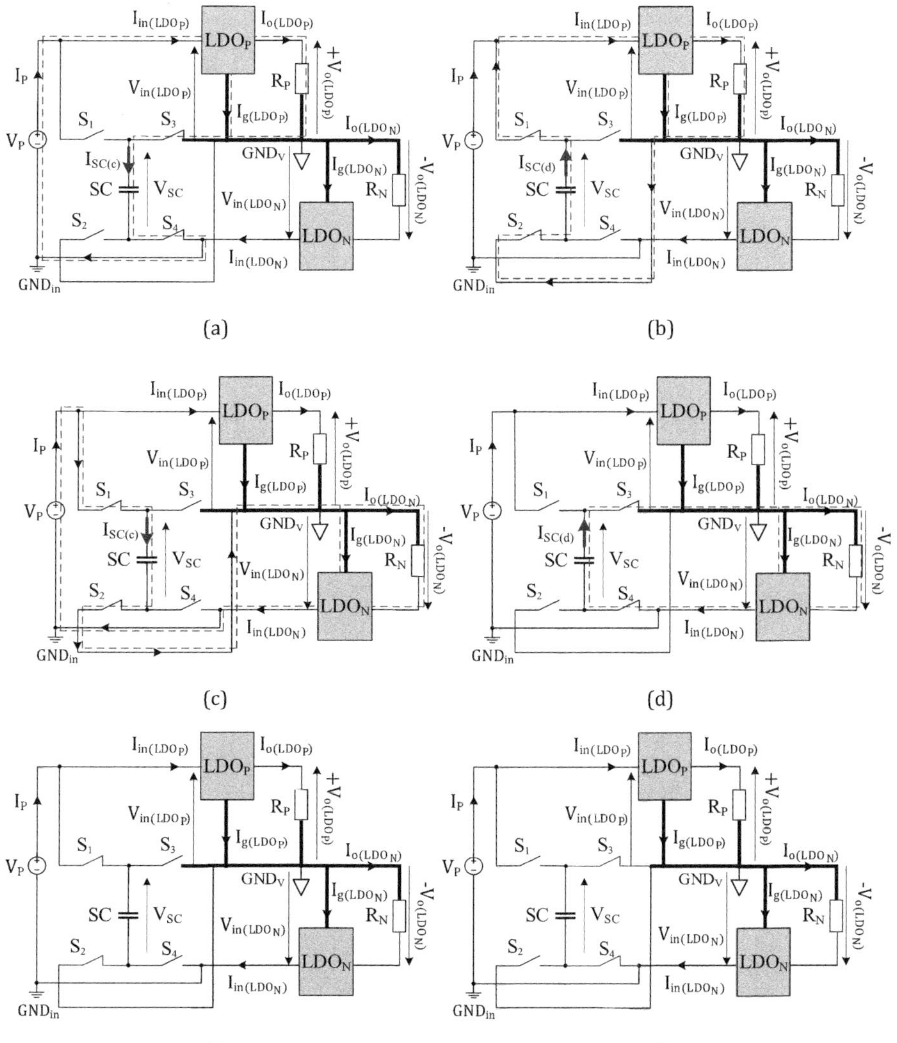

Fig. 11.9 DO-SCALDO operation under load imbalances: (A and B) Mode 2, (C) and (D) Mode 3, and (E and F) Mode 4.

11.4.5 Implementation aspects of the 12 to ±5 V DO-SCALDO prototype

The PCB and the informative schematic of the 12 to ±5 V DO-SCALDO proof-of-concept prototype are shown in Fig. 11.10 whereas the circuit parameters are provided in Table 11.2. The MCU is used to monitor the source and virtual ground voltages via two analog-to-digital (ADC) channels. The V_{min} value of the two LDO regulators is defined as 5.4 V in the converter MCU program which is 150 mV higher than that specified in the LDO datasheets. The control algorithm of the prototype is displayed in Fig. 11.11.

Table 11.1 Power and energy losses of the DO-SCALDO regulator.

Losses		Value	Remarks
1	The heat dissipation of the current limiting resistor (R_C)	$\frac{1}{2}C_{SC}V_{\min}^2$	This energy loss happens only once when Mode 1 is initiated
2	The ohmic losses due to the switches and the ESR of the SC	$I_{\text{diff}}^2(2R_{SW}+R_{ESR})$ where I_{diff} is the differential load current, R_{SW} is the resistance of a switch and R_{ESR} is the ESR of the SC	
3	The dynamic losses due to the switching transient events	≈ 0	Due to the very low switching frequency
4	The power consumption of the control circuit	$V_P I_{ct}$ where I_{ct} is the control circuit current	
5	The energy loss when the capacitors with different voltages are connected in parallel	$\frac{C_{in}\left(V_{SC}-V_{\text{in(cap)}}\right)^2}{T}$ where $V_{\text{in(cap)}}$ is the voltage of the input capacitor which comes in parallel with the SC at the start of a single switching cycle and T is the time of one switching cycle	Based on the following assumptions: • worst-case power estimation • input capacitors in both rails have the same capacitance (C_{in})
6	The power losses in the LDO regulators	$\approx \frac{V_P}{2}I_{g(\text{LDO}_\text{P})} + \frac{V_P}{2}I_{g(\text{LDO}_\text{N})} + \left(\frac{V_P}{2} - V_{o(\text{LDO}_\text{P})}\right)I_{o(\text{LDO}_\text{P})} + \left(\frac{V_P}{2} - V_{o(\text{LDO}_\text{N})}\right)I_{o(\text{LDO}_\text{N})}$	Depends on the quiescent current and the load current of the LDO regulators

In the case of integrated circuit (SCALDO IC) implementation, by using suitable voltage comparator stages and other analog circuit blocks combined, the requirement of a MCU can be removed.

11.4.6 Steady-state performance

Fig. 11.12 shows the resulting waveforms when the DO-SCALDO regulator is operating in the steady state. The source voltage is kept fixed at 12 V and the source current is measured using a current sensor of 100 mV/A. The input and output voltages of the

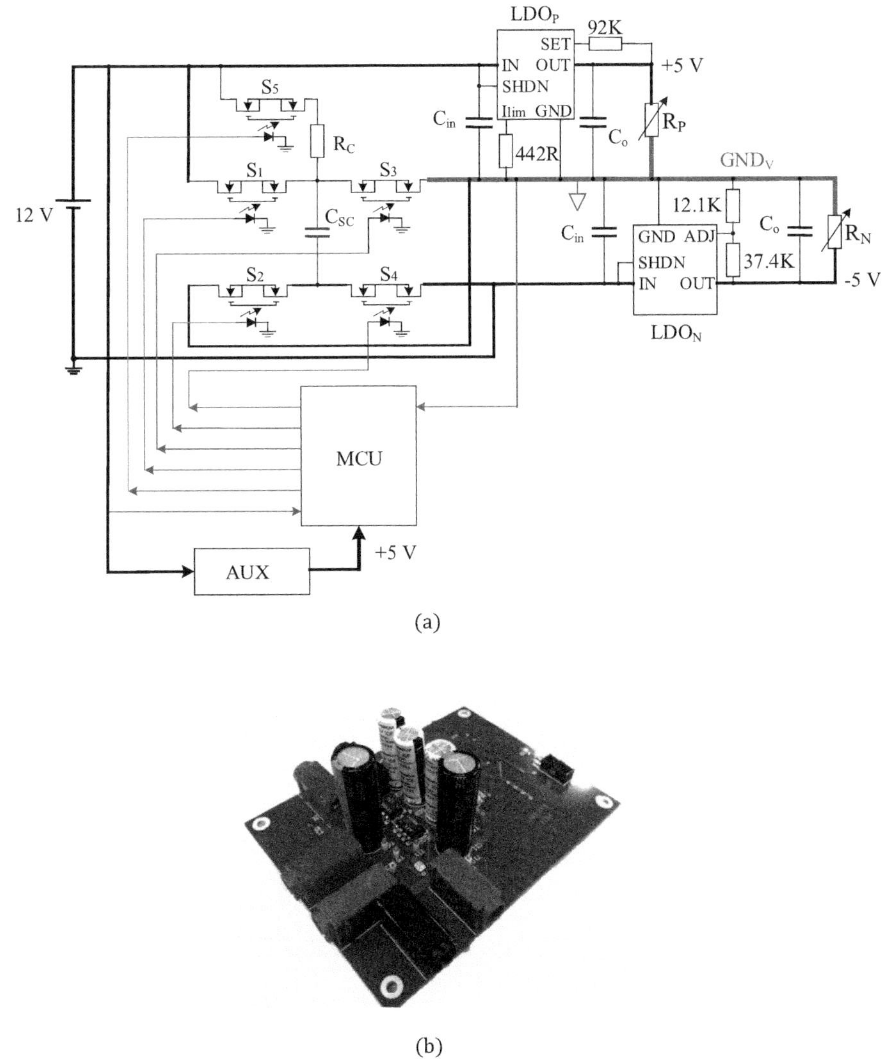

Fig. 11.10 The 12 to ±5 V DO-SCALDO proof-of-concept prototype: (A) schematic and (B) PCB design.

LDO regulators are measured with respect to GND_V. The input voltage of the negative-rail LDO regulator reflects the inverted GND_V. The key parameters for different levels of output currents are summarized in Table 11.3. Although the input voltages of the two LDO regulators change due to the charge balance of the SC switching network, their output voltages are regulated all the time.

Table 11.2 Circuit parameters of 12 to ±5 V DO-SCALDO prototype.

Component	Type	Properties
LDO_P	LT3086	Ground pin current (I_g) ≈ 5 mA (at 6 V and 1.5 A) Maximum output current = 1.5 A (Adjusted) Minimum input voltage = 5.4 V (MCU firmware defined)
LDO_N	LT3091	Ground pin current (I_g) ≈ 5 mA Maximum output current (Adjusted) = 1.5 A Minimum input voltage = −5.4 V (MCU firmware defined)
R_C (Charging resistor)	SMD Chip resistor	27 Ω (1%, 5 W)
C_{SC} (Three 10 F caps in series)	Electric double-layer capacitors (EDLC) (Manufacturer: NESSCAP)	C_{SC} = 3.3 F/8.1 V (±5%), ESR (R_{ESR}) = 90 mΩ
S_1, S_2, S_3, S_4, S_5	TLP3543	Photovoltaic relays On-resistance (R_{SW}) = 50 mΩ (each)
Control circuit auxiliary power	NCP1117	12–5 V linear regulator Quiescent current = 6 mA
Microcontroller (MCU)	ATMega16	Quiescent current = 25 mA
C_{in} (Two parallel input capacitors)	Electrolytic Ceramic	4700 μF/16 V (Buffer capacitor) 10 μF/16 V (Filter capacitor)
C_o (Output capacitor)	Ceramic	10 μF/16 V ESR = 400 mΩ at 1 kHz

11.4.7 Load-transient response

The load-transient response of individual LDO regulator is compared with the DO-SCALDO regulator. While disabling LDO_N, 6 V is applied to the input of LDO_P, and the load current is changed from 0 to 500 mA at a slew rate around 10 mA/μs to get the positive-rail regulator response. Similarly, −6 V is applied to LDO_N following the same procedure. This method keeps only the LDO regulator under test active while all the other parts of the circuit are isolated. On the other hand, the step load test is conducted when the SC switching is enabled to get the load-transient response of the DO-SCALDO regulator.

As shown in Fig. 11.13, the transient output voltage of LDO_P regulator is almost the same in all the three cases before and after applying the DO-SCALDO operation.

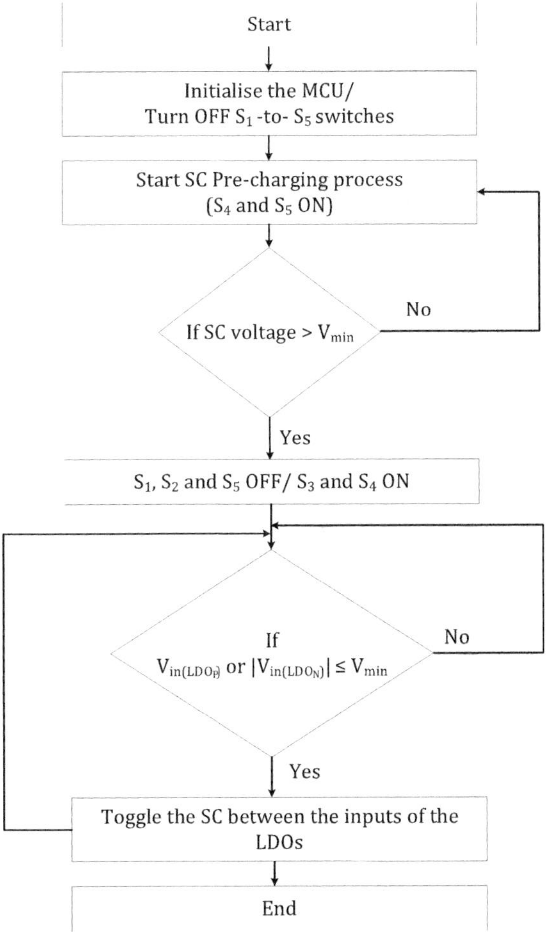

Fig. 11.11 The MCU algorithm of the DO-SCALDO prototype.

A similar result is achieved for negative-rail LDO regulator as shown in Fig. 11.14. Therefore, these step load results imply that the overall DO-SCALDO regulator produces the same quality voltage regulation as an LDO regulator for transient loads.

11.4.8 Power losses and efficiency estimation

Based on Table 11.1 and maximum differential load current (1.5 A) and rated input voltage (12 V), the power and energy losses of the DO-SCALDO prototype is calculated and presented in Table 11.4. The estimated total power loss is around 1.89 W if the loss contribution of the current-limiting resistor is neglected assuming that the regulator is run for a considerably longer period compared to the time duration of Mode 1.

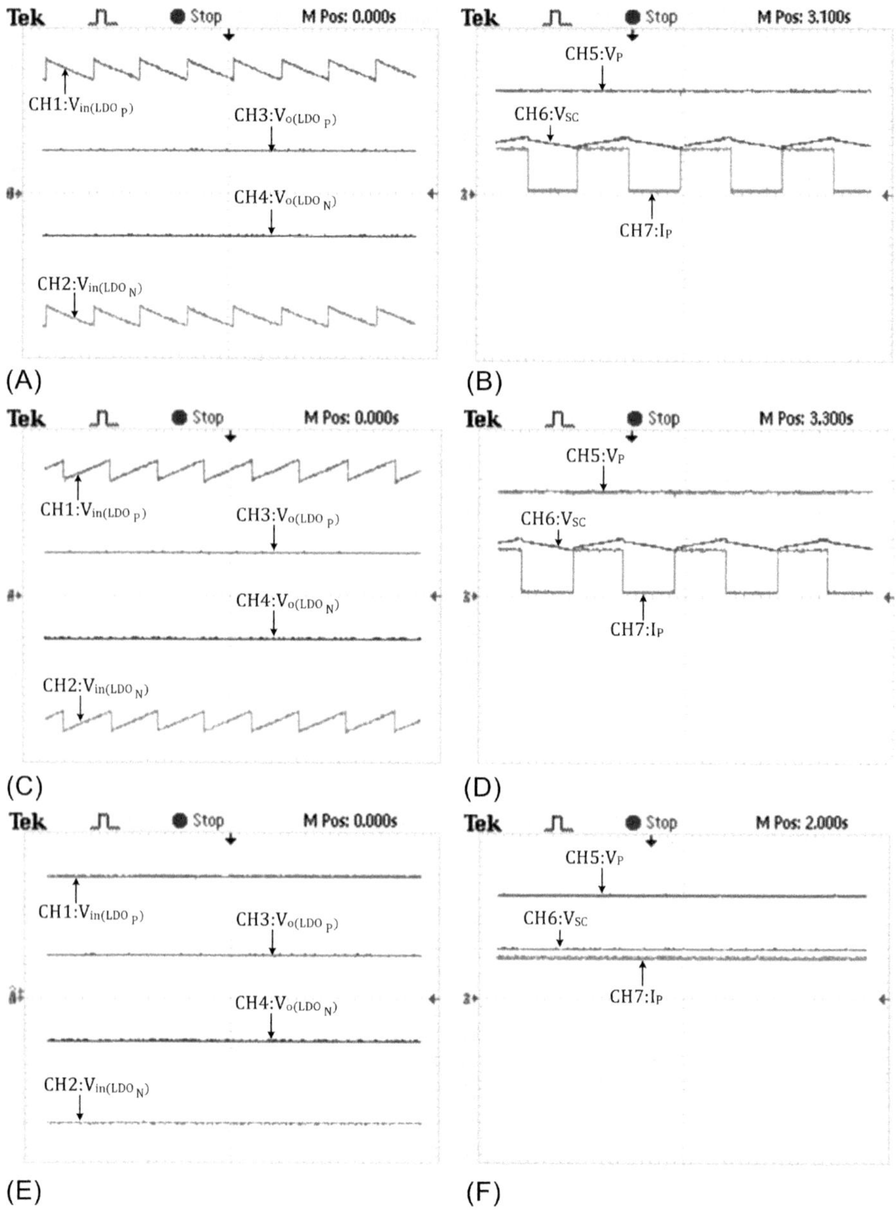

Fig. 11.12 Steady-state waveforms of the DO-SCALDO regulator: (A and B) when $I_{o(\mathrm{LDO_P})} = 1100\,\mathrm{mA}$, $I_{o(\mathrm{LDO_N})} = 100\,\mathrm{mA}$; (C and D) when $I_{o(\mathrm{LDO_P})} = 100\,\mathrm{mA}$, $I_{o(\mathrm{LDO_N})} = 1100\,\mathrm{mA}$; and (E and F) when $I_{o(\mathrm{LDO_P})} = 1000\,\mathrm{mA}$, $I_{o(\mathrm{LDO_N})} = 1000\,\mathrm{mA}$ (oscilloscope settings: CH1-CH2 (DC-coupling: 2 V/div), CH3-CH6(DC-coupling: 5 V/div), CH7(DC-coupling: 100 mV/div), time scale: 2.5 s/div).

Table 11.3 Summary of steady-state performance characteristics.

Parameter	Mode 2 $I_{o(LDO_P)} = 1100\,mA$ $I_{o(LDO_N)} = 100\,mA$		Mode 3 $I_{o(LDO_P)} = 100\,mA$ $I_{o(LDO_N)} = 1100\,mA$		Mode 4 $I_{o(LDO_P)} = 1000\,mA$ $I_{o(LDO_N)} = 1000\,mA$	
	Charge	**Discharge**	**Charge**	**Discharge**	**Charge**	**Discharge**
V_P (source voltage)	12 V	12 V	12 V	12 V	12 V	12 V
I_P (source current)	1130 mA	130 mA	1130 mA	130 mA	1030 mA	1030 mA
$T/2$ (half cycle)	2.8 s	2.8 s	2.8 s	2.8 s	$\ll$1 h	$\ll$1 h
$V_{in(LDO_P)}$	6.6 to 5.4 V		5.4 to 6.6 V		6 V	
$V_{o(LDO_P)}$	5 V		5 V		5 V	
$V_{in(LDO_N)}$	−5.4 to −6.6 V		−6.6 to −5.4 V		−6 V	
$V_{o(LDO_N)}$	−5 V		−5 V		−5 V	
I_{SC} (supercapacitor current)	1 A		1 A		0 A	

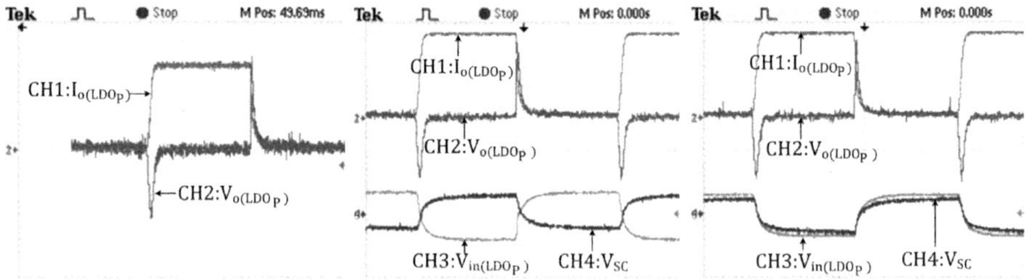

(A) (B) (C)

Fig. 11.13 Load-transient response of LDO_P: (A) stand-alone LDO_P regulator, (B) when running on DO-SCALDO method—SC charging phase, and (C) SC discharging phase (oscilloscope settings: CH1(DC-coupling: 20 mV/div), CH2(AC-coupling: 10 mV/div), CH3 (AC-coupling: 100 mV/div), CH4(AC-coupling: 50 mV/div), time scale: 250 μs/div).

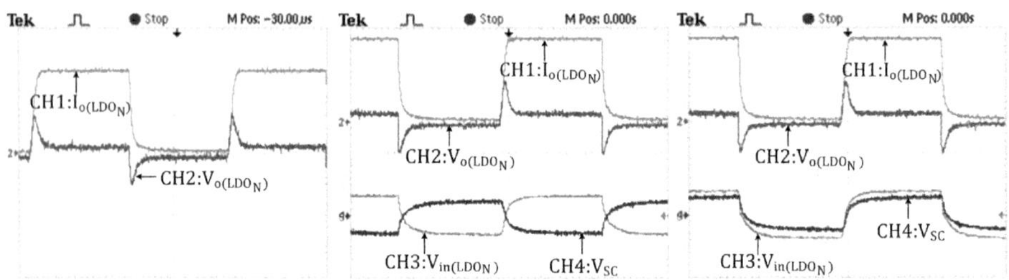

(A) (B) (C)

Fig. 11.14 Load-transient response of LDO_N: (A) stand-alone LDO_N regulator, (B) when running on DO-SCALDO technique—SC charging phase, and (C) SC discharging phase (oscilloscope settings: CH1(DC-coupling: 20 mV/div), CH2(AC-coupling: 10 mV/div), CH3 (AC-coupling: 100 mV/div), CH4(AC-coupling: 50 mV/div), time scale: 250 μs/div).

Table 11.4 Power and energy losses in the 12 to ±5 V DO-SCALDO regulator.

Loss factor	Description	Estimated value
1	The heat dissipation of the current limiting resistor	48.1 J
2	Ohmic losses due to the ESR of the SC and the switches	428 mW
3	Dynamic losses during the switching transients	≈0 W
4	Power consumption of the control unit	300 mW
5	Energy loss due to the paralleling of capacitors with different voltages	2.57 μW
6	Power losses in a single LDO regulators	
	(1) Due to the series-pass device	1.101 W
	(2) Due to the ground pin current	60 mW

Even though the maximum theoretical ETEE of the prototype should be 83.3%, according to Eq. (11.9), the measured efficiency is around 81% for 15 W output power when running at 12 V and driving 1.5 A load in each output. The ETEE against the load current of both rails is plotted in Fig. 11.15. As the control circuit power dominates at light loads, the ETEE becomes low. This issue can be minimized by using low quiescent current LDO regulators and replacing the MCU with low-power analog comparators and logic circuits designed to run the same switching algorithm.

The ETEE for ±5% deviation from the rated input voltage is plotted in Fig. 11.16. According to this diagram, the ETEE can reach up to 86.2% when the source voltage is reduced by 5% from the rated value. Similarly, the ETEE does not reduce lower than 78% if the source voltage is increased by 5%.

In summary, the DO-SCALDO regulator (12 to ±5 V version) runs at an end-to-end efficiency of around 81%–86%. The useful characteristics of the single-stage SCALDO regulator such as fast dynamic response, low noise, and low EMI/RFI are also maintained in this split-rail version making it applicable in a wide variety

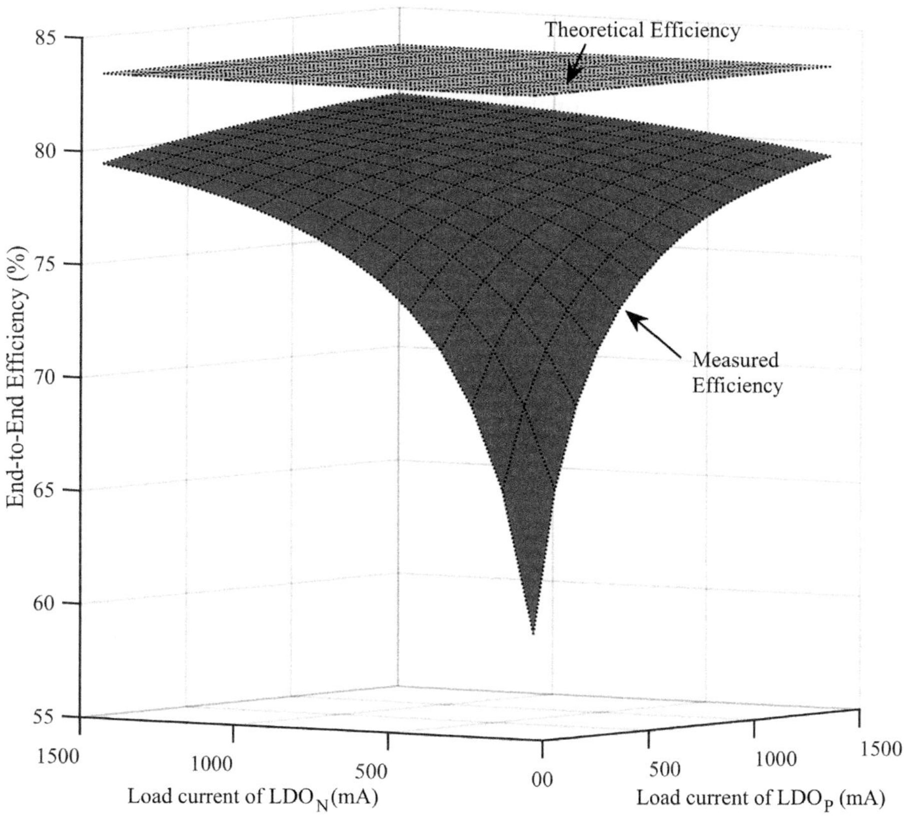

Fig. 11.15 Comparison of estimated and experimental end-to-end efficiency of 12 to ±5 V DO-SCALDO prototype.

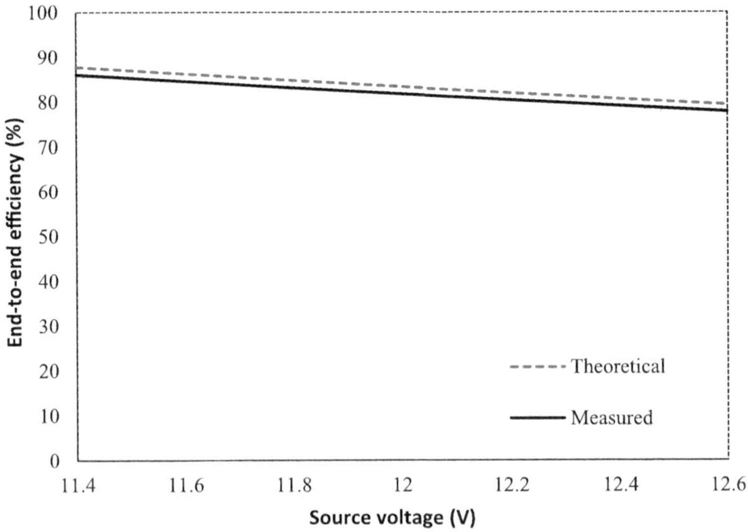

Fig. 11.16 End-to-end efficiency for ±5% deviation of the rated input voltage.

of noise-sensitive and high-current applications. In addition to the 12 to ±5 V proof-of-concept implementation, other common split-rail voltage combinations such as ±1.8 V, ±3 V can also be achieved due to the availability of adjustable voltage LDO regulators.

References

da Silva, E.R., dos Santos, W.R., Jacobina, C.B., Oliveira, A.C., 2011. Single-phase uninterruptible power system topology concepts: application to an universal active filter. In: 2011 IEEE Energy Conversion Congress and Exposition, pp. 3179–3185.

Fernandez, A., Sebastian, J., Martin-Ramos, J.A., Corral, J., Ruiz, F., 2002. Multiple output AC/DC converter with an internal DC UPS. In: 2002 IEEE 33rd Annual IEEE Power Electronics Specialists Conference. Proceedings (Cat. No.02CH37289). pp. 1095–1100 vol. 1093.

Gunawardane, K., Subasinghage, K., Kularatna, N., 2018. Efficiency enhanced linear DC-DC converter topology with integrated DC-UPS capability. In: 2018 IEEE International Conference on Industrial Technology (ICIT), pp. 712–717.

Hwang, J., Chen, J., Pan, J., Huang, Y., 2011. Measurement method for online battery early faults precaution in uninterrupted power supply system. IET Electr. Power Appl. 5 (3), 267–274.

Kwon, W., 2012. Design and modeling of highly power efficient node-level DC UPS. In: 2012 9th International Conference on Electrical Engineering/Electronics, Computer, Telecommunications and Information Technology, pp. 1–4.

Mondal, S., Keisling, E., 2012. Efficient data center design using novel modular DC UPS, server power supply with DC voltage and modular CDU cooling. In: 2012 IEEE International Conference on Power Electronics, Drives and Energy Systems (PEDES), pp. 1–6.

Pan, S., Wang, C., Jain, P., 2018. Structure and implementation of a hybrid 48V/380V DC UPS for IT datacenters. In: 2018 IEEE Energy Conversion Congress and Exposition (ECCE), pp. 3851–3856.

PCS100 UPS-I. https://new.abb.com/ups/systems/industrial-ups. (Accessed 18 December 2020). 2018. PCS100 UPS-I 150 kVA to 3000 kVA Industrial UPS.

Santos, W.R.N., Fernandes, E.d.M., Silva, E.R.C.d., Jacobina, C.B., Oliveira, A.C., Santos, P.M., 2016. Transformerless single-phase universal active filter with UPS features and reduced number of electronic power switches. IEEE Trans. Power Electron. 31 (6), 4111–4120.

Subasinghage, K., 2019. Advanced Analysis of Supercapacitor-Assisted Low-Dropout (SCALDO) Regulator. Auckland University of Technology.

Subasinghage, K., Gunawardane, K., Kularatna, N., Lie, T.T., 2019. Extending the supercapacitor-assisted low-dropout regulator (SCALDO) technique to a split-rail DC-DC converter application. IEEE Access 7, 124034–124047.

Wickramasinghe, T.C.P.K., 2016. Supercapacitor-Based Linear Converter for Voltage Regulator Modules. Electrical and Electronic Engineering. University of Waikato, Hamilton, New Zealand.

Wickramasinghe, T., Kularatna, N., Steyn-Ross, A., 2017. Reduced-switch SCALDO: an extra low-frequency DC-DC converter technique for VRM applications. IET Power Electron. 10 (15), 2180–2189.

Supercapacitor-assisted LED lighting technique and its applications in DC microgrids

12

Dilini Jayananda

12.1 Introduction to different lighting systems

Due to discoveries and advancements in new materials and lighting technologies, new generations of energy-efficient lighting devices have been developed over the past few decades. Even though there are various types of lamps available on the market, they differ in their operating principle, materials used, and importantly their energy efficiency. Different types of luminaires are being used in many day-today applications, such as outdoor lighting, indoor lighting, specialized lighting, etc.

Because of the rapid developments in materials and processing technologies, conventional lamps are disappearing and being replaced by newly developed solid-state lamps (Consumer NZ, 2013; Stardust, 2017; Kim et al., 2012). These lamps provide high efficiency and better quality than conventional lamps (Kim et al., 2012; DiodesIncorporated, n.d.). Fig. 12.1 presents the typical varieties of LED lamps and their applications. All new solid-state lighting are considered energy-efficient lighting families (Kim et al., 2012).

The rapid development of semiconductor technology has enabled the design of LED bulbs with lower energy consumption than the other available lighting technologies. LEDs have an extremely long life span, are energy efficient, and available in a variety of colors. Continuous development in this technology has enabled the improvement of LED lamps and their use in new applications. Colored LEDs are now commonly used commercially in exit signs and traffic signals, which can significantly reduce maintenance costs (Kim et al., 2012).

A comparison of energy savings and consumptions for common lighting techniques is provided in Fig. 12.2. In comparison to conventional incandescent lamps, the energy consumption of LED lamps is less than 20%, giving an energy saving of more than 80%. A detailed comparison of incandescent, fluorescent, and LED lighting technologies is provided in Table 12.1. Compared with compact fluorescent (CFL) or incandescent lamps, LED lamps have a longer life span, making them more reliable than other lighting technologies.

Color rendering index (CRI): This is a qualitative measure of the ability of a light source to reveal the color of an object in comparison with an ideal or natural light source (or standard reference light source). The CRI scale is from 0 to 100, with a

Energy Storage Devices for Renewable Energy-Based Systems. https://doi.org/10.1016/B978-0-12-820778-9.00006-1

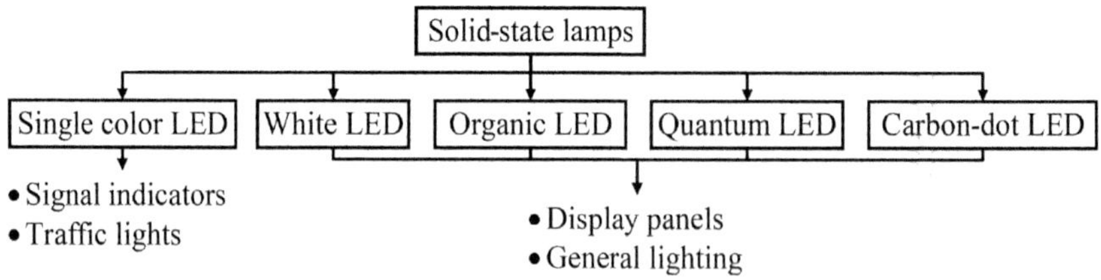

Fig. 12.1 Types of solid-state lightings and their applications.

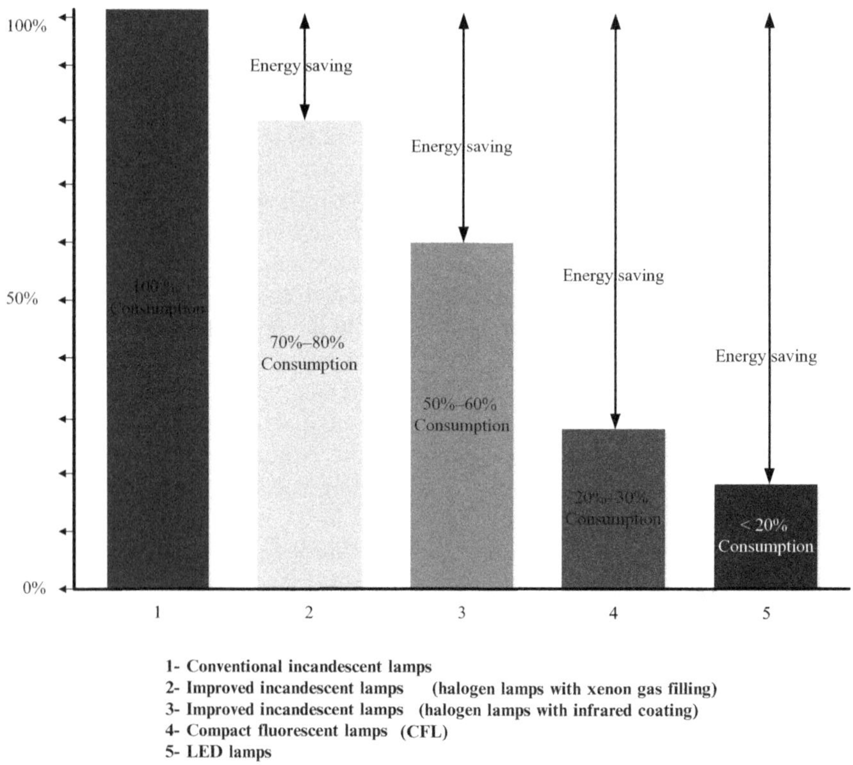

Fig. 12.2 Energy saving vs. consumption for common lighting technologies. Drawn from the data available from the European Commission 2009.

value of 100 indicating excellent color rendering. Typically sunlight and most incandescent lamps have a CRI of 100.

Color temperature: This is a way to compare the color of the light emitted from different types of lamps. Color temperature is measured in Kelvin (K) on a scale from 1,000 to 10,000. Colors with temperatures over 5,000 K are called cool colors (slightly blue), while color temperatures in the range of 2,700–3,000 K are known as warm (slightly orange) colors. For example, incandescent lamps give warm color temperatures, while some fluorescent lamps and sunlight emit cool color temperatures.

Table 12.1 Comparison of different lighting technologies.

Parameter	Incandescent	CFL	Halogen	LED
Initial cost	Low	Medium to low	Medium	Medium to high
Energy consumption	High	Low	High	Low
Operating temperature	High	Low	High	Low
Color rendering ability	High	Medium to high	High	Low to medium
Average rated lifetime (h)	750–1,500	6,000–10,000	2,000–4,000	Up to 100,000
Efficacy (lumen/W)	10–20	50–70	16–29	80–100+
Color rendering index (CRI)	100	80–90	100	65–90
Color temperature (K)	2,400–2,900	2,700–6,500	2,850–3,200	2,700–6,500

Typically, lamps with 2,000–6,500 K color temperatures are used for commercial and residential lighting applications.

Efficacy: This provides a measure of how much light output (in lumens) is delivered from a lamp per watt.

12.2 DC operation of LED units: 12 V and higher voltage DC operable flood lighting units

LEDs have been around for many decades but were restricted to indication lamps (red, green, orange, or yellow in color) and not used initially for general lighting. However, LED lamps have now become the newest and one of the fastest-developing energy-efficient lighting technologies in the category of general lighting. An LED lamp is a solid-state light-emitting source comprised of a substrate base, a $p-n$ junction diode, and a protective glass or plastic enclosure, which emits light through a phenomenon known as electroluminescence (Khan, 2013). Solid-state lighting means a semiconductor converts electricity to light without producing heat, usually within a small area of about 1mm^2 or less.

The structure of a single LED is much different from that of a regular semiconductor single diode. Single color light will be emitted from the LED when its $p-n$ junction is forward biased. Typically, the $p-n$ junction is covered by a transparent epoxy resin hemisphere, shaped in such a way that the photons emitted by the junction are reflected away from the surrounding substrate base and are focused through the top of the LED. The structure of an LED device is shown in Fig. 12.3.

In an LED lamp, there are multiple LEDs, and the light generated by each LED is projected in one direction, which creates the need to use diffusers or reflectors

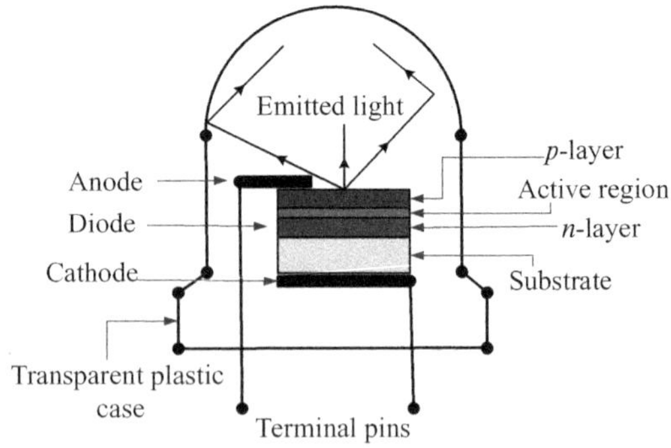

Fig. 12.3 Simple structure of an LED device.

(Matulka and Wood, 2013; Consumer NZ, 2013; Stardust, 2017). As most LED lamps are specified to last around 50,000 h versus 1,000 h for incandescent and 8,000 h for CFLs, low maintenance costs are involved for the replacement of lamps in large facilities. LEDs also give instant brightness without needing to warm up like conventional high-power lamps. In addition, they are more efficient, with very low energy consumption as shown in Fig. 12.2. Further, a reduction in CO_2 emissions by changing to LED alternatives is also a key driving factor for government environmental agencies. For example, a single 100 W incandescent bulb that is on for 4 h in a day produces 63 kg of carbon per year. Switching to a 12 W LED lamp, which has the equivalent light output to a 100 W incandescent, results in the emission of only 3 kg of carbon per year (Littelfuse, 2016). According to the data available for 2012, the United States saved about 675 million of dollars of annual energy costs by installing LED lamps (Matulka and Wood, 2013).

The savings achievable by switching from incandescent lamps to energy-efficient lighting technology are given in Table 12.2. This illustrates the value of the latest LED bulbs compared with CFLs and incandescent bulbs in terms of overall energy saving and cost-effectiveness.

Table 12.2 Cost comparison between different lighting technologies.

Parameter	Incandescent	CFL	LED
Wattage (equiv. 60 W incandescent) (W)	60	18	10
Cost per bulb (NZD)	1	10	11
Electricity per year (kWh/year) (5 h operation time per day, 10 bulbs per household)	1,095	329	183
Cost of electricity per year (at 0.2 per kWh) (NZD)	219	65.80	36.50
Cost for 10 bulbs (NZD)	10	100	110
Savings to household by switching to from incandescent (NZD)	0	63.20	82.50
Energy saving per year (kWh/year)	0	≈153	≈183

Note: Cost of electricity and bulbs vary, and bulb replacement cost is not considered in this comparison.

There are two categories of LED lamps commercially available categorized by their operating voltage as:

(a) Line voltage (230 or 120 V_{AC}) operable LED lamps
(b) Low-voltage LED lamps

12.2.1 Line voltage operable LED lamps

This type of LED lamp is the most commonly used in both commercial and residential buildings. These lamps can be directly connected to the AC mains. Line voltage operable LED lamps contain a power converter (AC/DC), a driver IC for the LEDs, and a heat sink for thermal management and optics to optimize the light quality. Since these lamps are intended to be factor-compatible with current incandescent and CFL bulbs, they contain an AC/DC converter circuit so they can operate from standard bulb "sockets." The construction of a typical residential LED lamp is shown in Fig. 12.4.

Since these lamps are directly connected to AC mains (e.g., 120/230 V AC), they can be damaged by lightning-induced surges and transient load switching (originating outside the bulb), can create voltage spikes or ring waves that can stress and damage the electronic components, and can ultimately destroy the LED lamp (Littelfuse, 2016). Therefore, it is crucial to use a transient voltage protection circuit within the package for reliability.

Fig. 12.5 provides a generalized schematic representation showing the various components and systems within the power and control circuits inside an LED lighting assembly. This figure specifically provides details on how to protect the LED array and the circuits, specially from transient surges, such as lightning induced or inductive dumps. While this schematic provides a wide set of protection schemes suggested by a surge absorber component provider (in this case Littelfuse Inc., US), in commercially available LED lamps, only a limited number of MOVs or TVS diodes may be used due to reasons of cost. As shown in Fig. 12.5, an AC fuse is used in series with the line to

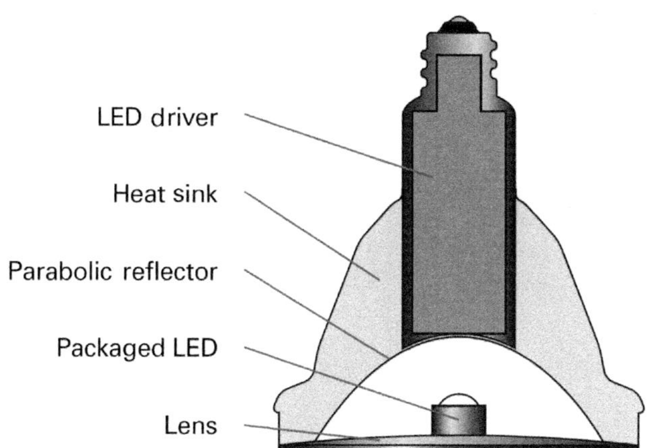

Fig. 12.4 Typical residential LED lamp construction (Littelfuse, 2016).

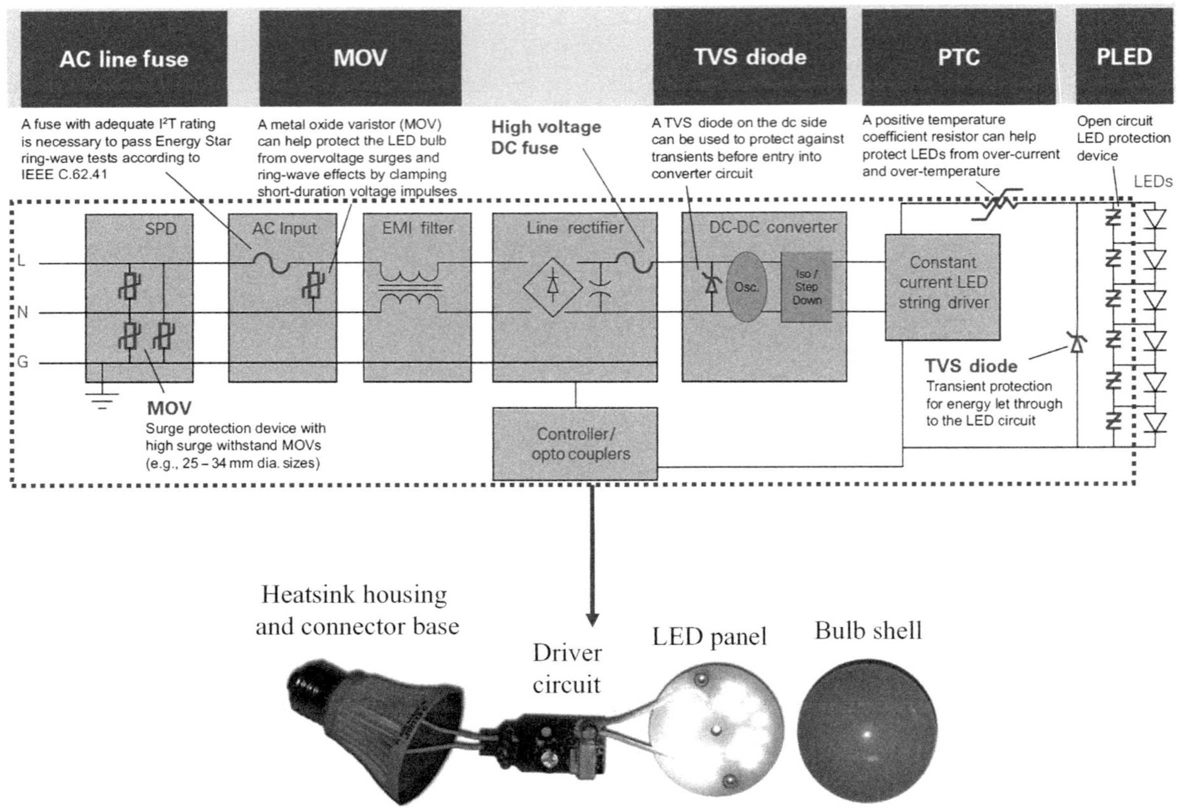

Fig. 12.5 Generalized line voltage operable LED luminaire driver circuit with transient and surge protection devices (Littelfuse, 2016).

protect against short circuit and overload conditions. Either isolated or nonisolated topologies can be used as power converters. A transient voltage suppression diode is used on the DC side as the transient protection circuitry. In the case of nonisolated LED drivers, which are more prominent in many retrofit lamps and low power applications, the LED array itself can be damaged by the surge energy as there is no transformer isolation between the AC input and the LEDs on the DC side (Littelfuse, 2016). Isolated LED drivers are more robust since the transformer in the DC-DC converter stage provides a significant degree of isolation and surge protection (Littelfuse, 2016). Nevertheless, surge protection is still required to limit higher surge events, especially in outdoor lamp and luminaire environments, because in the case of high-energy surge events, such as those caused by nearby lightning strikes, there can be a flashover from the primary to the secondary side of the transformer. A metal oxide varistor (MOV) across the AC line, as shown in Fig. 12.5, is used in the lamp to protect against lighting-related transients. For a detailed discussion on surge protection, please see Bibiliography at the end of the chapter.

Line operable LED light bulbs are available in different base types. The typical plug (E27) and pin (B22) are two common kinds of LED bulb bases. These are respectively called Edison Screw and the Bayonet Cap types. There are several other kinds of screw-in bases for LED light bulbs, which are categorized as candelabra and intermediate. It is important to choose a base type appropriate to the intended application.

12.2.2 Low-voltage LED luminaires

Tungsten-halogen and standard incandescent are the lamps of choice in low-voltage lighting systems. They typically operate from 12 to 24 V DC or AC, using step-down transformers (DiodesIncorporated, n.d.). This transformer is usually located within the fixture. However, it may also be remotely mounted, depending on the application. These luminaires are used in many retail and consumer lighting applications. Common applications include residential and commercial decorative/display lighting, task lighting, and retail product highlighting (Stardust, 2017; Mccallum, 2014; DiodesIncorporated, n.d.). Other special applications include swimming pool, fountain illumination, and landscape/garden/kitchen lighting. However, low efficiency, heat generation, and halogen capsule handling issues are among the disadvantages of using tungsten-halogen and incandescent lamps in low-voltage lighting applications.

Low-voltage LED lamps are a more energy-efficient and better solution to replace low-voltage tungsten-halogen and incandescent lamps, because these low-voltage LED luminaires are designed to retrofit to the same base connector type MR 16 used for tungsten-halogen and incandescent bulbs. Therefore, upgrading them to new LED lamps presents no issues.

The internal circuitry of a low-voltage LED lamp is shown in Fig. 12.6. The LED array needs DC power for the operation. Hence, an appropriate circuit, such as an LED driver, is required to convert AC voltage to DC voltage as shown in Fig. 12.6. Therefore, LED drivers are essential components of LED lamps. These drivers can be put inside the luminaire, which is called a built-in type (or integral) or be put outside, which is called an independent type (or remote). According to different applications, different types of LED drivers need to be applied; for example, an outdoor driver for a

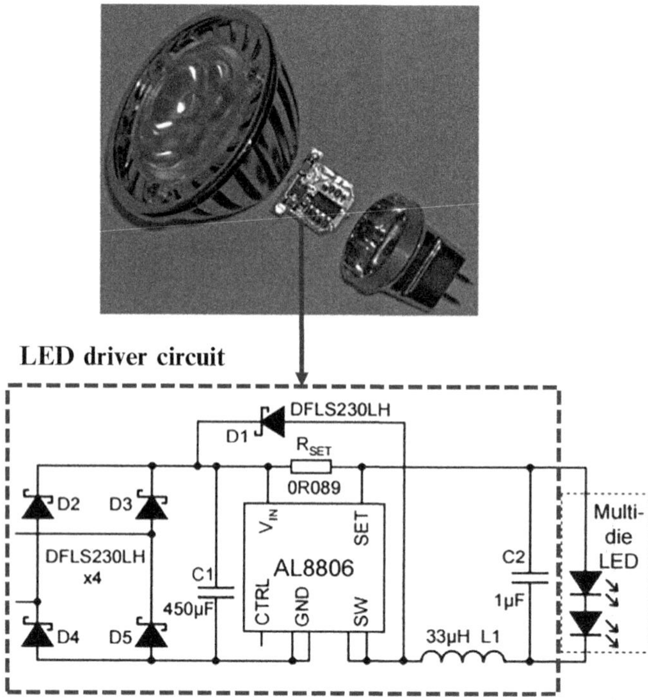

Fig. 12.6 Typical low-voltage operable LED luminaire internal circuitry (DiodesIncorporated, n.d.).

street light, an indoor point driver for a down-light, and an indoor linear driver for a panel light. It should be noted here that different LED lamp manufacturers use various LED driver circuits to drive a multi-die LED array. These low-voltage LED lamps also include protection circuits, as in-line voltage operable LED lamps. These low-voltage lamps also come in different wattages and color temperatures.

12.2.3 Practical operational region of 12 V LED luminaries

Commercially available low-voltage LED luminaries do not require the transformer to supply the exact nominal value of the voltage specified. In a simple laboratory test, as per Fig. 12.7, together with a low cost light output meter (such as EXTECH HD 450) an approximate measurement can be carried out to establish this range, in a dark room. If an exact brightness measurement versus AC/DC input voltage is to be carried out, a proper lamp testing apparatus will be needed.

Simple test setup shown in Fig. 12.7 requires only two multimeters and a DC power supply of 0–30 V DC for a case of 12 V LED.

12.2.4 Current-voltage characteristics

Based on the test set up of Fig. 12.7, several LED luminaries were tests for their constant brightness region as per Fig. 12.8. Fig. 12.8 shows the *I–V* characteristics curve,

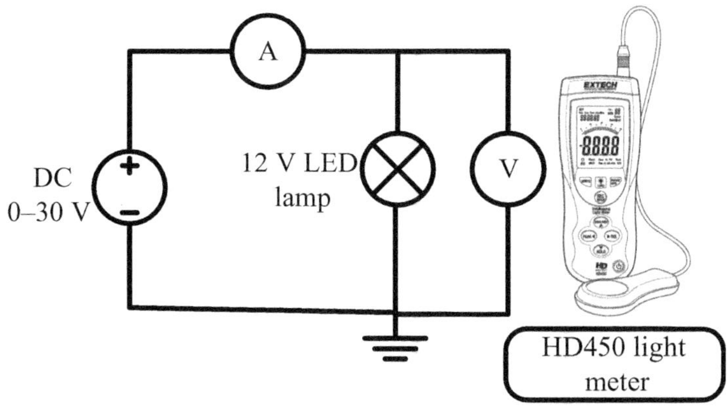

Fig. 12.7 Test setup used for the experiment.

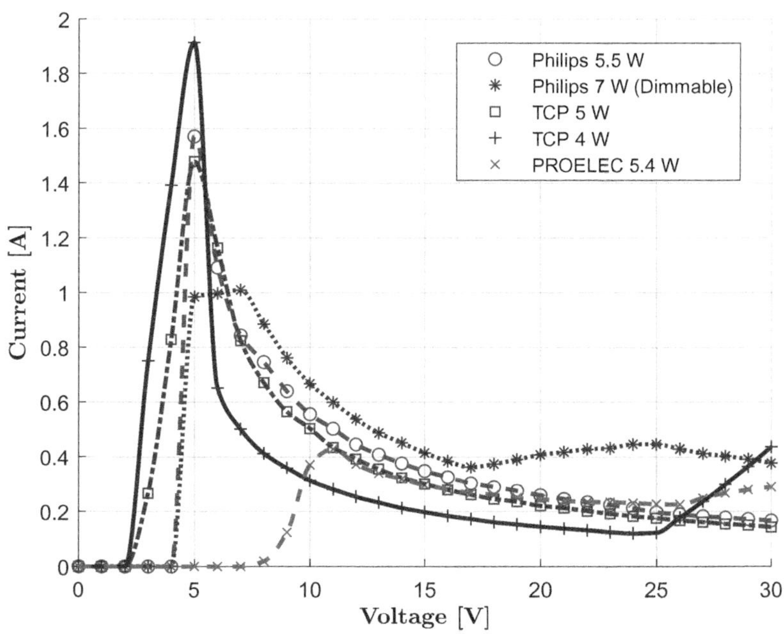

Fig. 12.8 Current-voltage characteristics for five commercially available LED lamps.

in which there is an approximately constant current region after initially drawing high current. This current is limited by the driver circuit, which avoids damage to the LED string due to excess power dissipation (Jayananda, 2020).

As per Fig. 12.8, LED lamps draw more current when the voltage across the LED lamp is reduced. With the exception of one LED lamp, all other LED lamps drew the highest current around 5 V. This maximum current depends on the wattage rating of the LED lamp and the driver circuit. For example, for the TCP 4 W and Philips 5.5 W

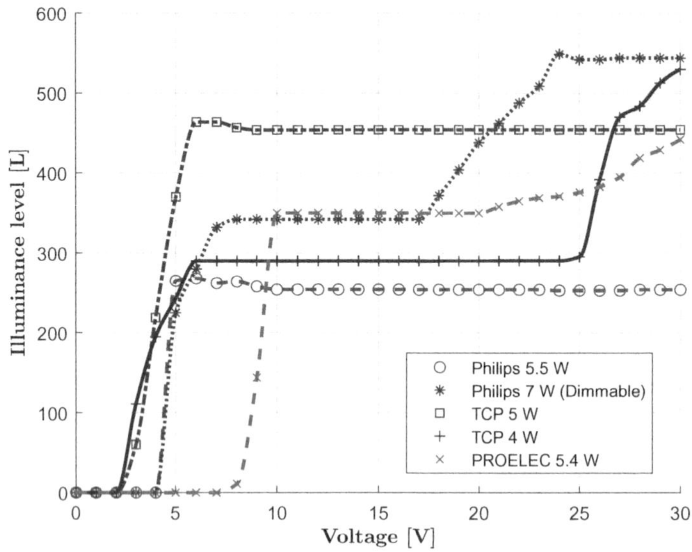

Fig. 12.9 Variation of illuminance with voltage for five commercially available LED lamps.

lamps, the highest current was 1.9 and 1.6 A, respectively. This confirms that each manufacturer uses different LED drivers inside their LED lamps. Fig. 12.8 shows that each brand has a different voltage, in which the lamp starts to draw current. From these current-voltage characteristics, it can be observed that basically, an LED lamp does not act as a constant resistive load.

12.2.5 Luminance-voltage characteristics

Fig. 12.9 shows the luminance versus the voltage curve, in which there is a constant brightness region. As the applied voltage increases (from much lower values than the nominal voltage), the lamp brightness increases proportionally until it reaches a certain voltage and then settles down to a constant brightness, regardless of any further increase in voltage. For the Philips 7 W dimmable LED lamp, there were two different constant brightness regions as shown in Fig. 12.9. One region is from 8 to 17 V and other region starts at 25 V. Both the TCP 5 W and Philips 5.5 W LED lamp showed constant brightness in the 9–30 V voltage region. However, the design process of supercapacitor-assisted LED (SCALED) converter detailed later only considered the 9–20 V region because higher voltages can reduce the lifetime of an LED lamp.

12.2.6 Power-voltage characteristics

Power dissipation in the tested LED lamps was calculated using the current-voltage measurements recorded during the experiment. The power-voltage characteristics for each lamp are shown in Fig. 12.10. The graph indicates that every LED lamp dissipated approximately constant power in the constant brightness region. Therefore,

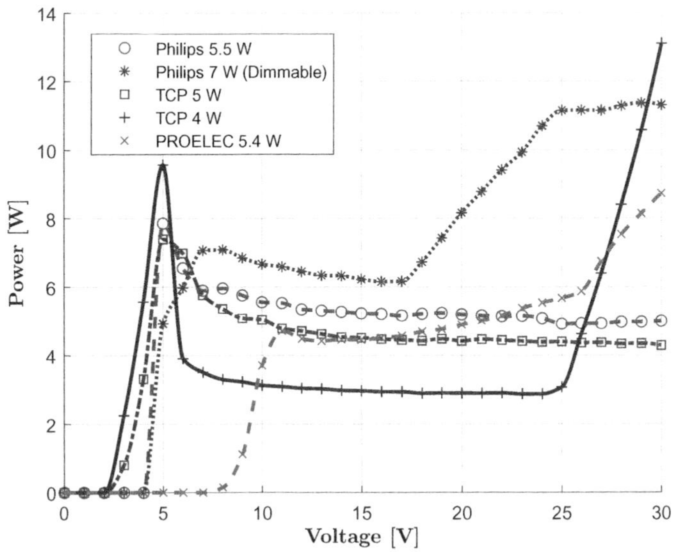

Fig. 12.10 Power-voltage characteristics for five commercially available LED lamps.

instead of considering an LED lamp as a constant resistive device, it is possible to consider it as a constant power device for a specific region of voltage.

Fig. 12.11 shows the DC voltage reverse capability for the Philips 5.5 W LED lamp. All five lamps were tested for DC reverse voltage capability. However, since Philips 5.5 W LED lamps were used in the development of SCALED, this capability in the Philips brand is shown here. From Fig. 12.11, it can be seen that some LED lamps can operate with either positive or negative DC voltages without compromising the brightness. The test results are based on our own laboratory measurements since this data are not provided by the lamp manufacturers.

In summary, commercially available 12-V LED lamps have of the following desirable features:

(i) ability to operate with positive or negative DC voltages;
(ii) ability to operate over a wide range of AC/DC voltages without compromising brightness; and
(iii) ability to operate as a constant power load in the constant brightness region.

These three features were used in developing a SCALED technique useful in DC-microgrid (DCMG) environments. Fig. 12.11 depicts the voltage reversal capability of the Phillips 5.5 W LED lamp, which was also used to reduce the switch counts while developing the SCALED converter.

12.2.7 Flood lights (high-bay light fixtures)

This kind of lighting is the go-solution in many industrial places to achieve bright and uniform illumination across large, such as high-ceilinged indoor spaces. These

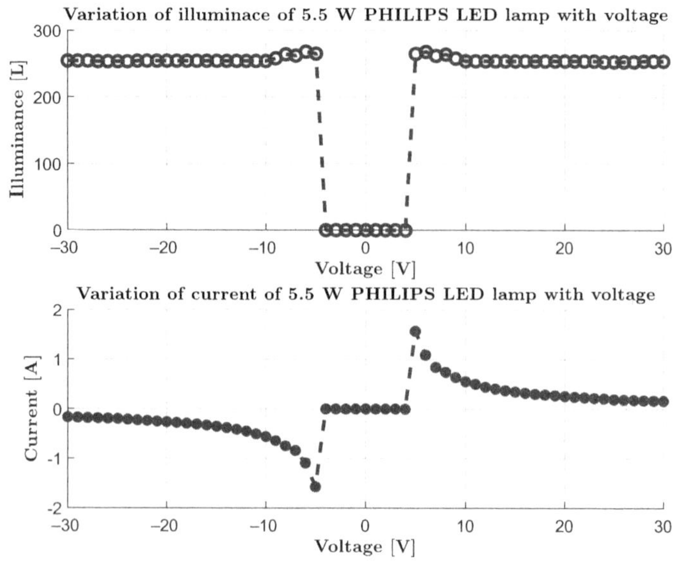

Fig. 12.11 DC voltage reversal capability of Philips 5.5 W LED lamps.

luminaries are most often found in warehouses, department stores, manufacturing areas, and sports halls. Rather than any single fixture style, the term "high-bay lighting" refers to the positioning of the luminaries, and as such it covers a broad and flexible range of lamp types and fittings. As a rule of thumb, high-bay lighting is usually the best option for any indoor setting where the floor-to-ceiling height measures 8 m or taller. Different types of high-bay lamps are available in the market depending on the wattage, voltage and lighting technology. Each has its own unique benefits and potential limitations depending on the demands of the space and types of the work being carried out. However, we pay more attention toward LED high-bay lighting.

By using LED high-bay lamps, it is possible to improve the energy efficiency vastly, resulting in anywhere from 50% to 80% lower bills (depending on the specifics of setup and intensity of use), and the lifespan of an LED bulb is typically many times longer than that of metal halide or fluorescent lamps.

High-bay LED fittings illuminate to full intensity instantly, rather than having to "warm-up" like many other lamp types. They are also suitable for use with microwave sensors without degrading the lifespan of the lamp, which allows for true wide-aspect motion detection over large areas.

Both AC and DC operable LED flood lamps are commercially available. However, for the testing we used DC operable LED flood lamps. In order to verify that the SCALED concept could be extended to flood lamps, a sample of 110 and 200 V DC rated lights were tested separately. For these tests, we used a variable AC (variac) power supply and a diode bridge to get the required DC voltage levels. During the testing, the variation in illuminance level with the applied voltage was recorded and is shown in Fig. 12.12. When testing 110 V flood lamp, it was not possible to

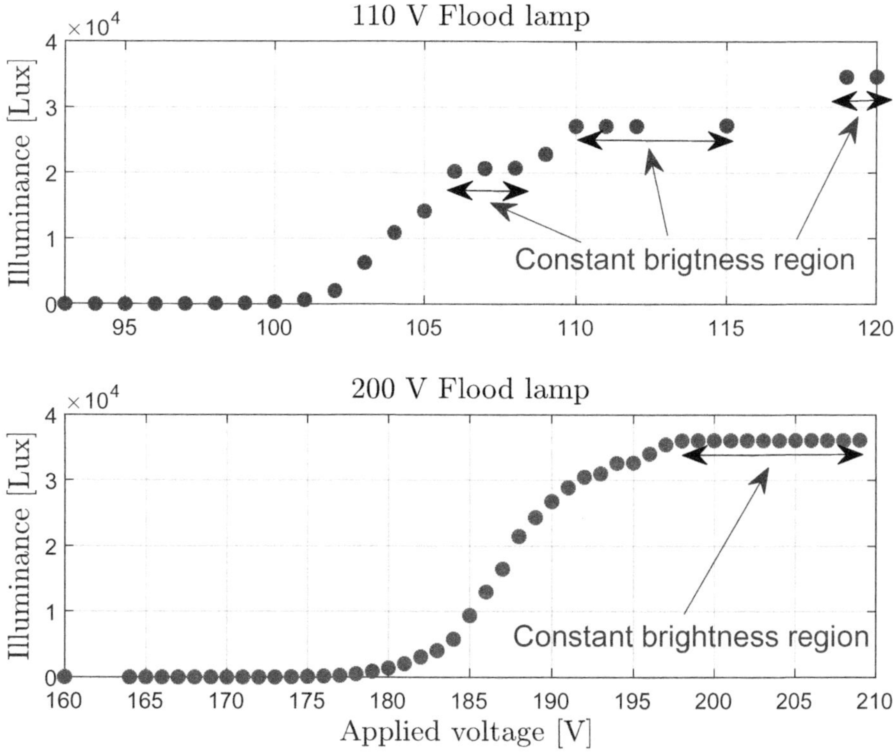

Fig. 12.12 Variation of brightness in flood lamps with applied voltage.

record the lamp brightness values for the voltages from 116 to 119 V, due to sudden spikes occurs in the LED current. This situation causes the variac to jump from 115 to 120 V. It can be seen from the top graph in Fig. 12.12 that the 110 V flood lamp has approximately constant brightness regions in three different voltage ranges:

1. Region 1: from 105 to 108 V
2. Region 2: from 110 to 115 V
3. Region 3: from 118 to 120 V

However, since these three regions have different brightness levels an operating voltage range can be selected depending on the brightness requirement. Therefore, the SCALED concept can easily be applied to 110 V flood lamp. However, although this seems easy, completely new design aspects will need to be considered during hardware design because 110 V rated flood lamps have three constant brightness regions. This application may need to use different switching arrangements to the one used in the SCALED converter.

Conversely, when considering the lower graph in Fig. 12.12, the 200 V flood lamp has only one constant brightness region. Therefore, the SCALED concept can be directly applied to 200 V flood lamps by adequately sizing the SC banks and selecting switches with appropriate ratings.

Accordingly, this section establishes that the SCALED technique is not only valid for 12 V indoor lighting, but also can be extended to outdoor lighting applications and the indoor lamps with various voltage and power ratings. In the next part of the design process and the related background information for the SCALED technique is presented.

12.3 Supercapacitors (SCs) for short-term DC-UPS capability to overcome solar energy fluctuations

Although power generation using solar energy is economical, we need to cope with the frequent fluctuations in irradiance and temperature and nonavailability of solar energy overnight. Therefore, in order to have a reliable operation, we need to use energy storage device (ESD) which can supply the load during fluctuating conditions. Currently, most common energy storage used in solar power based systems is rechargeable batteries, typically lead acid or lithium ion. These batteries act as both short-term and long-term energy storage supplying the load either in case of short-duration or long-duration fluctuations. However, all the rechargeable batteries have limited charge-discharge cycle as well as calendar life and are environmental unfriendly.

Nevertheless, with the continuous development of SC technology over the last decade, it is now possible to adopt SCs as short-term ESD in solar-powered DCMGs to replace the electrochemical battery packs. When SCs are to be used in solar power-based DCMG environment, there is a possibility of directly operating "white goods" (such as washing machines, refrigerators, dishwashers, and air conditioners) and lighting. With the proper sizing of the SC banks to be used in the DCMG, it is possible to overcome the short duration irradiance and temperature fluctuations.

12.4 Replacing battery banks with supercapacitors: Issue of MPPT implementation

Despite all the advantages related to energy generation using PV technology, the conversion efficiency is still at very low value in the range of 9%–20%. Also a PV module inherits nonlinear characteristics as shown in Fig. 12.13, which varies with temperature and solar irradiation level. According to Fig. 12.13A, typical characteristics of a solar panel show its near constant current region (A–B), the narrow maximum power point (MPP), area of operation (B–C), and the typical voltage-source type operational region (C–D). It is vital to operate at the MPP in order to achieve the maximum efficiency for the overall system (includes PV array, converters, etc.). Also, it is obvious from Fig. 12.13D power generated by the PV module varies with irradiance and temperature. Thus, it becomes necessary to use a technique in order to extract maximum power from the PV panels and deliver it to the load connected. Several methods to extract maximum power have been proposed in the literature (Babaa et al., 2014; Dolara et al., 2009; de Brito et al., 2011a, b; Adly et al., 2012; Subudhi and Pradhan, 2013; De Brito et al., 2013). Each of these tracking techniques has different

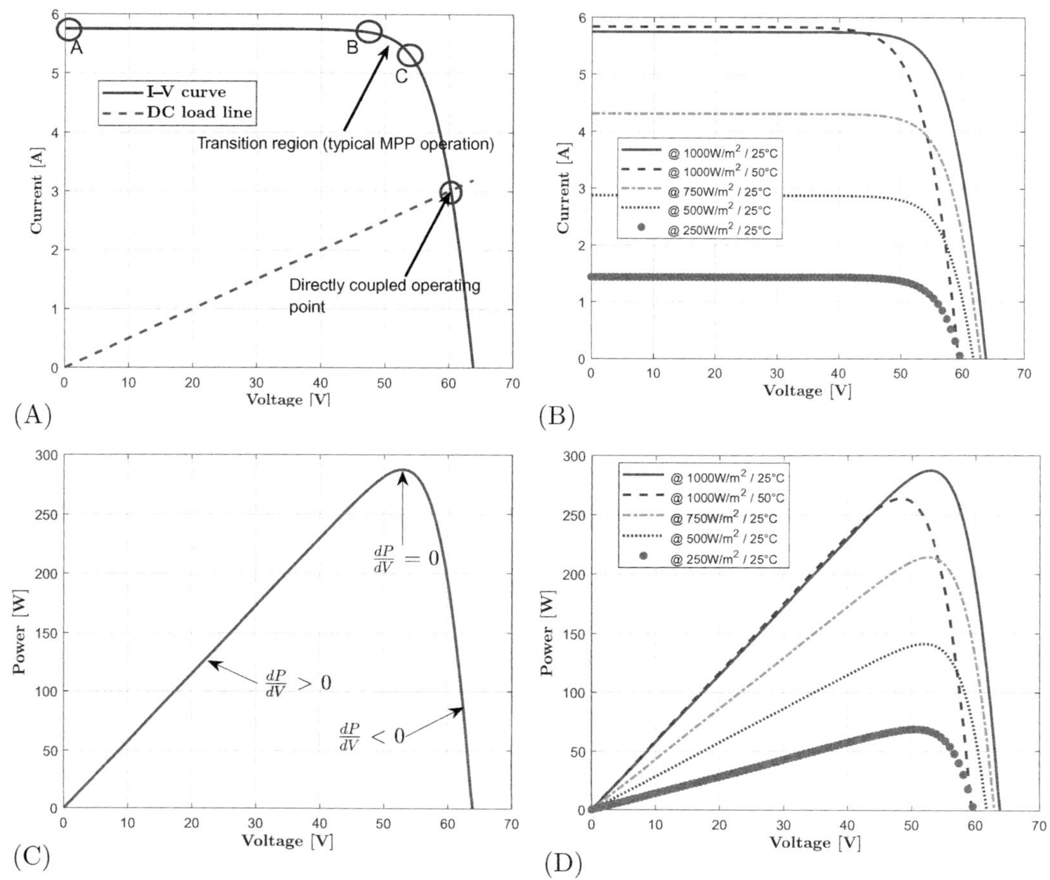

Fig. 12.13 Characteristic curves of solar PV panel: (A) PV current versus voltage characteristics with directly coupled operating point; (B) current-voltage variations with irradiance and temperature; (C) power versus voltage curve; and (D) power-voltage variation with irradiance and temperature.

efficiencies (De Brito et al., 2013). Theoretically, maximum possible load power occurs when the source resistance matches the load resistance (Babaa et al., 2014). This is achieved by changing the duty cycle of the switch-mode DC-DC converter in the maximum power point tracking (MPPT) charge controller.

If a PV array is directly connected to a load (i.e., direct-coupled systems), the operating point of the overall system will be at the intersection of DC load line and $I-V$ curve of PV array, as shown in Fig. 12.13A. This operating point is not the MPP for PV array; therefore, it is not possible to extract the maximum power out from the PV array. Thus, for a direct-coupled system, it is must to use oversized PV array in order to ensure that load's power requirements can be matched.

12.4.1 Summary of existing MPPT techniques

This fundamental theory of maximum power transfer is used in every technique or algorithm used for MPPT controllers. That is the impedance of the PV system must match that of the load in order to deliver maximum power to the connected load.

Commercially available main MPPT techniques are as follows.

12.4.1.1 Fixed duty cycle

This is the simplest method available and does not require any feedback. Load impedance is adjusted to match with PV module resistance only once to extract maximum power from the panel, being classified as off-line method. Variation in MPP due to the temperature and irradiance is disregarded in this method (Babaa et al., 2014).

12.4.1.2 Constant voltage (CV) method

In here it is assumed that $V_{MPP} = kV_{OC}$ where V_{MPP} is MPP voltage, $0.7 \leq k \leq 0.8$ and V_{OC} is the open circuit voltage of the PV module under standard test conditions (1,000 W m^{-2}, 25°C). Since value of k is based on an approximation, it is not always giving the proper MPP. This results in poor performance as the linear relationship is unable to capture the complexities of shading. Also the open circuit voltage of PV module varies a little with the solar radiation, and could significantly change due to temperature variations. So, this method is good for regions where the temperature does not vary widely (De Brito et al., 2013). Flowchart for this method is given in Fig. 12.14.

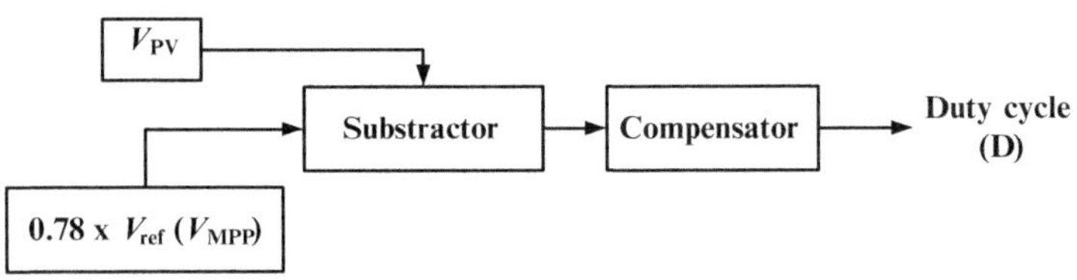

Fig. 12.14 Flowchart for constant voltage (CV) method.

12.4.1.3 MPP locus characterization

The objective of this method is to establish a linear relationship between the current of the PV array and the voltage at MPP. Eq. (12.1) is used in this method. Basically this relationship is given by the tangent line drawn to MPP locus curve for the PV current in which minimum irradiation condition satisfies the sensitivity of this method. The mathematical derivation for this method is discussed in De Brito et al. (2013). As the MPP locus varies with the irradiance and temperature, continuous update of the model for the tangent line is required. This is achieved by measuring the V_{OC} of the array periodically and plug in measured value to the following equation. It will require to disconnect the PV module from the loads during these measurements which will result power loss for the loads (De Brito et al., 2013).

$$T_L = \left(\frac{\eta V_T}{I_{MPP}} - NR_S\right) I_{MPP} + (V_{OC} - \eta[VD_o + V_T]) \tag{12.1}$$

where T_L is tangent line, N is number of cells, R_S is the series resistance of the PV module, V_{OC} is the open circuit voltage, I_{MPP} is current at the MPP, V_T is the thermal voltage, and $V\,D_o$ is differential voltage (De Brito et al., 2013).

12.4.1.4 Perturb and observe (P&O)

According to Fig. 12.13C, it is obvious that power produced by the PV module increases with the panel voltage before MPP and decreases after reaching MPP with the increment of voltage. The algorithm compares power of the previous step ($P[k-1]$) with that for the next step ($P[k]$). So that it can increase or decrease the reference voltage (V_{ref}) of the switched DC-DC converter, that is, if the current operating point is left to the MPP, the increment of voltage will result in the increment of power. Therefore, further voltage perturbation (increasing V_{ref}) is needed to reach the MPP. If current operating point is right to the MPP, the increment of voltage will result in decrement of power. Hence, reduction of V_{ref} is needed to reach the MPP. Therefore, the operating point is aligned toward the MPP by either increasing or decreasing the PV array voltage (De Brito et al., 2013). The flowchart representation for this method is shown in Fig. 12.15.

12.4.1.5 Incremental conductance (IC)

This method was proposed to overcome some limitations of P&O method. This is based on the fact that slope of the $P-V$ characteristics curve is zero at MPP $\left(\frac{dP}{dV}=0\right)$, positive in the left, and negative in the right as shown in Fig. 12.13C. Therefore, the MPP is related to the incremental change in the conductance of the PV array, and by manipulating the following equations, peak power delivery point can be located (De Brito et al., 2013). Flowchart for the algorithm used in this method is represented in Fig. 12.16.

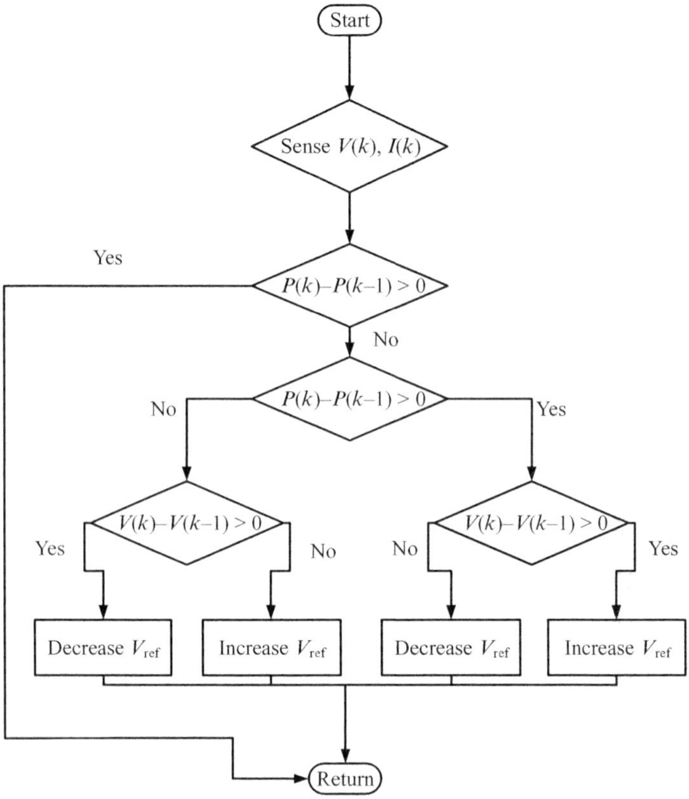

Fig. 12.15 Flowchart for perturb and observe (P&O) method.

$$\frac{dp}{dv} = \frac{d(vi)}{dv} = i + v\frac{di}{dv} = 0 \tag{12.2}$$

$$\frac{di}{dv} = -\frac{i}{v} \tag{12.3}$$

Based on the results of the comparison given below, the algorithm takes decision to either increase or decrease the voltage across the PV array by changing the V_{ref} of given to the switched mode DC-DC converter.

(a) $\Delta v/\Delta i = -i/v$: operating point lies with MPP;
(b) $\Delta v/\Delta i > -i/v$: operating point lies to the left of the MPP; and
(c) $\Delta v/\Delta i < -i/v$: operating point lies to the right of the MPP.

12.4.1.6 Beta method

The beta method provides an approximation of the MPP using an intermediate variable β (De Brito et al., 2013). The following equation is used to calculate this variable

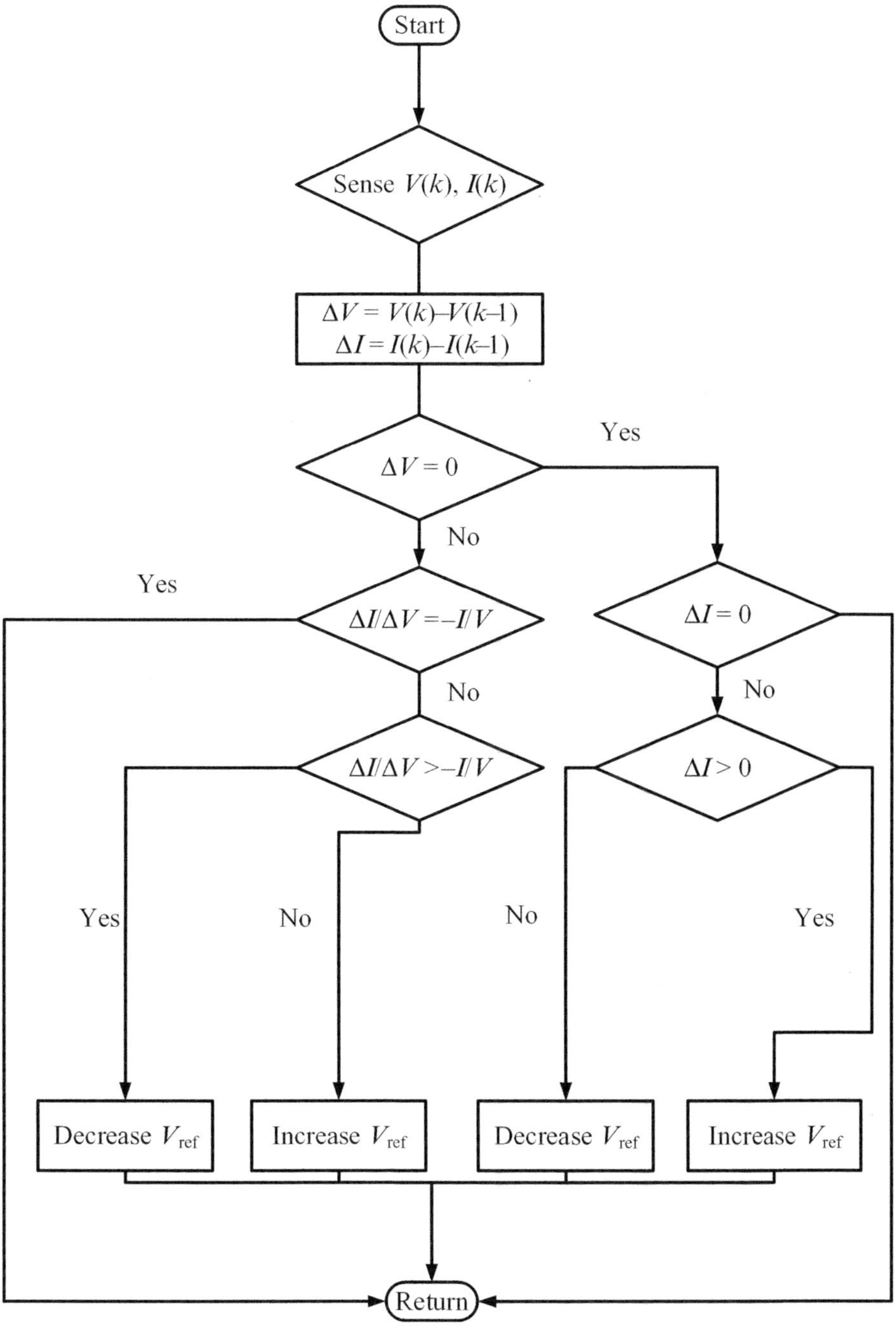

Fig. 12.16 Flowchart for incremental conductance (IC) method.

in which the value of β at optimum operating point remains constant. Therefore, β can be continuously calculated using the panel voltage and current.

$$\beta(V_{PV}, I_{PV}) = \ln\left(\frac{I_{PV}}{V_{PV}}\right) - cV_{PV}$$

where c is a constant depends on electron charge (q), temperature (T), quality factor (η), Boltzmann constant (K_B), and number of series cells (N_s). β can be continuously calculated using voltage and current of the panel. However, for optimal performance, electrical parameters of the PV module should be known to estimate the optimal performance point (De Brito et al., 2013).

12.4.1.7 Temperature method

For this technique, a low-cost temperature sensor is used and to modify the MPP algorithm function, because MPP strictly changes with temperature. However, the irregular temperature distribution on PV array can cause practical implementation problematic. The following equation is used in here:

$$V_{MPP}(T) = V_{MPP}(T_{ref}) + T_{Kvoc}(T - T_{ref})$$

where V_{MPP} is MPP voltage, T is the panel temperature, T_{Kvoc} is the temperature coefficient of V_{MPP}, and T_{ref} is standard test condition temperature.

All these methods simply used to match load impedance with the resistance of the solar panel. Major characteristics of earlier-mentioned methods are discussed in Table 12.3. More details regarding this can be found in the literature (De Brito et al., 2013).

Table 12.3 Major characteristics of MPPT algorithms (De Brito et al., 2013).

Method	Dependency of PV array	Tracking factor (TF)	Implementation complexity	Accuracy	Sensors
Fixed duty	No	0.87	Very simple	Very low	None
Constant voltage (CV)	Yes	0.92	Simple	Low	V
Perturb and observe	No	0.96	Simple	Very high	V, I
Incremental conductance	No	0.96	Medium	Very high	V, I
Beta	Yes	0.99	Medium	Very high	V, I
Temperature	Yes	0.98	Simple	Very high	V, T

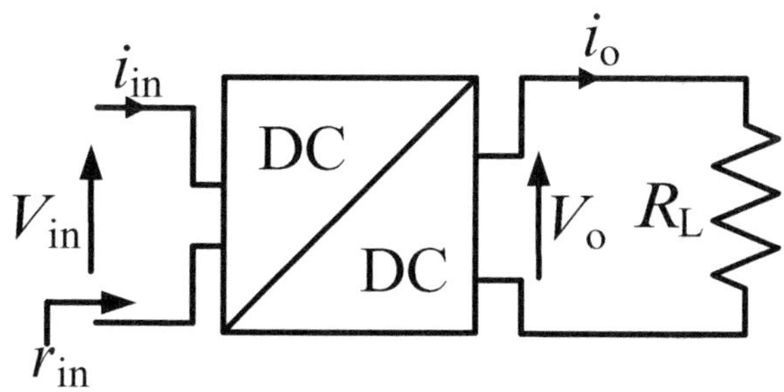

Fig. 12.17 Block diagram of loaded switched mode DC-DC converter.

The fluctuating nature of solar energy necessitates suitable energy storage systems (ESSs) in solar-powered DCMGs. Compared to typical battery banks, SCs offer longer cycle life eliminating the need to replace them regularly. However, compared to a typical MPPT controller, where the battery bank and resistive load fed by a switch-mode DC-DC converter allows impedance matching for maximum power transfer, an SC bank's significantly large capacitive load does not permit the typical impedance matching for maximum power transfer. This section discusses about this scenario.

12.4.2 Input impedance of switch mode DC-DC converters

A switch mode DC-DC converter, regardless of its topology, presents a negative incremental input impedance. Typical DC-DC converter in Fig. 12.17 works to maintain a constant output voltage V_o, at output power of P_0. For an ideal switch-mode DC-DC converter with 100% theoretical efficiency, following valuable relationships can be easily established (Panizza, 2009).

The converter under continuous conduction mode (CCM) acts as a DC transformer with a turns ratio of n which leads to the useful relationship of

$$r_{in} = \frac{V_{in}}{i_{in}} = n^2 R_L \tag{12.4}$$

where R_L is the load resistance $\left(R_L = \frac{V_0}{i_0}\right)$.

In accordance with Eq. (12.4) and Table 12.4, it can be observed that the input impedance of a switched mode DC-DC converter is a function of the duty cycle of the converter. Hence by varying D, it is possible to vary the effective input impedance of the converter. This approach is used in MPPT controllers to match the PV array resistance with the load resistance. Fig. 12.18 depicts the variation of input resistance with duty cycle ratio for the earlier converter types.

With the aim of exploring more about the variation of input resistance of switch mode DC-DC converter, following analysis was carried out with both battery pack and capacitor connected at the output of the converter separately. It should be noted that for this analysis, that an ideal component based buck converter was used as the DC-DC converter.

Table 12.4 Transformation coefficient n for switched mode converter.

Converter type	Transformation coefficient (n)
Buck converter	$1/D$
Boost converter	$1 - D$
Buck-boost converter	$(1 - D)/D$

D is the duty ration of the converter.

12.4.2.1 Battery connected to the output of the buck converter

Let us analyze a situation shown in Fig. 12.19, where a resistive load buffered by a battery pack is connected to the output of a buck converter.

By considering Fig. 12.19, the Thevenin resistance of the combined battery and load is

$$R_0 = \frac{R_L r_b}{R_L + r_b} \tag{12.5}$$

with r_b being the internal resistance of the battery. Therefore, r_{in} can be modified as

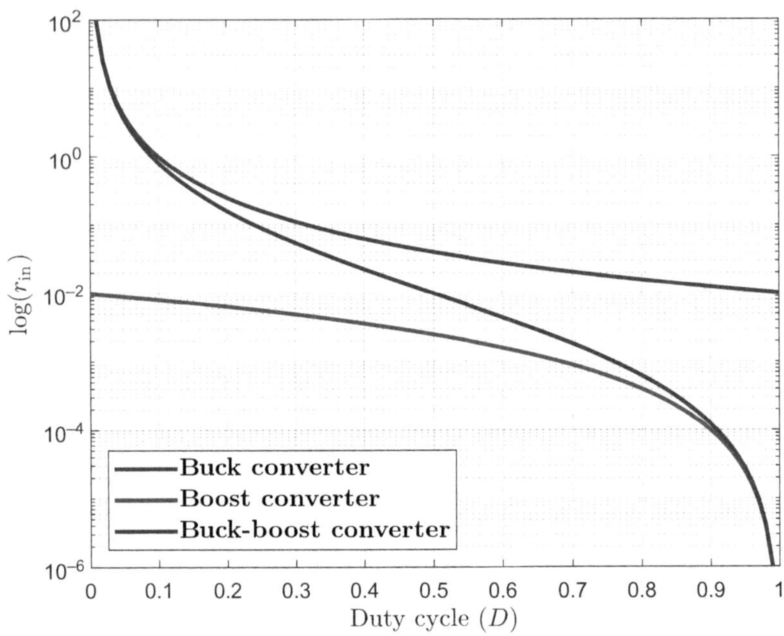

Fig. 12.18 Variation of input resistance r_{in} of an ideal switching converter under CCM with duty cycle D.

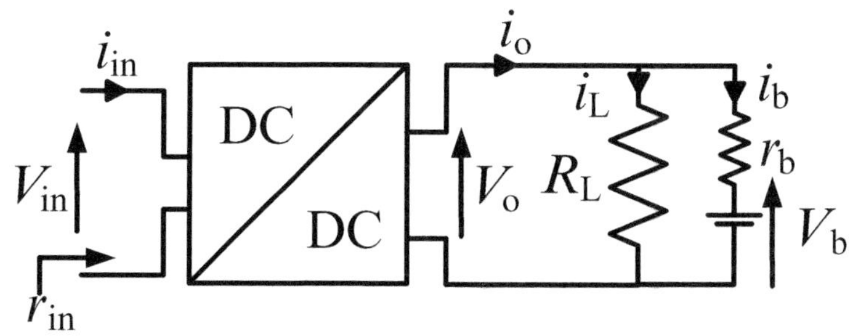

Fig. 12.19 Buck converter with resistive load is buffered by a battery bank.

$$r_{in}=\frac{R_0}{D^2}=\frac{R_L r_b}{D^2(R_L+r_b)} \tag{12.6}$$

showing that r_{in} varies from $\infty \rightarrow R_0$ as D goes from 0 to 1. With the variation of irradiance and temperature, the MPP of PV panel also changes. Hence, by varying the duty cycle, it is possible to match the load impedance with PV module impedance to extract maximum power.

12.4.2.2 Capacitor connected to the output of the buck converter

With recent advances such as hybrid SCs and CAPA batteries offering higher energy densities, battery packs may be easily replaced by the higher life cycle-based SC banks, with the secondary advantage of the constant ESR irrespective of the depth of discharge. Nevertheless, if a battery bank is replaced solely by an SC bank, existing MPPT schemes are not adaptable since the load becomes predominantly capacitive, compared to the case of a battery bank and resistive load which allows simple impedance matching based on the effective resistive load seen by the solar panel. Following analysis provides a summary of the two cases, namely parallel and series connection of the capacitor which is shown in Fig. 12.20.

(A) Parallel connection of the capacitor:

Consider Fig. 12.20A where load is buffered with a SC bank.

Using Ohm's law output current i_0 can be derived as

$$i_0=i_c+i_L=\frac{V_0}{R_L r_c}\left(r_c+R_L e^{\frac{-t}{r_c C}}\right) \tag{12.7}$$

where C is the capacitance and r_c is ESR of the capacitor. Capacitor is assumed to start with zero initial voltage.

Using Eq. (12.7), the general expression for Thevenin resistance R_0 can be obtained as

$$R_0=\frac{R_L r_c}{r_c+R_L e^{\frac{-t}{r_c C}}} \tag{12.8}$$

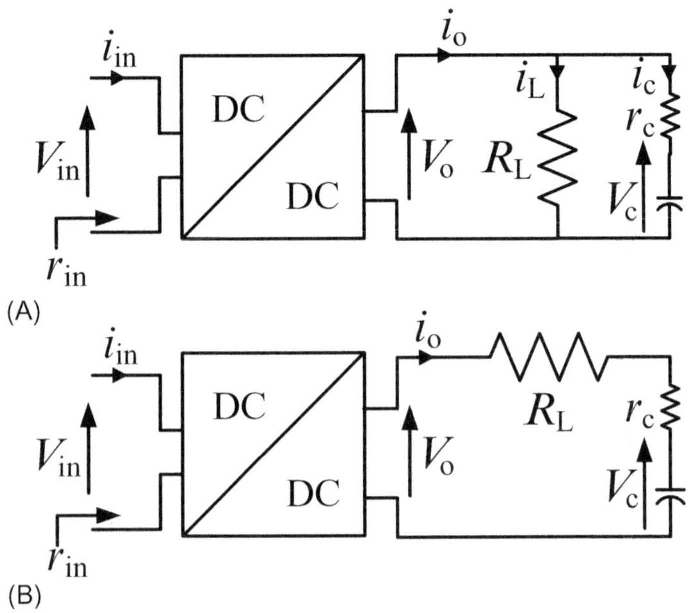

Fig. 12.20 Buck converter with a capacitor at the output: (A) parallel connection and (B) series connection.

Accordingly, r_{in} of the buck converter will be

$$r_{in} = \frac{R_L r_c}{D^2\left(r_c + R_L e^{\frac{-t}{r_c C}}\right)} \tag{12.9}$$

From Eq. (12.9), it can be seen that when $t = 0$ (initially) r_{in} of the DC-DC converter is $\frac{R_L r_c}{D^2(R_L + r_c)}$, and as $t \to \infty$ input impedance r_{in} approaches R_L/D^2.

(B) Series connection of the capacitor:

Considering Fig. 12.20B, i_0 can be derived as

$$i_0 = \frac{V_0}{R_L + r_c} \exp(-t/(r_c + R_L)C) \tag{12.10}$$

So we estimate the input resistance of the DC-DC converter r_{in} to be

$$r_{in} = \frac{V_0}{D^2 i_0} = \frac{(R_L + r_c)\exp(t/(r_c + R_L)C)}{D^2} \tag{12.11}$$

When considering Eq. (12.11), it is clear that in the case of series connection, input impedance r_{in} varies from (R_L + r_c) to ∞ as t goes from $0 \to \infty$.

Fig. 12.21 shows the behavior of input impedance of the buck converter when SC bank is used as the sole ESD for both parallel and series connection.

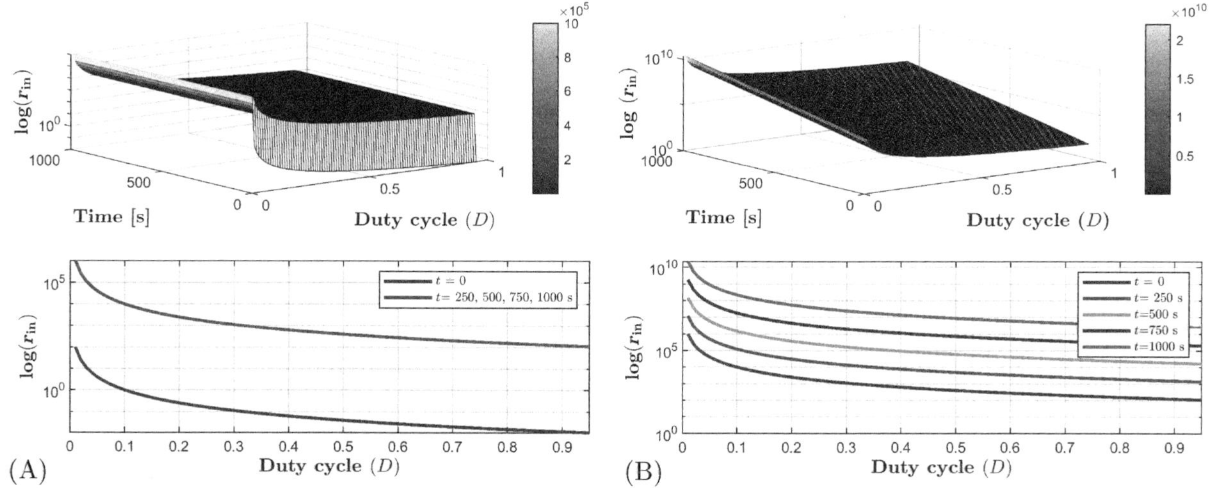

Fig. 12.21 Variation of r_{in} for buck converter with capacitor at output: (A) parallel connection of capacitor and (B) series connection of capacitor ($C = 1$ F, $R_{\mathrm{L}} = 100\Omega$, $r_{\mathrm{c}} = 0.01\Omega$).

The impedance of a PV panel is significantly varies with the irradiance (G) and temperature (T). Since G and T are continuously varying in real-world condition, panel resistance is also changing continuously. Therefore, in order to transfer maximum power, impedance of the connected load should also be varied continuously. When battery bank is connected as ESD in PV system, this is achieved by continuously varying the duty ratio D of the DC-DC converter according to Eq. (12.6). In the case of battery bank, input impedance of DC-DC converter is a function of duty ratio D only.

Compared to the case of a battery bank resulting in an input resistance given by Eq. (12.6), a SC bank makes the input resistance of the converter dependent on the state of the charge of the SC bank, which is also related to the time constant of the output circuit. This makes it extremely difficult to adapt the existing MPPT schemes into SC-based ESD systems. This is clear from Eqs. (12.9) to (12.11).

If a capacitor with an initial charge (or initial voltage V_{Cin}) was used in the case of series connection, we can rewrite Eq. (12.10) as

$$i_0 = \frac{V_0 - V_{\mathrm{Cin}}}{R_{\mathrm{L}} + r_{\mathrm{c}}} \exp\left(-t/(r_{\mathrm{c}} + R_{\mathrm{L}})C\right) \tag{12.12}$$

Defining $k = \frac{V_{\mathrm{Cin}}}{V_0}$ where $0 \leq k \leq 1$, Eq. (12.12) becomes

$$i_0 = \frac{V_0(1-k)}{R_{\mathrm{L}} + r_{\mathrm{c}}} \exp\left(-t/(r_{\mathrm{c}} + R_{\mathrm{L}})C\right) \tag{12.13}$$

So, input impedance r_{in} can be rewritten as

$$r_{\mathrm{in}} = \frac{(R_{\mathrm{L}} + r_{\mathrm{c}}) \exp\left(t/(r_{\mathrm{c}} + R_{\mathrm{L}})C\right)}{(1-k)D^2} \tag{12.14}$$

Fig. 12.22 shows the variation of r_{in} of buck converter considering Eq. (12.14) where $V_0 = 24$ V, $R_{\mathrm{L}} = 100\ \Omega$, $r_{\mathrm{C}} = 0.01\Omega$, $C = 1$ F, and $V_{\mathrm{Cin}} = 9$ V. It can be seen that from Fig. 12.22, with the state of charge of the capacitor, the input impedance of the converter changes making it difficult to match the PV panel impedance with r_{in} of the converter.

Table 12.5 summarizes the input impedance of a DC-DC converter for each case discussed in this section. It is obvious that existing MPPT techniques are not adaptable for a situation where an SC bank is used solely as the ESD for a PV system because capacitors store energy in its electrostatic field where in a battery, electrochemical reaction is related to the back electromotive force.

12.5 SCALED topology and its theory related to higher-efficiency LED lighting

There is a global interest in DCMGs with the proliferation of renewable energy sources. One good evidence was that the IEEE Power Electronics and Industry Application Society launched the biennial series of IEEE International Conference on DC

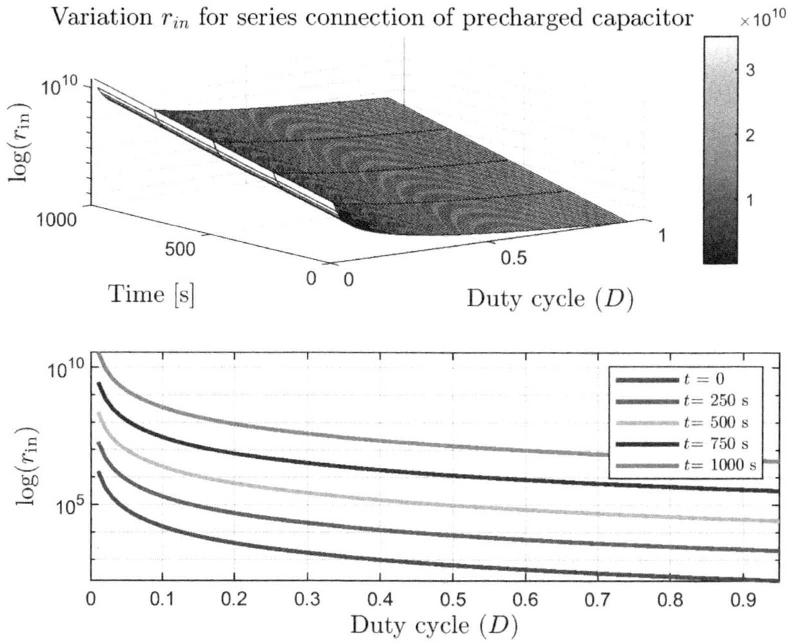

Fig. 12.22 Series connection of precharged capacitor.

Table 12.5 Input impedance of switch mode DC-DC converter (Jayananda, 2020).

Configuration	r_{in}
Battery connected to the output	$n^2 \dfrac{R_L r_b}{(R_L + r_b)}$
Parallel connection of the empty capacitor	$\dfrac{n^2 R_L r_c}{r_c + R_L \exp(-t/r_c C)}$
Series connection of the empty capacitor	$n^2(R_L + r_c)\exp(t/(r_c + R_L)C)$
Series connection of the precharged capacitor	$\dfrac{n^2(R_L + r_c)\exp(t/(r_c + R_L)C)}{(1-k)}$

Microgrids (ICDCM). With the third of the series held in Matsue Prefecture, Japan (2019), there is a rapidly growing interest in DC-powered products on a worldwide basis. The LED lighting is one of the key areas to harness the benefits of DCMG concepts and to identify the potential extension of the supercapacitor-assisted loss management (SCALOM) concept [summarized in Chapter 9] based on the extended *RC* concepts into the LED lighting area.

In DCMG environment with a renewable energy supply, DC-powered appliances eliminate the need for an inverter, to achieve a higher end-to-end efficiency (ETEE). However, this situation does not let us use the AC grid supply directly to power the appliances. In such a situation, an ESS is essential. The continuous advances of SC technologies have made the SCs to be used in electronic and electrical systems to

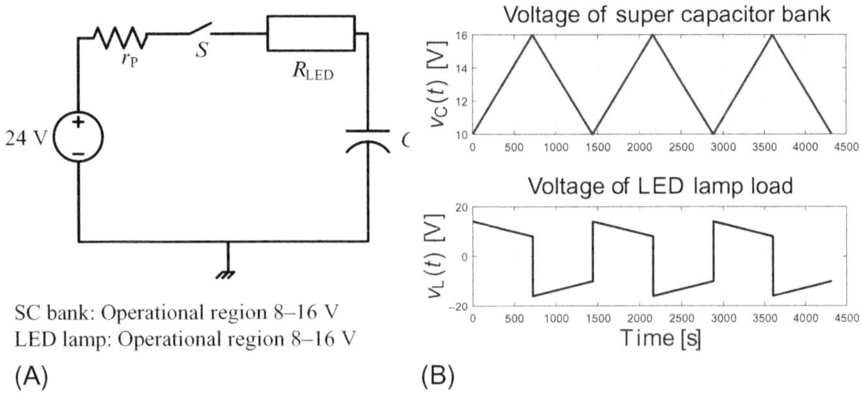

Fig. 12.23 Concept of proposed SCALED converter: (A) modified *RC* charging loop with LED lamp and (B) voltage waveforms for SC bank and LED lamp load.

overcome short-term fluctuations of the energy supply. Therefore, SC banks could be used to replace an environmentally unfriendly electrochemical battery pack used in a DCMG environment as short-term ESS.

12.5.1 Basic concept

The SCALED converter utilizes the same SCALOM concept discussed in Chapter 9. Here, 12-V DC operable LED lamps serve as the "useful load" in the *RC* loop, consuming the energy while charging the SC.

If the DC equivalent resistance of LED lamps (R_{LED}) is much higher than the parasitic resistance of the loop (r_P) in Fig. 12.23, most of the energy wasted in the *RC* charging loop will be utilized by the LED lamps. Fig. 12.23A depicts a case of nominal 24 V DC voltage source, switch, and the LED lamp working in series, charging a SC bank which is operating with a DC voltage range of 8–16 V. Fig. 12.23B shows the voltage across SC bank and LED lamp assuming that $R_{LED} \gg r_P$. The switch (*S*) should be operated in such a way that both LED lamp and SC bank are not exposed to any voltage higher than the maximum limit of the operational voltage region. When the SC bank reaches the upper limit, the excess energy within the bank should be released to the load. By using an adequately rated SC bank (approximately twice or larger DC voltage capability of LED bank) with LED lamp, SCALED technique helps to significantly reduce the losses in the *RC* loop.

12.5.2 Design considerations

In order to conserve energy, minimum conversion from DC to AC and AC to DC must take place. When considering existing solar-powered LED lighting system there are several conversion stages, which will lead to the reduction of the overall efficiency. The ETEE of a system will be the multiplication of efficiency of each conversion

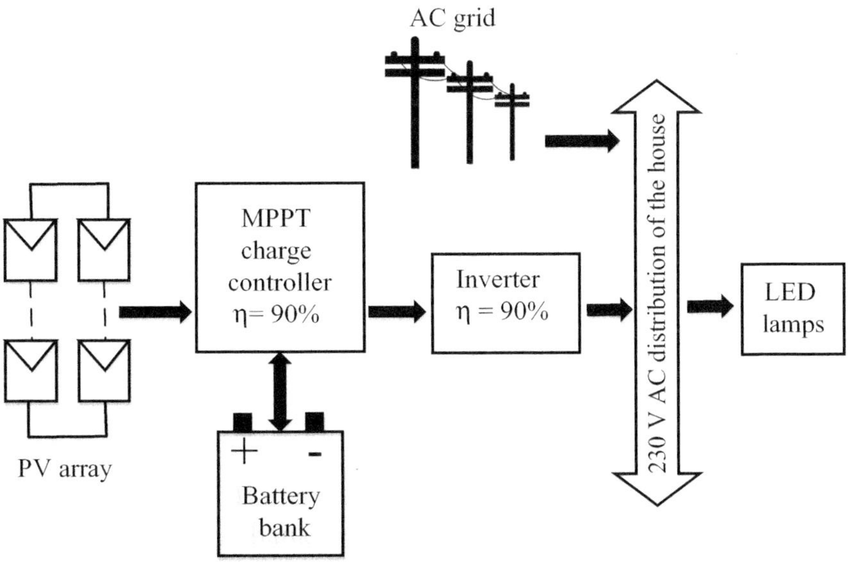

Fig. 12.24 Existing solar-powered LED lighting systems (with indicative converter efficiencies around 90%).

stage. For the case in Fig. 12.24 ETEE will be only 81%, which is lower than the individual efficiencies of the MPPT controller and inverter. Hence, by reducing the number of conversion stages, it is possible to improve the efficiency of the system.

Typically, existing solar-powered LED lighting system consists of three conversion stages as shown in Fig. 12.24. DC voltage generated by the solar panel is converted into higher DC voltage using DC-DC converter. Using a DC-AC converter (solar inverter), the output of the DC-DC converter is converted into AC which will power the LED lighting system. LED lamp consists of an LED driver circuit that comprises an AC-DC converter (Rectifier), which will convert the AC supplied by the solar inverter to DC to drive the internal LED array. With all three stages working together to power the LEDs, and if the AC-DC converter overall efficiency could be 73% (assuming that AC-DC converter is having an efficiency of 90%).

Since these 12 V LED lamps can be powered directly from 12 V DC supply, the author focused to implement a SCALED lighting system for DCMGs, which can operate directly from solar panels as shown in Fig. 12.25.

12.5.2.1 SCALED converter: Proof of concept

Three important properties of 12 V LED lamps: (a) ability to operate with positive or negative DC voltages, (b) ability to operate over a wide range of AC/DC voltages without compromising brightness, and (c) ability to operate as constant power load in the constant brightness region were used during the design phase of the SCALED converter.

As shown in Fig. 12.26, once switch S_1 is closed, the PV panel will charge the SC bank C while powering the LED lamp. Once SC bank C, charges until V_{LED} reaches to

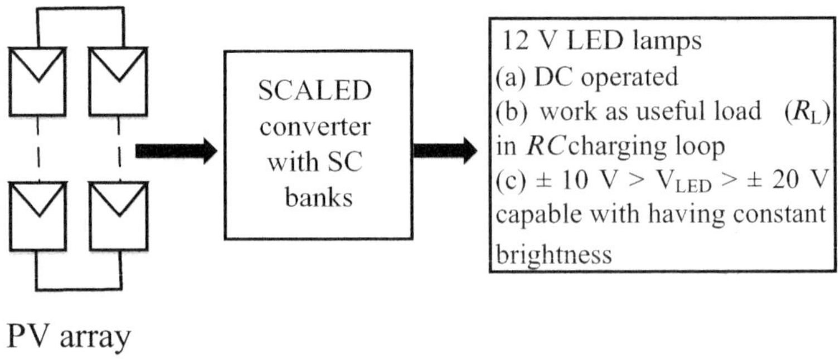

Fig. 12.25 Proposed concept of SCALED converter for solar-powered DCMGs: simplified block diagram.

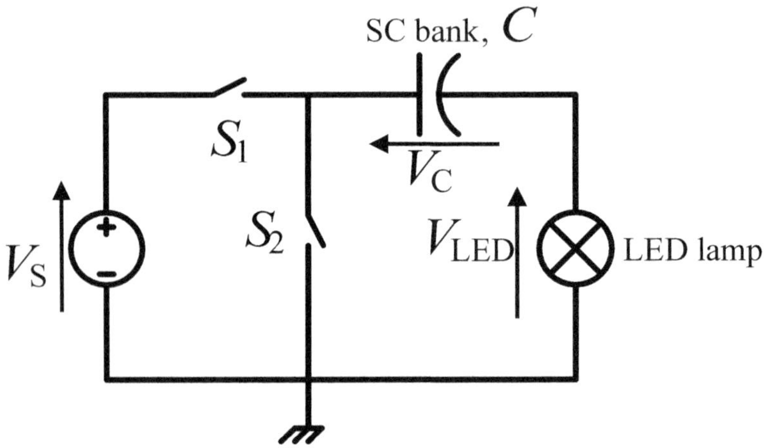

Fig. 12.26 Basic SCALED converter (proof of concept).

V_{LED}^{min} it will switch to discharging phase by closing S_2 and opening S_1. Then, C will discharge to the LED lamp until V_C drops to V_{LED}^{min}. This cycle repeats to keep the circuit working continuously. During initial proof of concept, V_{LED}^{max} was set as 14 V and V_{LED}^{min} was set as 10 V.

The initial circuit was designed to operate in following voltage ranges:

1. Operational voltage range of LED lamp: $10\text{ V} \geq V_{LED} \geq 14\text{ V}$
2. Operational voltage range of SC bank: $10\text{ V} \geq V_C \geq 14\text{ V}$
3. Input voltage range: $21\text{ V} \geq V_S \geq 24\text{ V}$

A digital controller to monitor the voltage of LED lamps and control the switches was implemented using a peripheral interface controller (PIC) microcontroller. This circuit was build and tested with laboratory power supply as the proof of concept.

Fig. 12.27A indicates the schematic diagram of the circuit used of proof of concept based on a PIC processor as the controller, and Fig. 12.27B depicts the oscilloscope waveforms of the circuit implemented (results obtained for 22 V input voltage is only

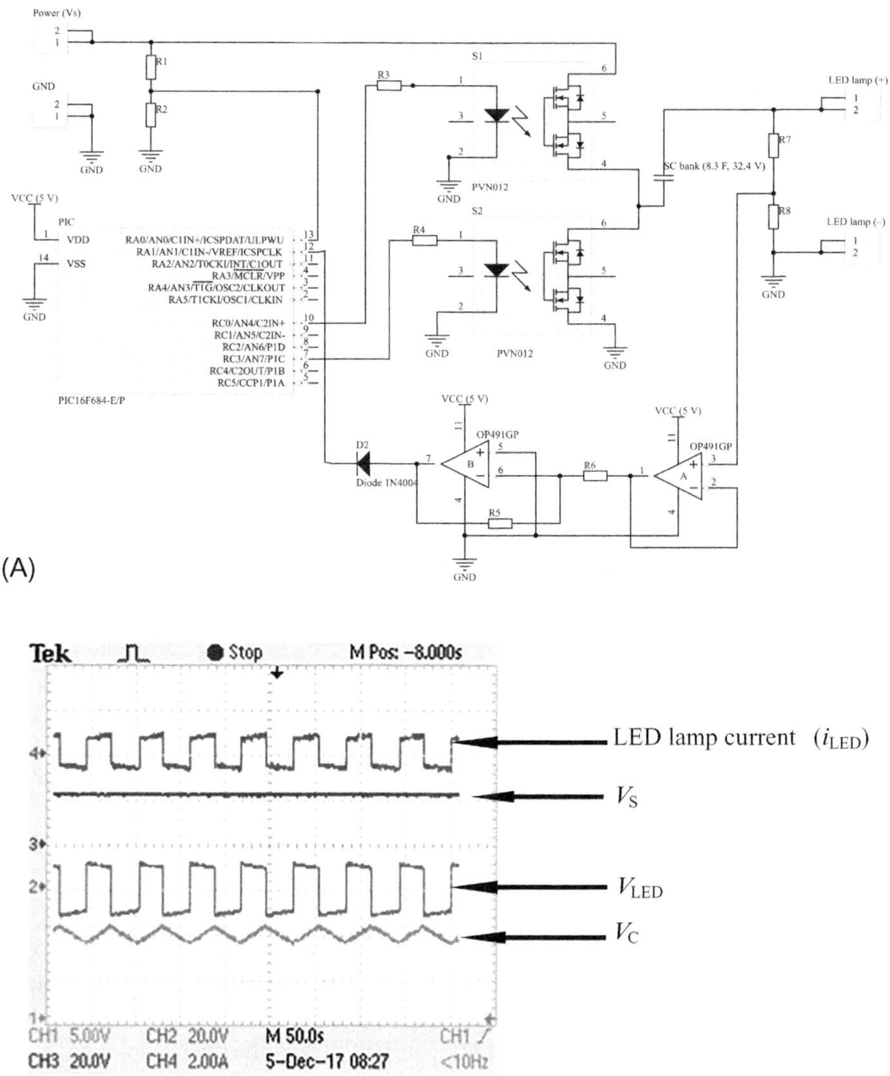

Fig. 12.27 Proof of concept: (A) schematic diagram of basic SCALED converter and (B) oscilloscope waveforms of the basic SCALED converter at input of 22 V.

shown here). When considering the current trace, during the charging phase of the SC bank, current is positive (at this stage current is drawn from the power supply) and during the discharging phase, current is negative (at this state SC bank is discharged to the LED lamp and positive terminal of the SC bank is connecting to the ground).

From this initial prototype, applicability of supercapacitor-assisted loss management (SCALOM) concept for LED lighting was verified. The overall efficiency of the circuit was calculated using the following equation:

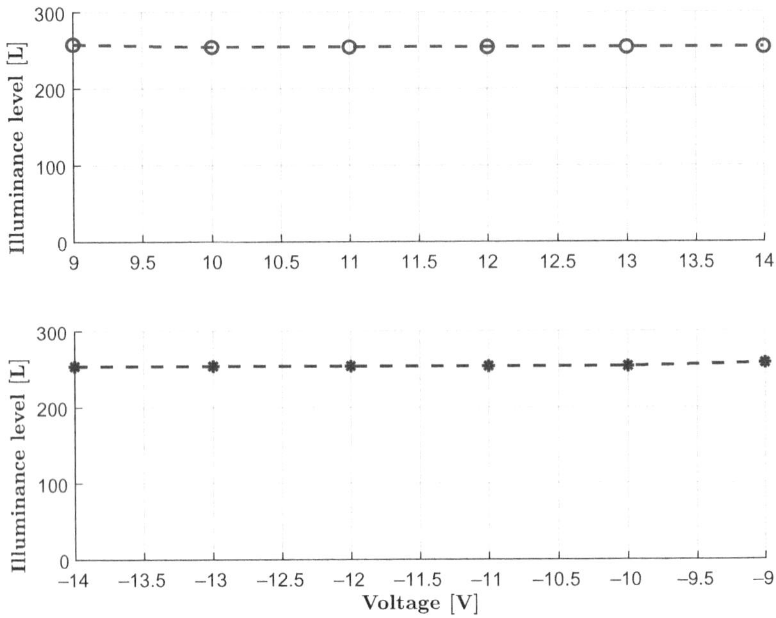

Fig. 12.28 Brightness of the Philips 5.5 W LED lamp during SCALED operation.

$$\eta = \frac{P_{\text{out}}}{P_{\text{in}}} = \frac{\overline{V_{\text{LED}}} \cdot \overline{i_{\text{LED}}}}{\overline{V_{\text{S}}} \cdot \overline{i_{\text{S}}}} \tag{12.15}$$

where overbar denotes average over time. The overall efficiency of the circuit was ≈89%.

Also, the switching frequency of the converter was well below 100 Hz. This frequency can be changed by changing the LED lamp load and size of the SC bank (i.e., capacitance of the SC bank) used. During testing of this prototype, brightness of the LED lamp was also recorded and shown in Fig. 12.28. It can be observed that brightness of the LED lamp was approximately constant during the operation of SCALED converter as expected.

12.5.3 Modified SCALED converter

However, in this basic SCALED topology, during the discharging phase of the SC bank, the PV panel is disengaging from the converter and won't supply until SC bank voltage reaches to $V_{\text{LED}}^{\min}$. Therefore, during the discharging phase, the power generated from the PV panel will be wasted. In a practically and commercially useful implementation, a large LED lighting load (12, 24, 48, or over 100 V DC cases) can be first divided into two halves, and one of these could be connected in series with a partly

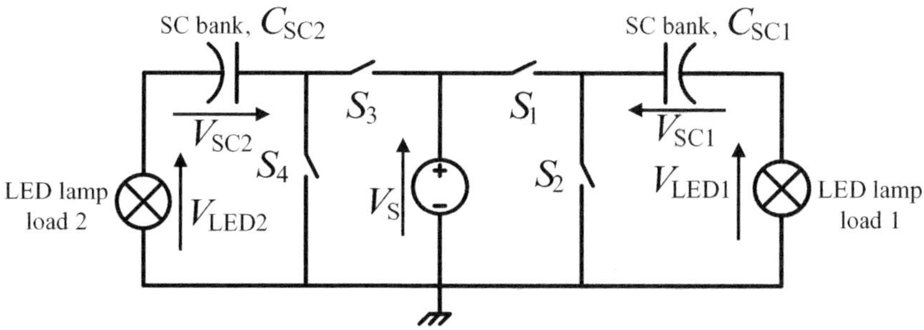

Fig. 12.29 SCALED converter (proof of concept).

charged SC module which is nominally around the DC rated voltage of the LED lamp bank. The other half is expected to operate from a fully charged identical SC module.

Hence, the basic SCALED converter was modified by dividing the LED lamp load into halves and using SC bank with each LED lamp load. In order to use SC banks as practical ESDs within an overall system with charge balance over each cycle, the concept shown in Fig. 12.29.

At any given time, we can look at the right or left side of the circuit, as a replica of the circuit in Fig. 12.26, where we have added S_2 and S_4 to discharge excess energy into the same LED lamps, acting as useful lighting loads. Since commercial LED lamps can operate with a negative DC source, S_2 and S_4 help maintain the charge balance of the respective SC banks, by alternatively switching S_1 and S_3 as a pair. A controller is required to maintain the sequence of switches accordingly. As explained in Section 12.5.1, in this topology the resistive losses in the well-known textbook case of the *RC* circuit fed by a voltage source is mostly absorbed into the LED lamp banks, for higher efficiency. One major advantage of this technique is that it does not require switching of the polarity of the SC bank during discharge, as the LED lamps can operate with either positive or negative polarity voltage.

As further improvement to topology, initial charging of the SC bank is also included in the actual SCALED design as shown in Fig. 12.30. The final converter incorporates three configurations, namely,

1. initial charging phase of SC1 and SC2;
2. charging phase of SC2 (or discharging phase of SC1); and
3. discharging phase of SC2 (or charging phase of SC1).

At the start-up phase (initial phase) of the converter, SC banks 1 and 2 will charge up to 20 and 12 V, respectively, by directly connecting PV module to SC banks and disconnecting the two LED lamp banks as shown in Fig. 12.30B. Once both SC banks are charged up to required voltage levels, the converter commences normal operation in which one SC bank charges to 20 V through LED lamps while the other SC bank discharges to the other LED lamps.

When first SC bank is charged to 20 V, it will switch to discharging phase while the other bank now switches to charging phase and cycle repeats. Due to the ability of LED lamps to operate within wide range of positive or negative voltages around

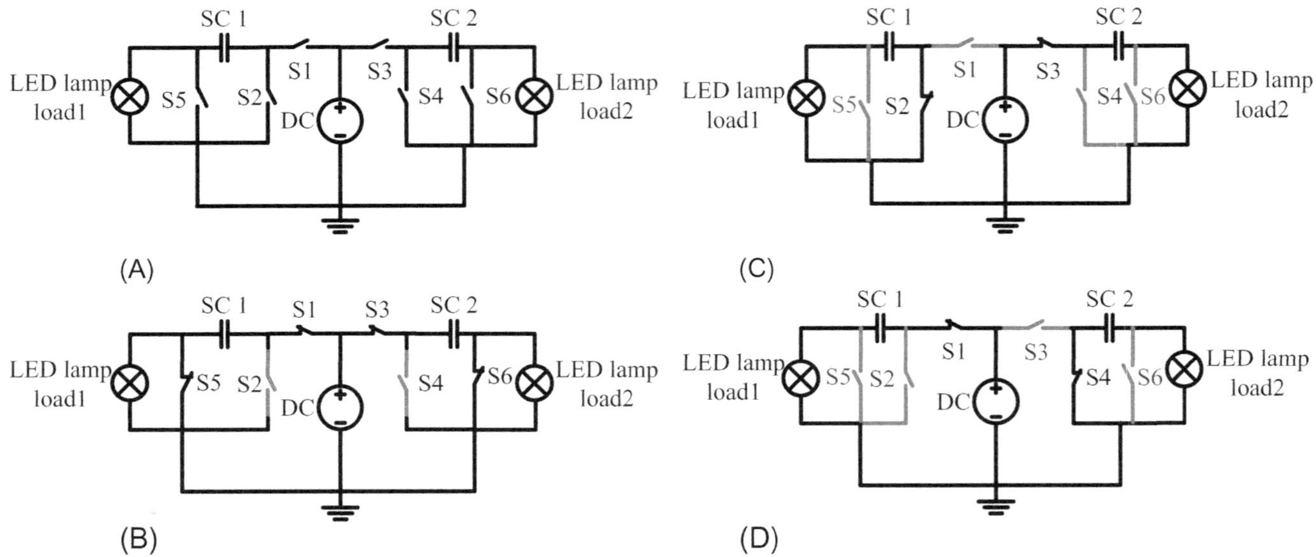

Fig. 12.30 The proposed SCALED converter with three operational configurations: (A) basic diagram of the converter; (B) initial charging phase of SC1 and SC2; (C) SC1 discharging, SC2 charging phase; and (D) SC1 charging, SC2 discharging phase.

Table 12.6 Switching arrangement of proposed SCALED converter.

Mode	S1	S2	S3	S4	S5	S6
Initial charging phase	On	Off	On	Off	On	On
SC2 charging and SC1 discharging	Off	On	On	Off	Off	Off
SC2 discharging and SC1 charging	On	Off	Off	On	Off	Off

the nominal value without degradation of brightness, this technique can cope with large fluctuations of solar irradiance. In addition, SC banks are able to provide power to LED lamp loads during low irradiance levels to provide short-term (in order of minutes) DC-uninterruptible power supply (DC-UPS) capability. In this converter, switches S2 and S4 help to preserve the charge balance of respective SC banks by alternately switching S1 and S3 as a pair. In order to maintain the switching sequence presented in Table 12.6, using a controller is compulsory.

Let us consider the charging phase of any of the SC bank, C_{SC}. After charging for a period t, the instantaneous voltage across the SC bank is given by

$$V_{SC} = V_{SC}(0) + \frac{I_{LED}t}{C_{SC}} \tag{12.16}$$

where I_{LED} is the average LED lamp current and $V_{SC}(0)$ is the initial voltage of the capacitor.

The PV panel voltage V_S equals the sum of the SC bank voltage and LED lamp voltage

$$V_S = V_{SC}(t) + V_{LED}(t) \tag{12.17}$$

The SC bank charges until voltage across LED lamp load 1 reaches V_{LED}^{min} while the voltage across the C_{SC} reaches $V_S - V_{LED}^{min}$ at the end of charging time (t_{ch}). Nevertheless, during the discharging phase in order to discharge this SC bank until voltage across LED lamp reaches V_{LED}^{min}, the criterion $V_S - V_{LED}^{min} > V_{LED}^{min}$ must be satisfied. This creates the condition

$$V_S > 2V_{LED}^{min} \tag{12.18}$$

However, to avoid the damage to the LED lamp during discharging, the final voltage of the SC bank should not be exceed the maximum voltage limit of the LED lamp load, V_{LED}^{min}. Hence, this creates another condition

$$V_S \le V_{LED}^{min} + V_{LED}^{max} \tag{12.19}$$

In order to maintain the charge balance of the SC bank, the same amount of charge stored during charging phase should delivered back to the LED lamp load during the discharging phase. Therefore, considering the charge balance of the SC during the charge-discharge cycle ($Q_{ch} = Q_{dch}$), it is possible to derive the following relationship

$$I_{LED1} t_{ch} = I_{LED1} t_{dch} \tag{12.20}$$

where t_{ch} and t_{dch} are charge and discharge time, respectively. These times will be dependent on the number of LED lamps used. This provides

$$t_{ch} = t_{dch} \tag{12.21}$$

The ETEE of SCALED converter can be calculated using the following equation:

$$\eta = \frac{P_{out}}{P_{in}} = \frac{\overline{V_{LED1}} \cdot \overline{i_{LED1}} + \overline{V_{LED2}} \cdot \overline{i_{LED2}}}{\overline{V_s} \cdot \overline{i_s}} \tag{12.22}$$

where overbar denotes average over time.

12.5.4 Theoretical foundation and analysis of SCALED converter

This converter has two main configurations as shown in Fig. 12.31 and is based on fact that LED lamps are capable of operating with both positive and negative rails. Analysis of charging phase and discharging phase of SCALED converter is given below under the more realistic assumption that the solar panels operate in the nonconstant current region, modeled as a constant DC voltage source with finite internal resistance.

12.5.4.1 Analysis of charging phase of SCALED converter

The simplified version of the SCALED charging phase is shown in Fig. 12.31A. The parasitic resistance r_{PC} includes the internal resistance of the supply source, ESR of the capacitor, and path resistance. The analysis of charging loop was done for two cases namely as zero and nonzero r_{PC}.

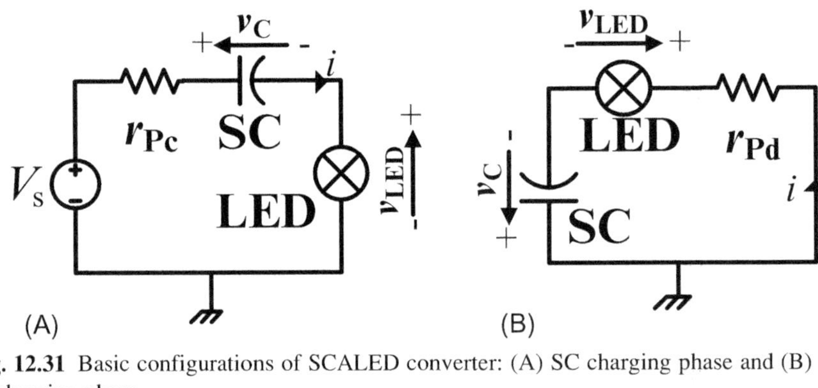

Fig. 12.31 Basic configurations of SCALED converter: (A) SC charging phase and (B) SC discharging phase.

(A) Case A: Neglecting parasitic resistance r_{Pc}

In here the supply was considered as an ideal CV source having voltage of V_s and the path resistance and the ESR of the capacitor is neglected. Since the power consumption of the LED lamp load P_0 is constant over the constant brightness region (discussed in Chapters 4 and 5), voltage across the LED lamp is

$$v_{LED} = \frac{P_0}{i} \tag{12.23}$$

where the second equality follows from Kirchoff's voltage law (KVL) as

$$v_{LED} = V_S - v_C \tag{12.24}$$

and the loop current of the circuit given by

$$i = C \cdot \frac{dv_C}{dt} \tag{12.25}$$

Using Eqs. (12.23)–(12.25), the general expression for the charging phase can be obtained as

$$(V_S - v_C)\frac{dv_C}{dt} = \frac{P_0}{C} \tag{12.26}$$

Since Eq. (12.26) is a nonlinear differential equation, the authors derived both analytical solution and numerical solutions for capacitor voltage v_C. The Euler method was used to derive the numerical solution. Integrating Eq. (12.26):

$$\int_0^{v_C(t)} (V_S - v_C) dv_C = \int_0^t \frac{P_0}{C} dt \tag{12.27}$$

leads to

$$\left(V_S - \frac{v_C(t)}{2}\right) \cdot v_C(t) = \frac{P_0}{C} t + K \tag{12.28}$$

a quadratic equation in $v_C(t)$ with solution

$$v_C(t) = V_S - \sqrt{{V_S}^2 - \frac{2 \cdot P_0}{C} t - 2K} \tag{12.29}$$

to satisfy the initial conditions with the condition that

$${V_S}^2 \geq 2 \cdot \left(\frac{P_0}{C} t + K\right) \tag{12.30}$$

Here K is a constant whose value can be found by substituting the initial conditions to Eq. (12.28). We find the charging time of the capacitor to reach supply voltage V_S to be

$$t_{\text{charging}} = \frac{C}{2 \cdot P_0} (V_S - v_C(0))^2 \tag{12.31}$$

(B) Case B: Including parasitic resistance r_{Pc}

In this case, a practical voltage source having finite internal resistance r_{S} was considered as the supply source to the SCALED circuit. When the effect of parasitic resistance is accounted, Eq. (12.26) will be modified as

$$\left(V_{\mathrm{S}} - v_{\mathrm{C}} - r_{\mathrm{Pc}} \cdot C\frac{dv_{\mathrm{C}}}{dt}\right)\frac{dv_{\mathrm{C}}}{dt} = \frac{P_0}{C} \tag{12.32}$$

Solving Eq. (12.32) for $\frac{dv_{\mathrm{C}}}{dt}$, we obtain a differential equation that can be evaluated using Euler method

$$\frac{dv_{\mathrm{C}}}{dt} = \frac{(V_{\mathrm{S}} - v_{\mathrm{C}}) - \sqrt{(V_{\mathrm{S}} - v_{\mathrm{C}})^2 - 4 \cdot r_{\mathrm{Pc}} \cdot P_0}}{2 \cdot r_{\mathrm{Pc}} \cdot C} \tag{12.33}$$

12.5.4.2 Analysis of discharging phase of SCALED converter

During the discharging phase of the SCALED topology, precharged SC bank is discharged to the LED lamp load. The simplified version of the SCALED discharging phase is shown in Fig. 12.31B. The parasitic resistance r_{Pd} includes both path resistance and ESR of the capacitor.

(A) Case C: Neglecting the effect of parasitic resistance r_{Pd}

To simplify the analysis the effect of the parasitic resistance r_{Pd} is neglected. Applying Ohm's law to Fig. 12.31B, we can derive an expression for capacitor voltage v_{C} in terms of capacitance C and power of the LED lamp load P_0 as follows:

$$v_{\mathrm{C}}\frac{dv_{\mathrm{C}}}{dt} = -\frac{P_0}{C} \tag{12.34}$$

which integrates to

$$\frac{v_{\mathrm{C}}(t)^2}{2} = -\frac{P_0}{C}t + K' \tag{12.35}$$

where the value of K' can be found by substituting initial conditions. Hence

$$v_{\mathrm{C}}(t) = \sqrt{2K' - \frac{2 \cdot P_0}{C}t} \tag{12.36}$$

with condition

$$K' \geq \frac{P_0}{C}t \tag{12.37}$$

By using Eq. (12.37) and substituting initial conditions to Eq. (12.35) the time taken to fully discharge the capacitor was found as

$$t_{\text{discharging}} = \frac{C}{2 \cdot P_0} v_C(0)^2 \tag{12.38}$$

(B) Case D: Including parasitic resistance r_{Pd}

When the effect of parasitic resistance was taken into account, Eq. (12.34) becomes

$$\left(v_C + C\frac{dv_C}{dt} \right) \frac{dv_C}{dt} = -\frac{P_0}{C} \tag{12.39}$$

which can be rewritten as

$$\frac{dv_C}{dt} = \frac{-v_C + \sqrt{v_C^2 - 4r_{Pd}P_0}}{2r_{Pd}C} \tag{12.40}$$

and can be solved numerically.

12.6 An overview of pilot project

The SCALED converter was developed with a pilot grant from Ports of Auckland Ltd (POAL) in New Zealand. An experimental shipping container used as an office cubicle in the jetty area was illuminated by commercially available 12 V LED lamps. Field testing of the full version of SCALED system was carried by using a *CSUN*250-60*P* PV panel as the energy source which were installed in the roof top of the DCMG container in POAL site. Field testing was carried out for 3 days continuously to observe the behavior of the topology in long run. The open circuit of the PV panel was 37.3 V, SC banks were charged to 20 V and discharged to 12 V to avoid the unnecessary higher voltage across LED lamp loads during the charging phase of SC banks. A GL240 data logger was used to acquire the experimental data at the POAL site.

Operational behavior of SCALED converter for 3 different days with average irradiance levels of 730, 643, and 744 W m^{-2} is shown in Fig. 12.32. Fig. 12.32A shows the voltage across two SC banks and two LED banks. 11 W LED lamp loads (three numbers of LED lamps) were used on both sides of the converter. Irradiance variations throughout the day (from 0900 to 1600 hours) was recorded using an irradiance meter and shown in Fig. 12.32B.

The ETEE of overall system was calculated using Eq. (12.22) which was discussed earlier in this chapter. Using this proposed system, it is possible to achieve higher ETEE levels of in the range of 90%–94%. It is important to highlight here that the above range of efficiencies were achieved without using any MPPT type DC-DC converter, since impedance matching required in MPPT is not practicable in this case, due to SC banks used for temporary energy storage. By examine Fig. 12.32, it is obvious that this configuration is capable of buffering against irradiance fluctuations. This is due to the fact that LED lamps can be operated across a wide range of positive and negative voltages without degrading the brightness with the SC banks acting as energy

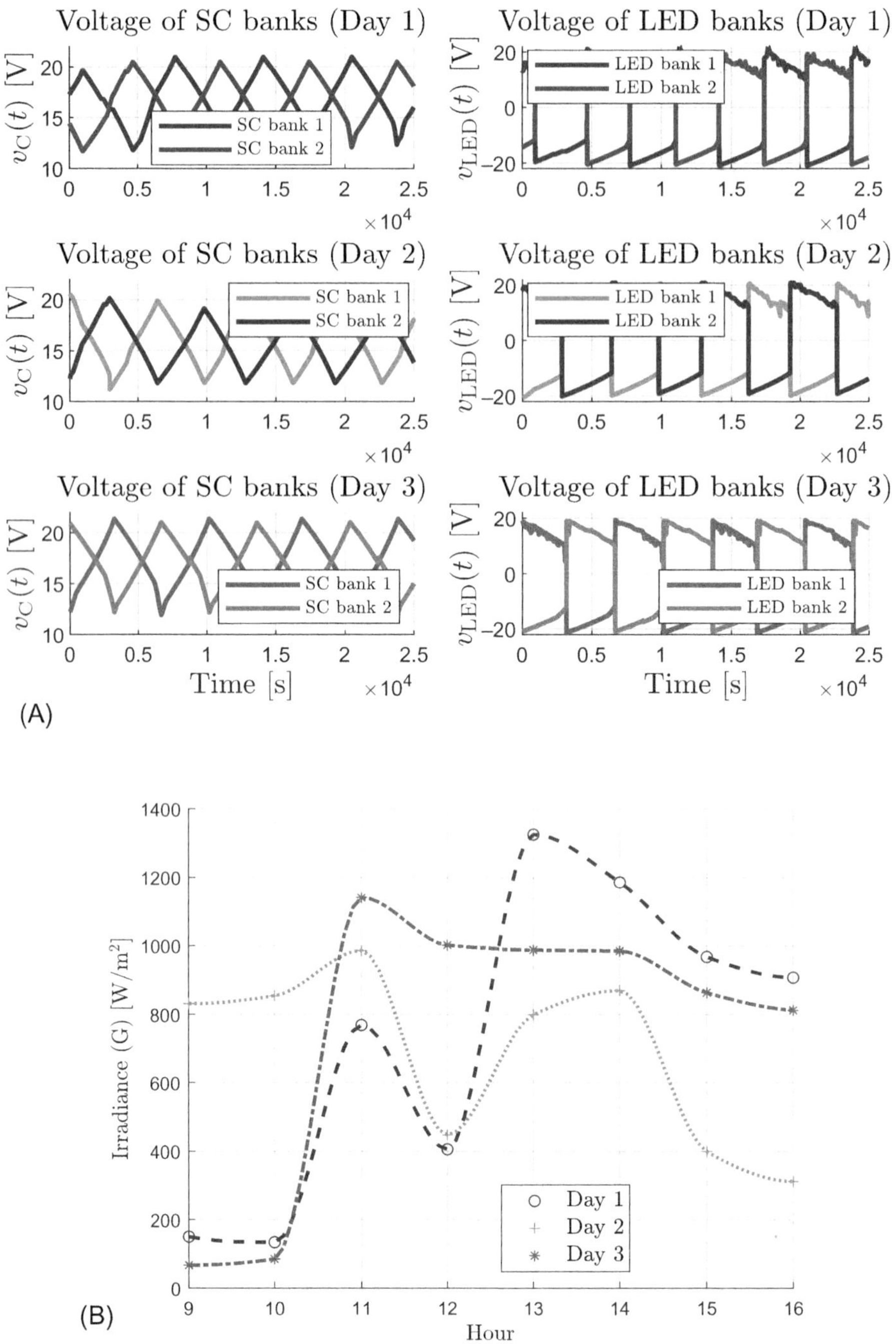

Fig. 12.32 Operational behavior of SCALED converter installed in POAL DCMG container: (A) supercapacitor and LED bank voltages vs. time for SCALED converter and (B) variation of irradiance throughout the day in POAL site.

Table 12.7 Summary of experimental results of SCALED converter.

Parameter	Specification
Operating voltage range of SC banks (LED bank)	20–12 V
Switching losses (for one cycle)	254 mW
ETEE	90%–94%
Switching frequency	0.15×10^{-3} Hz
Switching period	≈2 h
Autonomy time (depends on the lamp load)	15–30 min

buffers to smooth out irradiance fluctuations. In addition, this technique provides short-term DC-UPS capability in case of a sudden power loss. During testing, authors established that the switching frequency of the converter was well below 1 Hz for small loads. This frequency increases with the size of the LED lamp load. Results obtained from this testing are summarized in Table 12.7.

The SCALED converter is applicable for large LED lighting loads such as 24, 48, or over 100 V cases. Fluctuations in solar irradiance can be effectively accommodated by appropriate sizing of the SC banks.

By combining unique properties of commercially available LED lamps with SC banks, a novel loss circumventing SCALED converter for DC lighting in buildings is developed. In summary, the main features and advantages of this novel converter are

- higher ETEE with SC banks acting as lossless droppers;
- no high-frequency DC-DC converters required thus eliminating RFI/EMI issues; and
- short-term DC-UPS capability for irradiance fluctuations, by oversizing the SC banks, thus eliminating the need for battery banks.

Compared to the high-frequency switching converters coupled with a complex processor algorithm to achieve MPPT in a battery-based system achieving an ETEE around 90% (Chowdhury and Mourshed, 2016; Li and Lehman, 2004; Panov and Jovanovic, 2005; CharlesZhang, 2017; Hämmerle, 2017; Panizza, 2009; Middlebrook, 1976; De Brito et al., 2013), SCALED technique is able to achieve efficiencies in the range of 90%–94%, while maintaining the switching frequency in an extra-low range such as few hertz to fractional hertz. More details regarding this technique can be found in Jayananda et al. (2018a, b, 2019a, b, c, 2020) and Kularatna and Jayananda (2020).

References

Adly, M., Ibrahim, M., El Sherif, H., 2012. Comparative study of improved energy generation maximization techniques for photovoltaic systems. In: 2012 Asia-Pacific Power and Energy Engineering Conference, March, pp. 1–5.

Babaa, S., Armstrong, M., Pickert, V., 2014. Overview of maximum power point tracking control methods for PV systems. Power Energy Eng. 2, 59–72.

CharlesZhang, 2017. Analysis and Design of Input Filter for DC-DC Circuit. November. Texas Instruments.

Chowdhury, S.A., Mourshed, M., 2016. Off-grid electrification with solar home systems: an appraisal of the quality of components. Renew. Energy 97, 585–598.

Consumer, N.Z., 2013. LED bulb buying guide. @Online.

de Brito, M.A.G., Sampaio, L.P., Luigi, G., Canesin, C.A., 2011a. Evaluation of MPPT techniques for photovoltaic applications. In: 2011 IEEE International Symposium on Industrial Electronics, June, pp. 1039–1044.

de Brito, M.A.G., Sampaio, L.P., Luigi, G., e Melo, G.A., Canesin, C.A., 2011b. Comparative analysis of MPPT techniques for PV applications. In: 2011 International Conference on Clean Electrical Power (ICCEP), June, pp. 99–104.

De Brito, M.A.G., Galotto, L., Sampaio, L.P., e Melo, G.A., Canesin, C.A., 2013. Evaluation of the main MPPT techniques for photovoltaic applications. IEEE Trans. Ind. Electron. 60 (3), 1156–1167.

DiodesIncorporated, n.d. Cost effective LED driver, 2016. Available at: https://www.diodes.com/design/support/led-partners/cree/.

Dolara, A., Faranda, R., Leva, S., 2009. Energy comparison of seven MPPT techniques for PV systems. J. Electromagn. Anal. Appl. 1 (3), 152–162.

Hämmerle, F., 2017. Input Impedance Measurements for Stable Input-Filter Design. Omicorn Lab.

Jayananda, D., 2020. Supercapacitor Assisted LED Lighting (SCALED) Converter for Solar-Powered DC-Microgrids (Ph.D. thesis). University of Waikato.

Jayananda, D., Kularatna, N., Steyn-Ross, D.A., 2018a. Design approach for supercapacitor assisted LED lighting (SCALED) technique for DC-microgrids. In: 2018 IEEE International Conference on Industrial Electronics for Sustainable Energy Systems (IESES), January, pp. 27–31.

Jayananda, D., Kularatna, N., Steyn-Ross, D.A., 2018b. Powering 12-V LED luminiaries with supercapacitor-based energy storage in DC-microgrid systems. In: 44th Annual Conference of the IEEE Industrial Electronics Society 2018 (IECON), October, pp. 1922–1927.

Jayananda, D., Kularatna, N., Steyn-Ross, D.A., 2019a. Performance characteristics of energy-efficient LED lamps leading to supercapacitor assisted LED (SCALED) technique for DC-microgrid applications. In: 2019 IEEE International Conference on Industrial Technology (ICIT), February, pp. 515–520.

Jayananda, D., Kularatna, N., Steyn-Ross, D.A., 2019b. Supercapacitor assisted LED lighting (SCALED) for DC-micro grids. In: 2019 IEEE Third International Conference on DC Microgrids (ICDCM).

Jayananda, D., Kularatna, N., Steyn-Ross, D.A., 2019c. A validity of MPPT technique using supercapacitors as energy storage devices: example of the SCALED converter technique. In: IECON 2019—45th Annual Conference of the IEEE Industrial Electronics Society, October, 1, pp. 2301–2306. vol.

Jayananda, D., Kularatna, N., Steyn-Ross, D.A., 2020. Supercapacitor assisted LED (SCALED) technique for renewable energy systems: a very low frequency design approach with short-term DC-UPS capability eliminating battery banks. IET Renew. Power Gener. 14 (9), 1559–1570. https://doi.org/10.1049/iet-rpg.2019.1307.

Khan, M.N., 2013. Understanding LED Illumination. Boca Raton, CRC Press.

Kim, T.H., Wang, W., Li, Q., 2012. Advancement in materials for energy-saving lighting devices. Front. Chem. Sci. Eng. 6 (1), 13–26.

Kularatna, N., Jayananda, D., 2020. Supercapacitor-based long time-constant circuits: a unique design opportunity for new power electronic circuit topologies. IEEE Ind. Electron. Mag. 14 (2), 40–56. https://doi.org/10.1109/MIE.2019.2959199.

Li, P., Lehman, B., 2004. Performance prediction of DC-DC converters with impedances as loads. IEEE Trans. Power Electron. 19 (1), 201–209.

Littelfuse, 2016. LED design guide. Littelfuse, Inc., United States. Littelfuse LED design guide.

Matulka, R., Wood, D., 2013. The History of the Light Bulb. Department of Energy Facilities. Available at energy.gov/articles/history-light-bulb.

Mccallum, D., 2014. LED watt conversion and light replacement guide. Available at idavidmcallen.wordpress.com/2014/05/05/led-watt-conversion-light-replacement-guide/.

Middlebrook, R.D., 1976. Input filter considerations in design and application of switching regulators. In: IAS 76.

Panizza, M., 2009. Input Source Impedance and Its Effects on DC-DC Converter Performance and Characteristics. Vicor Corporation.

Panov, Y., Jovanovic, M., 2005. Practical issues of input/output impedance measurements in switching power supplies and application of measured data to stability analysis. In: Twentieth Annual IEEE Applied Power Electronics Conference and Exposition, 2005. APEC 2005, vol. 2, pp. 1339–1345.

Stardust, 2017. Light Bulbs. Available at: www.stardust.com/bulb.html.

Subudhi, B., Pradhan, R., 2013. A comparative study on maximum power point tracking techniques for photovoltaic power systems. IEEE Trans. Sustain. Energy 4 (1), 89–98.

Further reading

Kularatna, N., Steyn-Ross, A., Fernando, J., James, S., 2018. Design of Transient Protection Systems: Including Supercapacitor Based Design Approaches for Surge Protectors. Elsevier, USA 284 pp.

Supercapacitors for surge absorption: Supercapacitor assisted surge absorber (SCASA) technique

13

13.1 Introduction

Supercapacitors can be used in applications where they are not typically used as energy storage devices or in hybridized form with batteries. In Chapter 6, we used it as a theoretically lossless voltage dropper in a DC-DC converter, and in this chapter supercapacitor's ability to absorb high-voltage (HV) transient surges is discussed with a unique application.

If we consider a supercapacitor as an ideal device, the device will have an energy storage capacity of $\frac{1}{2}CV_C^2$, where C is the device capacitance and V_C is the rated DC voltage of the device. Now if this capacitor is used in a simple circuit where an ideal DC voltage source of value V_S and a resistor of value R are used, the overall RC circuit will have a time constant of $\tau = RC$. In Fig. 13.1A, this arrangement is shown and the graph in Fig. 13.1B indicates the exponential rise in the voltage across the capacitor, for the cases $V_S = V_C$ and $V_S = mV_C$ where m = 1 and 5.

For the circuit shown in Fig. 13.1, capacitor voltage is given by

$$v(t) = V_S\left(1 - e^{-\frac{t}{RC}}\right) \tag{13.1}$$

Eq. (13.1) shows that the capacitor can reach the value V_S after an approximate time of $T_{\text{final}} = 5RC$. However, if the rated voltage of the capacitor $V_C < V_S$, the capacitor may be destroyed due to overcharging, and in most cases we can expect a blast! However, if we can control the duration of the voltage source connected to the capacitor using a switch to disconnect it when the capacitor reaches its rated voltage, no destruction will occur. Also if the voltage source acts as a step voltage of duration T_{step} after which time the voltage abruptly drops to zero value, capacitor will not exceed its DC rating, provided T_{step} is adequately short. Figure 9.13 depicts this scenario. Figure 13.2 depicts the case of a normal capacitor versus a supercapacitor working in an RC circuit activated by a short duration pulse waveform (Silva, Kokuhennadige, Fernando, Kularatna & Steyn-Ross, 2021).

Based on the general validity of Eq. (13.1), if we can control the duration of occurrence of a HV source to a "short-enough" period, final voltage across the terminals of the capacitor could be kept within the limit of V_C. This discussion indicates that a capacitor with very large capacitance value, such as a supercapacitor in a circuit loop of finite series resistance, can be used to safely absorb energy from a HV (transient) source with a short-duration occurrence.

Energy Storage Devices for Renewable Energy-Based Systems. https://doi.org/10.1016/B978-0-12-820778-9.00013-9

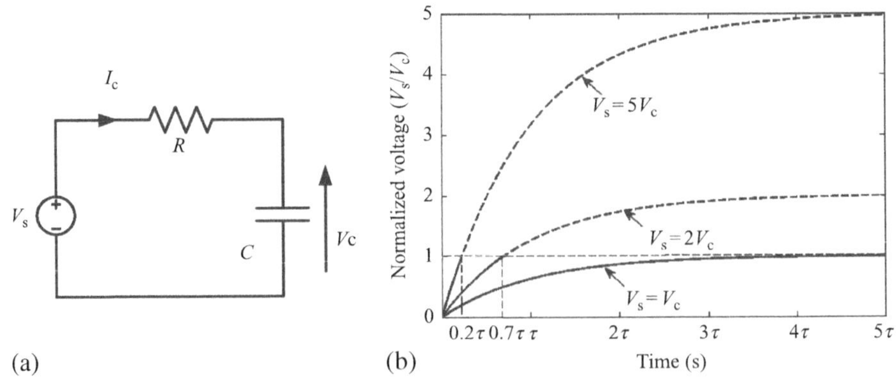

Fig. 13.1 A supercapacitor of capacitance value C charged from a simple DC voltage source. (A) Simple circuit and (B) graph of capacitor voltage vs time.

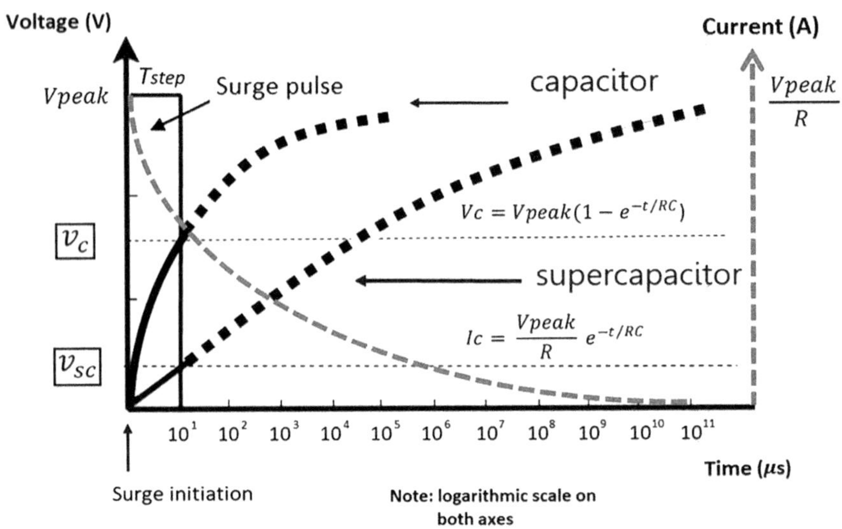

Fig. 13.2 Effect of 10 microsecond step pulse on a RC circuit based on a supercapacitor and a normal capacitor.

13.2 Lightning and inductive energy dumps in electric circuits and typical surge absorber techniques

In practical AC or DC power supplies, short-duration HV transient surges can get superimposed on the energy supply due to acts of Gods such as lightning, or due to switching off of large inductive loads such as high-power motor drives. Most electronic circuits designed with smaller feature size transistors such as the metal oxide

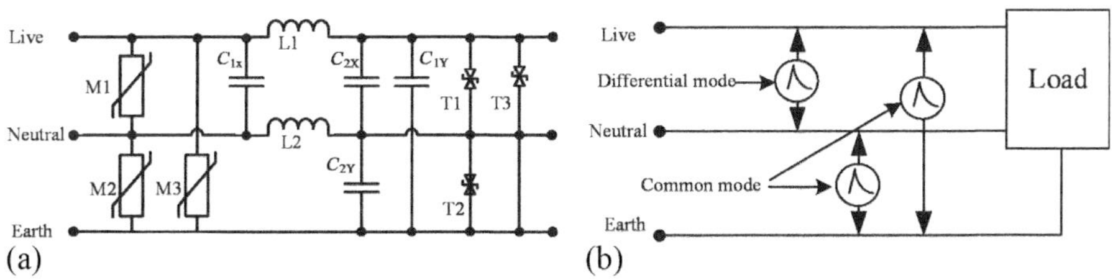

Fig. 13.3 Typical surge protection circuit and modes of surges. (A) Surge protector with MOVs and BBDs coupled with filter stages and (B) common and differential mode surges.

semiconductor field-effect transistors in high-performance processors and memories can be easily get destroyed due to these energy dumps. Superimposed voltages such as lightning-induced transients on incoming mains (or data and telecommunication circuits) can come in the form of differential and common mode surges, which become lethal to smaller feature size transistors in modern integrated circuits. Fig. 13.3 shows a typical circuit used for protecting electronic systems against surges and two modes of surges: (i) common mode and (ii) differential mode.

In a typical surge protector circuit, nonlinear devices (NLDs) such as metal oxide varistors (MOVs) and bidirectional breakover diodes (BBDs) are coupled with LC-type filter stages as shown in Fig. 13.3A. Depending on the level of protection required, different sizes of NLDs are used in varied versions of this general configuration. When a HV surge such as a lightning gets induced on the wire pairs (differential form on the live-neutral pair, common mode on neutral-earth or live-earth pairs), if the peak voltage of the induced voltage exceeds the firing voltage of the corresponding MOV, it fires and starts conducting a high instantaneous current. A maximum voltage termed as *clamping voltage* develops across the terminals of the MOV and the MOV starts absorbing the surge energy based on the voltage-current product over the period of the surge. This mitigates the danger of surge voltage propagating toward the critical load. In fact, the inductor capacitor filters on the path filter another part of the surge energy due to impedance $L\omega$ of the series inductor and the $\frac{1}{C\omega}$ impedance value of the parallel capacitor absorbing some more energy of the HV transient. Any remaining HV transient is absorbed by the firing of BBDs at the critical load end, provided the peak of the remaining surge is above the firing voltage of the BBD.

Overall, such a surge protector that absorbs the energy of the HV transient is designed on the basis of transient energy absorption capability of the NLDs used. Table 13.1 summarizes the properties of a typical MOV from the component manufacturer Littelfuse, United States. In general, these NLDs are characterized by their transient energy absorption rating given for a short duration such as few milliseconds, which is typically related to transients lasting less than about 100–200 μs. During this time, the NLD heats up due to the absorbed transient energy, while maintaining the clamping voltage across the device. However, if repeated high-energy transients keep firing the NLD, device's transient energy absorption rating (usually called the *Joule rating*) will be exceeded and device ends up in a failure. In most general cases of MOVs, every time a transient is absorbed by the device, it deteriorates gradually, and after a limited period of time, its surge absorption capability is totally lost.

Table 13.1 Summary of a typical MOV characteristics-device type TMOV14RP320 from Littelfuse.

Device specification	Value	Units	Remarks
Continuous AC voltage rating	320	V	
Continuous DC voltage rating	420	V	
Transient energy capability	136	J	For a period of 2 ms
Peak forward surge current	200	A	Single pulse (of 8/20 μs current wave)
	4500	A	Two pulses (of 8/20 μs current wave)
Continuous RMS current	100	mA	
Agency approvals	UL1449, IEC61051-1, IEC61051-2, IEC60950-1 (annex Q)		

Source: Littelfuse, May 2014. Varistor Products Data Sheet for TMOV and iTMOV Series.

Table 13.2 presents the characteristics of a BBD such as the 1.5KE440CA series from Littelfuse, where continuous power dissipation rating of back-to-back diode pair is in the order of 6.5 W, but its transient power rating, specified for a limited duration of 1 ms, is 1500 W. What this means is that surges can fire the junctions and absorb the transient energy and the absorbed energy dissipates as heat within a millisecond-order time duration, provided surge is just a short duration of several 100 microseconds only. If a longer duration surge hits the device, failure may result due to excessive energy dissipated over the safe duration.

Table 13.2 Summary of a typical BBD characteristics-device type 1.5KE440CA from Littelfuse.

Device specification	Value	Units	Remarks
Steady state power dissipation	6.5	W	
Peak power dissipation rating	1500	W	Tested using a 10/1000 μs test waveform as per data sheet (Littelfuse, 2014b)
Breakdown voltage	418–462	V	With a test current of 1 mA
Maximum clamping voltage	602	V	
Maximum peak pulse current	2.5	A	

Source: Littelfuse, January 2014. Transient Voltage Suppressor Data Sheets, 1.5KE Series.

Lightning or inductive energy dumps appearing on circuits are completely statistical. Standards related to surge protection such as IEEE C62.4X series (IEEE recommended practice, 2002a,b) define test waveforms with specific rise times, width, and peak amplitude with different value sets as suitable for different classes of protection. In general, IEEE C62 series standards define three different categories (or classes) of protection based on the distance from the power entry point, as shown in Fig. 13.4. Table 13.3 presents test data usable for different test processes as specified by the standards. For more details on surge protection devices, test processes, and

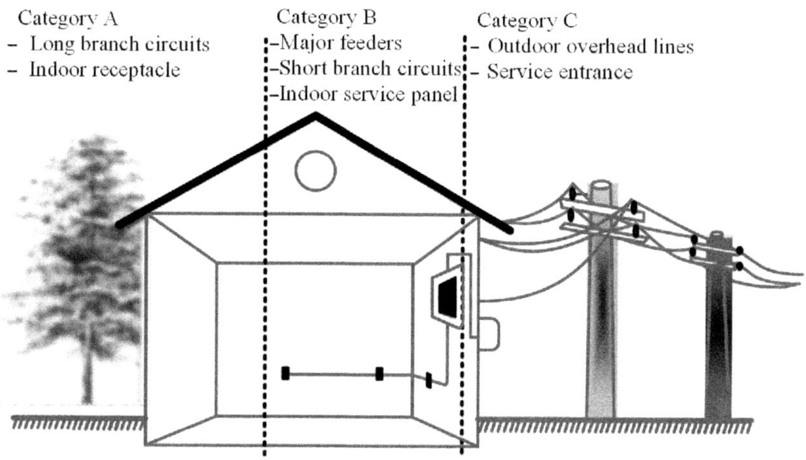

Fig. 13.4 IEEE surge categories: classes A, B, and C based on the distance from surge entry point.

Table 13.3 IEEE C62.41 location categories, frequency of occurrences, and surge waveform details.

		Ring-type test waveform details		
IEEE location category	**Level of exposure**	**2 kV, 70 A, 0.63 J**	**4 kV, 130 A, 2.34 J**	**6 kV, 200 A, 5.4 J**
A1	Low	0	0	0
A2	Medium	50	5	1
A3	High	1000	300	90
		Combination-type test waveform details		
IEEE location category	**Level of exposure**	**2 kV, 1 kA, 9 J**	**4 kV, 2 kA, 36 J**	**6 kV, 3 kA, 81 J**
B1	Low	0	0	0
B2	Medium	50	5	1
B3	High	1000	300	90

standards, the study by (Kularatna, Steyn-Ross, Fernando and James 2018) is suggested.

It is important to realize that the international standards define the test waveforms suitable for the categories A, B, and C based on double-exponential shapes or decaying sine waves as shown in Fig. 13.5. Usually, an open-circuit voltage waveform, short-circuit current waveform, and a ring wave are used as shown in Fig. 13.5A–C.

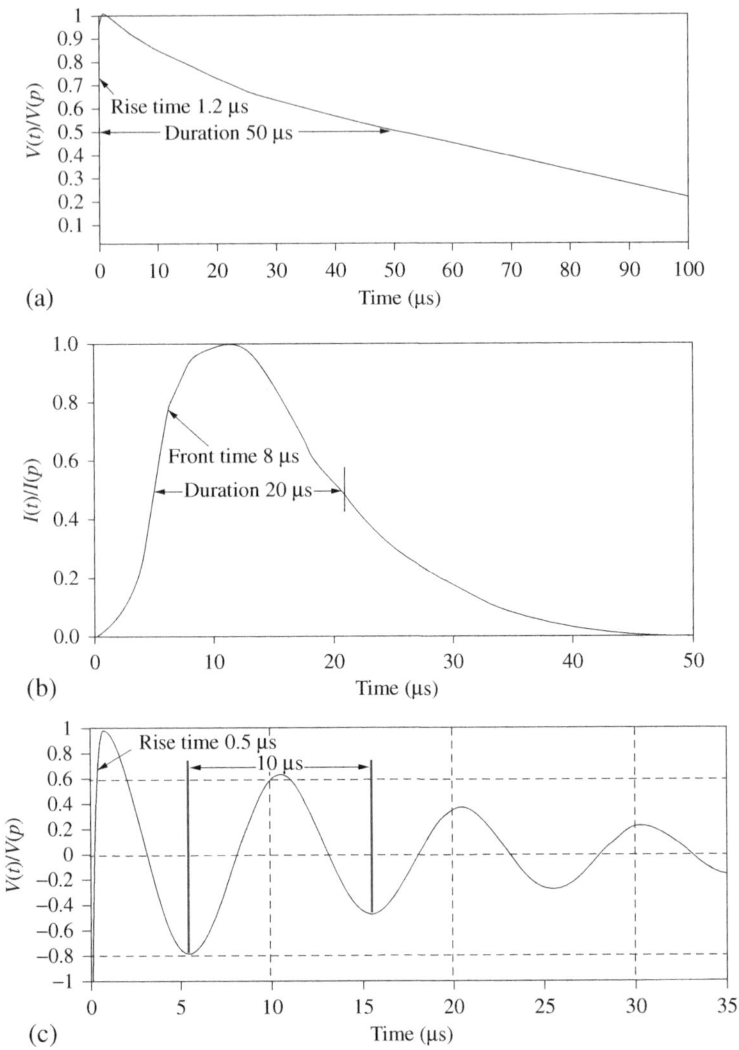

Fig. 13.5 Few examples of wave shapes defined in surge test standards. (A) Open-circuit voltage waveform, (B) short-circuit current waveform, and (C) ring waveform.

13.3 Supercapacitor as a surge absorption device: Summarized results of a preliminary investigation

In Chapter 9 we have discussed the general case of supercapacitor assisted (SCA) techniques, based on the new theoretical concept of SCA loss management (SCALoM). As indicated there, Supercapacitor Assisted Surge Absorber (SCASA) technique makes use of the advantageous property of SC based long time constant circuits, to develop a different kind of a surge protector.

In an early investigation conducted at University of Waikato where several supercapacitors were subjected to a single-shot HV surge as well as multiple surges of identical shape from a lightning surge simulator (LSS) [Types Noiseken LSS6230/LSS6130) indicated that symmetrical type ELDCs (common types of SCs) do not get destroyed by transient surges with 100 μs-order time durations. The waveforms used were as prescribed in standards such as IEEE C62-41 and IEC 61400-4-5. Fig. 13.6 indicates how three different supercapacitors developed their terminal voltage as we repeatedly surged them with identical HV transient waveforms with peak values from 1.5 to 6.5 kV. Devices used were (i) 120 F, 2.3 V device from Maxwell Technologies, (ii) 90 F 2.7 V device from NessCap, Korea, and (iii) 0.4 F 5.5 V device from Cap-XX, Australia. As shown in Fig. 13.6A, all what we could measure as voltage across the terminals was in the order of millivolts only.

Given this observation where the devices under test (DUTs) do not develop adequate DC terminal voltage even after 20 pulses that would exceed the rated DC voltage, it was decided to develop an automatic test interface for repeated application of standard surge waveforms to the DUT in order to determine their surge endurance. The purpose of this was to eliminate the tedious and time-consuming nature of repeated tests, allowing an LSS to recharge within about 20 s before each pulse was applied. In fact, the results shown in this preliminary test closely relate to the fact that an LSS output is a limited duration positive voltage source, which lasts less than 100 μs compared to the time constant of a commercial SC such as Cap-XX 0.4 F device with about 100 mΩ equivalent series resistance (ESR), giving a time constant of 40 ms.

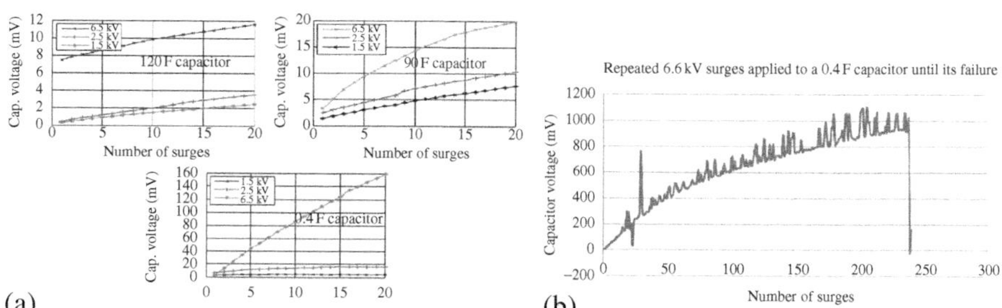

Fig. 13.6 Terminal voltage development vs number of surges at different peak values of a combined waveform as per IEC61400-4-5. (A) Testing of three capacitors, where vertical axis units are in millivolts and (B) repeatedly applying surge to a small supercapacitor (0.4 F) until its failure.

If we approximate the case of LSS output to be a constant 6.6 kV voltage, which is energized only for a period of less than 100 μs, we can expect only millivolts order voltages developed across the terminals of the device. This also indicates that the device still retains its capacitive behavior and is not adversely affected by the transient HV at the terminals. However, as shown in the case of repeated surge application into very small supercapacitors such as 0.4 F device failed after about 160 repeats, as shown in Fig. 13.6B.

Given the preliminary results showing that a supercapacitor is not destroyed by the repeated application of HV transients as specified in C62.4X standards, etc., and the gradual voltage rise across terminals after each hit is in the order of millivolts, we were encouraged to develop a more detailed test system and test process to confirm this observation supported by theory (Kularatna et al., 2011a). Based on the simple formula for a capacitor's energy storage capability of $\frac{1}{2}CV^2$, a supercapacitor of 10 F with a terminal DC voltage rating of 2 V gives an energy storage capability of 20 J. A typical computer power supply's smoothing capacitor within the rectifier stage of value 220 μF with a voltage rating of 400 V will have a maximum energy storage capacity of 17.6 J. This means that a small supercapacitor could absorb almost the same energy, which is capable of destroying a smoothing capacitor in an off-the-line DC power supply. Author's long working experience in an equatorial island with two monsoon seasons indicates that lightning induced on incoming mains destroys electronic components in rectifier stages and switch-mode power converter components, irrespective of the presence of (limited) surge protection capability on the desktop computer power supplies, which is sometimes called the silver box. James et al. (2010) have provided a summary of laboratory surge tests conducted on computer silver boxes with peak voltages up to 6.6 kV.

The above summary indicates that a supercapacitor can be used to absorb part of the surge energy during a transient traveling through the incoming AC mains, DC bus of a renewable energy source or the telecom/data circuits. From various models developed for supercapacitors, two common and useful models depicted in Fig. 13.7 based on the publications (Kularatna et al., 2011a; Spyker and Nelms, 2000) depict the equivalent circuit representing the combination of the LSS, connection cables, and the supercapacitor under test (IEEE recommended practice, 2002a).

Table 13.3 indicates a summary of test waveforms usable to test various surge protectors used at different IEEE location categories, with different statistical levels of exposure such as low, medium, and high. It also indicates that a laboratory instrument such as an LSS with a combination-type waveform or a ring waveform (as shown in Fig. 13.5A and C, respectively) with peak voltage capability up to about 6 kV and energy ratings up to about 90 J is used for testing surge protector devices.

The voltage and current values presented in Table 13.3 are the peak voltages applicable to Fig. 13.5A and the peak current values applicable to Fig. 13.5B, respectively. The current values presented in Table 13.3 also refer to the short-circuit current capabilities of test systems such as an LSS used to test industrial surge protectors. Fig. 13.8 shows the case of an open-circuit voltage waveform shown in Fig. 13.5A applied to an

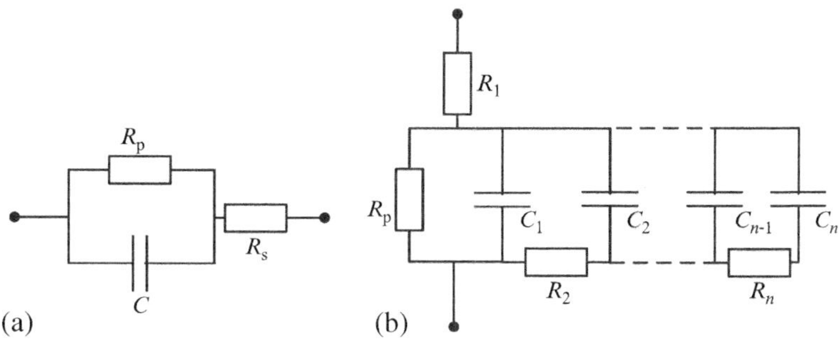

Fig. 13.7 Supercapacitor equivalent circuits usable in developing a surge protection circuit topology. (A) Classical equivalent circuit and (B) ladder circuit.
Adapted from Spyker, R.L., Nelms, R.M., 2000. Classical equivalent circuit parameters for a double-layer capacitor. IEEE Trans. Aerosp. Electron. Syst. 30 (3), 829–836.

SC with a simplified model shown in Fig. 13.7A. A surge waveform as shown in Fig. 13.3A can be represented mathematically as

$$v_{SG}(t) = \frac{\alpha\beta}{\alpha+\beta}\left(e^{-\alpha t} - e^{-\beta t}\right) \tag{13.2}$$

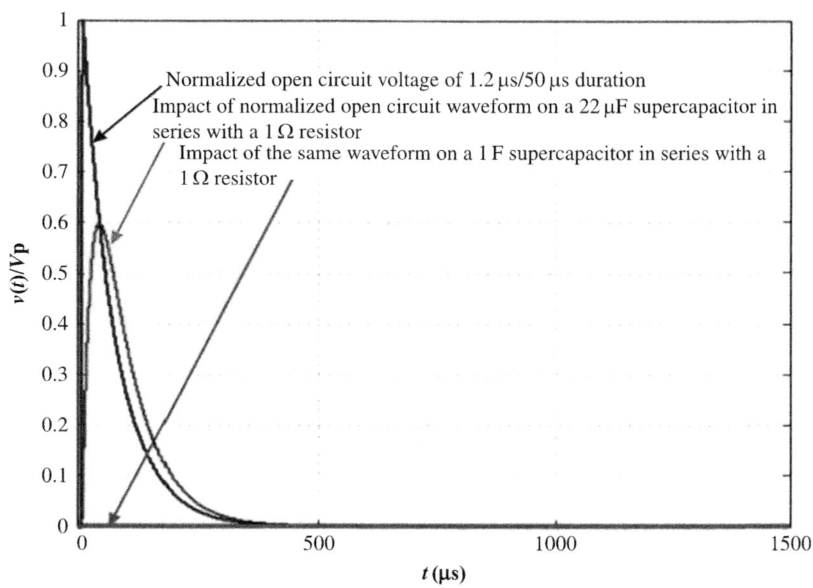

Fig. 13.8 MATLAB simulation results—applying an IEC61400-4-5 combined surge waveform to 220 μF capacitor and 1 F supercapacitor in series with 1 Ω.

where α represents the inverted value of the time constant related to the decaying part of the exponential waveform and β is the same for the rising part of the exponential waveform. By differentiation and estimating the maxima of the time-varying surge, we get the peak value of the surge waveform Vp as

$$V_{\mathrm{p}} = \frac{\alpha\beta}{\alpha+\beta}\left[\left(\frac{\alpha}{\beta}\right)^{\frac{\alpha}{\beta-\alpha}} - \left(\frac{\alpha}{\beta}\right)^{\frac{\beta}{\beta-\alpha}}\right] \tag{13.3}$$

From Eqs. (13.2), (13.3), we can arrive at a normalized open-circuit voltage similar to that shown in Fig. 13.5A. By suitable mathematical manipulations and approximations, we get the following relationship to approximate the normalized open-circuit voltage waveform as a function of time in microseconds:

$$v_{\mathrm{SG,nor}}(t) = 1.0203\left(\mathrm{e}^{-4.16t} - \mathrm{e}^{-0.0139t}\right) \tag{13.4}$$

Fig. 13.8 shows the MATLAB-generated waveform, which closely matches the shape of the normalized open-circuit voltage waveform together with its impact on an RC circuit comprising 220 μF and 1-F capacitors in series with a 1-Ω resistor. Fig. 13.8 clearly shows that due to very large time constants in supercapacitors, such an HV waveform, which has a total duration of about 100 μs, will not develop any adequate voltage across the capacitor. For example, if the peak voltage of the surge waveform is 1 kV, a 220-μF capacitor with 1-Ω resistor develops an approximate capacitor voltage of 600 V, while a 1-F capacitor's case does not show any measurable value.

Given the aforementioned situation, we simply predicted that an SC of farad-order value with 10- to 100-mΩ ESR may withstand HV transients of durations less than a few hundreds of microseconds. However, no publication indicated if this is the case, in the context of any other possible failure phenomenon which can occur due to a HV pulse. In analyzing the effect of a surge or a transient on an SC, similar to the case of an HV LSS output applied on the DUT, we can use the circuits shown in Fig. 13.9. In this case, we have simplified the SC using the equivalent circuit shown in Fig. 13.7A. Based on the Thevenin equivalent circuit seen at the two points A and B in Fig. 13.9A, the case can be simplified to the circuit as shown in Fig. 13.9B.

In this case

$$V_{\mathrm{EQ}} = V_{\mathrm{SG}}\left[\frac{1}{1+\left[\left(R_{\mathrm{SG}}+R_{\mathrm{S}}+r_{\mathrm{path}}\right)/R_{\mathrm{p}}\right]}\right] \tag{13.5}$$

$$R_{\mathrm{EQ}} = R_{\mathrm{Thevenin}} = \frac{R_{\mathrm{SG}}+R_{\mathrm{S}}+r_{\mathrm{path}}}{1+\dfrac{R_{\mathrm{SG}}+R_{\mathrm{S}}+r_{\mathrm{path}}}{R_{\mathrm{p}}}} \tag{13.6}$$

where R_{SG} and r_{path} are the internal resistance of the surge source and the path resistance of interconnections between the SC and the surge source, respectively. It can be

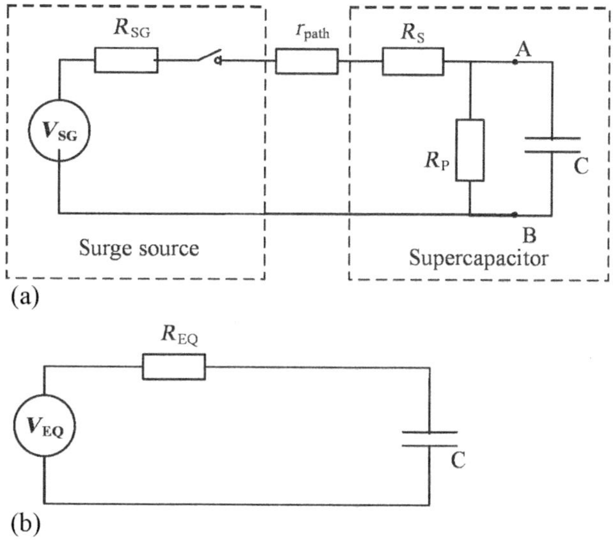

Fig. 13.9 Surge source connected to a supercapacitor. (A) Case with simplified equivalent circuit of an SC and (B) Thevenin equivalent circuit with simplifications.

easily seen that the time constants applicable to the combined effect will be always significantly higher than those with the case for $\tau = R_S C$, which would be the lowest possible time constant for the SC charging. This is based on the assumption that the value of R_p is significantly higher than other series path resistances. Given the aforementioned scenario, if a capacitor of value C is subjected to a step voltage source of duration T with maximum amplitude of V_{max}, the worst-case voltage developed across the capacitor at the end of the step voltage will be

$$V_{c,max} = V_{max}\left(1 - e^{\frac{-T}{R_S C}}\right) \tag{13.7}$$

which could be further simplified to

$$V_{c,max} = V_{max}\left[\frac{T}{\tau} - \frac{1}{2}\left(\frac{T}{\tau}\right)^2 + \frac{1}{3!}\left(\frac{T}{\tau}\right)^3 - \cdots\right] \tag{13.8}$$

This indicates that an approximate value of $\frac{T}{\tau}V_{max}$ when the time constant τ is much larger than the step surge voltage duration is T. When a step voltage source of 100-μs duration with a maximum value of V_{max} is applied to a supercapacitor of 10 mΩ ESR and a capacitance value of 100 F, the capacitor could be charged only up to $10^{-4}V_{max}$. This comfortably predicts that a supercapacitor will not acquire enough charge to exceed its DC voltage rating during the period of single-shot application of a combination surge waveform with even 10 kV. (Kularatna, Steyn-Ross, Fernando and James 2018) provided details of subjecting three different commercial supercapacitor families to surge waveforms up to 6.6 kV peak values, and it provided adequate confidence to use supercapacitors as surge-absorbing components. It is important to indicate here that only the symmetrical supercapacitors (or the common types of supercapacitors with two identical activated carbon electrodes) were used in the SCASA technique.

13.4 Design approaches to a supercapacitor-based surge protector

Supercapacitors in general have the energy absorption capabilities of MOVs and BBDs used in typical surge protectors, as shown in Fig. 13.1A. However, commercially available supercapacitors have very low DC voltage ratings, such as less than 4 V for single-cell devices, which are far below the instantaneous voltages occurring on the AC mains. Given this problem, a designer of a surge protector cannot just substitute a supercapacitor for a MOV or a BBD for the following two reasons:

- Low terminal voltage capability
- Their low impedance value at mains utility frequency.

Given this situation, completely different approaches should be used in designing a supercapacitor-based surge absorber circuit. Figure 9.14 and the related explanation in Chapter 9 provides the reasoning as to why a normal ceramic or other AC operable normal capacitor is not useful in absorbing a significant share of the surge energy, in comparison to a supercapacitor.

Supercapacitor-assisted surge absorber (SCASA) is a patented new technique, developed by the research team led by the author Kularatna and Fernando (2014), based on the concept in the circuit of Fig. 13.10A. In this circuit, typical NLDs such as a MOV or a BBD are combined with a magnetic component and a SC-based

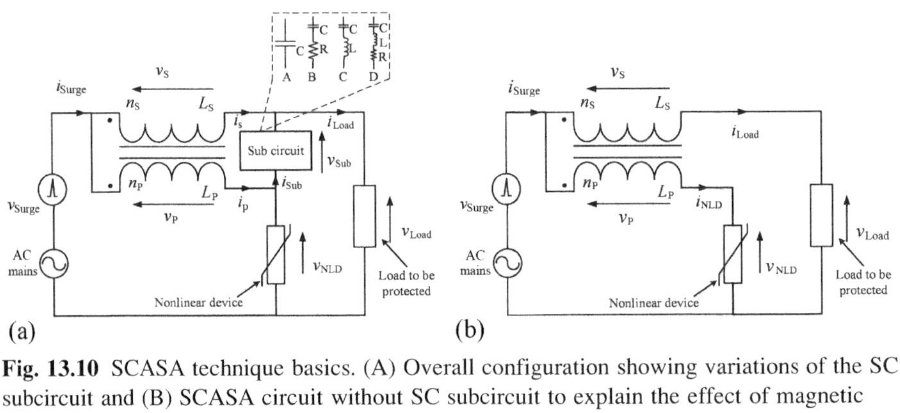

Fig. 13.10 SCASA technique basics. (A) Overall configuration showing variations of the SC subcircuit and (B) SCASA circuit without SC subcircuit to explain the effect of magnetic component.

subcircuit. However, compared to a typical surge arrester without supercapacitors, where NLDs are placed directly across the pairs of wires such as the neutral and the live (or neutral or live and earth), the NLD is placed in between the load side end of a primary coil of a transformer and the return wire as shown in Fig. 13.10A. Given this configuration, when a surge occurs at the AC input, and when the instantaneous voltage due to surge exceeds the firing voltage of the NLD, high instantaneous current flows through the primary coil, developing a voltage across the primary turns. This in turn develops an induced voltage across the secondary coil, and by winding the secondary coil in the opposite direction, secondary induced voltage can be generated to oppose the surge voltage. Result is to create a voltage across the critical load (to be protected), which could be less than the instantaneous surge voltage. By adjusting the turns ratio, we can adjust secondary voltage in such a way that the instantaneous voltage across the supercapacitor-based subcircuit can be varied.

Once a superimposed HV transient travels along the mains input, NLD fires and enters into conduction stage, developing a transient voltage across the connected winding (v_p). When the HV transient exceeds the firing voltage of the NLD, it conducts heavily creating a surge current through the primary coil. Due to induction, secondary coil also develops a voltage (v_s) and the two windings are configured to create this induced secondary voltage, which is higher than that of the primary winding, and to oppose the transient so that the critical load end sees the difference between these two voltages as shown in Fig. 13.11. Graphs in Fig. 13.11 are for a circuit with an MOV (type Q20K275), which has a maximum clamping voltage of 710V.

As indicated in Fig. 13.10A, a supercapacitor-based subcircuit configuration could have a few possible variations (A, B, C, and D). Voltage difference between the two coils appears across the SC subcircuit and creates a circulating current through the subcircuit, dissipating and absorbing the transient energy. Owing to the capability of the supercapacitor to absorb part of the surge energy, which will be dissipated in the closed loops formed, transient surge energy burden on the NLD is significantly

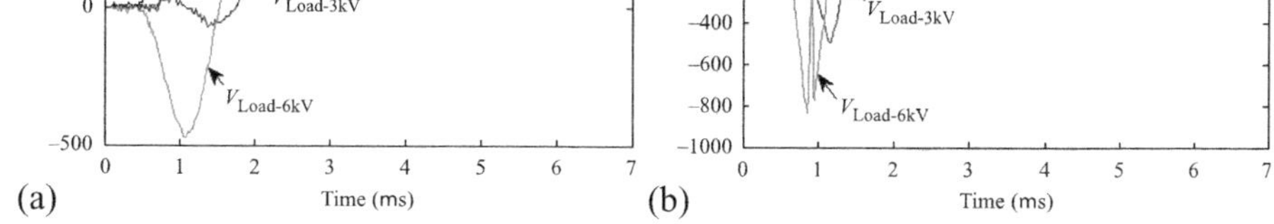

Fig. 13.11 Load and varistor voltage variation for 1, 3, and 6 kV surge input for different toroid with 6:30 turn ratio. (A) Powdered core ($\mu_r = 60$) and (B) ferrite core ($\mu_r = 10{,}000$).

reduced. In addition, supercapacitor subcircuit will perform a useful filter function to reduce the ringing waveform created by the surge.

If generalized impedance Z_{Sub} is used to represent the subcircuit options A–D as shown in Fig. 13.9A, we can establish the following mathematical relationships:

$$v_p = L_p \frac{di_p}{dt} + M \frac{di_s}{dt} \tag{13.9}$$

$$v_s = L_s \frac{di_s}{dt} + M \frac{di_p}{dt} \tag{13.10}$$

$$v_{sub} = \left(v_s - v_p\right) = i_{sub} Z_{sub} \tag{13.11}$$

where L_p and L_s are the self-inductances of the primary and secondary coils, respectively, and M is the mutual inductance between the two windings. Voltages v_p, v_s, and v_{Load} are the instantaneous values at the transformer primary, secondary, and at the load, respectively, when the superimposed surge, v_{Surge}, is applied to the circuit. v_{NLD} and i_{Sub} are the voltage across the NLD and the transient-related current flowing through the SC-based subcircuit.

13.4.1 Selection of the magnetic core

In the SCASA technique, overall performance is mainly governed by the capability of the magnetic component, where its leakage inductance combined with the transformer action assists in creating a lower effective clamping voltage across the critical load. Depending on the permeability of the selected core material, overall performance varies, since the secondary winding voltage due to the superimposed surge depends on the core's saturation behavior. Fig. 13.10B shows the test circuit used to estimate the transformer action in suppressing the overall surge when no subcircuit is connected.

Two transformers with primary to secondary turns ratio (n_p/n_s) of 6:30 with approximately equal cross sections but different core materials were used for this test and some essential data for two cores are presented in Table 13.4. Fig. 13.11 shows the voltage waveforms at the load when the overall circuit is subjected to peak surge voltages such as 1 kV ($V_{Load-1kV}$), 3 kV ($V_{Load-3kV}$), and 6 kV ($V_{Load-6kV}$). The $V_{NLD-6kV}$ line in each figure shows the voltage across the NLD for a 6-kV surge input. In Fig. 13.11A, results are for a powdered-core transformer with relative permeability

Table 13.4 Properties of different toroid cores used for testing (data sheet parameters).

Core type	Magnetics part no.	Relative permeability (μ_r)	A_L value (nH/T^2)
Powdered core	0077083A7	60	$81 \pm 8\%$
Ferrite core	ZW41605TC	10,000	0%

(μ_r) of 60. In Fig. 13.11B, results are for a ferrite core of relative permeability of 10,000. It can be seen from Fig. 13.11B that the secondary voltages generated due to surge are much narrower than the case in the powdered core shown in Fig. 13.11A for higher surge voltages. Hence, it stores less energy from the source, which will be absorbed and dissipated by the SC subcircuit. More details on these aspects are provided in studies by Kularatna and Fernando (2014) and Fernando and Kularatna (2013, 2014).

13.4.2 Impact of the supercapacitor subcircuit

In the SCASA technique, supercapacitor subcircuit plays the role of absorbing part of the surge energy superimposed on the incoming pair of wires. Since this is placed across the far end of the transformer-winding pair, difference in the instantaneous voltages across the two windings creates a voltage difference charging the supercapacitor. Overall effect is that instantaneous voltage across the critical load becomes much less than the surge-related voltage across the NLD.

Fig. 13.12 depicts the case of circuit arrangement used to measure the effect of different capacitance values within the subcircuit used in the SCASA technique. For performance comparison purposes, using the configuration as shown in Fig. 13.10A, authors used an ordinary 50-nF ceramic capacitor as well as a 5-F supercapacitor as the subcircuit and applied 1–6 kV lightning surges generated by the LSS. This is the type B subcircuit with a resistor value of 1 Ω.

When a ceramic capacitor of 50 nF is used as shown in Fig. 13.13A1–A3, voltage across the subcircuit starts oscillating at different frequencies varying from about 122 to 343 kHz. These waveforms are due to several situations occurring at the magnetic component such as

- core parameter nonlinearity variation (due to core saturation);
- transformer leakage inductance variation due to the rise of HV surge through the windings.

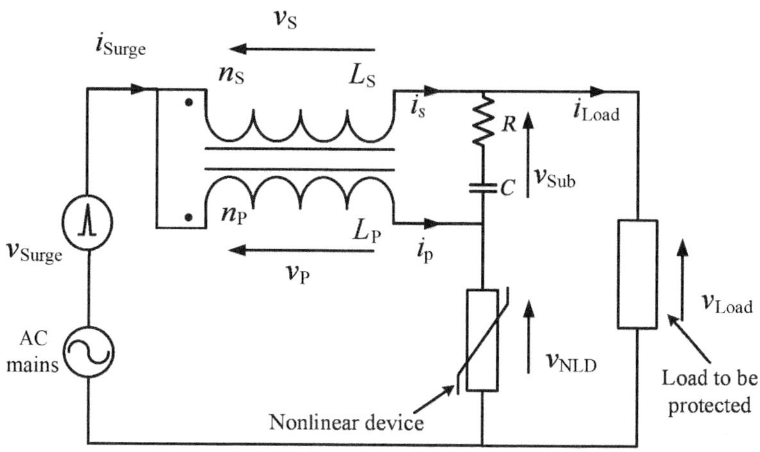

Fig. 13.12 Circuit developed for final testing of SCASA.

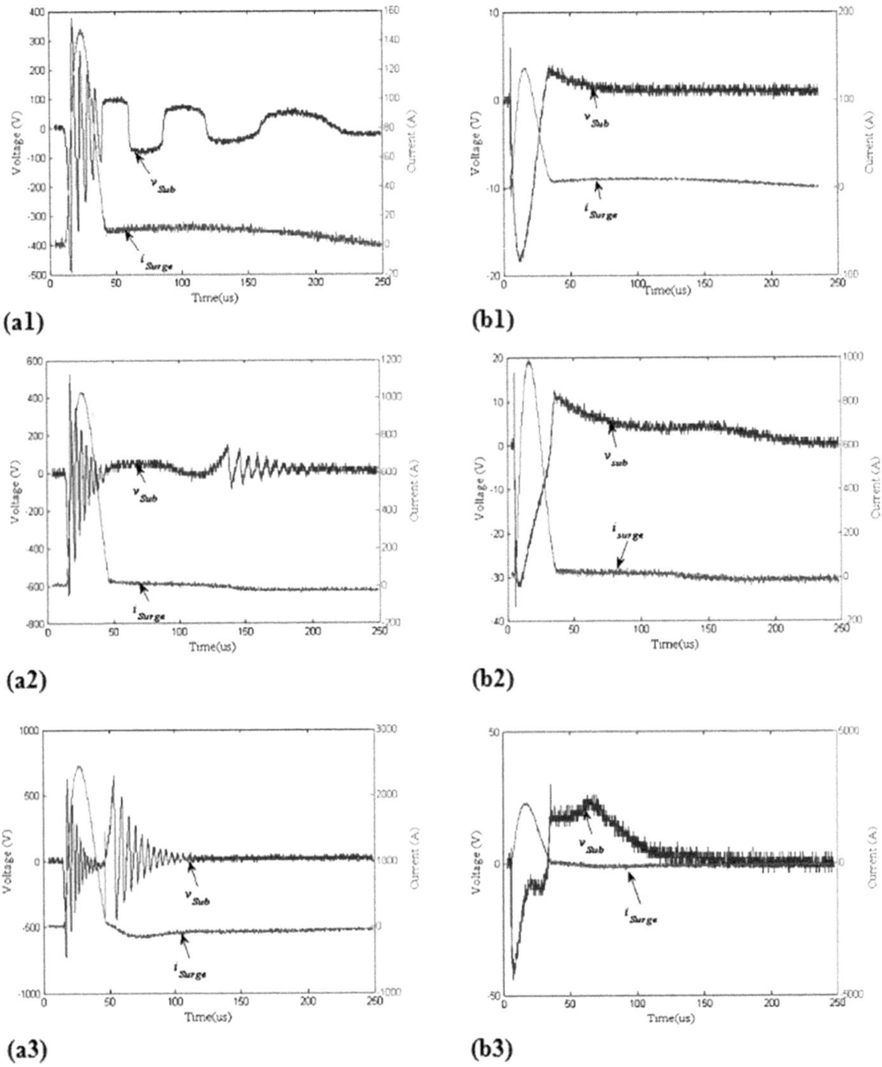

Fig. 13.13 Impact of the supercapacitor subcircuit in the SCASA technique compared with a ceramic capacitor used in lieu of supercapacitor subcircuit. (A1) Effect of 50 nF ceramic capacitor as the subcircuit for a 1 kV surge. (A2) Effect of 50 nF ceramic capacitor as the subcircuit for a 3 kV surge. (A3) Effect of 50 nF ceramic capacitor as the subcircuit for a 6 kV surge. (B1) Effect of SCASA subcircuit with a 5 F supercapacitor for a 1 kV surge. (B2) Effect of SCASA subcircuit with a 5 F supercapacitor for a 3 kV surge. (B3) Effect of SCASA subcircuit with a 5 F supercapacitor for a 6 kV surge.

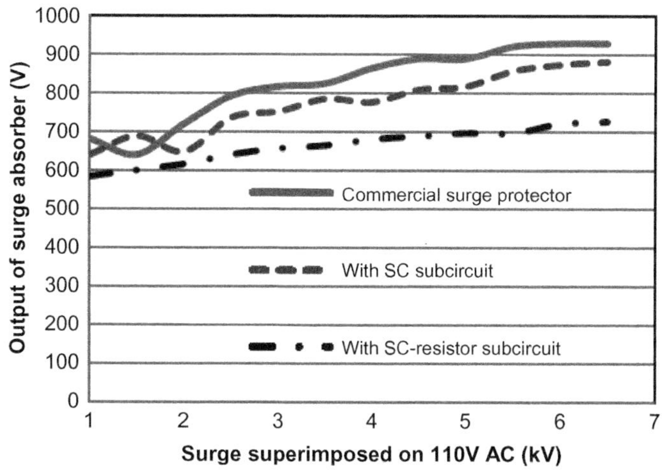

Fig. 13.14 Comparison of SCASA performance with a common commercial surge protector circuit.

Compared to this scenario, a supercapacitor of 5 F replacing the 50-nF ceramic shows very much lower voltages across the critical load. Fig. 13.13B1–B3 shows the scenario. With the 5-F supercapacitor, it can be clearly seen that there are no oscillations and a voltage generated across the subcircuit due to the current i_{Sub}, which opposes the voltage across NLD (v_{NLD}), creates lower voltage across the load than v_{NLD}. Another important design requirement of the SCASA technique is that there is minimal 50 or 60 Hz line frequency voltage components appear across the supercapacitor based subcircuit, given that the two coils are connected together at the live terminal of the mains supply. In this design, leakage impedance of the transformer plays a major role, and hence the selection of the right core for the transformer.

Mathematical analysis to detail this situation is beyond the scope of this chapter, but Eqs. (13.9)–(13.11) form the basis. More details can be found in the research thesis related to the project (Fernando, 2015).

In Fig. 13.14, results show the overall effect of the surge on the critical load end with a B-type subcircuit as shown in Fig. 13.12 and also depict the overall performance comparison of the SCASA configuration with a traditional commercial surge protector. A commercial version was successfully built and it was able to satisfy the conditions required by the UL1449 3rd Edition standard. A comprehensive discussion on design of surge protectors with a practical approach is available in (Kularatna, Steyn-Ross, Fernando and James 2018).

13.5 Conclusion

SCASA technique makes use of the capability of a supercapacitor to absorb transient HVs due to large time constants in a circuit loop, and forms the basis of an entirely new and nontraditional application of the ELDCs. This useful capability of supercapacitors can be extended to other major areas such as uninterruptible power supplies, and this is an entirely new applications research area at the University of Waikato, and early work under the concept of surge-resistant UPS is ongoing (Kularatna et al., 2011b; Kumaran, 2011). The subject area opens up another unique application set for EDLCs.

References

Fernando, J., 2015. Characterization of supercapacitors for special applications. Ph.D. thesis, The University of Waikato.

Fernando, J., Kularatna, N., 2013. A supercapacitor based enhancement for standalone surge protector circuits. In: Proceedings of ISIE-2013 Conference, Taipei, Taiwan, June, pp. 1–6.

Fernando, J., Kularatna, N., September 2014. Supercapacitor assisted surge absorber (SCASA) technique: selection of supercapacitor and magnetic components. In: Proceedings of ECCE 2014.

IEEE, 2002a. IEEE recommended practice on characterization of surges in low-voltage (1000V and less) AC power circuits, IEEE Standard C62.41.2.

IEEE, 2002b. IEEE recommended practice on surge testing for equipment connected to low-voltage (1000V and less) AC power circuit, IEEE Standard C62.45.

James, S., Kularatna, N., Steyn-Ross, A., Kunnemeyer, R., Tantrigoda, D., November 2010. Investigation of failure patterns of desk top computer power supplies using a lightning surge simulator and the generation of a database of comprehensive surge propagation study. In: Proceedings of IEEE IECON 2010 Conference, AZ, USA, pp. 1275–1280.

Kularatna, N., Steyn-Ross, A., Fernando, J., James, S., 2018. *Design of Transient Protection systems: Including Supercapacitor Based Design Approaches for Surge Protectors*. Elsevier, USA, p. 284.

Kularatna, N., Fernando, L.J., 2014. Power and Telecommunication Surge Protection Apparatus. New Zealand Patent Number 604332. March 21.

Kularatna, N., Fernando, J., Pandey, A., James, S., 2011a. Surge capability testing of supercapacitor families using a lightning surge simulator. IEEE Trans. Ind. Electron. 58 (10), 4942–4949.

Kularatna, N., Tilakaratne, L., Kumaran, P.K., 2011b. Design approaches to supercapacitor based surge resistant UPS techniques. In: Proceedings of IEEE-IECON 2011, November, pp. 4094–4099.

Kumaran, P.K., 2011. Development of Supercapacitor Based SRUPS. ME thesis.

Littelfuse, January 2014. Transient Voltage Suppressor Data Sheets, 1.5KE Series.

Silva, T.S.U., Kokuhennadige, S., Fernando, J., Kularatna, N. and Steyn-Ross, A.D., 2021. Supercapacitor assisted surge absorber (SCASA) technique: selection of magnetic components based on permeance, (accepted for IEEE-APEC 2021, To be published).

Spyker, R.L., Nelms, R.M., 2000. Classical equivalent circuit parameters for a double-layer capacitor. IEEE Trans. Aerosp. Electron. Syst. 30 (3), 829–836.

Supercapacitors in a rapid heat transfer application

14

14.1 Introduction

State-of-the-art supercapacitors (SCs) come in all sizes from about fractional farads to over 5000F in single-cell devices. For smaller value devices, their equivalent series resistance (ESR) values range from about 25 to 100 mΩ. The larger capacitance devices from about 300 to 5000F have much lower ESR values such as 0.3 to about 5 mΩ. Given this summary, and the discussion we had in Chapter 1 about the maximum power transfer capability of a voltage source given by $\frac{V^2}{4r_{int}}$ where V is the voltage source value and r_{int} is the internal resistance of the voltage source, provided that we terminate the voltage source with a load resistance, R_L equal to r_{int}. Now if we apply this fact to a typical battery and an SC to estimate the maximum possible output power, we could see that the internal resistance device is the primary determining factor. In this chapter we discuss the case where we could use a bank of SCs to fast-heat water flowing in domestic water systems and to minimize the waste of treated water due to the stored cold water in domestic pipes between the central water heater and the individual faucet. The discussion will also demonstrate how a bank of SCs with an appropriate control system could circumvent the problem of developing a direct AC mains-based high-power fast-acting water heating solution, due to power delivery limits of an electrical subcircuit with a maximum limitation of about 2.3 kW in a 230 V AC, 50 Hz case where a 10 A limit is there for an individual sub-circuit.

14.2 Problem of wasted water in day-to-day situations at home

In domestic hot water supply systems and the central heating systems, whether it is gas or electrical, the hot water unit is usually placed in a central location and small diameter pipes carry hot water from the central unit to individual faucets. Table 14.1 provides a typical scenario in New Zealand residences.

Based on information in Table 14.1, one can estimate the wasted amount of water on a daily basis, in addition to the frustration due to initial cold water flushed through the hot water supply activation. Although fast-acting gas heater systems are becoming popular, the wasted water due to cold water stored in the buried pipes between the central gas unit and the individual faucet is unavoidable in domestic situations. In New Zealand with a small population of approximately 4.5 million, more than 15 million liters of treated water is wasted annually due to this problem.

Energy Storage Devices for Renewable Energy-Based Systems. https://doi.org/10.1016/B978-0-12-820778-9.00010-3

Table 14.1 Typical hot water distribution scenario in New Zealand residences.

Parameter	Average value
Average distance from central hot water unit to individual faucets	3–50 m
Water flow rate	4–8 L/min
Hot water temperature	35–55°C
Time delay for hot water due to water storage in wall-buried pipes	3–40 s
Wasted amount of water due to delay in hot water	0.3–3 L/use
Average number of times for faucet usage	5–20/day

14.3 Problem of traditional heating from direct AC mains supply and heating system specifications

The first requirement in this situation is to supply the energy to flowing water at an adequate rate. Following heat transfer relationship is the starting point to analyze the requirement.

$$P = \rho f_{\mathrm{w}} c\, \Delta\theta \tag{14.1}$$

where P is the power (W), ρ is the density of water (kg/L), f_{w} is the water flow rate (L/s), C is the specific heat of water (J/kg/°C), and $\Delta\theta$ is the temperature rise (°C).

Based on the above relationship, if the water flow rate is 8 L/min, the power delivery should happen at a rate of approximately 11.2 kW for an average temperature rise of nearly 20°C. Given the 10 A maximum current limit from a domestic AC mains wall outlet of 230 V AC/50 Hz the maximum power delivery is now limited to 2.3 kW (RCCB, 2011), which is very much below the requirement. However, total energy requirement to heat the water with an average 20°C temperature rise within 30 s at a flow rate of 8 L/min, is only about 150 Wh (540 kJ), which is compared with the energy required for an mid-size electric car to drive about 1 km distance.

Considering the following key design criteria in Table 14.2, we can start working on a different solution based on prestored energy.

14.4 Commercial solutions for eliminating water wastage due to storage in buried plumbing

There are a few commercially available solutions such as tank-less systems (Laing, 2013), under the sink tanks (Permenter, 2004), and recirculation systems (Fazekas, 1984), to reduce or eliminate this delay, but costs range from 500 to 6000 NZ dollars. However, these solutions do not eliminate the problem completely and economically. The tank-less system heats water as it passes through the unit. This option saves energy as only the water used is heated. However, this solution requires replacement of the existing faucets and heating system (Laing, 2013). Under-sink tanks are an alternative option, but this system contains a mini tank installed under the sink and

Table 14.2 Essential specifications for fast-heating solution based on energy storage.

Specification	
Total maximum temperature rise	20°C
Average water flow rate	6 L/min
Estimated time to receive central hot water from the gas heater	20 s
Total energy requirement (with losses neglected)	47 Wh
Power capability of the heating element	8.4 kW
Maximum voltage allowed for safety reasons	60 V DC or 42.4 V_{peak} (based on AS/NZS 60950.1:2003 Standard)

a mixing faucet. This solution is expensive and requires regular maintenance (Permenter, 2004). Another solution is a recirculation system; this system is made up of a closed loop that connects to each hot water usage point and a pump for recirculating water. These systems cannot be easily retrofitted and running costs are expensive (Fazekas, 1984). More details are given in Kulartna et al. (2014).

14.5 Practical requirements for a localized solution

The above power and energy requirement estimated as a typical worst case requirement indicated that the required energy can be stored in a suitable storage device and can be converted to a short-term high power delivery at the required rate into flowing water, without any additional water tanks, etc. However, several prime requirements as listed below are essential in such a system to be acceptable to regulatory authorities.

14.5.1 Electrical safety and isolation

Due to safety reasons, a power source over 60 V DC cannot be utilized based on safety extra low voltage requirement for electrical shock safety. If the mains power is directly used, galvanic isolation via a transformer supplying less than 60 V (rms) may be required in case a high-power heater coil is placed inside pipe carrying domestic water, in addition to the overload safety provided via a residual type circuit breaker (Datasheet, n.d.; Particular requirement for instantaneous water heaters, 2004).

14.5.2 Domestic installation regulations

Short-term high power delivery as discussed in an application such as the fast water heating requires special wiring and protection devices (Datasheet, n.d.). It is impractical to use an existing electrical subcircuit to deliver this rate of power. In case high-power 30 A subcircuit is to be used, maximum power level can be increased to about 7.5 kW, but this is still beyond the power requirement of about 10–15 kW.

However, since the system required for inline water heating system such as this requires to be powered for a period of less than 30 s in general (due to this the central gas or electrical heating system can kick in within 2–5 s in general) and the statistical nature of water usage at a domestic faucet, prestored energy storage solution is a viable and cost-effective solution.

14.6 SC-based solution with prestored energy

Developing a SC-based energy storage system with an energy storage capacity in the range of 25–200 Wh is practically feasible based on large single-cell capacitors from multiple sources of manufacturers. For example, Maxwell Technologies, United States (Maxwell Technologies, n.d.) and LS Mtron, Korea (LS Mtron, n.d.) are representative companies that manufacture traditional symmetrical double layer type supercapacitor devices suitable for an application like this, and their important specifications for the application are summarized in Table 14.3.

As we see from Table 14.3, these large single devices, usually available up to about 5000 F in general, have extremely low ESR values, i.e., less than 1 mΩ, and current delivery capabilities of 500–3000 A. Based on the simple theoretical considerations in Chapter 1 based on Eq. (1.9) and the related discussion, a series capacitor bank of ten 3000 F from either manufacturer will allow an initial maximum power of 54 kW at a voltage of 25 V across the 10-element series bank. This is nearly impossible from a 230 V/50 Hz or 120 V/60 V single-phase domestic socket outlet. In fact, this is the maximum power rating of the single 3000 F SC cell charged to 2.5 V, suitably terminated by a load equal to its DC ESR.

A 10-element bank charged up to 25 V will be able to hold an energy content of 93 kJ or 26 Wh. To reach the energy stored level of 150 Wh, we require six parallel strings of 10-element 3000 F capacitor banks in general, but the designer was challenged with the prohibitive cost of the energy storage bank for a domestic application such as inline water heating. Fig. 14.1 shows the representative arrangement of the SC bank of 60 elements. Fig. 14.1B shows the general case of Thevenin equivalent circuit

Table 14.3 Comparison of device specifications for fast energy delivery system such as inline water heating.

Manufacturer	Capacitance (F)	Rated voltage (V)	DC ESR (mΩ)	Maximum current (A)	Stored energy (Wh)
Maxwell Technologies	3000	2.7	0.29	2396	3.04
	1500	2.7	0.47	1426	1.52
	650	2.7	0.8	640	0.66
Samwha Electric	70,000 F (battery-capacitor)	2.7	1.1		43 (EOD at 1.6 V)
	7500 (Hybrid capacitor)	2.8	0.8		7.3 (EOD at 0.9 V)

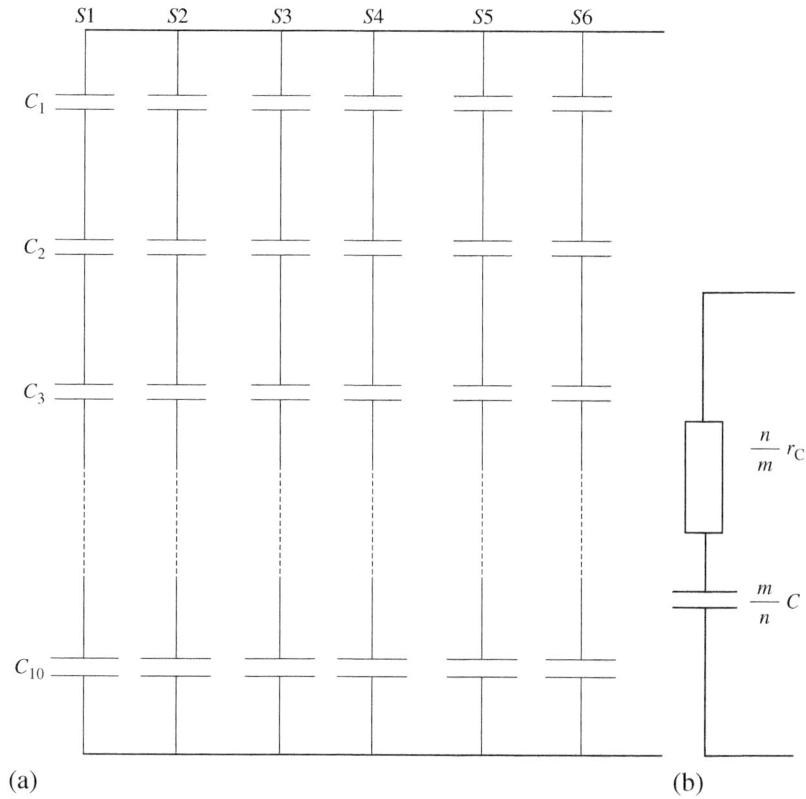

Fig. 14.1 A supercapacitor bank of 60 3000 F single cells. (A) Configuration and (B) Thevenin equivalent circuit.

for a bank of n series elements and m strings of n series elements combined. This gives a case of total capacitance of $\frac{m}{n}C$ with an overall ESR of $\frac{n}{m}r_C$, where C and n are the capacitance and ESR of the capacitor used to build the array, respectively.

Given the case of such a capacitor bank connected to a load resistance of R_L via a switch as shown in Fig. 14.2A, we can show that the instantaneous voltage across the load is given by

$$v_{RL}(t) = \frac{V_{C0}R_L}{R_L + r_{eq}} e^{-\frac{t}{(R_L + r_{eq})C_{eq}}} \tag{14.2}$$

and the energy transferred to the load and the ESR is given by

$$E(0) - E(t) = \frac{1}{2}C_{eq}V_{C0}^2\left[1 - e^{-\frac{2t}{(R_L + r_{eq})C_{eq}}}\right] \tag{14.3}$$

where V_{C0} is the initial capacitor bank voltage, $C_{eq} = \frac{m}{n}C$, and $r_{eq} = \frac{n}{m}r_C$.

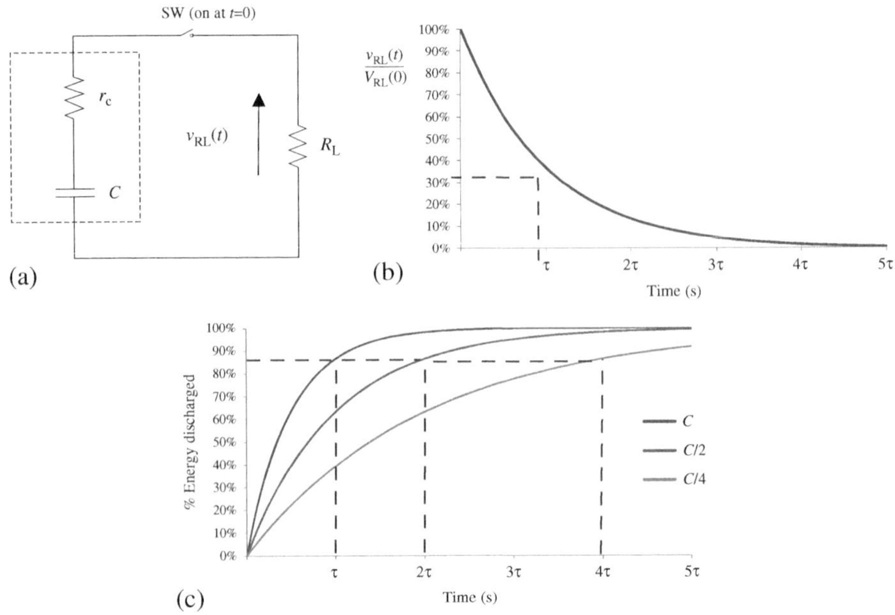

Fig. 14.2 A supercapacitor bank discharging into a resistive load. (A) Equivalent circuit, (B) capacitor voltage vs time, and (C) percentage of delivered energy vs time.

The above discussion gives rise to the graphs in Fig. 14.2B and C for normalized capacitor voltage and the percentage energy discharged over time.

In a case like this, the discharge time constant is given by

$$\tau = \frac{mC}{n}\left(\frac{n}{m}r_C + R_L\right) \tag{14.4}$$

and the approximate average power, P_{ave}, delivered by the bank into the load is given by (with individual capacitors with maximum voltage of V per capacitor)

$$P_{ave} = \frac{nV^2}{10\left(\frac{n}{m}r_C + R_L\right)} \tag{14.5}$$

In general, it is quite easy to show that an SC-based solution is practicable, compared to any other energy source such as a battery bank. In the case of a battery bank, since the life cycle for full depth of discharge (DOD) is in the range of 500–2000 for lead-acid or Li-ion chemistry, in addition to the gradual rise of the internal resistance with the DOD (Coleman et al., 2007), it does not become a very commercially attractive solution. However, in order to reduce the cost of the overall solution, it is possible to combine a suitable battery bank with a smaller SC bank. A discussion on this solution is beyond the scope of the chapter. In this discussion, we have used the most simplified model for an SC (Miller, 2010). Since the state-of-the-art SCs have relatively low leakage currents, an SC bank made of single cells without any balancing circuits can be fully charged and about 80%–85% energy can be reused overnight, allowing the bank to recharge slowly based on a small capacity DC power supply suitable for an SC bank. In case of a SC bank with no balancing circuits (for lower cost), we need to maintain maximum voltage per supercapacitor with a reasonable margin (of at least 0.2 to 0.3 V cell).

14.7 Results from an ongoing prototype development exercise

A complete system concept suitable for an application such as this is given in Fig. 14.3. In order to maintain and control the temperature rise within the range of 20–30°C, an overall system was built by combining three heating coils (in the range of 70–200 mΩ) within the last 50 cm length of the plumbing before the faucet. As shown in Fig. 14.3, in order to develop the control system temperature sensors were placed at the input and the output ends of this short piece of retrofit plumbing, in addition to a flow rate sensor. Multiple energy storage banks were used in the early development stages, which can independently feed independent heating elements under the control of a processor-based system.

14.7.1 Early experimental results

Based on the above design approach, the research team developed a proof-of-concept laboratory system based on SC banks having different capacities and rated voltages. Thermocouples were inserted into both the inlet and outlet sides of the tube. The

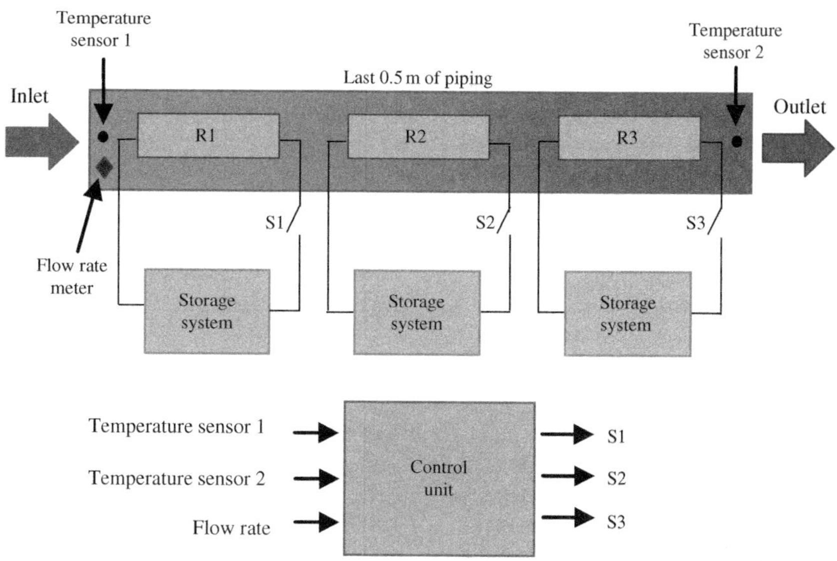

Fig. 14.3 Complete system architecture of an SC-based inline water heating system.

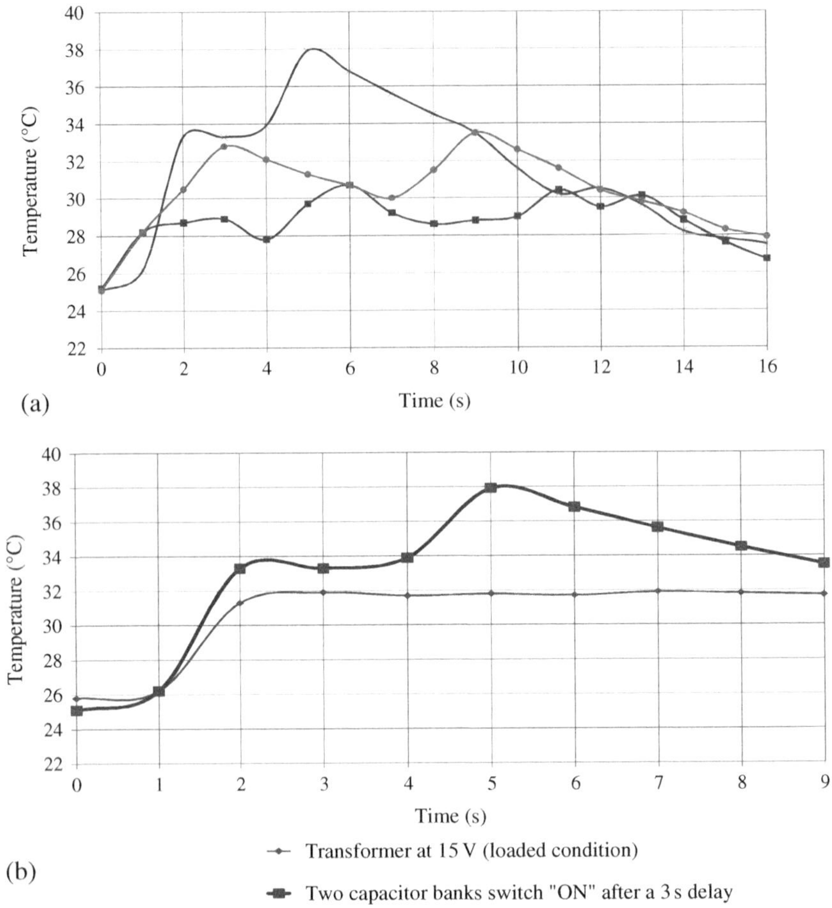

Fig. 14.4 Temperature rise vs time in the inline water heating system configurations. (A) Switching 15 V capable SC banks with different change over delays and (B) SC bank delivery compared with an isolation transformer with loaded terminal voltage of 15 V into a 100 mΩ heater coil.

temperature vs time plots shown in Fig. 14.4 are for various sizes of SC banks, switched with different delays.[a] Early experimental data indicated that it was a feasible solution.

Fig. 14.4B shows one such SC bank switched into heater coils compared with the case of energy fed from the 230 V AC mains via an isolation transformer capable of maintaining approximately 15 V at the terminals of the heater coils, fed from a 2.3-kW rated socket outlet. The same terminal voltage capability SC bank pair was used on the same heater elements, with 3 s delay between the banks. As we clearly see, while the

[a] The details are commercially sensitive and hence not presented here.

transformer can provide a steady state of temperature rise, its internal losses were prohibitive for a commercial solution.

In addition, the transformer's initial start-up delay due to the LR circuit formed with the heating element was clearly observed as shown in Fig. 14.4B. SC banks can be upscaled to reach the required average power level (hence temperature rise) based on suitably prestored energy. In order to optimize the time constants (by modifying the resistance value of the inline heating element), different capacitor heater coil combinations were attempted.

Fig. 14.5 depicts the temperature of the outlet water vs time at different flow rates based on a series-connected SC bank energy storage with a terminal voltage rating of 30 V and an overall equivalent capacitance of 58 F. Given the examples with 90 and 180 mΩ heating coils at 4–8 L/min flow rates, we can see the possibility of tuning the circuit time constants to manage the heating process, which usually completes within less than 30 s, before the central gas heater kicks in.

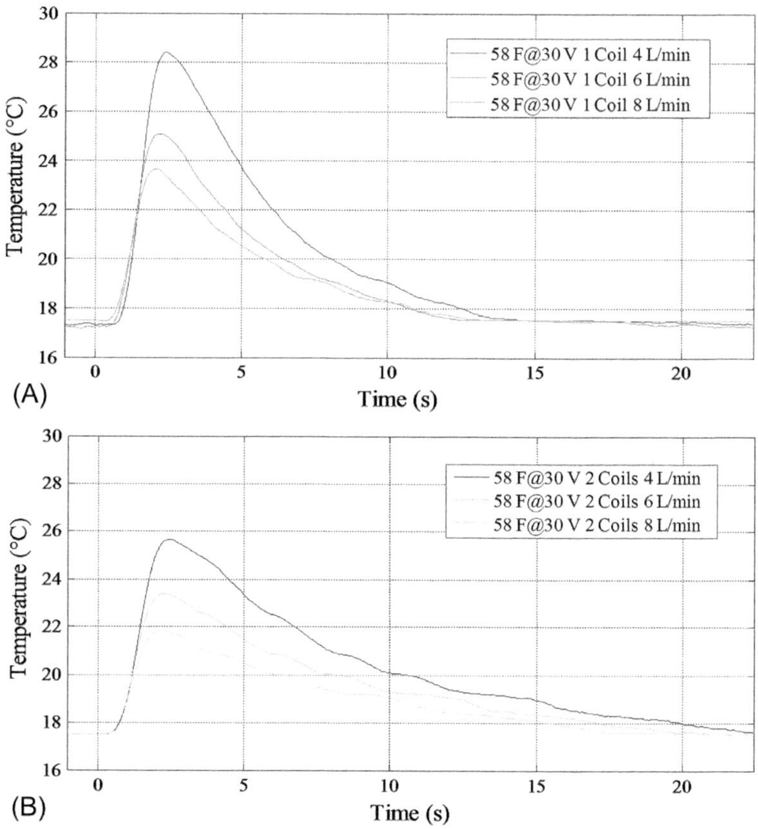

Fig. 14.5 The outlet temperature vs time at different flow rates. (A) 90 mΩ heating element and (B) 180 mΩ heating element.

14.8 Specific advantages of SC energy storage

The above proof-of-concept work clearly proves that the SC energy storage is an elegant solution due to the following reasons.

(1) With the relatively low leakage of commonly available SCs, it is possible to precharge a bank and maintain almost 85% of the total energy at the end of an 8-h period. (i.e., night charging for early morning hot water usage).
(2) Due to the statistical nature of the water usage, solution is practical with prestored energy.
(3) Since the heater coil delivers the power continuous to the cold water only for a period of less than 30 s, overheating of system components due to excessive power levels does not occur.
(4) Practical time constants associated with typical SC banks and a heating coil (as per Fig. 14.2A) are in the range of 1–10 s, which is comparable with the delays associated with centrally heated hot water.

14.9 Implementation challenges

Several implementation challenges were identified during the course of the initial prototyping stage as given below:

(1) Optimizing the SC bank with the required energy storage capability
(2) Limiting the energy transfer during the switching ON to prevent overheating or bubble forming in flowing water
(3) Developing an inline version of a high-power heater coil
(4) Sizing the SC bank with a cost consideration and current trends in SC prices (Brouji et al., 2009)
(5) System integration aspects

An interesting research challenge set from this work is to develop a fast SC bank-charging system based on flowing options:

(1) Dividing the SC bank into three smaller subunits and to develop an energy circulating system
(2) Developing a fast charging system based on three concepts:
    (a) Limited energy transfer into an SC bank at an extremely rapid charging rate
    (b) Investigating the possibility of extracting higher energy from AC mains for a short period of time, using the advantage of in-rush current allowance built into the residual current circuit breakers and other circuit breakers used in domestic systems
    (c) Developing special circuit topologies for fast but energy-limited charge transfers into SC banks

A fast charger technique suitable for this kind of an application is summarized in Chapter 9. New supercapacitor families such as the hybrid types and the battery capacitors (or capa-batteries) entering the market after 2015 with much higher energy densities than symmetrical EDLCs allows this type techniques to be realistically possible, without mixing with battery packs. With the supercapacitor prices expected to drop heavily within the next 5 years (Ultracapacitor Market Forecast, 2015–2020), developing this technique and other nontraditional applications further may open up new research and development areas in power electronics.

14.10 Recent developments of commercial supercapacitor families and their impact on this technique

The design approach to rapid heat transfer systems discussed above was fully implemented using a 30V SC bank supplemented by a lead-acid automotive battery by the year 2016. The details of these systems are available in Ref (Gurusinghe, N., 2016).

Within the last 5 years after this experimental work completed, new supercapacitor families as discussed in Chapter 8 have nearly reached the energy density of lead-acid batteries. A particularly impressive development are the capa-batteries with single cell devices of up to about 70,000 F. With these new devices this Supercapacitor Assisted Temperature Modification Apparatus (SCATMA) technique (NZ Patent, 2016), is really feasible to provide commercial rapid heating solutions.

References

Brouji, E., et al., 2009. Impact of calendar life and cycling ageing on supercapacitor performance. IEEE Trans. Veh. Technol. 58 (8), 3917–3929.

Coleman, M., et al., 2007. State-of-charge determination from EMF voltage estimation: using impedance, terminal voltage, and current for lead-acid and lithium-ion batteries. IEEE Trans. Ind. Electron. 54 (5), 2550–2557.

Datasheet, System Pro M Compact Miniature Circuit Breaker S 200/S200M [Online]. Available from: http://www05.abb.com/global/scot/scot209.nsf/veritydisplay/44c5b5ac208f3f25c1257ad7004ec86a/$file/2CDC002157D0202_view.pdf.

Fazekas, D.J., 1984. Control Means and Process for Domestic Hot Water Re-Circulating System. US Patent 06613452, May 24.

Gurusinghe, N., 2016. Supercapacitor-Assisted Temperature Modification Apparatus (SCATMA) and Fast Supercapacitor Charger, PhD thesis, University of Waiktao.

Kulartna, N., Gattuso, A., Gurusinghe, N., Jayasuriya, T., du Toit, J., November 2014. Pre-stored supercapacitor energy as a solution for burst energy requirements in domestic in-line fast water heating systems. In: Proceedings of IECON 2014, USA.

Laing, O., 2013. Water Delivery System and Method for Making Hot Water Available in Domestic Hot Water Installation. US Patent 20130279891, April 11.

LS Mtron. Available from: http://www.lsmtron.com/page/productMain.asp. [online].

Maxwell Technologies. Available from: http://www.maxwell.com/ultracapacitors/. [online].

Miller, J.R., 2010. Introduction to electrochemical capacitor technology. IEEE Electr. Insul. Mag. 26 (4), 40–47.

NZ Patent 625359, Fluid Temperature Modification Apparatus, Kularatna, A.D.V.N., 30 March 2016.

Particular requirement for instantaneous water heaters, 2004. AS/NZS 60335.2.35.

Permenter, J.B., 2004. Insulated Hot Water Storage Tank for Sink. US Patent 10788150, February 2.

RCCB, 2011. Residual Current Operated Circuit Breakers Without Integral Overcurrent Protection for Household and Similar Uses (RCCBs)—Part 1: General Rules. AS/NZS 61008.1:2011.

Ultracapacitor Market Forecast 2015–2020. Available from: http://www.marketresearchmedia.com/?p=912. [online].

Appendix: Capacitors and AC line filtering

Co-author: Nicoloy Gurusinghe

One of the most common applications of capacitors is to apply them as AC line frequency filters in AC to DC converters. Usually a half-wave or a full-wave bridge rectifier is used together with an electrolytic capacitor of a sufficient value, to smoothen out the rectified waveform so that the output is an average DC voltage with a limited peak to peak 50/60 or 100/120 Hz ripple waveform superimposed on the DC average. In this short discussion we will compare the performance of electrolytic capacitors and supercapacitors in AC-DC converters to show why the present-day supercapacitors are not suitable for the specific application.

Fig. A.1 depicts a simple full bridge rectifier and a capacitor connected to a load and the output voltage versus time. In a simplified case, we can show that,

$$V_{DC} \approx V_{peak} - 2V_D - \frac{I_L}{4fC} \tag{A.1}$$

where V_{DC} is the approximate DC output voltage, V_D is the diode forward voltage drop, I_L is the average DC load current, f is the AC line frequency, and C is the smoothing capacitor value. In this case we assume that the two diodes in series conduct for only a very short time within each half cycle of the waveform, and, the approximate time the capacitor discharges in each half cycle is very close to the half-cycle period of the AC input mains.

However, in practice, the capacitor used is not ideal and it comes with a fixed equivalent series resistance (ESR) in series with the capacitance value. Given this situation, we can analyze the situation using the case in Fig. A.2, where the AC input source (and the diodes in series) presents an internal resistance of R_S and this resistor

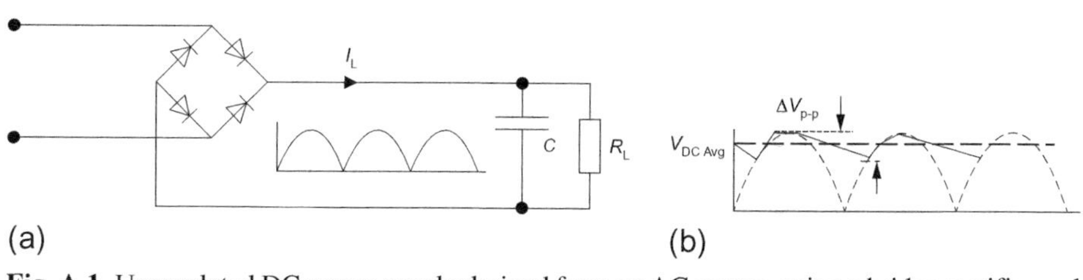

Fig. A.1 Unregulated DC power supply derived from an AC source, using a bridge rectifier and a smoothing capacitor. (A) Simple case without a transformer and (B) approximate case of DC output with superimposed ripple.

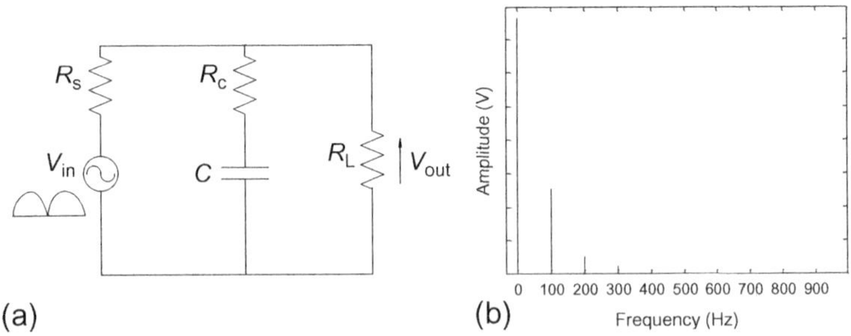

Fig. A.2 Conceptual view of the rectification process considering nonideal case of capacitor and the AC source-rectifier combination. (A) Conceptual circuit with AC source rectifier and (B) spectrum of the full-wave rectified sine waveform.

comes together with the capacitor which has an ESR of R_C in series with a capacitance value of C. Fig. A.2A depicts the circuit, and Fig. A.2B shows the Fourier series components of the rectified DC source (before smoothing by the capacitor), which presents a set of harmonic components of the rectified AC waveform, together with an average DC component.

Given the above nonideal situation, the capacitor is expected to filter out the ripple, and the output of the circuit is expected to provide a DC average voltage with minimized ripple frequency components. In this situation, the transfer function applied to the general case will be,

$$\frac{V_o(s)}{V_{in}(s)} = \frac{R_L(R_C C s + 1)}{[(R_L + R_C)R_S + R_L R_C]Cs + (R_L + R_S)} \tag{A.2}$$

The purpose of the capacitor is to act as a low-pass filter to minimize the harmonic components generated by the rectification process, and to achieve a higher DC average value. Fig. A.3A depicts the rectified voltage waveform and the current drawn from the AC source. In this case we have used components with details as in Table A.1, with a bridge rectifier. Fig. A.3B indicates fast Fourier transform (FFT) waveform of the rectified output, in this case without any capacitor. As we see from Fig. A.3C with the electrolytic capacitor added to the circuit, DC output becomes smooth with only a small ripple superimposed. Fig. A.3D indicates the FFT plot where DC component has significantly increased (note that FFT plot is in dB scale) compared to the case in Fig. A.3B, with most higher order harmonics are suppressed due to capacitor acting as an effective low-pass filter.

Fig. A.3E indicates the case of a 25-F supercapacitors in series used instead of an electrolytic capacitor. As seen here input current drawn by the circuit is significantly increased, since at 100 Hz and higher order harmonics, the impedance shown by the capacitor is gradually decreased, compared to its approximately constant ESR. Additional current drawn from the input creates heat in the two resistors (output resistance of the AC source, R_s and the capacitor ESR).

It is important to note here that the electrolytic capacitor acts as a ripple filter due to its low time constant compared with the time of the AC waveform's half period

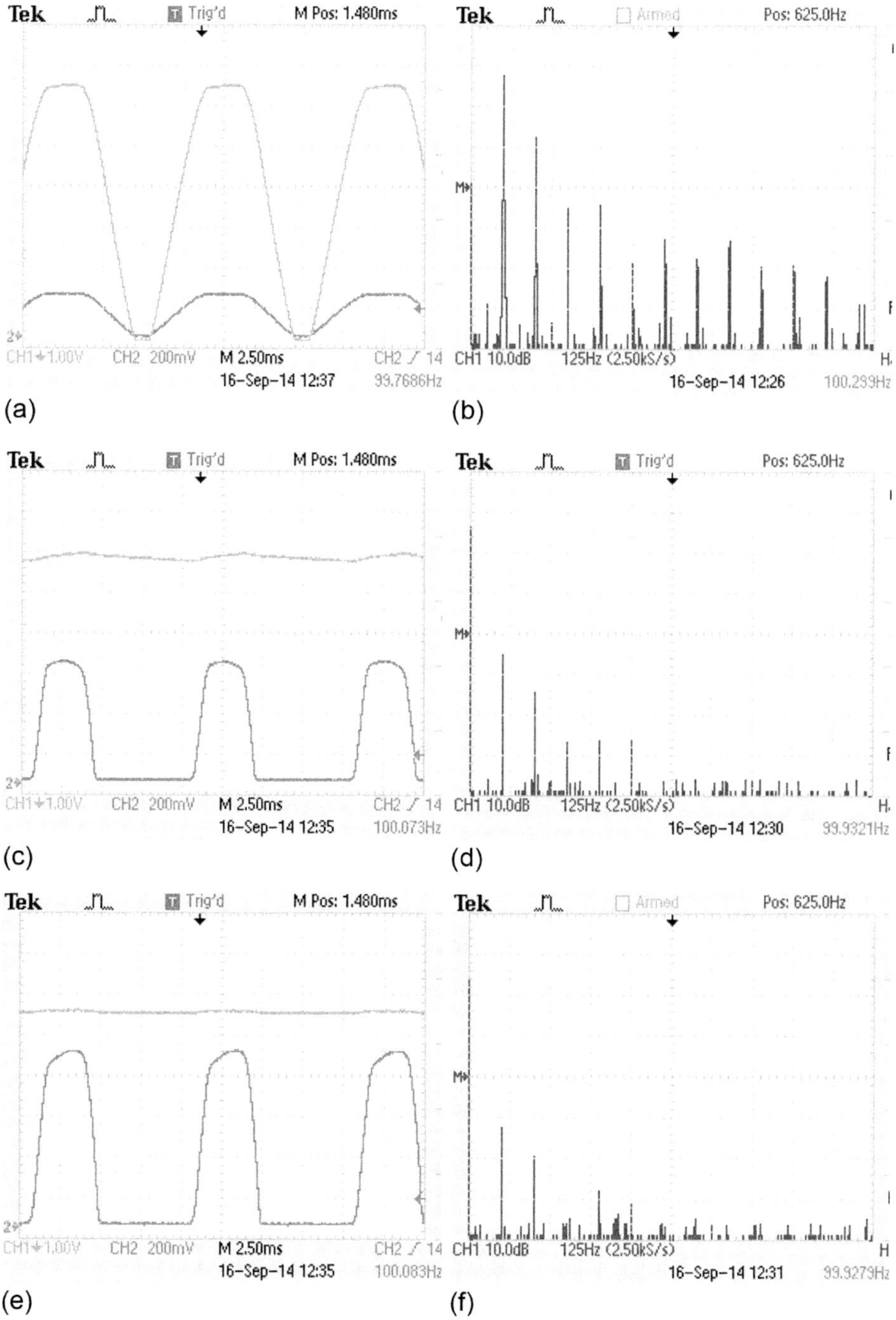

Fig. A.3 Time-domain plots (oscillographs) and FFT plots of lightly loaded full-wave rectified case and filtered output. (A) Time-domain plots with no capacitor connected, (B) FFT plot with no capacitor, (C) time-domain plots with an electrolytic capacitor, (D) FFT plot with electrolytic capacitor, (E) time-domain plot with a supercapacitor, and (F) FFT plot with supercapacitor.

Table A.1 Test setup data used to show the impact of different capacitors as AC line filters.

Parameter	Value	Remarks
AC output from the isolation transformer	5 V rms	Approx. 200 W
Load resistance	20 Ω	±10% tolerance
Electrolytic capacitor	3300 μF, ESR of 20 mΩ	Nichicon corporation
Supercapacitors (two in series)	25 F, 42 mΩ	Maxwell technologies

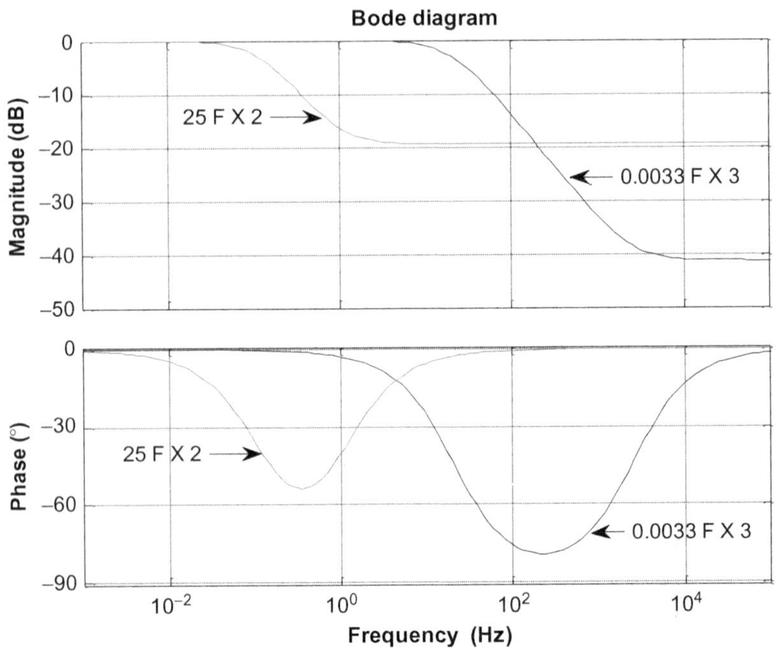

Fig. A.4 Gain and phase plots for two cases of filters.

duration of 10 ms. If we look at the FFT plots carefully, we can see that the DC component at the output is increased by about 25 dB compared to the case of rectifier bridge only.

A careful look at the DC component in Fig. A.3F for the supercapacitors indicates that it is about 2 dB lower than the case of electrolytic capacitor in Fig. A.3D. Bode plots in Fig. A.4 indicates the two cases where the electrolytic capacitor providing a useful low-pass filtering action with about a phase angle of around −60 degree, whereas the supercapacitors showing a phase angle of almost zero degrees around 100 Hz frequency.

Due to supercapacitor impedance reduced at 100 Hz and above significantly, current drawn from the rectified AC source increases due to the simple voltage divider

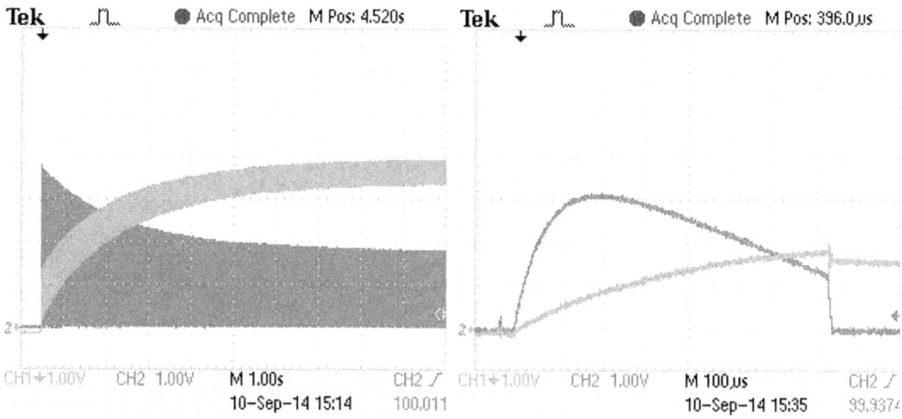

Fig. A.5 Time delays to stabilize the DC output of capacitor filters. (A) 25-F supercapacitor and (B) 3300-μF electrolytic capacitor.

formed due to R_S (combined with R_L) and R_C and without the additional impedance of the capacitor. This means that the supercapacitor circuit wastes energy as heat inside these two components. This is the impact of large time constant of the supercapacitor in general compared to the electrolytic capacitor.

Another issue with supercapacitors is the initial time to stabilize the DC output. This is indicated in Fig. A.5. In the case of supercapacitor in Fig. A.5A it takes about 6 s to stabilize the DC output, whereas in Fig. A.5B it takes about 750 μS only to stabilize the DC output.

In summary, this discussion details the case of present-day supercapacitors' inability to perform AC line filtering in AC-DC converters, due to their long-time constants compared to electrolytic capacitors. This issue is discussed in Chapter 5. Miller et al. (2011) and Burke and Miller (2008) provide more details and present approaches to develop better supercapacitors suitable for AC ripple filtering.

References

Burke, A.F., Miller, J.R., 2008. Electrochemical capacitors: challenges and opportunities for real-world applications. Electrochem. Soc. Interface 17, 53.

Miller, J.R., Outlaw, R.A., Holloway, B.C., 2011. Graphene electric double layer capacitor with ultra-high-power performance. Electrochim. Acta 56, 10443–10449.

Index

Note: Page numbers followed by *f* indicate figures and *t* indicate tables.

F

G

H

I

V

W

Z

9780128207789